Digital Filter Design and Realization

RIVER PUBLISHERS SERIES IN SIGNAL, IMAGE AND SPEECH PROCESSING

Series Editors

Indexing: All books published in this series are submitted to Thomson Reuters Book Citation Index (BkCI), CrossRef and to Google Scholar.

The "River Publishers Series in Signal, Image and Speech Processing" is a series of comprehensive academic and professional books which focus on all aspects of the theory and practice of signal processing. Books published in the series include research monographs, edited volumes, handbooks and textbooks. The books provide professionals, researchers, educators, and advanced students in the field with an invaluable insight into the latest research and developments.

Topics covered in the series include, but are by no means restricted to the following:

- Signal Processing Systems
- Digital Signal Processing
- Image Processing
- Signal Theory
- Stochastic Processes
- Detection and Estimation
- Pattern Recognition
- Optical Signal Processing
- Multi-dimensional Signal Processing
- Communication Signal Processing
- Biomedical Signal Processing
- Acoustic and Vibration Signal Processing
- Data Processing
- Remote Sensing
- Signal Processing Technology
- Speech Processing
- Radar Signal Processing

For a list of other books in this series, visit www.riverpublishers.com

Digital Filter Design and Realization

Takao Hinamoto
Hiroshima University
Japan

Wu-Sheng Lu
University of Victoria
Canada

River Publishers

Routledge
Taylor & Francis Group
LONDON AND NEW YORK

Published 2017 by River Publishers
River Publishers
Alsbjergvej 10, 9260 Gistrup, Denmark
www.riverpublishers.com

Distributed exclusively by Routledge
4 Park Square, Milton Park, Abingdon, Oxon OX14 4RN
605 Third Avenue, New York, NY 10017, USA

Digital Filter Design and Realization / by Takao Hinamoto, Wu-Sheng Lu.

Routledge is an imprint of the Taylor & Francis Group, an informa business

ISBN 978-87-93519-64-0 (print)

Contents

Preface

Analysis, design, and realization of digital filters have experienced major developments since the 1970s, and have now become an integral part of the theory and practice in the field of contemporary digital signal processing. This book is written to present an up-to-date and comprehensive account of the analysis, design, and realization of digital filters. It is intended to be used as a text for graduate students as well as a reference book for practitioners in the field. Prerequisites for this book include basic knowledge of calculus, linear algebra, signal analysis, and linear system theory.

The text is organized into seventeen chapters which are outlined as follows:

Chapter 1 presents introductory materials on digital signal processing. Chapter 2 describes several fundamental sequences, the z-transforms of commonly encountered discrete-time functions, and basic properties of linear discrete-time systems. Chapter 3 studies stability of recursive digital filters and their coefficient sensitivity. Chapter 4 deals with mathematical properties of linear discrete-time dynamical systems, studies transfer functions of linear systems and their relation to state-space descriptions.

Chapter 5 presents the fundamentals of FIR digital filters and several methods for the design of FIR digital filters. The next five chapters are related to the design of digital filters. In Chapter 6, we are concerned with the design of recursive digital filters using analog filter theory. Chapter 7 presents several methods for the design of recursive digital filters in the frequency domain while Chapter 8 investigates several methods for the design of recursive digital filters in the time domain. Chapter 9 deals with efficient techniques for the design of interpolated and frequency-response-masking (FRM) FIR digital filters. Chapter 10 addresses the design of a class of composite digital filters by an alternating convex optimization strategy to achieve equiripple passband and least-squares stopband subject to peak-gain constraint.

Chapter 11 studies the finite-word-length effects in the implementation of recursive digital filters. Chapter 12 deals with the l_2-sensitivity analysis and minimization of state-space digital filters. Chapter 13 explores an pole and zero sensitivity analysis and minimization of state space digital filters.

Chapter 14 studies error spectrum shaping in the recursive digital filters that are described by transfer functions or state-space models. Chapter 15 examines an roundoff noise analysis and minimization of state-space digital filters, and develops a technique for jointly optimizing high-order error feedback and realization to minimize the roundoff noise gain at the filter's output. Chapter 16 presents roundoff noise and l_2-sensitivity analyses of the generalized transposed direct-form II structure and its equivalent state-space realization, and describes a procedure for synthesizing the optimal filter structure or equivalent state-space realization. In Chapter 17, we consider block-state realization of an IIR digital filter, and examine several properties of the block-state realization. Analysis of roundoff noise and minimization of average roundoff noise gain subject to l_2-scaling constraints for block-state realization are also examined. Moreover, a quantitative analysis on l_2-sensitivity is performed, and two techniques for minimizing a sensitivity measure known as average l_2-sensitivity subject to l_2-scaling constraints are presented.

We wish to express our sincere gratitude to Professor Akimitsu Doi of Hiroshima Institute of Technology, Hiroshima, Japan, for his kindness and significant contributions in terms of extensive computer simulations and drawing many figures for the book.

List of Figures

List of Tables

List of Abbreviations

1-D	One-dimensional
2-D	Two-dimensional
A/D	Analog-to-digital
BFGS	Broyden-Fletcher-Goldfarb-Shanno
BP	Bandpass
BS	Bandstop
CCF	Complementary comb filter
CCP	Convex-concave procedure
C-filter	Composite filter
D/A	Digital-to-analog
DFT	Discrete Fourier transform
DSP	Digital signal processor
ECG	Electrocardiograph
EEG	Electroencephalogram
EPLSS	Equiripple passbands and least-squares stopbands
FIR	Finite Impulse Response
FRM	Frequency response masking
FWL	Finite-word-length
HP	Highpass
IDFT	Inverse discrete Fourier transform
IFIR	Interpolated FIR
IIR	Infinite Impulse Response
KKT	Karush-Kuhn-Tucker
LBR	Lossless bounded real
(L, L) system	L-input/L-output state-space model
LP	Lowpass
M-D	Multidimensional
P-M filter	Parks-McClellan filter
P-wave	Primary wave
QP	Quadratic programming
RGB	Red, green, and blue
SDP	Semidefinite programming
S/H	Sample-and-hold
SISO	Single-input/single-output
SOCP	Second-order cone programming
S-wave	Secondary wave
VLSI	Very large scale integrated

1

Introduction

1.1 Preview

This is a book that is primarily concerned with basic concepts and methods in digital filter design and realization. The recent advances in the theory and practice of digital signal processing have made it possible to design sophisticated high-order digital filters and to carry out the large amounts of computations required for their design and realization. In addition, these advances in design and realization capability can be achieved at low cost due to the widespread availability of inexpensive, powerful digital computers and related hardware. Briefly put, therefore, the focus of this book is the design and realization of digital filters.

To begin with, we introduce basic terminology for signal analysis and present an overview of digital signal processing, explaining its advantages and disadvantages. We then examine the sampling of an analog signal and that of a continuous-time sinusoidal signal in connection with aliasing. Finally, the sampling theorem is presented and also the method to recover an analog signal from its discrete-time samples is explained. It is shown that if the bandwidth of an analog signal is finite, in principle the analog signal can be reconstructed from the samples, provided that the sampling rate is sufficiently high to avoid aliasing.

1.2 Terminology for Signal Analysis and Typical Signals

1.2.1 Terminology for Signal Analysis

A one-dimensional (1-D) signal is a function of a single scalar variable. A speech signal is an example of 1-D signals where the variable is time. For 1-D signals, the variable is usually labeled as *time*. If the variable is continuous, the signal is called a *continuous-time signal*, which is defined at very time instantly. When the variable is discrete, the signal is called a *discrete-time signal*, which is defined at discrete instants of time. A continuous-time signal with continuous amplitude is called an *analog signal*. A speech signal is an example of analog signals. A discrete-time signal with discrete-valued amplitudes represented by a finite number of digits is called a *digital signal*. A digitized music signal stored on a CD-ROM disk is an example of digital signals. A discrete-time signal with continuous-valued amplitude is called a *sampled-data signal*. Therefore, a digital signal is a quantized sampled-data signal. A continuous-time signal with discrete-valued amplitudes is called a *quantized boxcar signal*. Four types of these signals are illustrated in Figure 1.1, where the abscissa and ordinate axes are time and the amplitude of signals, respectively.

A two-dimensional (2-D) signal is a function of two independent variables. An image signal is an example of 2-D signals where the two independent variables are two spatial variables. A multidimensional (M-D) signal is a function of more than one variable. A black-and-white video signal is an example of 3-D signals where the three independent variables are two spatial variables and time. A color video signal is a 3-channel signal composed of

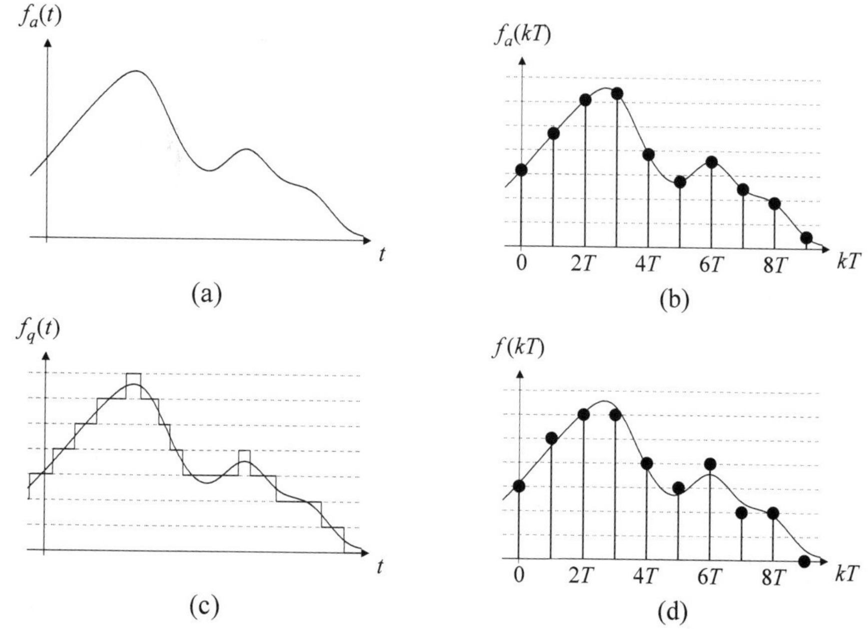

Figure 1.1 Four types of signals: (a) Analog signal, (b) Sampled-data signal, (c) Quantized boxcar signal, (d) 3-bit quantized digital signal.

three 3-D signals representing the three primary colors, namely, red, green, and blue (RGB). In this book, our focus will be on the processing of 1-D signals.

1.2.2 Examples of Typical Signals

A. Electrocardiograph (ECG) Signals

Electrical activity of the heart may be represented by ECG signals which are essentially periodic waveforms.

B. Electroencephalogram (EEG) Signals

Several times the overall effect of the electrical activity due to random firing of billions of individual neurons in the brain is represented by EEG signals.

C. Seismic Signals

Seismic signals are generated by the movement of rocks resulting from an earthquake, a volcanic eruption, or an underground explosion. Specifically, ground movement causes elastic waves in terms of primary wave (P-wave), secondary wave (S-wave), and surface wave that propagate through the body of the earth in all directions from the source of movement.

D. Diesel Engine Signals

In the precision adjustment of diesel engines during production, signal processing plays an important role. For the efficient operation of the engine, accurate determination of the topmost point of piston's movement inside the cylinder of the engine is required.

E. Speech Signals

A speech signal is formed by exciting the vocal tract and is composed of two types of sounds, namely, voiced and unvoiced.

F. Musical Sounds

The sound generated by most musical instruments is produced by mechanical vibrations caused by activating some form of mechanical oscillator that in turn causes other parts of the instruments to vibrate. All these vibrations together in a single instrument generate the musical sound.

G. Time Series

A *time series* is a sequence of data points in a successive order. Time series occurs in business, economics, physical sciences, social sciences, engineering, medicine, and many other fields. Examples of time series abound, for instance, the yearly average number of sunspots, daily stock prices, the value of total monthly exports of a country, the yearly population of animal species in a certain geographical area, the annual yields per acre of crops in a country, and the monthly totals of international airline passengers over certain periods.

H. Images and Video Signals

An image is a 2-D signal whose intensity is a function of two spatial variables. Typical examples are still images, photographs, radar and sonar images, and medical X-rays. An image sequence such as that seen in a television, is a 3-D signal whose image intensity at any point is a function of three variables, i.e., two spatial variables and time.

1.3 Digital Signal Processing

1.3.1 General Framework for Digital Signal Processing

Most signals of practical interest, such as speech signals, biological signals, seismic signals, radar signals, sonar signals, audio and video signals, etc. are analog. Digital signal processing techniques can be utilized to process analog signals. In general, digital processing of analog signals consists of three basic steps:

(1) An analog signal is converted into a digital signal by an A/D converter.
(2) This digital signal is then processed by a digital signal processor, resulting in a processed digital signal.
(3) The processed digital signal is finally converted into an analog signal by a D/A converter.

The digital processing of analog signals is illustrated in a block diagram form in Figure 1.2. Since the amplitude of an analog signal varies with time, a *sample-and-hold* (S/H) circuit is employed at first to sample the analog signal at periodic intervals, and hold the sampled value constant at the input of the *analog-to-digital* (A/D) converter to allow accurate digital conversion. The input to the A/D converter is a staircase-type analog signal, and the output of the A/D converter is a binary data stream which is processed by a digital signal processor where the desired signal processing algorithm is implemented. The output of the digital signal processor is another binary data stream which is

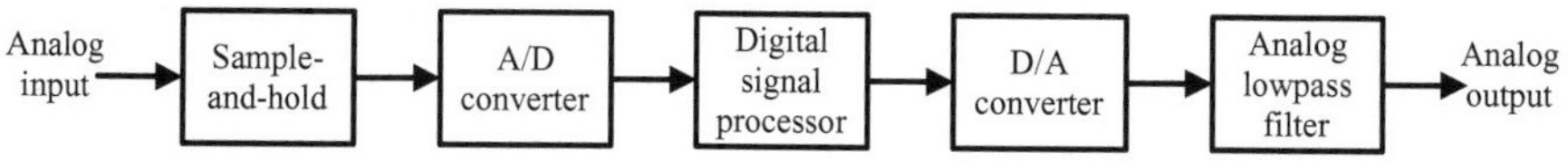

Figure 1.2 General framework for the digital processing of an analog signal.

converted into a staircase-type analog signal by a *digital-to-analog* (D/A) converter. A lowpass filter is then used at the output of the D/A converter to remove all undesired high-frequency components and to deliver a processed analog signal to the output.

1.3.2 Advantages of Digital Signal Processing

Digital signal processing offers several advantages:

(1) Operations of digital circuits do not depend on precise values of the digital signals. Hence a digital circuit is less sensitive to tolerance of component values, and is fairly independent of external parameters such as temperature, aging, etc.
(2) Amenable to full integration. In particular, advances in very large scale integrated (VLSI) circuits have made it possible to integrate highly sophisticated DSP systems on a single chip.
(3) Since the signals and coefficients describing a processing operation are represented as binary values, desirable accuracy can be achieved by simply increasing the wordlength, subject to cost constraint. Moreover, using floating-point arithmetic can further increase the dynamic range for signals and coefficients.
(4) Digital implementation permits easy adjustment of processor characteristics during the processing, such as in adaptive filtering.
(5) Digital implementation allows realization of certain characteristics, which are impossible with analog implementation, such as exact linear phase and multirate processing.
(6) Digital circuits can be cascaded without loading problems.
(7) Digital signals can be stored almost indefinitely without loss of information on various storage media such as magnetic tapes and disks, and optical disks.

1.3.3 Disadvantages of Digital Signal Processing

There are also disadvantages when digital signal processing techniques are applied:

(1) Increased system complexity in digital processing of analog signals.
(2) Frequencies available for sampling and digital processing are often limited.
(3) Digital systems are constructed by active devices that consume electrical power.

1.4 Analysis of Analog Signals

1.4.1 The Fourier Series Expansion of Periodic Signals

We now consider a periodic signal $f(t)$ that is a periodic function of time. The Fourier series allows one to express a given periodic function of time as sum of an infinite number of sinusoids whose frequencies are harmonically related. That is,

$$f(t) = \frac{1}{2}a_0 + \sum_{n=1}^{\infty}\left(a_n \cos n\Omega_0 t + b_n \sin n\Omega_0 t\right) \tag{1.1}$$

where $\Omega_0 = 2\pi/T_0$ with T_0 the period of the signal is called the *fundamental frequency*, and the expansion coefficients a_n and b_n are given by

$$a_n = \frac{2}{T_0}\int_{-T_0/2}^{T_0/2} f(t)\cos n\Omega_0 t\,dt \quad \text{for } n = 0, 1, 2, \cdots$$

$$b_n = \frac{2}{T_0}\int_{-T_0/2}^{T_0/2} f(t)\sin n\Omega_0 t\,dt \quad \text{for } n = 1, 2, 3, \cdots$$

Equation (1.1) is called the *sine-cosine-form of the Fourier series.* Using Euler's formula, we can write

$$a_n \cos n\Omega_0 t + b_n \sin n\Omega_0 t = c_n e^{jn\Omega_0 t} + c_{-n} e^{-jn\Omega_0 t} \tag{1.2}$$

where

$$c_n = \frac{1}{2}\left(a_n - jb_n\right), \qquad c_{-n} = \frac{1}{2}\left(a_n + jb_n\right)$$

By substituting (1.2) into (1.1), we obtain

$$\begin{aligned} f(t) &= c_0 + \sum_{n=1}^{\infty}\left(c_n e^{jn\Omega_0 t} + c_{-n} e^{-jn\Omega_0 t}\right) \\ &= \sum_{n=-\infty}^{\infty} c_n e^{jn\Omega_0 t} \end{aligned} \tag{1.3}$$

This expression is called the *complex form of the Fourier series.* From (1.1) and (1.2), it follows that

$$c_n = \frac{1}{T_0}\int_{-T_0/2}^{T_0/2} f(t) e^{-jn\Omega_0 t} dt \quad \text{for } n = 0, \pm 1, \pm 2, \cdots \tag{1.4}$$

The magnitude $|c_n|$ and phase angle $\angle c_n = -\tan^{-1}(b_n/a_n)$ of c_n are called the *magnitude spectrum* and *phase spectrum*, respectively. In the sequel, we shall use a two-way arrow to represent a Fourier series pair, for instance

$$f(t) \longleftrightarrow \{c_n\} \tag{1.5}$$

1.4.2 The Fourier Transform

A signal that is a function of time can be represented by a combination of sinusoids and co-sinusoids of various frequencies. In such a depiction, an infinite number of terms are usually employed. It is called a *frequency domain* representation and known as the Fourier transform, which has been a powerful tool for the analysis and design of filters.

The Fourier transform consists of a pair of integral relations

$$F(\Omega) = \mathcal{F}[f(t)] = \int_{-\infty}^{\infty} f(t)e^{-j\Omega t}dt \quad \text{for} \quad -\infty < \Omega < \infty \tag{1.6}$$

and

$$f(t) = \mathcal{F}^{-1}[F(\Omega)] = \frac{1}{2\pi}\int_{-\infty}^{\infty} F(\Omega)e^{j\Omega t}d\Omega \tag{1.7}$$

Equations (1.6) and (1.7) are called the *Fourier transform* and *inverse Fourier transform*, respectively. The magnitude and phase angle of $F(\Omega)$, namely,

$$|F(\Omega)| \text{ and } \angle F(\Omega) = \tan^{-1}\left(\frac{\text{Im}\{F(\Omega)\}}{\text{Re}\{F(\Omega)\}}\right)$$

are called the *magnitude spectrum* and *phase spectrum*, respectively. In the sequel, we shall use a two-way arrow to represent a Fourier transform pair, for instance

$$f(t) \longleftrightarrow F(\Omega) \tag{1.8}$$

We now consider the energy contained in a signal and then relate that energy to the Fourier transform of the signal.

The total energy contained in a continuous-time signal $f(t)$ is given by

$$E = \int_{-\infty}^{\infty} |f(t)|^2 dt \tag{1.9}$$

From (1.6) and (1.7), it follows that

$$\int_{-\infty}^{\infty} |f(t)|^2 dt = \frac{1}{2\pi}\int_{-\infty}^{\infty} |F(\Omega)|^2 d\Omega \tag{1.10}$$

This is called *Parseval's theorem* that relates the energy contained in a continuous-time signal to the Fourier transform of that signal.

1.4.3 The Laplace Transform

We start by obtaining the Fourier transform of an one-sided function of time, multiplied by a function that decays as time increases, that is, $f(t) = 0$ for $t < 0$ and $e^{-\sigma t} f(t)$ for $t \geq 0$ where σ is a real number. The Fourier transform pair in this case can be written as

$$F_\sigma(\Omega) = \int_0^\infty f(t)e^{-(\sigma+j\Omega)t}dt \quad \text{for} \quad -\infty < \Omega < \infty \tag{1.11}$$

and

$$f(t) = \frac{1}{2\pi}\int_{-\infty}^\infty F_\sigma(\Omega)e^{(\sigma+j\Omega)t}d\Omega \tag{1.12}$$

where the factor $e^{-\sigma t}$ has been moved from the left side of (1.12) to the right.

By defining

$$s = \sigma + j\Omega \tag{1.13}$$

we obtain

$$\frac{ds}{d\Omega} = j \tag{1.14}$$

provided that σ is constant. Substituting (1.13) and (1.14) into (1.11) and (1.12) yields

$$F(s) = \mathcal{L}[f(t)] = \int_0^\infty f(t)e^{-st}dt \tag{1.15}$$

and

$$f(t) = \mathcal{L}^{-1}[F(s)] = \frac{1}{2\pi j}\int_{\sigma-j\infty}^{\sigma+j\infty} F(s)e^{st}ds \tag{1.16}$$

respectively. Function $F(s)$ is called the *Laplace transform* of $f(t)$ and, conversely, $f(t)$ is called the *inverse Laplace transform* of $F(s)$. In the sequel, we shall use a two-way arrow to represent a Laplace transform pair, for instance

$$f(t) \longleftrightarrow F(s) \tag{1.17}$$

Final-value theorem:

The *final-value theorem* can be stated as follows [7]: If $f(t)$ and $df(t)/dt$ are both Laplace transformable, if $F(s)$ is the Laplace transform of $f(t)$, and if $\lim_{t\to\infty} f(t)$ exists, then

$$\lim_{t \to \infty} f(t) = \lim_{s \to 0} sF(s) \tag{1.18}$$

Initial-value theorem:
The *initial-value theorem* can be stated as follows [7]: If $f(t)$ and $df(t)/dt$ are both Laplace transformable and if $\lim_{s \to \infty} sF(s)$ exists, then

$$f(0) = \lim_{t \to +0} f(t) = \lim_{s \to \infty} sF(s) \tag{1.19}$$

Some common Laplace transform pairs can be found in Table 1.1.

Table 1.1 Laplace transform pairs

$f(t)$	$F(s)$
Unit impulse $\delta(t)$	1
Unit step $u_o(t)$	$\frac{1}{s}$
e^{-at}	$\frac{1}{s+a}$
t	$\frac{1}{s^2}$
$\cos \Omega t$	$\frac{s}{s^2 + \Omega^2}$
$\sin \Omega t$	$\frac{\Omega}{s^2 + \Omega^2}$
$e^{-at} \cos \Omega t$	$\frac{s+a}{(s+a)^2 + \Omega^2}$
$e^{-at} \sin \Omega t$	$\frac{\Omega}{(s+a)^2 + \Omega^2}$
$\frac{df(t)}{dt}$	$sF(s) - f(0)$
$\int_0^t f(\tau) d\tau$	$\frac{F(s)}{s}$
$e^{-at} f(t)$	$F(s+a)$
$\int_0^t f_1(\tau) f_2(t-\tau) d\tau$	$F_1(s) F_2(s)$

1.5 Analysis of Discrete-Time Signals

1.5.1 Sampling an Analog Signal

We now consider the process of sampling an analog signal and holding this value. A discrete-time signal can be obtained by sampling an analog signal at times kT for $k = 0, 1, 2, \cdots$ where T is the *sampling period*. The *sampling frequency* is given by

$$F_s = \frac{1}{T} \text{ Hz} \tag{1.20}$$

The sampling process is illustrated in Figure 1.3.

Suppose $f_a(t)$ is an analog signal input to the sampler, then the sampled output signal $\hat{f}_a(t)$ is the product

$$\hat{f}_a(t) = f_a(t)\Delta_T(t) \tag{1.21}$$

where the subscript a of $f_a(t)$ in (1.21) is used to indicate an analog signal. The modulating function $\Delta_T(t)$ is a train of uniformly spaced impulse functions given by

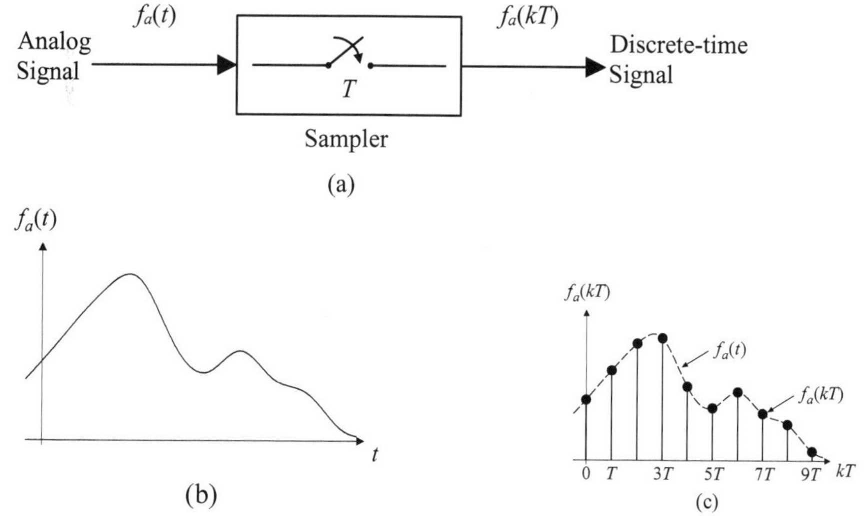

Figure 1.3 Periodic sampling of an analog signal: (a) Sampler, (b) Analog signal, (c) Sampled-data signal.

$$\Delta_T(t) = \sum_{k=-\infty}^{\infty} \delta_a(t - kT) \tag{1.22}$$

where $\delta_a(t)$ is the *Dirac delta function* or simply the *impulse function* defined by

$$\delta_a(t) = \begin{cases} \infty, & t = 0 \\ 0, & t \neq 0 \end{cases} \qquad \int_{-\infty}^{\infty} \delta_a(t)dt = 1$$

$$\int_{-\infty}^{\infty} f_a(t)\delta_a(t - t_0)dt = f_a(t_0)$$

If $\Delta_T(t)$ is replaced by its series representation, the sampled output signal $\hat{f}_a(t)$ in (1.21) can be expressed as

$$\begin{aligned} \hat{f}_a(t) &= f_a(t) \sum_{k=-\infty}^{\infty} \delta_a(t - kT) \\ &= \sum_{k=-\infty}^{\infty} f_a(t)\delta_a(t - kT) \end{aligned} \tag{1.23}$$

1.5.2 The Discrete-Time Fourier Transform

We shall start by finding the Fourier transform of a sampled analog signal $\hat{f}_a(t)$, which has been discussed in the previous section. The Fourier transform of $\hat{f}_a(t)$ is given by

$$\begin{aligned} \mathcal{F}[\hat{f}_a(t)] &= \int_{-\infty}^{\infty} \hat{f}_a(t)\, e^{-j\Omega t} dt \\ &= \int_{-\infty}^{\infty} \sum_{k=-\infty}^{\infty} f_a(t)\delta_a(t - kT)\, e^{-j\Omega t} dt \\ &= \sum_{k=-\infty}^{\infty} \int_{-\infty}^{\infty} f_a(t)\delta_a(t - kT)\, e^{-j\Omega t} dt \end{aligned} \tag{1.24}$$

Using a property of the Dirac delta function, this expression can be reduced to

$$\mathcal{F}[\hat{f}_a(t)] = \sum_{k=-\infty}^{\infty} f_a(kT)\, e^{-j\Omega kT} \tag{1.25}$$

where $f_a(kT)$ denotes the value of the kth sample of $f_a(t)$.

If we let $f(k) = f_a(kT)$, then (1.25) induces the Fourier transform of $\{f(k)\}$ as $\mathcal{F}[f(k)]$. In general, the Fourier transform of the discrete-time signal is given by

$$F(\omega) = \mathcal{F}[f(k)] = \sum_{k=-\infty}^{\infty} f(k)\, e^{-j\omega k} \tag{1.26}$$

where $\omega = \Omega T$. $F(\omega)$ in (1.26) is called the *discrete-time Fourier transform* of a sequence $f(k)$. The Fourier transform $F(\omega)$ is periodic with period 2π, as can easily be verified from (1.26). The *inverse discrete-time Fourier transform* of $F(\omega)$ is found to be

$$f(k) = \mathcal{F}^{-1}[F(\omega)] = \frac{1}{2\pi}\int_0^{2\pi} F(\omega)e^{j\omega k} d\omega \tag{1.27}$$

The magnitude and phase angle of $F(\omega)$, namely,

$$|F(\omega)| \text{ and } \angle F(\omega) = \tan^{-1}\left(\frac{\text{Im}\{F(\omega)\}}{\text{Re}\{F(\omega)\}}\right)$$

are called the *magnitude spectrum* and *phase spectrum*, respectively. In the sequel, we shall use a two-way arrow to represent a discrete-time Fourier transform pair, for instance

$$f(k) \longleftrightarrow F(\omega) \tag{1.28}$$

We now derive *Parseval's theorem* for discrete-time signals. The total energy contained in a discrete-time signal $f(k)$ is given by

$$E = \sum_{k=-\infty}^{\infty} |f(k)|^2 \tag{1.29}$$

From (1.26) and (1.27), it follows that

$$\sum_{k=-\infty}^{\infty} |f(k)|^2 = \frac{1}{2\pi}\int_0^{2\pi} |F(\omega)|^2 d\omega \tag{1.30}$$

where $|F(\omega)|^2$ is the *power spectrum* of the signal $f(k)$. This is the discrete-time version of Parseval's theorem.

1.5.3 The Discrete Fourier Transform (DFT)

Given a finite length sequence $\{f(n)\}$ which is defined only in the interval $0 \leq n \leq N-1$, the *discrete Fourier transform* (DFT) is defined by

$$F(k) = \sum_{n=0}^{N-1} f(n)\, W_N^{kn} \quad \text{for } k = 0, 1, \cdots, N-1 \tag{1.31}$$

where $W_N = e^{-\frac{2\pi}{N}}$ and W_N is called the *twiddle factor*. Conversely, the *inverse discrete Fourier transform* (IDFT) is given by

$$f(n) = \frac{1}{N} \sum_{k=0}^{N-1} F(k)\, W_N^{-kn} \quad \text{for } n = 0, 1, \cdots, N-1 \tag{1.32}$$

The magnitude and phase angle of $F(k)$, namely,

$$|F(k)| \text{ and } \angle F(k) = \tan^{-1}\left(\frac{\text{Im}\{F(k)\}}{\text{Re}\{F(k)\}}\right)$$

are called the *magnitude spectrum* and *phase spectrum*, respectively. In the sequel, we shall use a two-way arrow to represent a DFT pair, for instance

$$f(n) \longleftrightarrow F(k) \tag{1.33}$$

We now examine *Parseval's theorem* for an N-point sequence $f(n)$ and its N-point DFT $F(k)$. The total energy contained in an N-point sequence $f(n)$ is given by

$$E = \sum_{n=0}^{N-1} |f(n)|^2 \tag{1.34}$$

From (1.31) and (1.32), it follows that

$$\sum_{n=0}^{N-1} |f(n)|^2 = \frac{1}{N} \sum_{k=0}^{N-1} |F(k)|^2 \tag{1.35}$$

This is commonly referred to as Parseval's theorem for the DFT.

1.5.4 The *z*-Transform

In the study of discrete-time signals and systems, the z-transform plays an important role similar to that of the Laplace transform for continuous-time signals and systems. We shall start by obtaining the Laplace transform of a sampled analog signal $\hat{f}_a(t)$, which has been discussed in Section 1.5.1. From (1.23), the Laplace transform of $\hat{f}_a(t)$ is described by

$$\begin{aligned}\mathcal{L}[\hat{f}_a(t)] &= \int_0^{\infty} \hat{f}_a(t)\, e^{-st} dt \\ &= \int_0^{\infty} \sum_{k=-\infty}^{\infty} f_a(t)\delta_a(t-kT)\, e^{-st} dt \\ &= \sum_{k=-\infty}^{\infty} \int_0^{\infty} f_a(t)\delta_a(t-kT)\, e^{-st} dt\end{aligned} \tag{1.36}$$

By virtue of a property of the Dirac delta function, this expression can be deduced to

$$\mathcal{L}[\hat{f}_a(t)] = \sum_{k=-\infty}^{\infty} f_a(kT)\, e^{-skT} \tag{1.37}$$

If we let $f(k) = f_a(kT)$ and

$$e^{sT} = z \tag{1.38}$$

then (1.37) and (1.38) induce the Laplace transform of $\{f(k)\}$ as $\mathcal{Z}[f(k)]$. In general, the z-transform of the discrete-time signal $f(k)$ is given by

$$F(z) = \mathcal{Z}[f(k)] = \sum_{k=-\infty}^{\infty} f(k)\, z^{-k} \tag{1.39}$$

The reader is referred to Sections 2.3, 2.4 and 2.5 for further details.

1.6 Sampling of Continuous-Time Sinusoidal Signals

We now consider a continuous-time sinusoidal signal of the form

$$\begin{gathered}x_a(t) = A\cos(\Omega t + \theta) = A\cos(2\pi F t + \theta) \\ -\infty < t < \infty, \qquad -\infty < \Omega, F < \infty\end{gathered} \tag{1.40}$$

where A is the *amplitude* of the sinusoid, Ω is the *frequency* in radians per second (rad/s), F is the *frequency* in hertz (Hz), θ is the *phase* in radian, and $\Omega = 2\pi F$.

By acquiring samples of the signal $x_a(t)$ every T seconds, the discrete-time sinusoidal signal can be obtained as

$$x(k) \triangleq x_a(kT) = A\cos(2\pi FkT + \theta) = A\cos(k\omega + \theta) \qquad (1.41)$$

where

$$\omega = 2\pi f = 2\pi FT = \Omega T, \qquad f = FT = \frac{F}{F_s}$$

Here, f and F_s are called the *normalized frequency* and the *sampling frequency*, respectively.

Assuming that

$$\omega_i = \omega_0 + 2\pi i \ \text{ for } \ i = 0, 1, 2, \cdots \qquad (1.42)$$

we have

$$\begin{aligned} x_i(k) &= A\cos(k\omega_i + \theta) = A\cos\{k(\omega_0 + 2\pi i) + \theta\} \\ &= A\cos(k\omega_0 + \theta) = x_0(k) \end{aligned} \qquad (1.43)$$

hence discrete-time sinusoids, whose frequencies are separated by an integer multiple of 2π, are identical. Alternatively, the sequences of any two sinusoids with frequencies in the range $-\pi \le \omega \le \pi$ or $-\frac{1}{2} \le f \le \frac{1}{2}$ are distinct. In other words, discrete-time sinusoidal signal with frequencies in the range $|\omega| \le \pi$ or $|f| \le \frac{1}{2}$ are unique, and all frequencies in the range $|\omega| > \pi$ or $|f| > \frac{1}{2}$ are aliases. As a result, periodic sampling of a continuous-time signal can be viewed as a mapping of an infinite frequency range $-\infty < F < \infty$ into a finite frequency range $-\frac{1}{2} \le f \le \frac{1}{2}$.

From $f = F/F_s$ and $|f| \le \frac{1}{2}$, it follows that

$$-\frac{F_s}{2} \le F \le \frac{F_s}{2} \qquad (1.44)$$

Since the highest frequency in a discrete-time signal is $\omega = \pi$, i.e. $f = \frac{1}{2}$ with a sampling rate $F_s = 1/T$, the corresponding highest value of the continuous-time frequency is given by

$$F_{max} = \frac{F_s}{2} = \frac{1}{2T} \qquad (1.45)$$

1.7 Aliasing

We now examine what happens to the frequencies $\{F\}$ with $F > F_{max} = F_s/2$.

With sampling rate $F_s = 1/T$, the sampling of a continuous-time sinusoidal signal

$$x_a(t) = A\cos(2\pi F_0 t + \theta) \tag{1.46}$$

yields a discrete-time signal of the form

$$x(k) \triangleq x_a(kT) = A\cos(2\pi F_0 kT + \theta) = A\cos(2\pi f_0 k + \theta) \tag{1.47}$$

where $f_0 = F_0/F_s$ is the normalized frequency of the sinusoid. Suppose the frequency range for a continuous-time signal $x_a(t)$ is $-F_s/2 \leq F_0 \leq F_s/2$, the frequency range for a discrete-time signal $x(k)$ is $-1/2 \leq f_0 \leq 1/2$ which is a one-to-one relationship between F_0 and f_0. Hence, the analog signal $x_a(t)$ can be reconstructed from the samples $x(k)$.

Alternatively, if the sinusoid signals described by

$$x_a(t) = A\cos(2\pi F_i t + \theta) \tag{1.48}$$

are sampled at rate $F_s = 1/T$ where

$$F_i = F_0 + iF_s, \qquad i = \pm 1, \pm 2, \cdots$$

then the frequency F_i is outside the frequency range $-F_s/2 \leq F_0 \leq F_s/2$. In this case, the sampled signal is described by

$$\begin{aligned} x(k) \triangleq x_a(kT) &= A\cos(2\pi F_i kT + \theta) \\ &= A\cos\left(2\pi \frac{F_0 + iF_s}{F_s} k + \theta\right) \\ &= A\cos(2\pi f_0 k + \theta) \end{aligned} \tag{1.49}$$

Equation (1.49) coincides with (1.47) which is derived from (1.46) by sampling. Hence, there exist an infinite number of continuous-time sinusoids that produce the same discrete-time signal after sampling. As a result, the frequencies $F_i = F_0 + iF_s$ for $-\infty < i < \infty$ are indistinguishable from the frequency F_0 after sampling and hence they are aliases of F_0. The relation between the frequency variables of the continuous-time and discrete-time signals is illustrated in Figure 1.4. An example of aliasing is shown in

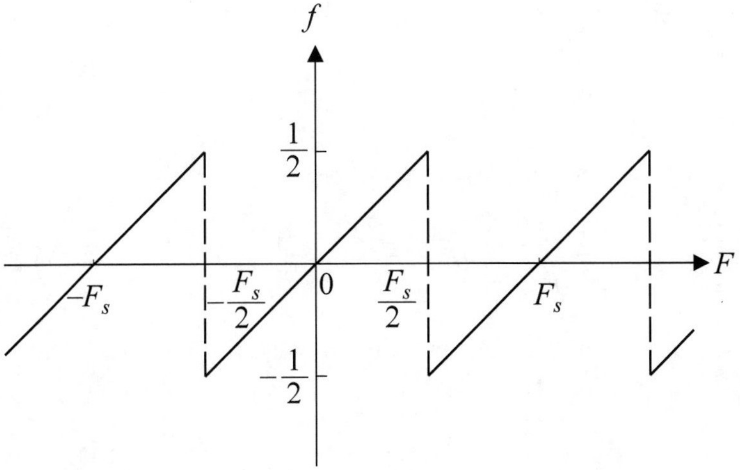

Figure 1.4 Relationship between the continuous-time and discrete-time frequency variables in periodic sampling.

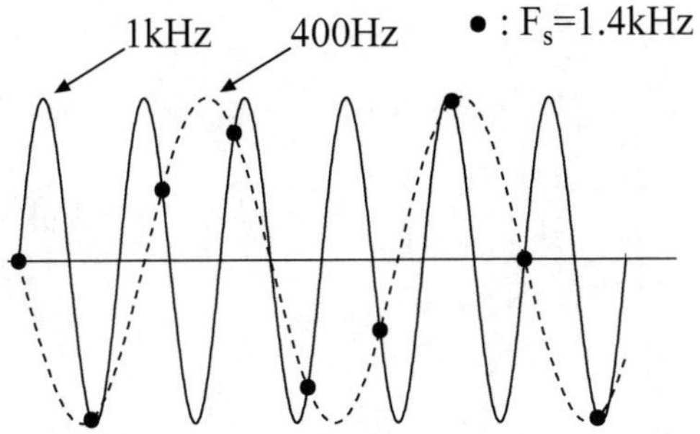

Figure 1.5 Illustration of aliasing.

Figure 1.5 where two sinusoids with frequencies $F_0 = 400$ Hz and $F_{-1} = -1000$ Hz yield identical samples when a sampling rate of $F_s = 1400$ Hz is employed.

1.8 Sampling Theorem

The periodic function $\Delta_T(t)$ in (1.22) can be written using the Fourier series expansion as

$$\Delta_T(t) = \sum_{k=-\infty}^{\infty} c_k \, e^{j\frac{2\pi}{T}kt} \tag{1.50}$$

where the Fourier series expansion coefficients are given by

$$c_k = \frac{1}{T}\int_{-\frac{T}{2}}^{\frac{T}{2}} \Delta_T(t)\, e^{-j\frac{2\pi}{T}kt} dt \tag{1.51}$$

Substituting (1.22) into (1.51), we obtain

$$c_k = \frac{1}{T}\int_{-\frac{T}{2}}^{\frac{T}{2}} \delta_a(t)\, e^{-j\frac{2\pi}{T}kt} dt = \frac{1}{T}e^0 = \frac{1}{T} \tag{1.52}$$

Hence, by using (1.50), the sampled output signal $\hat{f}_a(t)$ in (1.21) can be expressed as

$$\hat{f}_a(t) = f_a(t)\,\Delta_T(t) = \frac{1}{T}\sum_{k=-\infty}^{\infty} f_a(t)\, e^{j\frac{2\pi}{T}kt} \tag{1.53}$$

Taking the Laplace transform on the both sides of (1.53) then gives

$$\begin{aligned}\hat{F}_a(s) = \mathcal{L}[\hat{f}_a(t)] &= \frac{1}{T}\int_0^{\infty} \sum_{k=-\infty}^{\infty} f_a(t)\, e^{j\frac{2\pi}{T}kt} e^{-st} dt \\ &= \frac{1}{T}\sum_{k=-\infty}^{\infty} \int_0^{\infty} f_a(t)\, e^{-(s-j\frac{2\pi}{T}k)t} dt \\ &= \frac{1}{T}\sum_{k=-\infty}^{\infty} F_a\left(s - j\frac{2\pi}{T}k\right)\end{aligned} \tag{1.54}$$

where $F_a(s) = \mathcal{L}[f_a(t)]$. Substituting $s = j\Omega$ into (1.54), we obtain

$$\begin{aligned}\hat{F}_a(j\Omega) &= F_s \sum_{k=-\infty}^{\infty} F_a\big[j(\Omega - k\Omega_s)\big] \\ &= F_s \sum_{k=-\infty}^{\infty} F_a\big[j2\pi(F - kF_s)\big]\end{aligned} \tag{1.55}$$

where $\Omega = 2\pi F$ and $\Omega_s = 2\pi/T = 2\pi F_s$. This explains the relationship between the spectrum $\hat{F}_a(j\Omega)$ of the sampled discrete-time signal $f_a(kT)$ and the spectrum $F_a(j\Omega)$ of the analog signal $f_a(t)$. The right side of (1.55) is periodic with repetition of the spectrum $F_sF_a(j\Omega)$ and period Ω_s.

We now consider the case where the spectrum $F_a(j\Omega)$ of an analog signal $f_a(t)$ is band-limited as shown in Figure 1.6(a) and the spectrum is zero for $|F| \geq B$. If the sampling frequency F_s is chosen as $F_s > 2B$, the spectrum $\hat{F}_a(j\Omega)$ of the sampled discrete-time signal $f_a(kT)$ appears as shown in Figure 1.6(b). Therefore, if the sampling frequency F_s is chosen so that $F_s \geq 2B$, where the frequency $F_s = 2B$ is usually referred as the *Nyquist frequency*, then it follows that

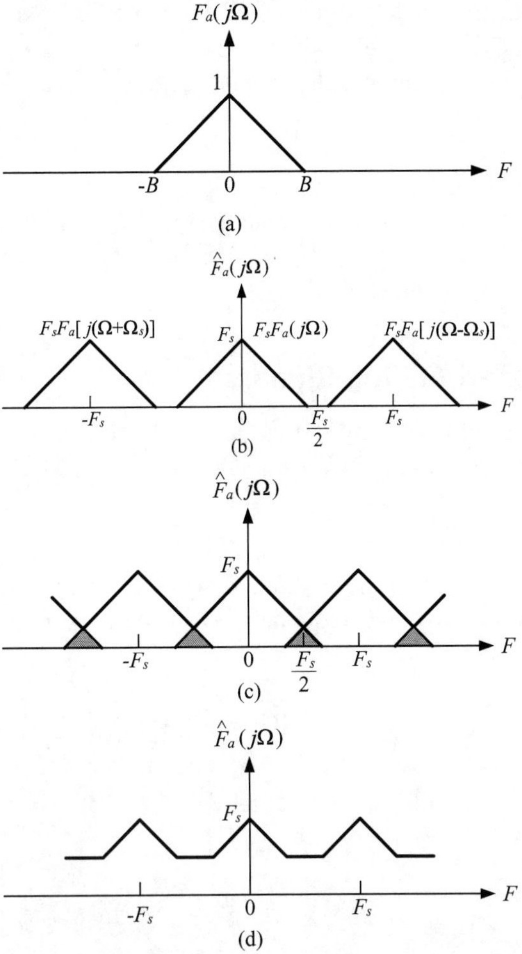

Figure 1.6 Aliasing of spectral components: (a) Spectrum of a band-limited analog signal, (b) Spectrum of the discrete-time signal, (c) (d) Spectrum of the discrete-time signal with spectral overlap.

$$\hat{F}_a(j\Omega) = F_s F_a(j\Omega), \qquad |F| \leq F_s/2 \tag{1.56}$$

In this case, aliasing does not exist and hence the spectrum of the sampled discrete-time signal $f_a(kT)$ is essentially identical with that of the analog signal $f_a(t)$ in the frequency range $|F| \leq F_s/2$.

If the sampling frequency F_s is chosen so that $F_s < 2B$, the periodic continuation of $F_a(j\Omega)$ generates spectral overlap as shown in Figures 1.6(c) and 1.6(d). Hence, the spectrum $\hat{F}_a(j\Omega)$ of the sampled discrete-time signal $f_a(kT)$ contains aliased frequency components of the analog signal spectrum $F_a(j\Omega)$. As a result, the analog signal $f_a(t)$ cannot be recovered from its sample values $\{f(kT)\}$ due to aliasing.

From the above observations, the *sampling theorem* can be stated as follows.

Sampling Theorem:
If the highest frequency contained in an analog signal $f_a(t)$ is $F_{max} = B$ and the signal is sampled at a rate $F_s \geq 2F_{max} = 2B$, then $f_a(t)$ can be exactly recovered from its sample values $\{f_a(kT)\}$.

1.9 Recovery of an Analog Signal

For a band-limited signal with highest frequency B, it is possible to avoid aliasing by sampling the signal at a frequency $F_s = 1/T$ ($\Omega_s = 2\pi/T$) satisfying

$$F_s = 1/T \geq 2B \qquad (\Omega_s = 2\pi/T \geq 4\pi B) \tag{1.57}$$

Namely, aliasing is avoidable if the signal is sampled at a frequency F_s which is at least as high as the Nyquist frequency $2B$. If the condition in (1.57) is satisfied, then

$$\hat{F}_a(j\Omega) = F_s F_a(j\Omega) \quad \text{for } |\Omega| \leq \frac{\Omega_s}{2} \quad \left(|F| \leq \frac{F_s}{2}\right) \tag{1.58}$$

holds. Since $F_a(j\Omega) = 0$ outside the interval $|\Omega| \leq \frac{\Omega_s}{2}$, by using the inverse Fourier transform of $F_a(j\Omega)$, we obtain

$$f_a(t) = \frac{1}{2\pi} \int_{-\infty}^{\infty} F_a(j\Omega) e^{j\Omega t} d\Omega = \frac{T}{2\pi} \int_{-\Omega_s/2}^{\Omega_s/2} \hat{F}_a(j\Omega) e^{j\Omega t} d\Omega \tag{1.59}$$

If $\hat{F}_a(j\Omega)$ in (1.59) is represented by the discrete-time Fourier transform

$$\hat{F}_a(j\Omega) = \sum_{k=-\infty}^{\infty} f_a(kT)e^{-j\Omega kT} \tag{1.60}$$

then it follows that

$$\begin{aligned} f_a(t) &= \frac{T}{2\pi}\int_{-\Omega_s/2}^{\Omega_s/2}\sum_{k=-\infty}^{\infty} f_a(kT)e^{-j\Omega kT}e^{j\Omega t}d\Omega \\ &= \frac{T}{2\pi}\sum_{k=-\infty}^{\infty} f_a(kT)\int_{-\Omega_s/2}^{\Omega_s/2} e^{j\Omega(t-kT)}d\Omega \\ &= \frac{T}{2\pi}\sum_{k=-\infty}^{\infty} f_a(kT)\left[\frac{e^{j\Omega(t-kT)}}{j(t-kT)}\right]_{-\Omega_s/2}^{\Omega_s/2} \\ &= \sum_{k=-\infty}^{\infty} f_a(kT)\,\frac{\sin(\pi/T)(t-kT)}{(\pi/T)(t-kT)} \end{aligned} \tag{1.61}$$

The reconstruction formula in (1.61) is known as an *interpolation formula* for reconstructing $f_a(t)$ from its sample values. This interpolation formula is built on the *interpolation function*

$$g(t) = \frac{\sin(\pi/T)t}{(\pi/T)t} \tag{1.62}$$

by summing up appropriately shifted $g(t)$ by kT, namely $g(t-kT)$, for $k = 0, \pm 1, \pm 2, \cdots$ with the sample values $f_a(kT)$ as the weights. It is noted that at $t = iT$, the interpolation function $g(t-kT)$ is zero except at $i = k$. As a result, $f_a(t)$ evaluated at $t = iT$ is simply equal to the sample value $f_a(iT)$. At all other times, the weighted sum of the time shifted versions of the interpolation function combines to produce exactly $f_a(t)$. The ideal band-limited interpolation process is illustrated in Figure 1.7.

1.10 Summary

This chapter has introduced terminology for signal analysis and examples of typical signals, and presented an overview of digital signal processing with the explanation of its advantages and disadvantages. This chapter has also analyzed analog and discrete-time signals, and examined the sampling of analog signals and that of continuous-time sinusoidal signals in connection

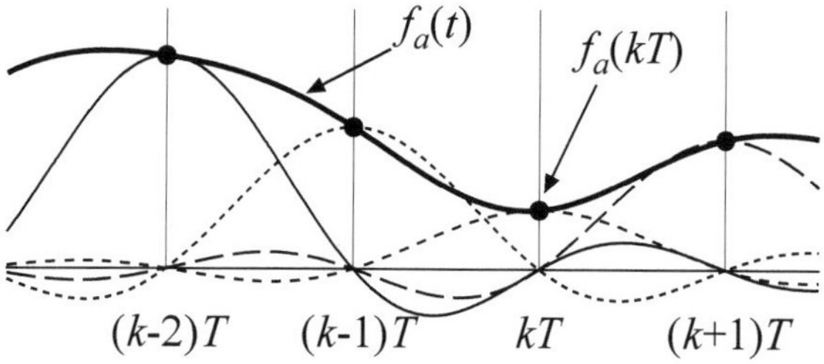

Figure 1.7 Ideal band-limited reconstruction by interpolation.

to aliasing. Moreover, the sampling theorem has been presented with the explanation of how to recover an analog signal from its discrete-time samples.

References

[1] A. V. Oppenheim and R. W. Schafer, *Digital Signal Processing*, NJ: Prentice-Hall, 1975.

[2] S. K. Mitra, *Digital Signal Processing*, 3rd ed. NJ: McGraw-Hill, 2006.

[3] J. G. Proakis and D. J. Manolakis, *Digital Signal Processing: Principles, Algorithms, and Applications*, 2nd ed. New York, Macmillan Publishing Company, 1992.

[4] J. R. Johnson, *Introduction to Digital Signal Processing*, NJ: Prentice Hall, 1989.

[5] T. Higuchi, *Fundamentals of Digital Signal Processing*, Tokyo, Japan, Shokodo, 1986.

[6] P. M. Chirlian, *Signals and Filters*, New York: Van Nostrand Reinhold, 1994.

[7] K. Ogata, *Modern Control Engineering*, 3rd ed., New Jersey: Prentice-Hall, Inc. 1997.

2

Discrete-Time Systems and *z*-Transformation

2.1 Preview

This chapter covers fundamental concepts relating to discrete-time signals and systems. In Section 2.2, we introduce common and important discrete-time sequences such as *unit pulse sequence, unit step sequence, unit ramp sequence, exponential sequence, and cosine sequence*. Most of these sequences consist of evenly space samples of familiar continuous-time functions. In Section 2.3, the one-side z-transform of a discrete-time sequence is defined, and then fundamental z-transforms as well as z-transform properties are explained. In Section 2.4, we present three methods for computing inverse z-transform using partial fraction expansion, power series expansion, and contour integration, respectively.

Several concepts that are most relevant to discrete-time systems, such as *linearity, time-invariance, stability*, and *causality*, are studied in Section 2.5. In Sections 2.6 and 2.7, linear time-invariant discrete-time systems are described in terms of difference equations as well as state-space descriptions. The block diagrams of digital filters and two methods for constructing state-space descriptions from a difference equation are presented. In Section 2.8, transfer functions in the frequency domain for causal linear time-invariant discrete-time systems are introduced, and several analysis and design issues for all-pass, notch, and doubly complementary digital filters are examined.

2.2 Discrete-Time Signals

A discrete-time signal $f(k)$ is a function of a discrete variable k. It is a sequence of numbers, called *samples*, indexed by the sample number k. In the rest of the book, unless specified otherwise, it is always assumed that a discrete-time signal is a sequence of samples obtained by sampling a continuous-time signal with a constant sampling frequency. Below are several special sequences that are found useful in the analysis of discrete-time signals and systems.

1. *Unit Pulse Sequence*
   The unit pulse sequence has a unit sample for $k = 0$ and all subsequent samples are zero, i.e.,

$$\delta(k) = \begin{cases} 1, & k = 0 \\ 0, & \text{otherwise} \end{cases} \tag{2.1}$$

2. *Unit Step Sequence*
   The unit step sequence has samples that are unity for $k = 0$ and thereafter, namely,

$$u_o(k) = \begin{cases} 1, & k = 0, 1, 2, \cdots \\ 0, & \text{otherwise} \end{cases} \tag{2.2}$$

3. *Unit Ramp Sequence*
   The unit ramp sequence is defined as

$$f(k) = ku_o(k) \tag{2.3}$$

   where $u_o(k)$ is the unit step sequence.
4. *Exponential Sequence*
   An exponential sequence is evenly spaced samples of an exponential function, that is,

$$f(k) = \begin{cases} e^{-\alpha kT}, & k = 0, 1, 2, \cdots \\ 0, & \text{otherwise} \end{cases} \tag{2.4}$$

   where $\alpha > 0$ and constant T, termed the *sampling interval* or *sampling period*, is the time interval between samples.
5. *Cosine Sequence*
   A cosine sequence is spaced samples of a cosine continuous-time function, that is,

$$f(k) = \begin{cases} \cos(\omega kT + \theta), & k = 0, 1, 2, \cdots \\ 0, & \text{otherwise.} \end{cases} \tag{2.5}$$

These sequences are shown in Figure 2.1.

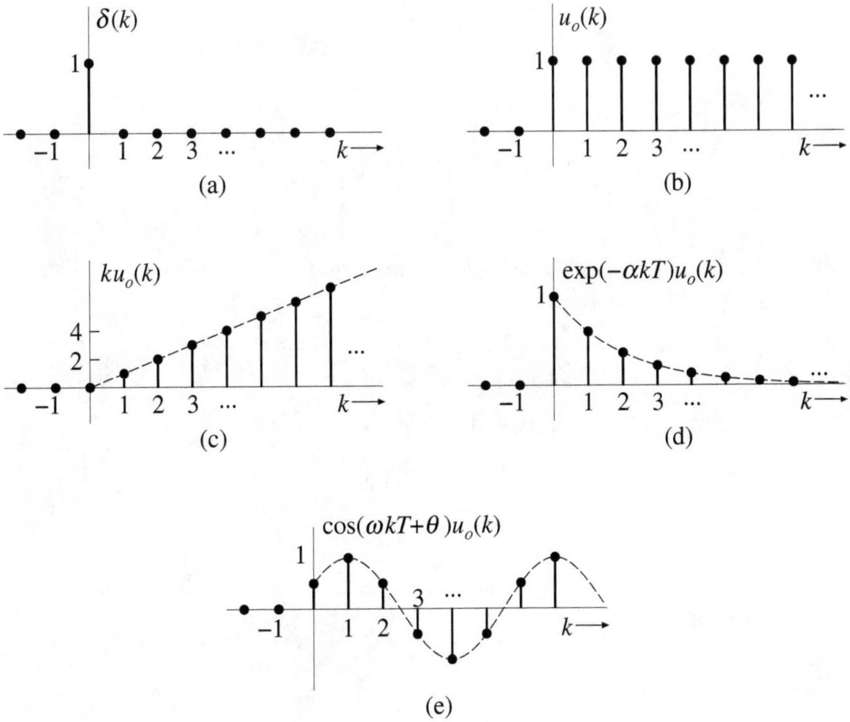

Figure 2.1 Several fundamental sequences. (a) Unit pulse. (b) Unit step. (c) Unit ramp. (d) Exponential. (e) Sinusoidal.

2.3 *z*-Transform of Basic Sequences

The *one-side* z-transform of a sequence $f(k)$ is defined by

$$\mathcal{Z}[f(k)] = F(z) = \sum_{k=0}^{\infty} f(k) z^{-k} \tag{2.6}$$

The z-transform is denoted by $\mathcal{Z}[\cdot]$ or $F(z)$. Note that the one-sided z-transform excludes samples before step zero in the transform.

2.3.1 Fundamental Transforms

1. The z-transform of the unit pulse is

$$\mathcal{Z}[\delta(k)] = \sum_{k=0}^{\infty} \delta(k) z^{-k} = \delta(0) = 1 \tag{2.7}$$

2. The z-transform of the unit step sequence becomes

$$\mathcal{Z}[u_o(k)] = \sum_{k=0}^{\infty} z^{-k} = \frac{1}{1 - z^{-1}} = \frac{z}{z - 1} \tag{2.8}$$

3. The z-transform of a geometrical progression $f(k) = a^k$ is

$$\mathcal{Z}[a^k] = \sum_{k=0}^{\infty} a^k z^{-k} = \frac{1}{1 - az^{-1}} = \frac{z}{z - a} \tag{2.9}$$

4. The z-transform of a ramp function $f(k) = ku_o(k)$ is

$$\begin{aligned} \mathcal{Z}[ku_o(k)] &= 0 + z^{-1} + 2z^{-2} + 3z^{-3} + 4z^{-4} + \cdots \\ z^{-1}\mathcal{Z}[ku_o(k)] &= 0 \ + \ z^{-2} + 2z^{-3} + 3z^{-4} + \cdots \end{aligned} \tag{2.10}$$

where $u_o(k)$ is the unit step sequence in (2.2). By subtracting the second equation from the first equation in (2.10), we obtain

$$(1 - z^{-1})\mathcal{Z}[ku_o(k)] = z^{-1}(1 + z^{-1} + z^{-2} + z^{-3} + \cdots) = \frac{z^{-1}}{1 - z^{-1}} \tag{2.11}$$

which yields

$$\mathcal{Z}[ku_o(k)] = \frac{z^{-1}}{(1 - z^{-1})^2} = \frac{z}{(z - 1)^2} \tag{2.12}$$

5. The z-transform of a cosine function $f(k) = \cos(k\omega T)$ becomes

$$\begin{aligned} Z\left[\cos k\omega T\right] &= \sum_{k=0}^{\infty} \cos k\omega T z^{-k} = \sum_{k=0}^{\infty} \frac{e^{jk\omega T} + e^{-jk\omega T}}{2} z^{-k} \\ &= \frac{1}{2} \sum_{k=0}^{\infty} \{(e^{j\omega T} z^{-1})^k + (e^{-j\omega T} z^{-1})^k\} \\ &= \frac{1}{2} \left\{ \frac{1}{1 - e^{j\omega T} z^{-1}} + \frac{1}{1 - e^{-j\omega T} z^{-1}} \right\} \\ &= \frac{1 - z^{-1} \cos \omega T}{1 - 2z^{-1} \cos \omega T + z^{-2}} = \frac{z(z - \cos \omega T)}{z^2 - 2z \cos \omega T + 1} \end{aligned} \tag{2.13}$$

The z-transform of several important sequences are listed in Table 2.1.

Table 2.1 z-Transform Pairs

$f(k)$	$F(z)$
$\delta(k)$	1
$u_o(k)$	$\frac{z}{z-1}$
a^k	$\frac{z}{z-a}$
k	$\frac{z}{(z-1)^2}$
$\cos k\omega T$	$\frac{z(z-\cos\omega T)}{z^2-2z\cos\omega T+1}$
$\sin k\omega T$	$\frac{z\sin\omega T}{z^2-2z\cos\omega T+1}$
$r^k\cos k\omega T$	$\frac{z(z-r\cos\omega T)}{z^2-2z\,r\cos\omega T+r^2}$
$r^k\sin k\omega T$	$\frac{z\,r\sin\omega T}{z^2-2z\,r\cos\omega T+r^2}$

2.3.2 Properties of *z*-Transform

Several important properties of the z-transform are given below.

1. *Linearity*
   The z-transform of a linear combination of sequences is the linear combination of the individual z-transforms, that is,

$$\begin{aligned}\mathcal{Z}[a_1 f_1(k) + a_2 f_2(k)] &= \sum_{k=0}^{\infty} a_1 f_1(k) z^{-k} + \sum_{k=0}^{\infty} a_2 f_2(k) z^{-k} \\ &= a_1 \mathcal{Z}[f_1(k)] + a_2 \mathcal{Z}[f_2(k)] \\ &= a_1 F_1(z) + a_2 F_2(z)\end{aligned} \tag{2.14}$$

   where a_1 and a_2 are arbitrary constants. Equation (2.14) shows that the z-transform is a linear transform.

2. *Step-Shifted Relations*
   A sequence that is delayed i steps has the z-transform

$$\begin{aligned}\mathcal{Z}[f(k-i)] &= \sum_{k=0}^{\infty} f(k-i) z^{-k} = z^{-i} \sum_{k=0}^{\infty} f(k-i) z^{-(k-i)} \\ &= z^{-i} \sum_{k=0}^{\infty} f(k) z^{-k} = z^{-i} F(z)\end{aligned} \tag{2.15}$$

where $f(k) = 0$ for $k < 0$. Similarly, a sequence that is i-step advance has the z-transform

$$\begin{aligned} \mathcal{Z}[f(k+i)] &= \sum_{k=0}^{\infty} f(k+i)z^{-k} = z^i \sum_{k=0}^{\infty} f(k+i)z^{-(k+i)} \\ &= z^i \left[\sum_{k=0}^{\infty} f(k)z^{-k} - \sum_{k=0}^{i-1} f(k)z^{-k} \right] \\ &= z^i \left[F(z) - \sum_{k=0}^{i-1} f(k)z^{-k} \right] \end{aligned} \tag{2.16}$$

3. *Multiplication by a geometrical progression* a^k
The z-transform of sequence $a^k f(k)$ with a nonzero constant a is given by

$$\mathcal{Z}[a^k f(k)] = \sum_{k=0}^{\infty} a^k f(k)z^{-k} = \sum_{k=l}^{\infty} f(k)(a^{-1}z)^{-k} = F(a^{-1}z) \tag{2.17}$$

4. *Differentiation*
A sequence multiplied by the step index k has the z-transform

$$\begin{aligned} \mathcal{Z}[kf(k)] &= \sum_{k=0}^{\infty} kf(k)z^{-k} = \sum_{k=0}^{\infty} f(k)(kz^{-k}) \\ &= \sum_{k=0}^{\infty} f(k) \left[-z\frac{d}{dz}z^{-k} \right] = -z\frac{d}{dz} \left[\sum_{k=0}^{\infty} f(k)z^{-k} \right] \\ &= -z\frac{dF(z)}{dz} \end{aligned} \tag{2.18}$$

5. *Convolution*
The z-transform of the convolution of two sequences $h(k)$ and $f(k)$ is given by

$$\begin{aligned} \mathcal{Z}\left[\sum_{l=0}^{\infty} h(l)f(k-l) \right] &= \sum_{k=0}^{\infty}\sum_{l=0}^{\infty} h(l)f(k-l)z^{-k} \\ &= \sum_{l=0}^{\infty} h(l)z^{-l} \sum_{k=0}^{\infty} f(k-l)z^{-(k-l)} = H(z)F(z) \end{aligned} \tag{2.19}$$

where $f(k) = 0$ for $k < 0$.

6. *Initial-Value Theorem*
The initial value of a sequence $f(k)$ is the value of the sequence at $k = 0$. It can be found from the z-transform of the sequence as

$$f(0) = \lim_{z \to \infty} \left[f(0) + f(1)z^{-1} + f(2)z^{-2} + \cdots \right] = \lim_{z \to \infty} F(z) \tag{2.20}$$

7. *Final-Value Theorem*
The Final value of a sequence $f(k)$ is the value of the sequence at $k = \infty$. By virtue of

$$\sum_{k=0}^{N} f(k)z^{-k} = f(0) + f(1)z^{-1} + \cdots + f(N)z^{-N}$$

$$\begin{aligned} \sum_{k=0}^{N} f(k-1)z^{-k} &= z^{-1}[f(0) + f(1)z^{-1} + \cdots \\ &\quad + f(N-1)z^{-(N-1)}] \\ &= z^{-1} \sum_{k=0}^{N-1} f(k)z^{-k} \end{aligned} \tag{2.21}$$

it follows that

$$f(N) = \lim_{z \to 1} \left[\sum_{k=0}^{N} f(k)z^{-k} - z^{-1} \sum_{k=0}^{N-1} f(k)z^{-k} \right] \tag{2.22}$$

As $N \to \infty$, the final-value theorem is found from (2.22) as

$$\lim_{k \to \infty} f(k) = \lim_{z \to 1} \left[F(z) - z^{-1}F(z) \right] = \lim_{z \to 1} (1 - z^{-1})F(z) \tag{2.23}$$

2.4 Inversion of *z*-Transforms

The z-transform of a sequence $f(k)$ is defined in (2.6) as

$$F(z) = \sum_{k=0}^{\infty} f(k)z^{-k} \tag{2.24}$$

By multiplying both sides of (2.24) by z^{i-1} and integrating with a contour integral for which the contour of integration encloses the origin and lies entirely in the region of convergence of $F(z)$, we obtain

$$\begin{aligned}\frac{1}{2\pi j}\oint_C F(z)z^{i-1}dz &= \frac{1}{2\pi j}\oint_C \sum_{k=0}^{\infty} f(k)z^{-k+i-1}dz \\ &= \sum_{k=0}^{\infty} f(k)\frac{1}{2\pi j}\oint_C z^{-k+i-1}dz\end{aligned} \tag{2.25}$$

where C is a counterclockwise closed contour in the region of convergence of $F(z)$ and encircling the origin of the z-plane. Applying *Cauchy's integral theorem*

$$\frac{1}{2\pi j}\oint_C z^{i-1}dz = \begin{cases} 1, & i = 0 \\ 0, & i \neq 0 \end{cases} \tag{2.26}$$

hence the right-hand side of (2.25) is equal to $f(i)$. Therefore, the inverse z-transform can be described by the contour integral as

$$f(k) = \frac{1}{2\pi j}\oint_C F(z)z^{k-1}dz \tag{2.27}$$

We now examine some methods for obtaining the inverse z-transform from a given z-transform.

2.4.1 Partial Fraction Expansion

A useful method for obtaining the inverse z-transform from a given rational z-transform is to perform a partial-fraction expansion and then compute the inverse z-transform of each individual term. Suppose a rational z-transform $F(z)$ has been expressed in a partial-fraction expansion of the form

$$F(z) = \frac{c_1}{1-\lambda_1 z^{-1}} + \frac{c_2}{1-\lambda_2 z^{-1}} + \cdots + \frac{c_N}{1-\lambda_N z^{-1}} \tag{2.28}$$

then the inverse z-transform of $F(z)$ is given by

$$f(k) = c_1\lambda_1^k + c_2\lambda_2^k + \cdots + c_N\lambda_N^k \quad \text{for } k = 0, 1, 2, \ldots \tag{2.29}$$

where the coefficients c_i for $i = 1, 2, \cdots, N$ can be derived from

$$c_i = (1-\lambda_i z^{-1})F(z)\big|_{z=\lambda_i}$$

As an example, consider the problem of computing the inverse z-transform of

$$F(z) = \frac{-z^2(z+1)}{6z^3 - 11z^2 + 6z - 1}$$

We now express $F(z)$ as

$$F(z) = \frac{-(1+z^{-1})}{6(1-z^{-1})(1-\frac{1}{2}z^{-1})(1-\frac{1}{3}z^{-1})}$$

$$= \frac{c_1}{1-z^{-1}} + \frac{c_2}{1-\frac{1}{2}z^{-1}} + \frac{c_3}{1-\frac{1}{3}z^{-1}}$$

and compute

$$c_1 = (1-z^{-1})F(z)\Big|_{z=1} = -1, \qquad c_2 = \left(1-\frac{1}{2}z^{-1}\right)F(z)\Big|_{z=1/2} = \frac{3}{2}$$

$$c_3 = \left(1-\frac{1}{3}z^{-1}\right)F(z)\Big|_{z=1/3} = -\frac{2}{3}$$

which yields

$$f(k) = -1 + \frac{3}{2}\left(\frac{1}{2}\right)^k - \frac{2}{3}\left(\frac{1}{3}\right)^k \quad \text{for } k = 0, 1, 2, \cdots$$

2.4.2 Power Series Expansion

For a rational z-transform, a power series expansion can be obtained by long division. Suppose a rational z-transform $F(z)$ can be expressed in a power series expansion of the form

$$F(z) = \frac{b_m z^m + \cdots + b_1 z + b_0}{z^n + a_{n-1}z^{n-1} + \cdots + a_0} = \sum_{k=0}^{\infty} f(k)z^{-k}, \qquad m \le n \tag{2.30}$$

then the inverse z-transform $f(k)$ of $F(z)$ is readily obtained from (2.30). Consider for example

$$F(z) = \frac{2z^2 + z}{z^2 - 2z + 1}$$

which admits a power series expansion as

$$F(z) = 2 + 5z^{-1} + 8z^{-2} + 11z^{-3} + 14z^{-4} + 17z^{-5} + 20z^{-6} + \cdots$$

Hence we obtain

$$\{f(0), f(1), f(2), f(3), f(4), f(5), f(6), \cdots\} = \{2, 5, 8, 11, 14, 17, 20, \cdots\}$$

2.4.3 Contour Integration

For a rational z-transform, the contour integrals in (2.27) can be obtained using the residue theorem, that is,

$$\begin{aligned} f(k) &= \frac{1}{2\pi j}\oint_C F(z)z^{k-1}dz \\ &= \sum \Big[\text{residues of } F(z)z^{k-1} \text{ at the poles inside } C\Big] \\ &\qquad k = 0, 1, 2, \cdots \end{aligned} \tag{2.31}$$

In the case where $z = a$ is a first-order pole, the residue of $F(z)z^{k-1}$ becomes

$$\text{Res}(a) = (z-a)F(z)z^{k-1}\Big|_{z=a} \tag{2.32}$$

If $z = a$ is a multiple pole with multiplicity n, the residue of $F(z)z^{k-1}$ is given by

$$\text{Res}(a) = \frac{1}{(n-1)!}\frac{d^{n-1}}{dz^{n-1}}(z-a)^n F(z)z^{k-1}\Big|_{z=a} \tag{2.33}$$

As an example, consider

$$F(z) = \frac{1}{8}\frac{z^3}{(z-1)(z-\frac{1}{2})(z-\frac{1}{4})}$$

We now compute

$$\begin{aligned} \text{Res}(1) &= (z-1)F(z)z^{k-1}\Big|_{z=1} = \frac{1}{3} \\ \text{Res}(\tfrac{1}{2}) &= \Big(z-\frac{1}{2}\Big)F(z)z^{k-1}\Big|_{z=1/2} = -\frac{1}{4}\Big(\frac{1}{2}\Big)^k \\ \text{Res}(\tfrac{1}{4}) &= \Big(z-\frac{1}{4}\Big)F(z)z^{k-1}\Big|_{z=1/4} = \frac{1}{24}\Big(\frac{1}{4}\Big)^k \end{aligned}$$

which gives

$$f(k) = \frac{1}{3} - \frac{1}{4}\Big(\frac{1}{2}\Big)^k + \frac{1}{24}\Big(\frac{1}{4}\Big)^k \quad \text{for } k = 0, 1, 2, \cdots$$

2.5 Parseval's Theorem

In Section 1.5.2, Parseval's theorem is explored in terms of Fourier transforms. In this section, this theorem is extended to the domain of z-transforms. For simplicity, we consider two real sequences $f(k)$ and $g(k)$ for $k \geq 0$.

Theorem 2.1: *Parseval's Theorem*
Let $f(k)$ and $g(k)$ be two sequences from l_2 and let $F(v)$ and $G(v)$ be their z-transforms. Then the following equality holds true:

$$\sum_{k=0}^{\infty} f(k)g(k) = \frac{1}{2\pi j} \oint_C F(v)G(1/v)v^{-1}dv \tag{2.34}$$

where the contour of integration is taken in the overlap of the regions of convergence of $F(v)$ and $G(1/v)$.

Proof
Define a sequence $w(k)$ as

$$w(k) = f(k)g(k) \tag{2.35}$$

so that

$$W(z) = \sum_{k=0}^{\infty} f(k)g(k)z^{-k} \tag{2.36}$$

From (2.27), we can write

$$f(k) = \frac{1}{2\pi j} \oint_{C_1} F(v)v^{k-1}dv \tag{2.37}$$

where C_1 is a counterclockwise contour within the region of convergence of $F(v)$. Substituting (2.37) into (2.36) yields

$$\begin{aligned} W(z) &= \sum_{k=0}^{\infty} g(k)\frac{1}{2\pi j} \oint_{C_1} F(v)(z/v)^{-k}v^{-1}dv \\ &= \frac{1}{2\pi j} \oint_{C_1} F(v)\left[\sum_{k=0}^{\infty} g(k)(z/v)^{-k}\right] v^{-1}dv \\ &= \frac{1}{2\pi j} \oint_{C} F(v)G(z/v)v^{-1}dv \end{aligned} \tag{2.38}$$

Note that

$$\sum_{k=0}^{\infty} w(k) = W(z)\Big|_{z=1} \tag{2.39}$$

which in conjunction with (2.38) gives

$$\sum_{k=0}^{\infty} f(k)g(k) = \frac{1}{2\pi j}\oint_C F(v)G(1/v)v^{-1}dv \tag{2.40}$$

This completes the proof of Theorem 2.1. ■

Suppose $F(z)$ and $G(z)$ converge on the unit circle, we can choose $v = e^{j\omega}$ so that (2.34) becomes

$$\sum_{k=0}^{\infty} f(k)g(k) = \frac{1}{2\pi}\int_0^{2\pi} F(e^{j\omega})G(e^{-j\omega})d\omega \tag{2.41}$$

In particular, if $f(k)$ and $g(k)$ are two complex sequences, then (2.34) is changed to

$$\sum_{k=0}^{\infty} f(k)g^*(k) = \frac{1}{2\pi j}\oint_C F(v)G^*(1/v^*)v^{-1}dv \tag{2.42}$$

because $\mathcal{Z}[g^*(k)] = G^*(z^*)$ where $g^*(k)$ denotes the conjugate complex number of $g(k)$.

2.6 Discrete-Time Systems

A discrete-time system can be considered as a unique transformation or operator that maps an input sequence $u(k)$ into an output sequence $y(k)$, i.e.,

$$y(k) = S[u(k)] \tag{2.43}$$

where $S[\cdot]$ represents a transformation or an operator. Discussed below are several basic notion and concepts related to discrete-time systems.

1. *Linearity*

The class of *linear systems* is defined by the principle of superposition. Supposing that $y_1(k) = S[u_1(k)]$ and $y_2(k) = S[u_2(k)]$, the system in (2.43) is said to be *linear* if

$$\begin{aligned} S[au_1(k) + bu_2(k)] &= aS[u_1(k)] + bS[u_2(k)] \\ &= ay_1(k) + by_2(k) \end{aligned} \tag{2.44}$$

for arbitrary constants a and b.

2. *Time-Invariance*
The system in (2.43) is said to be a *time-invariant* system provided that

$$y(k - k_o) = S[u(k - k_o)] \tag{2.45}$$

always holds for an arbitrary integer k_o.

By using the unit pulse sequence $\delta(k)$ defined in (2.1), an arbitrary sequence $u(k)$ can be represented as

$$u(k) = \sum_{i=-\infty}^{\infty} u(i)\delta(k - i) \tag{2.46}$$

Therefore, the system in (2.43) can be written as

$$y(k) = S\left[\sum_{i=-\infty}^{\infty} u(i)\delta(k - i)\right] \tag{2.47}$$

Under the assumption that the system in (2.43) is *linear* and *time-invariant*, (2.47) can be expressed as

$$y(k) = \sum_{i=-\infty}^{\infty} u(i)S\big[\delta(k - i)\big] = \sum_{i=-\infty}^{\infty} u(i)h(k - i) \tag{2.48}$$

where $h(k) = S[\delta(k)]$. The sequence $h(k)$ in (2.48) is called *unit-sample response* or *impulse response*. From (2.48), we see that a linear time-invariant system is completely characterized by its impulse response.

Equation (2.48) is commonly referred to as the *convolution sum*, and is said to be the *convolution of* $u(k)$ *with* $h(k)$ which is often denoted by

$$y(k) = u(k) * h(k) \tag{2.49}$$

An alternative expression for the system in (2.48) is given by

$$y(k) = \sum_{i=-\infty}^{\infty} h(i)u(k - i) = h(k) * u(k) \tag{2.50}$$

The system described by (2.48) or (2.50) is called a *digital filter*.

As illustrated in Figure 2.2, two linear time-invariant systems connected in cascade form constitute a linear time-invariant system whose impulse response is the convolution of the two impulse responses. Two linear time-invariant

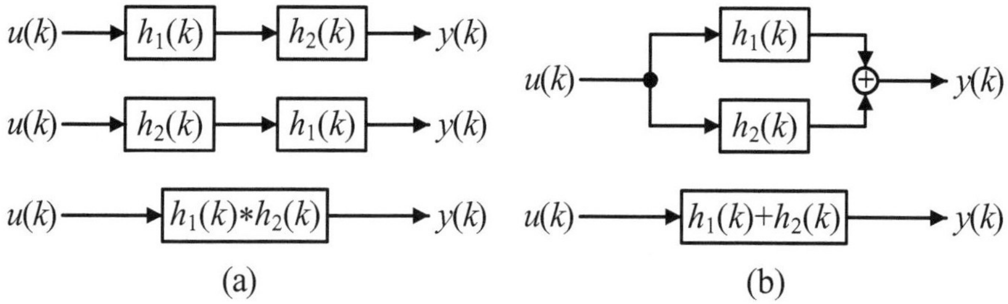

Figure 2.2 Typical linear time-invariant systems with identical unit-pulse responses. (a) Cascade forms. (b) Parallel forms.

systems connected in parallel form are equivalent to a single system whose impulse response is the sum of the individual impulse responses.

A more restricted class of linear time-invariant systems of practical significance obeys stability and causality.

3. *Stability*

A system is said to be *stable* if every bounded input produces a bounded output. A linear time-invariant system is *stable* if and only if

$$\sum_{k=-\infty}^{\infty} |h(k)| < \infty \tag{2.51}$$

4. *Causality*

A system is said to be *causal* if the output $y(k)$ for any $k = k_o$ depends on the input $u(k)$ for $k \leq k_o$ only. A linear time-invariant system is *causal* if and only if the impulse response $h(k)$ is zero for $k < 0$. Therefore, for linear time-invariant causal systems, (2.50) becomes

$$y(k) = \sum_{i=0}^{\infty} h(i)u(k-i) = \sum_{i=-\infty}^{k} h(k-i)u(i) \tag{2.52}$$

By taking the z-transform of both sides in (2.52), we obtain

$$Y(z) = \left[\sum_{i=0}^{\infty} h(i)z^{-i} \right] U(z) \tag{2.53}$$

where $U(z) = \mathcal{Z}[u(k)]$ and $Y(z) = \mathcal{Z}[y(k)]$. Therefore,

$$H(z) = \frac{Y(z)}{U(z)} = h(0) + h(1)z^{-1} + h(2)z^{-2} + \cdots \tag{2.54}$$

The $H(z)$ in (2.54) is called the *transfer function* of a linear time-invariant causal system in (2.52) whose *impulse response* is an infinite-length sequence. The digital filter in (2.52) is realizable and of practical importance.

2.7 Difference Equations

Consider a difference equation of the form

$$y(k) = -\sum_{i=1}^{n} a_i y(k-i) + \sum_{i=0}^{n} b_i u(k-i) \tag{2.55}$$

where $u(k)$ is a scalar input, $y(k)$ is a scalar output, a_i's and b_i's are scalar coefficients, and the initial conditions are chosen as $u(k) = y(k) = 0$ for $k < 0$. To illustrate the structure of a digital filter in a block diagram, we introduce several basic block-diagram symbols for adder, constant multiplier, and unit delay etc. as shown in Figure 2.3.

Using unit delay z^{-1} which is defined by

$$z^{-1}u(k) = u(k-1), \qquad z^{-1}y(k) = y(k-1) \tag{2.56}$$

Equation (2.55) can be expressed as

$$\left(1 + \sum_{i=1}^{n} a_i z^{-i}\right) y(k) = \left(\sum_{i=0}^{n} b_i z^{-i}\right) u(k) \tag{2.57}$$

Moreover, by introducing an appropriate intermediate variable $v(k)$, (2.57) is decomposed into two parts

$$\begin{aligned} y(k) &= (b_0 + b_1 z^{-1} + \cdots + b_n z^{-n}) v(k) \\ v(k) &= \frac{u(k)}{1 + a_1 z^{-1} + \cdots + a_n z^{-n}} \end{aligned} \tag{2.58}$$

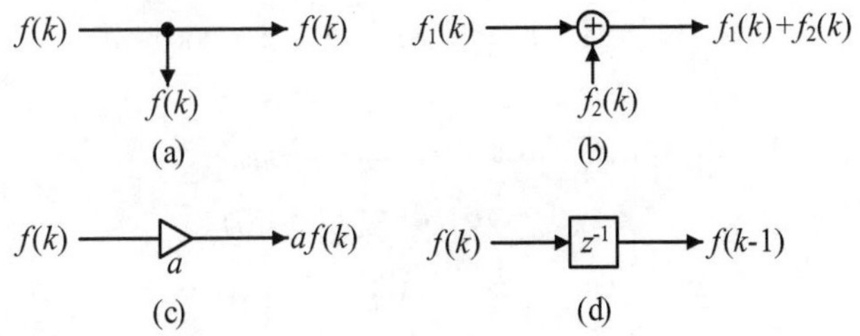

Figure 2.3 Block-diagram symbols for digital filters. (a) Drawer point. (b) Adder. (c) Constant multiplier. (d) Unit delay.

hence (2.57) is equivalent to

$$\begin{aligned} y(k) &= b_0 v(k) + b_1 v(k-1) + \cdots + b_n v(k-n) \\ v(k) &= u(k) - a_1 v(k-1) - \cdots - a_n v(k-n) \end{aligned} \tag{2.59}$$

A block diagram of the system in (2.59) is depicted in Figure 2.4. This figure is referred to as the *direct form II structure* and has the minimum possible number of delays [1]. Since feedback loops are involved in the filter structure, systems with such a structure are called *IIR* (*Infinite Impulse Response*) or *recursive digital filters.* In the case where feedback loop does not exist, i.e., $a_i = 0$ for $i = 1, 2, \cdots, n$, the difference equation in (2.55) becomes

$$y(k) = b_0 u(k) + b_1 u(k-1) + \cdots + b_n u(k-n) \tag{2.60}$$

The block diagram of the system in (2.60) is drawn in Figure 2.5. Since there exist only feedforward paths in Figure 2.5, systems with such a structure are called *FIR* (*Finite Impulse Response*), *nonrecursive*, or *transversal digital filters*.

Using (2.56), (2.57) is transformed into

$$\begin{aligned} y(k) &= b_0 u(k) + [b_1 u(k) - a_1 y(k)]z^{-1} + \cdots + [b_n u(k) - a_n y(k)]z^{-n} \\ &= b_0 u(k) + [b_1 u(k-1) - a_1 y(k-1)] + \cdots \\ &\quad + [b_n u(k-n) - a_n y(k-n)] \end{aligned} \tag{2.61}$$

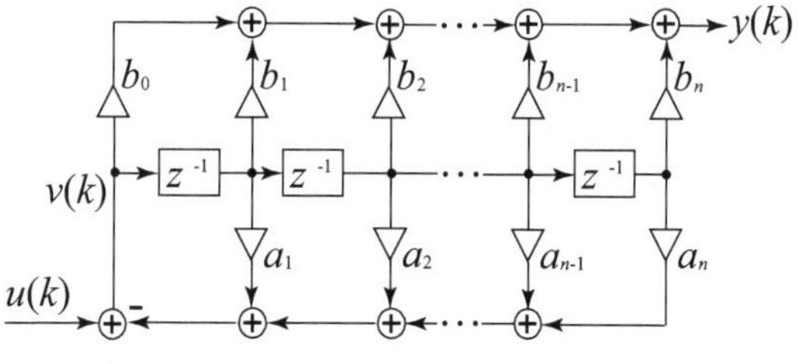

Figure 2.4 Direct form II structure of IIR digital filters.

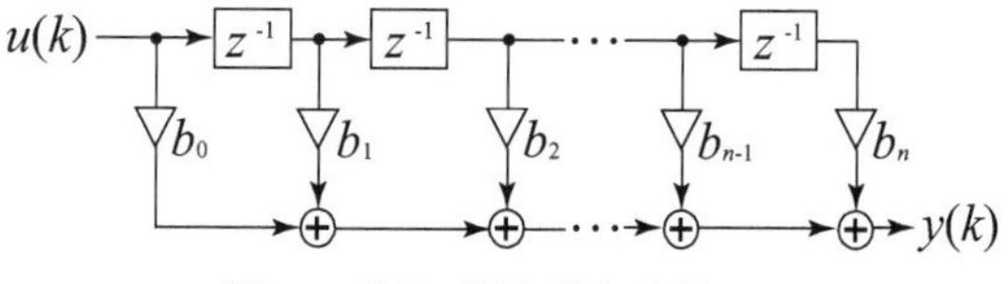

Figure 2.5 FIR digital filters.

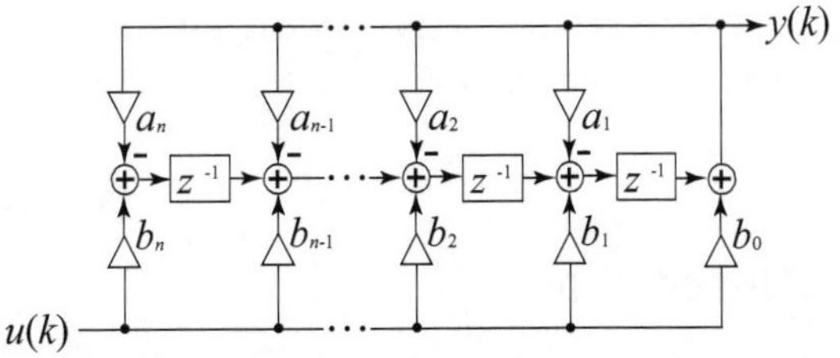

Figure 2.6 Transposed direct form II structure of IIR digital filters.

The block diagram of the system in (2.61) is illustrated in Figure 2.6. This figure is referred to as the *transposed direct form II structure* [1] and is generated by reversing the directions of all branches in the network of Figure 2.4. Such a procedure is called *flow-graph reversal* or *transposition*, that leads to a set of transposed filter structures. In addition, in the case where feedback loop does not exist, i.e., $a_i = 0$ for all $i = 1, 2, \cdots, n$, (2.61) becomes

$$\begin{aligned} y(k) &= b_0 u(k) + b_1 u(k) z^{-1} + \cdots + b_n u(k) z^{-n} \\ &= b_0 u(k) + b_1 u(k-1) + \cdots + b_n u(k-n) \end{aligned} \tag{2.62}$$

The block diagram of the system in (2.62) is shown in Figure 2.7. The block diagram in Figure 2.7 which is generated by reversing the directions of all branches in thenetwork of Figure 2.5 is called the *transposed form* of an FIR digital filter.

By taking the z-transform on the both sides of (2.55), we obtain

$$\Big(1 + \sum_{i=1}^{n} a_i z^{-i}\Big) Y(z) = \Big(\sum_{i=0}^{n} b_i z^{-i}\Big) U(z) \tag{2.63}$$

which leads to

$$H(z) = \frac{Y(z)}{U(z)} = \frac{b_0 + b_1 z^{-1} + \cdots + b_n z^{-n}}{1 + a_1 z^{-1} + \cdots + a_n z^{-n}} \tag{2.64}$$

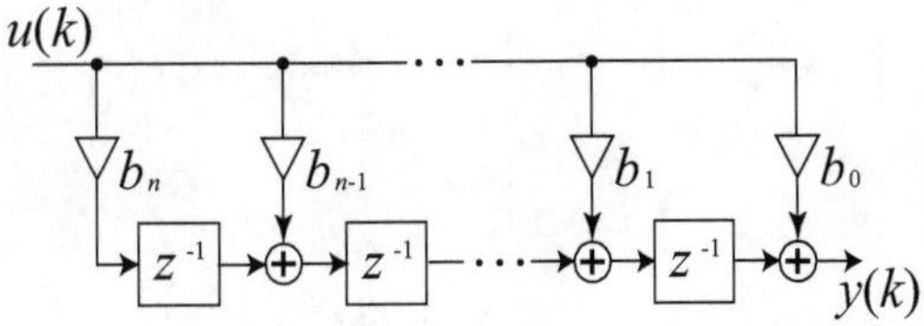

Figure 2.7 Transposed form of FIR digital filters.

The $H(z)$ in (2.64) is the transfer function of a difference equation in (2.55) whose impulse response is a sequence of infinite length unless $a_i = 0$ for all $i = 1, 2, \cdots, n$.

2.8 State-Space Descriptions

We now examine several procedures for realizing state-space models from difference equations.

2.8.1 Realization 1

By substituting the second equation into the first equation in (2.59), we obtain

$$y(k) = b_0 u(k) + (b_1 - a_1 b_0)v(k-1) + \cdots + (b_n - a_n b_0)v(k-n)$$
$$v(k) = u(k) - a_1 v(k-1) - \cdots - a_n v(k-n) \tag{2.65}$$

The block-diagram of the system in (2.65) is depicted in Figure 2.8 with definition of

$$\tilde{b}_i = b_i - a_i b_0 \quad \text{for } i = 1, 2, \cdots, n$$

Defining a state-variable vector $\boldsymbol{x}(k)$ by

$$\boldsymbol{x}(k) = [v(k-n), v(k-n+1), \cdots, v(k-1)]^T \tag{2.66}$$

the difference equation in (2.65) can be realized by a state-space model as

$$\begin{aligned} \boldsymbol{x}(k+1) &= \boldsymbol{A}\boldsymbol{x}(k) + \boldsymbol{b}u(k) \\ y(k) &= \boldsymbol{c}\boldsymbol{x}(k) + du(k) \end{aligned} \tag{2.67}$$

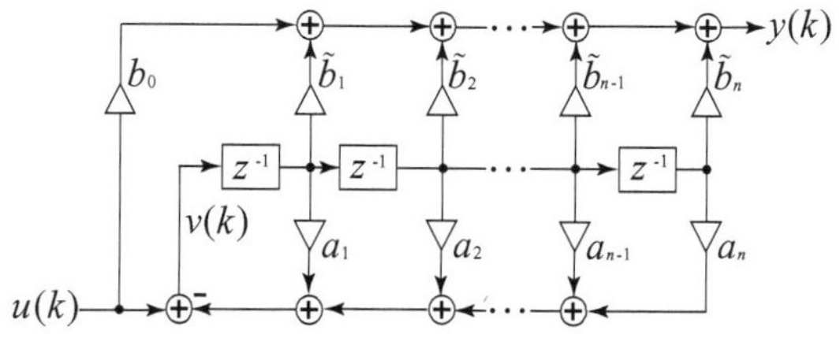

Figure 2.8 Transformation of IIR digital filters.

where

$$
\boldsymbol{A} = \begin{bmatrix} 0 & 1 & 0 & \cdots & 0 \\ 0 & 0 & 1 & \cdots & 0 \\ \vdots & \vdots & \vdots & \ddots & \vdots \\ 0 & 0 & 0 & \cdots & 1 \\ -a_n & -a_{n-1} & -a_{n-2} & \cdots & -a_1 \end{bmatrix}, \qquad \boldsymbol{b} = \begin{bmatrix} 0 \\ \vdots \\ 0 \\ 1 \end{bmatrix}
$$

$$
\boldsymbol{c} = \begin{bmatrix} \tilde{b}_n & \cdots & \tilde{b}_2 & \tilde{b}_1 \end{bmatrix}, \qquad d = b_0
$$

2.8.2 Realization 2

By decomposing (2.61) into two parts, we have

$$
\begin{aligned}
y(k) &= b_0 u(k) + v(k) \\
v(k) &= [b_1 u(k) - a_1 y(k)] z^{-1} + [b_2 u(k) - a_2 y(k)] z^{-2} + \cdots \\
&\quad + [b_n u(k) - a_n y(k)] z^{-n}
\end{aligned} \tag{2.68}
$$

which is equivalent to

$$
\begin{aligned}
y(k) &= b_0 u(k) + v(k) \\
v(k) &= [(b_1 - a_1 b_0) u(k) - a_1 v(k)] z^{-1} + \cdots \\
&\quad + [(b_n - a_n b_0) u(k) - a_n v(k)] z^{-n}
\end{aligned} \tag{2.69}
$$

The block diagram of the system in (2.69) is illustrated in Figure 2.9 with definition of

$$
\tilde{b}_i = b_i - a_i b_0 \quad \text{for } i = 1, 2, \cdots, n
$$

Defining the output of each unit delay z^{-1} by state variables $\tilde{x}_1(k)$, $\tilde{x}_2(k), \cdots, \tilde{x}_n(k)$ in order from the left to the right in Figure 2.9, the difference equation in (2.69) can be realized by a state-space model as

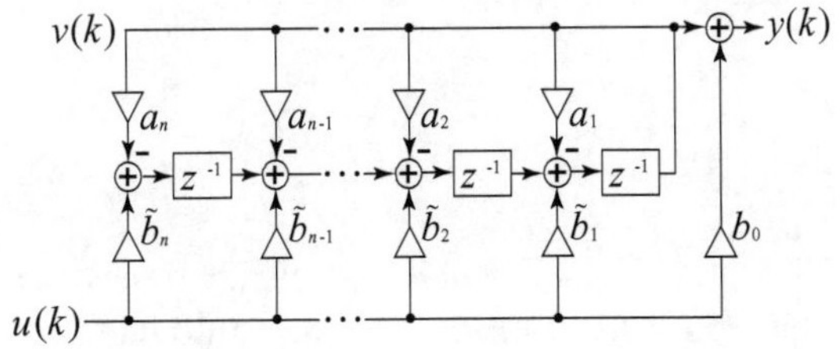

Figure 2.9 Transformation of the transposed form of IIR digital filters.

$$\tilde{\boldsymbol{x}}(k+1) = \tilde{\boldsymbol{A}}\tilde{\boldsymbol{x}}(k) + \tilde{\boldsymbol{b}}u(k)$$
$$y(k) = \tilde{\boldsymbol{c}}\,\tilde{\boldsymbol{x}}(k) + du(k) \tag{2.70}$$

where $\tilde{\boldsymbol{x}}(k) = [\tilde{x}_1(k), \tilde{x}_2(k), \cdots, \tilde{x}_n(k)]^T$, $v(k) = \tilde{x}_n(k)$ and

$$\tilde{\boldsymbol{A}} = \begin{bmatrix} 0 & \cdots & 0 & -a_n \\ 1 & \ddots & \vdots & \vdots \\ \vdots & \ddots & 0 & -a_2 \\ 0 & \cdots & 1 & -a_1 \end{bmatrix}, \qquad \tilde{\boldsymbol{b}} = \begin{bmatrix} \tilde{b}_n \\ \vdots \\ \tilde{b}_2 \\ \tilde{b}_1 \end{bmatrix}$$

$$\tilde{\boldsymbol{c}} = \begin{bmatrix} 0 & \cdots & 0 & 1 \end{bmatrix}, \qquad d = b_0$$

The coefficient matrices in (2.67) are related to those in (2.70) as

$$\boldsymbol{A} = \tilde{\boldsymbol{A}}^T, \qquad \boldsymbol{b} = \tilde{\boldsymbol{c}}^T, \qquad \boldsymbol{c} = \tilde{\boldsymbol{b}}^T \tag{2.71}$$

The system in (2.70) is called a *dual system* of the system in (2.67).

2.9 Frequency Transfer Functions

2.9.1 Linear Time-Invariant Causal Systems

We now consider a linear time-invariant causal system in (2.52) with the input given by a complex exponential function

$$u(k) = e^{j\omega kT} \tag{2.72}$$

With such an input, we can write

$$y(k) = \sum_{i=0}^{\infty} h(i)e^{j\omega(k-i)T} = \left[\sum_{i=0}^{\infty} h(i)e^{-j\omega iT}\right] e^{j\omega kT} \tag{2.73}$$

from which it immediately follows that

$$H(e^{j\omega T}) = \sum_{i=0}^{\infty} h(i)e^{-j\omega iT} = \sum_{i=0}^{\infty} h(i)z^{-i}\Bigg|_{z=e^{j\omega T}} \tag{2.74}$$

The $H(e^{j\omega T})$ in (2.74) is called the *frequency response* or *frequency transfer function*, which is obtained by substituting $z = e^{j\omega T}$ into its transfer function $H(z)$.

The $H(e^{j\omega T})$ in (2.74) can be expressed using polar coordinates form as

$$H(e^{j\omega T}) = \left|H(e^{j\omega T})\right| e^{j\theta(\omega)} \tag{2.75}$$

where

$$\left|H(e^{j\omega T})\right| = \sqrt{\text{Re}\{H(e^{j\omega T})\}^2 + \text{Im}\{H(e^{j\omega T})\}^2}$$

$$\theta(\omega) = \tan^{-1}\left(\frac{\text{Im}\{H(e^{j\omega T})\}}{\text{Re}\{H(e^{j\omega T})\}}\right)$$

are called the *amplitude characteristic* and *phase characteristic*, respectively. It is obvious that $|H(e^{j\omega T})|$ is an even function and $\theta(\omega)$ is an odd function, that is,

$$\left|H(e^{j\omega T})\right| = \left|H(e^{-j\omega T})\right|, \qquad \theta(-\omega) = -\theta(\omega) \tag{2.76}$$

2.9.2 Rational Transfer Functions

Consider a rational transfer function described by

$$H(z) = \frac{B(z)}{A(z)} = \frac{b_0 + b_1 z^{-1} + \cdots + b_m z^{-m}}{1 + a_1 z^{-1} + \cdots + a_n z^{-n}} \tag{2.77}$$

with frequency response

$$H(e^{j\omega T}) = \frac{B(e^{j\omega T})}{A(e^{j\omega T})} = \frac{b_0 + b_1 e^{-j\omega T} + \cdots + b_m e^{-jm\omega T}}{1 + a_1 e^{-j\omega T} + \cdots + a_n e^{-jn\omega T}} \tag{2.78}$$

Using polar coordinates form, $H(e^{j\omega T})$ in (2.78) can be expressed as

$$H(e^{j\omega T}) = \left|H(e^{j\omega T})\right| e^{j\theta(\omega)} \tag{2.79}$$

where

$$\left|H(e^{j\omega T})\right| = \frac{\sqrt{\left(\sum_{i=0}^{m} b_i \cos i\omega T\right)^2 + \left(\sum_{i=1}^{m} b_i \sin i\omega T\right)^2}}{\sqrt{\left(\sum_{i=0}^{n} a_i \cos i\omega T\right)^2 + \left(\sum_{i=1}^{n} a_i \sin i\omega T\right)^2}}$$

$$\theta(\omega) = \tan^{-1}\left(\frac{\sum_{i=1}^{n} a_i \sin i\omega T}{\sum_{i=0}^{n} a_i \cos i\omega T}\right) - \tan^{-1}\left(\frac{\sum_{i=1}^{m} b_i \sin i\omega T}{\sum_{i=0}^{m} b_i \cos i\omega T}\right)$$

with $a_0 = 1$ and $-\pi \le \omega T \le \pi$. By virtue of (2.76), we have

$$\begin{aligned} H(e^{j\omega T}) &= \left|H(e^{j\omega T})\right| e^{j\theta(\omega)} \\ H(e^{-j\omega T}) &= \left|H(e^{j\omega T})\right| e^{-j\theta(\omega)} \end{aligned} \tag{2.80}$$

This leads to

$$\theta(\omega) = \frac{1}{2j} \ln \left[\frac{H(e^{j\omega T})}{H(e^{-j\omega T})} \right] \tag{2.81}$$

The group delay of the phase characteristic is defined by

$$\tau(\omega) = -\frac{d\theta(\omega)}{d\omega} \tag{2.82}$$

which implies that

$$\tau(\omega) = -\left.\frac{d\theta(\omega)}{dz}\frac{dz}{d\omega}\right|_{z=e^{j\omega T}} = -\left. jTz\frac{d\theta(\omega)}{dz}\right|_{z=e^{j\omega T}} \tag{2.83}$$

Using (2.81) and (2.83), we compute

$$\begin{aligned} \tau(\omega) &= -\frac{T}{2}\left[z\left(\frac{d\ln H(z)}{dz} - \frac{d\ln H(z^{-1})}{dz} \right)\right]_{z=e^{j\omega T}} \\ &= -\frac{T}{2}\left[z\frac{1}{H(z)}\frac{dH(z)}{dz} + z^{-1}\frac{1}{H(z^{-1})}\frac{dH(z^{-1})}{dz^{-1}} \right]_{z=e^{j\omega T}} \\ &= -T\,\mathrm{Re}\left[z\frac{1}{H(z)}\frac{dH(z)}{dz} \right]_{z=e^{j\omega T}} \end{aligned} \tag{2.84}$$

where we have utilized

$$\begin{aligned} \frac{d\ln H(z^{-1})}{dz} &= \frac{d\ln H(z^{-1})}{dz^{-1}}\frac{dz^{-1}}{dz} \\ &= -z^{-2}\frac{d\ln H(z^{-1})}{dz^{-1}} \end{aligned} \tag{2.85}$$

By substituting (2.77) into (2.84), we obtain

$$\begin{aligned}\tau(\omega) &= -T\,\mathrm{Re}\left[z\frac{A(z)}{B(z)}\cdot\frac{1}{A^2(z)}\left\{A(z)\frac{dB(z)}{dz} - B(z)\frac{dA(z)}{dz}\right\}\right]_{z=e^{j\omega T}} \\ &= -T\,\mathrm{Re}\left[z\frac{1}{B(z)}\frac{dB(z)}{dz} - z\frac{1}{A(z)}\frac{dA(z)}{dz}\right]_{z=e^{j\omega T}} \\ &= -T\,\mathrm{Re}\left[\frac{\sum_{i=1}^{m} ib_i z^{-i}}{\sum_{i=0}^{m} b_i z^{-i}} - \frac{\sum_{i=1}^{n} ia_i z^{-i}}{\sum_{i=0}^{n} a_i z^{-i}}\right]_{z=e^{j\omega T}}\end{aligned} \tag{2.86}$$

2.9.3 All-Pass Digital Filters

All-pass digital filters form a particularly important class of IIR digital filters in which the numerator polynomial is generated from the denominator polynomial by reversing the order of the coefficients, hence the transfer function of an all-pass digital filter assumes the form

$$\begin{aligned}H(z) &= \frac{a_n + a_{n-1}z^{-1} + \cdots + a_1 z^{-(n-1)} + z^{-n}}{1 + a_1 z^{-1} + a_2 z^{-2} + \cdots + a_n z^{-n}} \\ &= \frac{1 + a_1 z + \cdots + a_n z^n}{1 + a_1 z^{-1} + \cdots + a_n z^{-n}}\, z^{-n}\end{aligned} \tag{2.87}$$

Hence the frequency response of the all-pass digital filter is given by

$$H(e^{j\omega}) = \frac{1 + a_1 e^{j\omega} + \cdots + a_n e^{jn\omega}}{1 + a_1 e^{-j\omega} + \cdots + a_n e^{-jn\omega}}\, e^{-jn\omega} = e^{j\theta(\omega)} \tag{2.88}$$

where

$$\theta(\omega) = -n\omega + 2\tan^{-1}\left(\frac{\sum_{i=1}^{n} a_i \sin i\omega}{1 + \sum_{i=1}^{n} a_i \cos i\omega}\right)$$

$$\theta(\omega) = \begin{cases} 0 & \text{if } \omega = 0 \\ -n\pi & \text{if } \omega = \pi \end{cases}$$

and without loss of generality, the sampling interval T has been set to unity.

From (2.88), it is seen that the amplitude response of all-pass digital filters is equal to unity over the entire baseband. As a result, by connecting a discrete-time system with a properly designed all-pass digital filter in cascade, the phase response of the system can be altered in a desired manner without affecting the amplitude response of the original discrete-time system. Moreover, as a signal processing building block all-pass digital filter admits computationally efficient implementation that renders it useful in many signal processing applications [8]. To see this, we express $H(z)$ in (2.87) as $H(z) = Y(z)/U(z)$. By applying the inverse z-transform to the equation, the following nth-order difference equation is found as the time-domain representation of the filter in (2.87)

$$y(k) = -\sum_{i=1}^{n} a_i \left[y(k-i) - u(k-n+i)\right] + u(k-n) \tag{2.89}$$

where only n multiplications are required to compute each output sample, whereas $2n$ delay (or storage) elements are required to realize the filter.

Another useful structure for realizing all-pass digital filters is the Gray and Markel lattice filter [8, 9]. For illustration purpose, we consider two multiplier lattice two-pair for all-pass digital filter implementation, shown in Figure 2.10.

From the figure, we can write

$$\begin{aligned} X_i(z) &= U_i(z) - k_i z^{-1} H_{i-1}(z) X_i(z) \\ Y_i(z) &= \left[z^{-1} H_{i-1}(z) + k_i\right] X_i(z) \end{aligned} \tag{2.90}$$

which yields

$$H_i(z) = \frac{Y_i(z)}{U_i(z)} = \frac{z^{-1} H_{i-1}(z) + k_i}{1 + k_i z^{-1} H_{i-1}(z)} \tag{2.91}$$

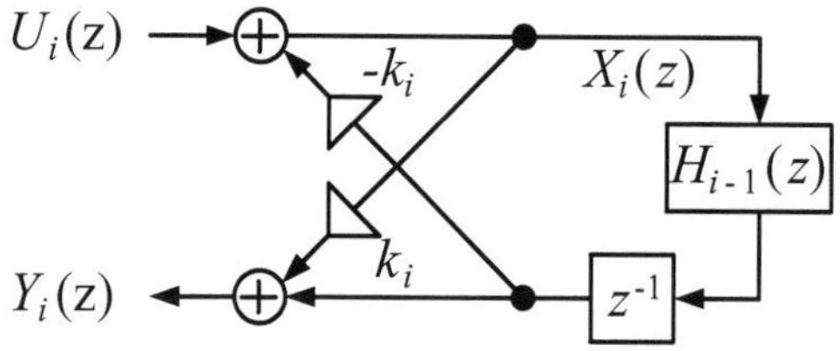

Figure 2.10 Two multiplier lattice two-pair for all-pass digital filter implementation.

where $H_0(z) = 1$ and $k_i = H_i(\infty)$. For example, from (2.91) it follows that

$$\begin{aligned} H_1(z) &= \frac{z^{-1}H_0(z) + k_1}{1 + k_1 z^{-1}H_0(z)} = \frac{z^{-1} + k_1}{1 + k_1 z^{-1}} \\ H_2(z) &= \frac{z^{-1}H_1(z) + k_2}{1 + k_2 z^{-1}H_1(z)} = \frac{z^{-2} + k_1(1 + k_2)z^{-1} + k_2}{1 + k_1(1 + k_2)z^{-1} + k_2 z^{-2}} \end{aligned} \tag{2.92}$$

which correspond to the transfer functions of first-order and second-order all-pass digital filters, respectively. Cascaded lattice realization of an nth-order all-pass digital filter using two multiplier lattice two-pair modules is depicted in Figure 2.11 where $H_n(z) = Y_n(z)/U_n(z)$. It is known [8] that the nth-order all-pass digital filter is stable if $|k_i| < 1$ for $i = 1, 2, \cdots, n$.

Alternatively, two multiplier lattice two-pair in Figure 2.10 can also be implemented as per Figures 2.12 or 2.13 without altering the overall transfer function.

We remark that Figure 2.12 is the single multiplier lattice two-pair [9], which requires the fewest number of multipliers. Figure 2.13 is the normalized lattice two-pair [10] where all internal nodes are automatically scaled in the l_2 sense [11]. By employing single multiplier lattice two-pair in Figure 2.12, the lattice structure of Figure 2.11 can be transformed into that of Figure 2.14.

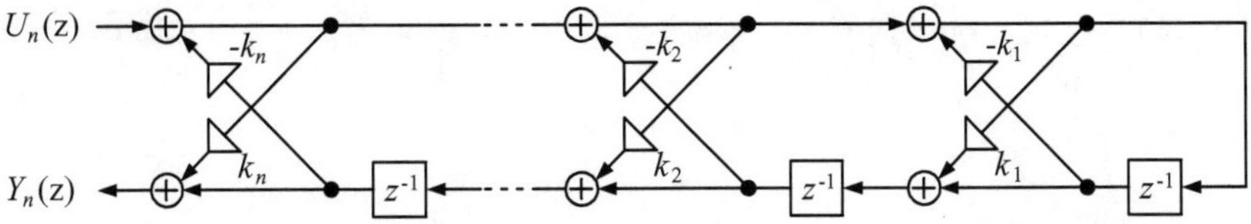

Figure 2.11 Cascaded lattice realization of an nth-order all-pass digital filter.

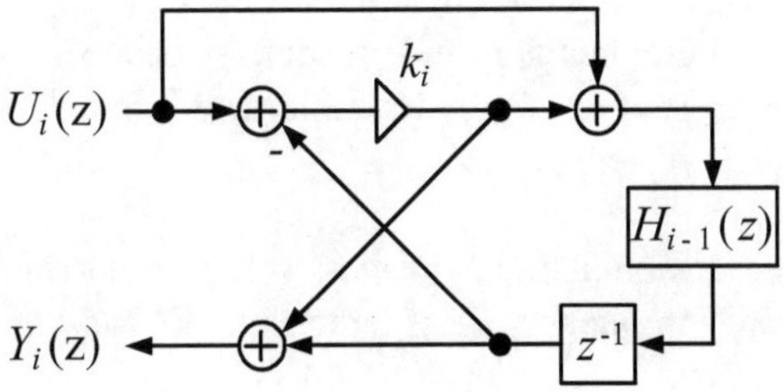

Figure 2.12 Single multiplier lattice two-pair.

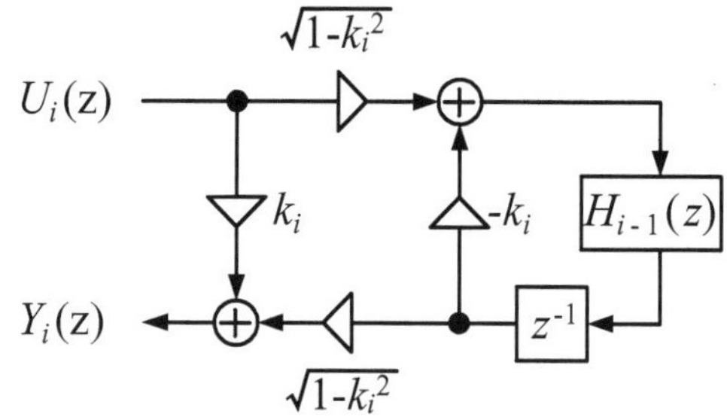

Figure 2.13 Normalized lattice two-pair.

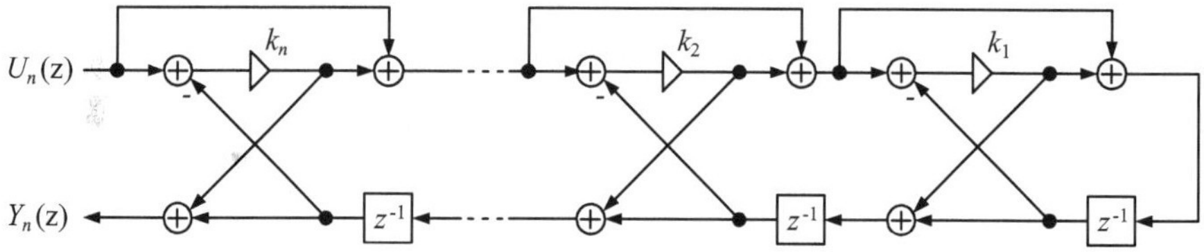

Figure 2.14 Another cascaded lattice realization of an nth-order all-pass digital filter.

2.9.4 Notch Digital Filters

Notch digital filters are useful for removing a single-frequency component from a signal such as an unmodulated carrier in communication systems or power supply hum from a sampled analog signal, and so on.

The magnitude response of an ideal notch digital filter $H_{notch}(z)$ satisfies

$$|H_{notch}(e^{j\omega})| = \begin{cases} 0, & \text{for } \omega = \omega_o \\ 1, & \text{otherwise} \end{cases} \tag{2.93}$$

where ω_o is the *notch frequency.* However, (2.93) can only hold in theory because zero bandwidth cannot be realized in practice. A realistic requirement at ω_o is to satisfy the specified 3 dB rejection bandwidth B. Note that the frequencies ω_1 and ω_2 where the magnitude response goes to $1/\sqrt{2}$ are called the *3 dB cutoff frequencies*, and if the magnitude response is less than $1/\sqrt{2}$ for $\omega_1 < \omega < \omega_2$, their difference $B = \omega_2 - \omega_1$ is called *3 dB rejection bandwidth.*

It can be shown that the actual transfer function of a single-frequency notch digital filter with a notch frequency ω_o of bandwidth B can be expressed in the form [8]

$$H(z) = \frac{1 + H_2(z)}{2} \tag{2.94a}$$

where

$$H_2(z) = \frac{k_2 - k_1(1+k_2)z^{-1} + z^{-2}}{1 - k_1(1+k_2)z^{-1} + k_2 z^{-2}} \tag{2.94b}$$

is a second-order all-pass digital filter with

$$k_1 = \cos\omega_o, \qquad k_2 = \frac{1 - \tan\frac{B}{2}}{1 + \tan\frac{B}{2}} \tag{2.94c}$$

The implementation of a single-frequency notch digital filter is depicted in Figure 2.15 where $H(z) = Y(z)/U(z)$.

The lattice structure of a second-order digital all-pass filter in (2.94b) is shown in Figure 2.16 where $H_2(z) = Y_2(z)/U_2(z)$.

By substituting (2.94b) into (2.94a), the transfer function of the single-frequency notch digital filter is found to be

$$\begin{aligned} H(z) &= \frac{1+k_2}{2} \cdot \frac{1 - 2k_1 z^{-1} + z^{-2}}{1 - k_1(1+k_2)z^{-1} + k_2 z^{-2}} \\ &= \frac{1+k_2}{2} \cdot \frac{z - 2k_1 + z^{-1}}{z - k_1(1+k_2) + k_2 z^{-1}} \end{aligned} \tag{2.95}$$

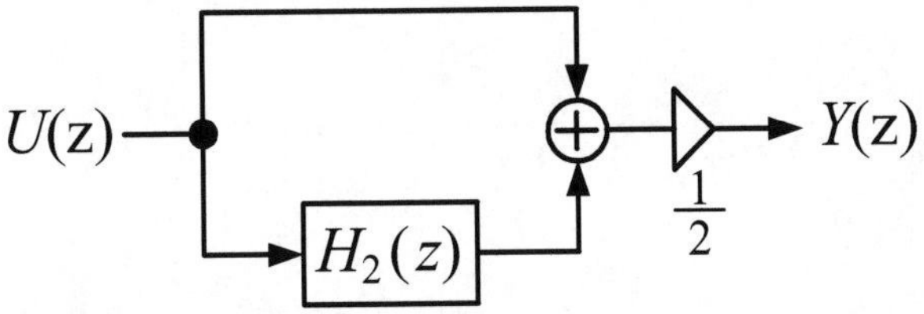

Figure 2.15 Implementation of a single-frequency notch digital filter.

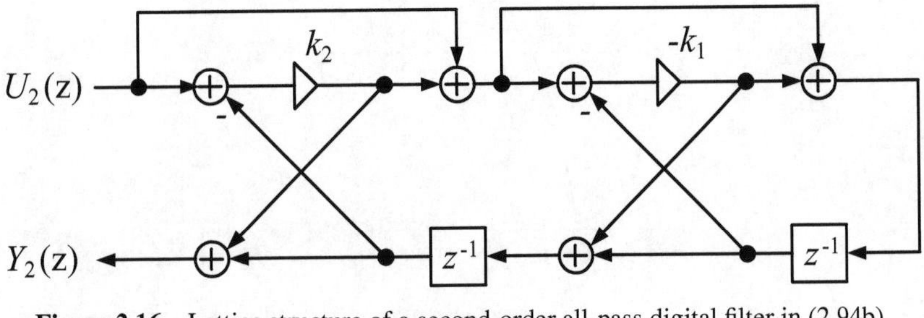

Figure 2.16 Lattice structure of a second-order all-pass digital filter in (2.94b).

and the frequency response of the filter in (2.95) is given by

$$H(e^{j\omega}) = \frac{1+k_2}{2}\frac{e^{j\omega} - 2k_1 + e^{-j\omega}}{e^{j\omega} - k_1(1+k_2) + k_2e^{-j\omega}}$$
$$= \frac{1}{1 + j\dfrac{1-k_2}{1+k_2}\cdot\dfrac{\sin\omega}{\cos\omega - k_1}} \tag{2.96}$$

whose magnitude response is described by

$$|H(e^{j\omega})| = \frac{1}{\sqrt{1 + \left(\dfrac{1-k_2}{1+k_2}\cdot\dfrac{\sin\omega}{\cos\omega - k_1}\right)^2}} \tag{2.97}$$

The notch frequency ω_o, cutoff frequencies ω_1, ω_2 $(\omega_1 < \omega_2)$ and bandwidth $B = \omega_2 - \omega_1$ of the magnitude response in (2.97) are illustrated in Figure 2.17.

Evidently, since $k_1 = \cos\omega_o$, the notch frequency is given by

$$\omega = \omega_o \tag{2.98}$$

Also, the cutoff frequencies ω_i for $i = 1, 2$ must satisfy $|H(e^{j\omega_i})| = 1/\sqrt{2}$. Hence,

$$\left(\frac{1-k_2}{1+k_2}\cdot\frac{\sin\omega_i}{\cos\omega_i - \cos\omega_o}\right)^2 = 1 \;\text{ for }\; i = 1, 2 \tag{2.99a}$$

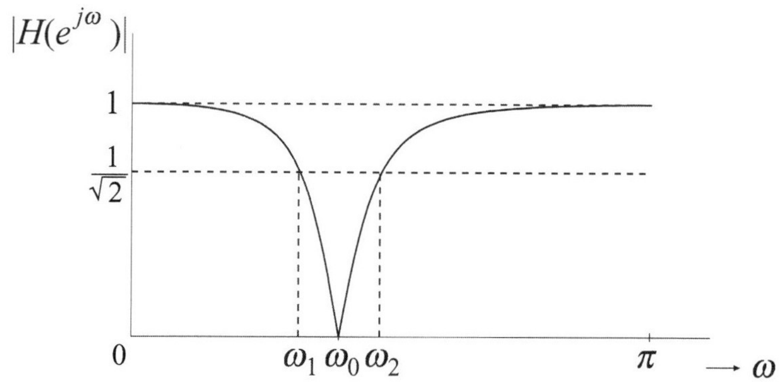

Figure 2.17 Notch frequency ω_o, cutoff frequencies ω_1, ω_2 and bandwidth $B = \omega_2 - \omega_1$ of the magnitude response in (2.97).

or equivalently,

$$\left(\frac{\sin\omega_i}{\cos\omega_i - \cos\omega_o}\right)^2 = \left(\frac{1+k_2}{1-k_2}\right)^2 \quad \text{for } i = 1, 2 \tag{2.99b}$$

must hold. By defining $A = (1+k_2)/(1-k_2)$, it can be derived from (2.99b) that

$$\frac{\sin\omega_1}{\cos\omega_1 - \cos\omega_o} = A, \qquad \frac{\sin\omega_2}{\cos\omega_2 - \cos\omega_o} = -A \tag{2.100}$$

where $A > 0$. By solving (2.100), we obtain

$$A = \frac{\sin\omega_1 + \sin\omega_2}{\cos\omega_1 - \cos\omega_2} = \frac{1}{\tan\dfrac{\omega_2 - \omega_1}{2}} = \frac{1}{\tan\frac{B}{2}} \tag{2.101}$$

From (2.101) and $A = (1+k_2)/(1-k_2)$, it follows that

$$\tan\frac{B}{2} = \frac{1}{A} = \frac{1-k_2}{1+k_2} \iff k_2 = \frac{1-\tan\frac{B}{2}}{1+\tan\frac{B}{2}} \tag{2.102}$$

Hence, if k_2 approaches unity, then we have

$$\tan\frac{B}{2} \simeq \frac{B}{2} \quad \text{and} \quad B \simeq 1 - k_2 \tag{2.103}$$

This reveals that if $k_2 \simeq 1$, then bandwidth B approaches zero, and (2.95) approaches an ideal notch filter. For the sake of stability, $|k_1| < 1$ and $|k_2| < 1$ have been assumed in the above analysis.

There are some variants of a single-frequency notch filter in (2.95). By employing an arithmetic-geometric mean inequality, we obtain

$$\frac{1+k_2}{2} \geq \sqrt{k_2} \tag{2.104}$$

where equality holds provided that $k_2 = 1$. Hence, if we assume $k_2 \simeq 1$ and set $k_2 = \rho^2$ where $0 \ll \rho < 1$, then it follows from (2.95) that [12, 13]

$$\begin{aligned} H(z) &\simeq \rho\,\frac{1 - 2k_1 z^{-1} + z^{-2}}{1 - 2k_1\rho z^{-1} + \rho^2 z^{-2}} \\ &\simeq \frac{1 - 2k_1 z^{-1} + z^{-2}}{1 - 2k_1\rho z^{-1} + \rho^2 z^{-2}} \end{aligned} \tag{2.105}$$

or [14]

$$H(z) \simeq \rho \frac{1 - 2k_1 z^{-1} + \rho(2-\rho)z^{-2}}{1 - 2k_1 \rho z^{-1} + \rho^2 z^{-2}} \tag{2.106}$$

The magnitude response of the notch digital filter in (2.106) is shown in Figure 2.18 where $\omega_o = 0.3\pi$ and $\rho = 0.985$.

The notch digital filter in (2.106) can be realized by a state-space model as [14]

$$\begin{aligned} \boldsymbol{x}(k+1) &= \boldsymbol{A}\boldsymbol{x}(k) + \boldsymbol{b}u(k) \\ y(k) &= \boldsymbol{c}\boldsymbol{x}(k) + du(k) \end{aligned} \tag{2.107}$$

where $\boldsymbol{x}(k)$ is a 2×1 state-variable vector, $u(k)$ is a scalar input, $y(k)$ is a scalar output, and $\boldsymbol{A}, \boldsymbol{b}$, $\boldsymbol{c}$ and d are real constant matrices defined by

$$\boldsymbol{A} = \rho \begin{bmatrix} \cos\omega_o & -\sin\omega_o \\ \sin\omega_o & \cos\omega_o \end{bmatrix}, \qquad \boldsymbol{c} = \rho \begin{bmatrix} 1 & 1 \end{bmatrix}$$

$$\boldsymbol{b} = (\rho - 1) \begin{bmatrix} \cos\omega_o - \sin\omega_o \\ \cos\omega_o + \sin\omega_o \end{bmatrix}, \qquad d = \rho$$

From (2.107), it is observed that $\boldsymbol{A}\boldsymbol{A}^T = \boldsymbol{A}^T\boldsymbol{A}$ holds, i.e., matrix $\boldsymbol{A}$ is normal [15]. Without loss of generality, the coefficient matrices $\boldsymbol{A}$ and $\boldsymbol{b}$ of the state-space model in (2.107) can be written as [14]

$$\boldsymbol{A} = \frac{\rho}{\sqrt{\alpha^2+\beta^2}} \begin{bmatrix} \alpha & -\beta \\ \beta & \alpha \end{bmatrix}, \qquad \boldsymbol{b} = \frac{\rho - 1}{\sqrt{\alpha^2+\beta^2}} \begin{bmatrix} \alpha - \beta \\ \alpha + \beta \end{bmatrix} \tag{2.108}$$

where α and β are real numbers.

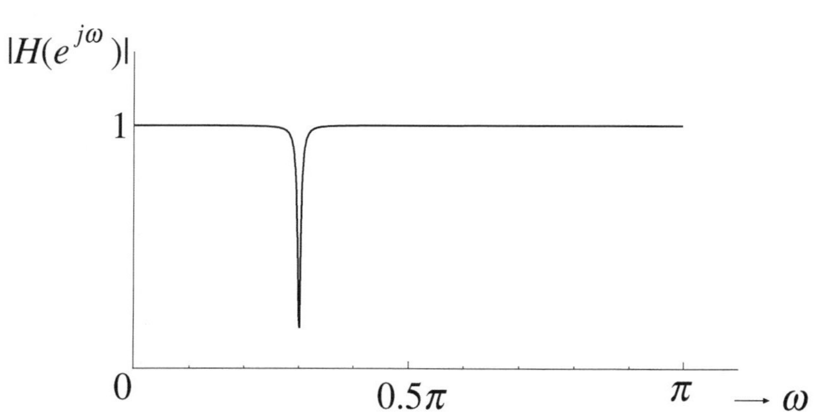

Figure 2.18 Magnitude response of a notch filter in (2.106) where $\omega_o = 0.3\pi$ and $\rho = 0.985$.

2.9.5 Doubly Complementary Digital Filters

All-pass digital filters play an important role in doubly complementary filters which find applications in various signal processing systems [8]. In this section, two stable transfer functions $G(z)$ and $H(z)$ which are both *all-pass complementary*, i.e.,

$$\left|G(e^{j\omega}) + H(e^{j\omega})\right| = 1 \text{ for all } \omega \tag{2.109}$$

and *power complementary*, i.e.,

$$\left|G(e^{j\omega})\right|^2 + \left|H(e^{j\omega})\right|^2 = 1 \text{ for all } \omega \tag{2.110}$$

are considered. Such transfer function pairs are termed *doubly complementary* [16].

We now examine a parallel structure of all-pass digital filters $A_1(z)$ and $A_2(z)$, shown in Figure 2.19.

The frequency responses of transfer functions $G(z) = Y_1(z)/U(z)$ and $H(z) = Y_2(z)/U(z)$ specified in Figure 2.19 can be expressed as

$$\begin{aligned} G(e^{j\omega}) &= \frac{1}{2}\left[A_1(e^{j\omega}) + A_2(e^{j\omega})\right] \\ &= \frac{1}{2}\left[e^{j\theta_1(\omega)} + e^{j\theta_2(\omega)}\right] \\ &= e^{j\frac{\theta_1(\omega)+\theta_2(\omega)}{2}} \cos\left[\frac{\theta_1(\omega)-\theta_2(\omega)}{2}\right] \end{aligned} \tag{2.111}$$

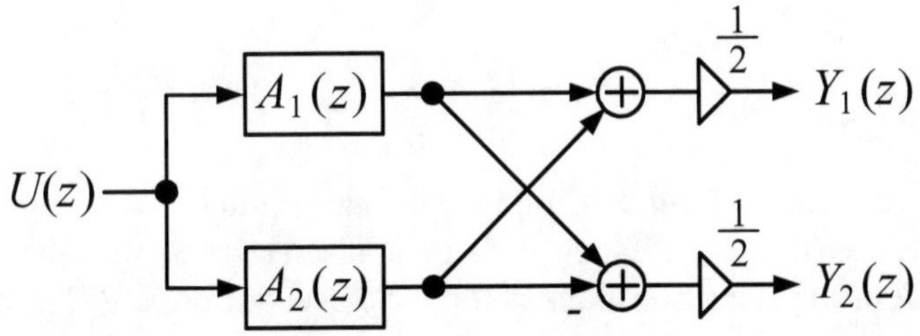

Figure 2.19 Implementation of the doubly complementary filter pair as the sum and difference of all-pass digital filters.

$$\begin{aligned} H(e^{j\omega}) &= \frac{1}{2}\left[A_1(e^{j\omega}) - A_2(e^{j\omega})\right] \\ &= \frac{1}{2}\left[e^{j\theta_1(\omega)} - e^{j\theta_2(\omega)}\right] \\ &= j\, e^{j\frac{\theta_1(\omega)+\theta_2(\omega)}{2}} \sin\left[\frac{\theta_1(\omega)-\theta_2(\omega)}{2}\right] \end{aligned} \tag{2.112}$$

respectively. Evidently, the transfer function $G(z)$ in (2.111) is a lowpass filter and the transfer function $H(z)$ in (2.112) is a highpass filter provided that

$$\theta_1(\omega) = \begin{cases} \theta_2(\omega), & \text{for } 0 \le |\omega| \le \omega_p \\ \theta_2(\omega) \pm \pi, & \text{for } \omega_s \le |\omega| \le \pi \end{cases} \tag{2.113}$$

where ω_p and ω_s denote the passband and stopband edges, respectively. It is straightforward to show that the transfer functions $G(z) = Y_1(z)/U(z)$ and $H(z) = Y_2(z)/U(z)$ satisfy both Equations (2.109) and (2.110) simultaneously. Hence, they are doubly complementary.

When designing a lowpass (highpass) digital filter using all-pass digital filters $A_1(z)$ and $A_2(z)$, it is a good choice to use delay elements $z^{-(n-1)}$ as an all-pass filter $A_1(z)$, i.e., $A_1(z) = z^{-(n-1)}$. The phase specification of the second all-pass filter $A_2(z)$ is given by

$$\theta_{desired}(\omega) = \begin{cases} -(n-1)\omega, & \text{for } 0 \le |\omega| \le \omega_p \\ -(n-1)\omega - \pi, & \text{for } \omega_s \le |\omega| \le \pi \end{cases} \tag{2.114}$$

The phase specification of an all-pass digital filter $A_2(z)$ is depicted together with the magnitude response in Figure 2.20.

Efficient techniques are available to design an all-pass digital filter $A_2(z)$ approximating a given phase response. The reader is referred to Section 7.3 for further details.

2.10 Summary

This chapter has covered fundamental concepts relating to discrete-time signals and systems. First, the concepts of a discrete-time sequence, the z-transformation, and z-transform inversion have been introduced. The properties of discrete-time systems have then been discussed, and discrete-time systems have been described in terms of difference equations as well as state-space descriptions. Frequency responses have also been induced from the

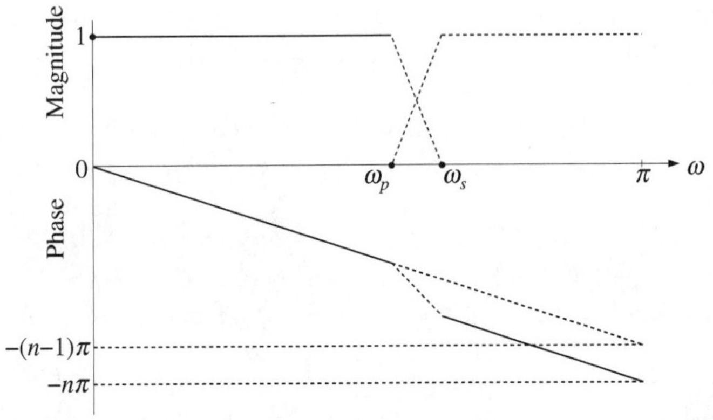

Figure 2.20 The phase specification of an all-pass digital filter $A_2(z)$.

transfer functions. In addition, analysis and design issues of several useful types of IIR digital filters, including all-pass digital filters, notch digital filters and doubly complementary digital filters, are examined.

References

[1] A. V. Oppenheim and R. W. Schafer, *Digital Signal Processing*, NJ: Prentice-Hall, 1975.

[2] Y. Sakawa, *Linear System Control Theory*, Tokyo, Japan, Asakura Publishing, 1979.

[3] H. Kogo and T. Mita, *Introduction to System Control Theory*, Tokyo, Japan, Jikkyo Shuppan, 1979.

[4] T. Higuchi, *Fundamentals of Digital Signal Processing*, Tokyo, Japan, Shokodo, 1986.

[5] M. S. Santina, A. R. Stubberud and G. H. Hostetter, *Digital Control System Design*, 2nd ed. Orlando, FL, Saunders College Publishing, Harcourt Brace College Publishers, 1994.

[6] S. Takahashi and M. Ikehara, *Digital Filters*, Tokyo, Japan, Baifukan, 1999.

[7] M. Hagiwara, *Digital Signal Processing*, Tokyo, Japan, Morikita Publishing, 2001.

[8] P. A. Regalia, S. K. Mitra and P. P. Vaidynathan, "The digital all-pass filter: A versatile signal processing building block," *Proc. IEEE*, vol. 76, no. 1, pp. 19–37, Jan. 1988.

[9] A. H. Gray, Jr., and J. D. Markel, "Digital lattice and ladder filter synthesis," *IEEE Trans. Audio Electroacoust.*, vol. AU-21, no. 6, pp. 491–500, Dec. 1973.
[10] A. H. Gray, Jr., and J. D. Markel, "A normalized filter structure," *IEEE Trans. Acoust., Speech, Signal Process.*, vol. ASSP-23, no. 3, pp. 268–270, June 1975.
[11] A. H. Gray, Jr., "Passive cascaded lattice digital filters," *IEEE Trans. Circuits Syst.*, vol. CAS-27, no. 5, pp. 337–344, May 1980.
[12] T. S. Ng, "Some aspects of an adaptive digital notch filter with constrained poles and zeros," *IEEE Trans. Acoust., speech, Signal Process.*, vol. ASSP-35, no. 2, pp. 158-161, Feb. 1987.
[13] M. V. Dragosevic and S. S. Stankovic, "An adaptive notch filter with improved tracking properties," *IEEE Trans. Signal Process.*, vol. 43, no. 9, pp. 2068–2078, Sep. 1995.
[14] Y. Hinamoto and S. Nishimura, "Normal-form state-space realization of single frequency IIR notch filters and its application to adaptive notch filters," in *Proc. APCCAS 2016*, pp. 599–602, Oct. 2016.
[15] R. E. Skelton and D. A. Wagie, "Minimal root sensitivity in linear systems," *J. Guidance Contr.*, vol. 7, pp. 570–574, Sep.–Oct. 1984.
[16] P. P. Vaidynathan, S. K. Mitra and Y. Neuvo, "A new method of low sensitivity filter realization," *IEEE Trans. Acoust., Speech, Signal Process.*, vol. ASSP-34, no. 2, pp. 350–361, Apr. 1986.

3

Stability and Coefficient Sensitivity

3.1 Preview

A dynamic system can hardly be useful in practice unless it is stable, meaning that it always produces a reasonable output in response to a reasonable input. For the sake of quantitative analysis of digital filters and algorithmic development of their designs, it is of fundamental importance to understand the notion of stability in rigorous terms and various criteria for verifying stability. In Section 3.2, after defining the bounded-input bounded-output stability for IIR digital filters, several stability criteria in terms of impulse response and system poles are presented. These include Schur-Cohn criterion, Schur-Cohn-Fujiwara criterion, Jury-Marden criterion, and Lyapunov criterion. Another issue addressed in this chapter is coefficient sensitivity. When the coefficients of an IIR digital filter are quantized for implementation purposes, the coefficient error may lead to substantial changes in the filter characteristics such as frequency response and stability. In Section 3.3, these changes are evaluated by examining the changes in the locations of poles due to the changes in the filter's coefficients.

3.2 Stability

3.2.1 Definition

An IIR digital filter is said to be stable if every bounded input produces a bounded output. A necessary and sufficient condition on the impulse response for stability is given in the following theorem.

Theorem 3.1
A necessary and sufficient condition for an IIR digital filter to be stable is that its impulse response $\{h(n)\}$ is absolutely summable, namely,

$$\sum_{n=-\infty}^{\infty} |h(n)| < \infty \tag{3.1}$$

Proof

Assume that (3.1) does not hold, i.e., the sum on the left-hand side of (3.1) is not bounded. Consider the sequence $x(n)$ defined by

$$x(n) = \begin{cases} 1 & \text{if } h(-n) \geq 0 \\ -1 & \text{if } h(-n) < 0 \end{cases}$$

which is obviously a bounded sequence. With this $\{x(n)\}$ as an input, the output of the filter at $n = 0$ is given by

$$y(0) = \sum_{m=-\infty}^{\infty} x(m)h(-m) = \sum_{m=-\infty}^{\infty} |h(m)| = \infty$$

which shows that (3.1) is a necessary condition. To prove the sufficiency, assume that (3.1) holds and $\{x(n)\}$ is bounded. Then we have

$$\begin{aligned} |y(n)| &= \left| \sum_{m=-\infty}^{\infty} x(m)h(n-m) \right| \\ &\leq \sum_{m=-\infty}^{\infty} |x(m)|\,|h(n-m)| \leq M \sum_{m=-\infty}^{\infty} |h(n-m)| < \infty \end{aligned}$$

where M is a bound of $x(n)$. Hence (3.1) is also a sufficient condition and the proof is complete. ■

3.2.2 Stability in Terms of Poles

For a causal IIR digital filters whose transfer function is a rational function of the form

$$H(z) = \frac{\sum_{i=0}^{M} a_i z^{-i}}{\sum_{i=0}^{N} b_i z^{-i}} \quad \text{with } b_0 = 1 \tag{3.2}$$

its stability can be characterized in terms its poles which are defined as the roots $\{p_i | \, i = 1, 2, \cdots, N\}$ of the equation $\sum_{i=0}^{N} b_i z^{-i} = 0$ in the z plane.

Theorem 3.2
The IIR filter in (3.2) is stable if and only if all its poles are located strictly inside the unit circle of the z plane, namely,

$$|p_i| < 1 \quad \text{for } i = 1, 2, \cdots, N \tag{3.3}$$

Proof
We begin by writing the transfer function in terms of its impulse response as

$$H(z) = \sum_{n=-\infty}^{\infty} h(n)z^{-n}$$

Because the filter is causal, $h(n)$ vanishes for every $n < 0$, hence we have

$$H(z) = \sum_{n=0}^{\infty} h(n)z^{-n}$$

If the filter is stable, then (3.1) implies that for any $z = re^{j\omega}$ with $r \geq 1$ we have

$$\left|H(re^{j\omega})\right| \leq \sum_{n=0}^{\infty} \left|h(n)r^{-n}e^{-jn\omega}\right| = \sum_{n=0}^{\infty} r^{-n}\left|h(n)\right| \leq \sum_{n=0}^{\infty} |h(n)| < \infty$$

which implies that any pole of $H(z)$ cannot be in the region $\{z\colon |z| \geq 1\}$, i.e. (3.3) must hold. Next, assume that (3.3) is satisfied. For $n > 0$, the impulse response $h(n)$ can be expressed as

$$h(n) = \frac{1}{2\pi j}\oint_{\Gamma} H(z)z^{n-1}dz = \sum_{i=1}^{N} p_i^{n-1}\,\text{res}_{z=p_i}H(z)$$

where the residue $\text{res}_{z=p_i}H(z)$ is known to be finite, hence $|\text{res}_{z=p_i} H(z)| \leq \bar{R}$ for some constant $\bar{R}$ and $p_i = r_i e^{j\psi_i}$ with $r_i \leq \bar{r}$ for some $\bar{r} < 1$. It follows that

$$\sum_{n=0}^{\infty} |h(n)| \leq |h(0)| + N\bar{R}\sum_{n=1}^{\infty} \bar{r}^{n-1} < \infty$$

hence the filter is stable which completes the proof. ■

3.2.3 Schur-Cohn Criterion

Given an Nth-order polynomial $B(z)$ of the form

$$B(z) = \sum_{i=0}^{N} b_i z^{N-i} \tag{3.4}$$

the Schur-Cohn criterion [1] examines whether or not all its zeros are inside the unit circle of the z plane by evaluating the determinants of N matrices whose sizes vary from 2×2 to $2N \times 2N$.

Theorem 3.3: *Schur-Cohn*
All zeros of polynomial $B(z)$ are strictly inside the unit circle of the z plane if and only if, for $k = 1, 2, \cdots, N$,

$$\det \boldsymbol{S}_k < 0 \text{ if } k \text{ is odd and } \det \boldsymbol{S}_k > 0 \text{ if } k \text{ is even}$$

where

$$\boldsymbol{S}_k = \begin{bmatrix} \boldsymbol{A}_k & \boldsymbol{B}_k \\ \boldsymbol{B}_k^T & \boldsymbol{A}_k^T \end{bmatrix}$$

$$\boldsymbol{A}_k = \begin{bmatrix} b_N & 0 & 0 & \cdots & 0 \\ b_{N-1} & b_N & 0 & \cdots & 0 \\ \vdots & \vdots & \vdots & & \vdots \\ b_{N-k+1} & b_{N-k+2} & b_{N-k+3} & \cdots & b_N \end{bmatrix}$$

$$\boldsymbol{B}_k = \begin{bmatrix} b_0 & b_1 & b_2 & \cdots & b_{k-1} \\ 0 & b_0 & b_1 & \cdots & b_{k-2} \\ \vdots & \vdots & \vdots & & \vdots \\ 0 & 0 & 0 & \cdots & b_0 \end{bmatrix}$$

3.2.4 Schur-Cohn-Fujiwara Criterion

A stability criterion based on the Schur-Cohn criterion with improved efficiency was described by Fujiwara [1].

Theorem 3.4: *Schur-Cohn-Fujiwara*
All zeros of polynomial $B(z)$ are strictly inside the unit circle of the z plane if and only if the matrix $\boldsymbol{F} = (f_{ij})$ of size $N \times N$ is positive definite, where

$$f_{ij} = \sum_{k=1}^{\min(i,j)} \left(b_{i-k}b_{j-k} - b_{N-i+k}b_{N-j+k}\right) \tag{3.5}$$

3.2.5 Jury-Marden Criterion

An efficient and easy-to-use stability criterion was developed by Jury [1] using a result of Marden [1] that considerably simplifies the calculations involved in Schur-Cohn criterion. In this criterion, an array of numbers known as the Jury-Marden array is constructed as follows: The first two rows of the array are just the coefficients of polynomial $B(z)$ in ascending and descending orders, respectively, see Table 3.1.

The elements of the third and fourth rows are calculated as

$$c_i = \begin{vmatrix} b_0 & b_{N-i} \\ b_N & b_i \end{vmatrix} \qquad \text{for } i = 0,\ 1,\ \cdots,\ N-1$$

and those of fifth and sixth rows as

$$d_i = \begin{vmatrix} c_0 & c_{N-1-i} \\ c_{N-1} & c_i \end{vmatrix} \qquad \text{for } i = 0,\ 1,\ \cdots,\ N-2$$

and so on until a total of $2N-3$ rows are computed. There will be three components in the last row, which are denoted as r_0, r_1, and r_2. The criterion can now be stated as

Theorem 3.5: *Jury-Marden*
All zeros of polynomial $B(z)$ are strictly inside the unit circle of the z plane if and only if the following conditions are satisfied:

Table 3.1 The Jury-Marden array [1]

Row	Coefficients			
1	b_0	b_1	$\cdots$	b_N
2	b_N	b_{N-1}	$\cdots$	b_0
3	c_0	c_1	$\cdots$	c_{N-1}
4	c_{N-1}	c_{N-2}	$\cdots$	c_0
5	d_0	d_1	$\cdots$	d_{N-2}
6	d_{N-2}	d_{N-3}	$\cdots$	d_0
$\cdots$	$\cdots\cdots$	$\cdots\cdots$	$\cdots\cdots$	
$2N-3$	r_0	r_1	r_2	

(i) $D(1) > 0$
(ii) $(-1)^N D(-1) > 0$
(iii) $b_0 > |b_N|$
$|c_0| > |c_{N-1}|$
$|d_0| > |d_{N-2}|$
$\vdots$
$|r_0| > |r_2|$

3.2.6 Stability Triangle of Second-Order Polynomials

By applying the Jury-Marden criterion to a second-order polynomial of the form

$$B(z) = z^2 + b_1 z + b_2 \tag{3.6}$$

it can be concluded that $B(z)$ is stable if and only if coefficients $\{b_1, b_2\}$ satisfy

$$b_2 < 1, \quad b_1 + b_2 > -1, \quad \text{and} \quad b_1 - b_2 < 1 \tag{3.7}$$

The region defined by the three linear constraints in (3.7) is depicted in Figure 3.1, which is often referred to as the stability triangle [2, 3].

3.2.7 Lyapunov Criterion

An IIR digital filter characterized in state space as

$$\begin{aligned} \boldsymbol{x}(k+1) &= \boldsymbol{A}\boldsymbol{x}(k) + \boldsymbol{b}u(k) \\ y(k) &= \boldsymbol{c}\boldsymbol{x}(k) + du(k) \end{aligned} \tag{3.8}$$

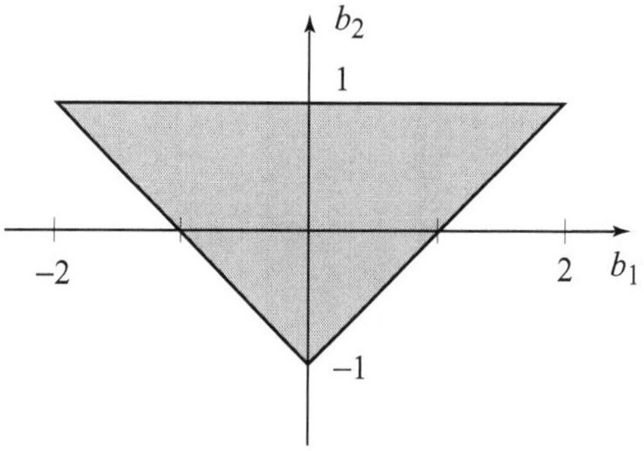

Figure 3.1 Stability triangle.

is stable if and only if the magnitudes of the eigenvalues of system matrix $\boldsymbol{A}$ are strictly less than unity.

Theorem 3.6: *Lyapunov*
The state-space digital filter in (3.8) is stable if and only if for a positive definite matrix $\boldsymbol{Q}$ (say, $\boldsymbol{Q} = \boldsymbol{I}$) there exists a unique positive definite matrix $\boldsymbol{P}$ that satisfies the Lyapunov equation

$$\boldsymbol{A}^T\boldsymbol{P}\boldsymbol{A} - \boldsymbol{P} = -\boldsymbol{Q} \tag{3.9}$$

Proof
Suppose there exists a positive definite $\boldsymbol{P}$ that satisfies the equation in (3.9). Let λ be any one of the eigenvalues of $\boldsymbol{A}$. By definition, there exists an eigenvector $\boldsymbol{x} \neq \boldsymbol{0}$ such that $\boldsymbol{A}\boldsymbol{x} = \lambda\boldsymbol{x}$. Multiplying (3.9) by $\boldsymbol{x}$ from right and $\boldsymbol{x}^T$ from left, we obtain $\boldsymbol{x}^T\boldsymbol{A}^T\boldsymbol{P}\boldsymbol{A}\boldsymbol{x} - \boldsymbol{x}^T\boldsymbol{P}\boldsymbol{x} = -\boldsymbol{x}^T\boldsymbol{Q}\boldsymbol{x}$ which implies that

$$(|\lambda|^2 - 1)\boldsymbol{x}^T\boldsymbol{P}\boldsymbol{x} = -\boldsymbol{x}^T\boldsymbol{Q}\boldsymbol{x} \tag{3.10}$$

Hence

$$|\lambda|^2 - 1 = -\frac{\boldsymbol{x}^T\boldsymbol{Q}\boldsymbol{x}}{\boldsymbol{x}^T\boldsymbol{P}\boldsymbol{x}} < 0 \tag{3.11}$$

because both $\boldsymbol{x}^T\boldsymbol{P}\boldsymbol{x}$ and $\boldsymbol{x}^T\boldsymbol{Q}\boldsymbol{x}$ are strictly positive. This shows that the magnitudes of the eigenvalues of $\boldsymbol{A}$ are strictly less than unity, therefore the system is stable.

Conversely, suppose the system is stable. For a given positive definite $\boldsymbol{Q}$, we can construct a matrix $\boldsymbol{P}$ as

$$\boldsymbol{P} = \sum_{k=0}^{\infty}(\boldsymbol{A}^T)^k\boldsymbol{Q}\boldsymbol{A}^k \tag{3.12}$$

Because the system is stable, the magnitudes of all eigenvalues of $\boldsymbol{A}$ are strictly less than unity, hence the above series is well defined. Moreover, $\boldsymbol{P}$ is positive definite because it is the sum of a positive define matrix $\boldsymbol{Q}$ plus many other terms each of which is at least positive semidefinite. Furthermore, we can compute

$$\boldsymbol{A}^T\boldsymbol{P}\boldsymbol{A} - \boldsymbol{P} = \sum_{k=0}^{\infty}(\boldsymbol{A}^T)^{k+1}\boldsymbol{Q}\boldsymbol{A}^{k+1} - \sum_{k=0}^{\infty}(\boldsymbol{A}^T)^k\boldsymbol{Q}\boldsymbol{A}^k = -\boldsymbol{Q} \tag{3.13}$$

Thus the positive definite matrix $\boldsymbol{P}$ constructed above satisfies (3.9), which completes the proof. ■

3.3 Coefficient Sensitivity

The difference equation associated with the IIR digital filter in (3.2) is given by

$$y(k) = \sum_{i=0}^{M} a_i x(k-i) - \sum_{i=1}^{N} b_i y(k-i) \tag{3.14}$$

When the filter coefficients $\{a_i\}$ and $\{b_i\}$ are quantized for implementation purposes, the coefficient error may lead to substantial changes in the filter characteristics such as frequency response and stability. These changes can be investigated by examining the changes in the locations of poles due to the changes in the filter's coefficients.

The transfer function in (3.2) can be expressed as

$$H(z) = \frac{z^{N-M}\left(a_0 z^M + a_1 z^{M-1} + \cdots + a_M\right)}{z^N + b_1 z^{N-1} + \cdots + b_N} \triangleq \frac{\tilde{A}(z)}{\tilde{B}(z)} \tag{3.15}$$

The polynomials $\tilde{A}(z)$ and $\tilde{B}(z)$ in (3.15) can be written as

$$\tilde{A}(z) = z^{N-M}\left(a_0 z^M + a_1 z^{M-1} + \cdots + a_M\right) = a_0 \prod_{i=1}^{N} (z - z_i) \tag{3.16}$$

and

$$\tilde{B}(z) = \sum_{k=0}^{N} b_k z^{N-k} = \prod_{i=1}^{N} (z - p_i) \tag{3.17}$$

where $\{z_i\}$ and $\{p_i\}$ are the zeros and poles of the IIR filter, respectively. To examine how changes in coefficients $\{b_k\}$ lead to change in pole p_i, we assume that the N poles are distinct and denote by dp_i the infinitesimal change in p_i due to infinitesimal changes $db_1, db_2, \ldots, db_n$ in $\{b_k\}$. The rule of total differentiation gives

$$dp_i = \sum_{k=1}^{n} \left.\frac{\partial p_i}{\partial b_k}\right|_{z=p_i} db_k$$

where

$$\left.\frac{\partial p_i}{\partial b_k}\right|_{z=p_i} = \left.\frac{\partial \tilde{B}(z) \big/ \partial b_k}{\partial \tilde{B}(z) \big/ \partial p_i}\right|_{z=p_i} = -\frac{p_i^{N-k}}{\prod_{j=1,\, j\neq i}^{N} (p_i - p_j)}$$

Therefore, we have

$$dp_i = -\frac{1}{\prod\limits_{j=1, j\neq i}^{N} (p_i - p_j)} \sum_{k=1}^{n} p_i^{N-k} db_k \tag{3.18}$$

Based on (3.18), several observations can be made as follows [3]:

1. Coefficient sensitivity increases when the poles are close together.
2. Coefficient sensitivity increases when a pole p_i moves closer to the unit circle because the magnitude of each p_i^{N-k} gets larger.
3. The filter is most sensitive to the variations of the last coefficient b_N because in (3.18) it is associated with $p_i^0 = 1$ that is the largest among all p_i^{N-k}. In general, the filter's sensitivity to the variations of coefficient b_i is greater than that of coefficient b_j as long as $i > j$.

For a high-order IIR filter with sharp transitions in its frequency response, typically the poles are not well-separated, hence the coefficient sensitivity of the filter can be high. A less sensitive realization of a high-order filter is to break the transfer function into lower-order sections and connect these section in cascade or parallel. In this way, the poles within each section can be well-separated and reduced overall coefficient sensitivity can be achieved. Although the use of fourth- or higher-order sections may be justified in some applications, second-order section has been the most-often employed building block in cascade and parallel structures.

3.4 Summary

This chapter presents several criteria that can be used to verify the stability of an IIR digital filters. The Schur-Cohn criterion, Schur-Cohn-Fujiwara criterion, and Jury-Marden criterion are of use when the transfer function of the IIR filter is given. The stability triangle defined by (3.7) is particularly convenient when the denominator of the IIR filter is given in terms of a product of second-order sections. The Lyapunov criterion is suitable for IIR filters that are modeled in state space. This chapter also presents a brief study on coefficient sensitivity. This is done by examining the changes in the locations of poles due to the changes in the filter's coefficients.

References

[1] E. I. Jury, *Theory and Application of the z-Transform Method*, John Wiley, New York, 1964.

[2] A. Antoniou, *Digital Filters: Analysis, Design, and Applications*, 2nd ed., McGraw-Hill, New York, 1993.

[3] R. A. Roberts and C. T. Mullis, *Digital Signal Processing*, Addison-Wesley, Reading MA, 1987.

4

State-Space Models

4.1 Preview

We have introduced the external (input-output) description and state-space description of dynamical linear systems in Chapter 2. Unlike the external description, by adequately defining a state-variable vector and establishing dynamical relations between the state variables and system's input and output, a state-space description reveals a great deal of internal structure of a dynamical linear system. In this chapter, the state-space models of dynamical linear systems are studied more systematically. In Section 4.2, a necessary and sufficient condition for a state-space model to be controllable and observable is described and proved. In Section 4.3, Faddeev's formula for deriving its transfer function from a given state-space model is presented. In Section 4.5, equivalent transformation is defined with its application to the derivation of various representations such as *canonical form, balanced form, input-normal form, output-normal form* and so on. In Section 4.6, Kalman's canonical structure theorem is introduced. In Section 4.6, the problems of minimal realization and minimal partial realization are addressed by means of the Hankel matrix. Finally, a passivity property of discrete-time linear systems, known as lossless bounded-real lemma, is examined in Section 4.7.

4.2 Controllability and Observability

Consider a state-space model $(\boldsymbol{A}, \boldsymbol{b}, \boldsymbol{c}, d)_n$ described by

$$\begin{aligned} \boldsymbol{x}(k+1) &= \boldsymbol{A}\boldsymbol{x}(k) + \boldsymbol{b}u(k) \\ y(k) &= \boldsymbol{c}\boldsymbol{x}(k) + du(k) \end{aligned} \tag{4.1}$$

where $\boldsymbol{x}(k)$ is an $n \times 1$ state-variable vector, $u(k)$ is a scalar input, $y(k)$ is a scalar output, and $\boldsymbol{A}, \boldsymbol{b}, \boldsymbol{c}$ and d are $n \times n$, $n \times 1$, $1 \times n$, and 1×1

real constant matrices, respectively. The state space of the model in (4.1) is an n-dimensional real vector space and is denoted by $\mathbf{\Sigma}$. The first and second equations in (4.1) are called the *state equation* and *output equation*, respectively. A block-diagram of the state-space model in (4.1) is depicted in Figure 4.1.

From (4.1), it follows that

$$\begin{aligned} \boldsymbol{x}(k) &= \boldsymbol{A}^k\boldsymbol{x}(0) + \boldsymbol{A}^{k-1}\boldsymbol{b}u(0) + \cdots + \boldsymbol{A}\boldsymbol{b}u(k-2) + \boldsymbol{b}u(k-1) \\ y(k) &= \boldsymbol{c}\boldsymbol{A}^k\boldsymbol{x}(0) + \boldsymbol{c}\boldsymbol{A}^{k-1}\boldsymbol{b}u(0) + \cdots + \boldsymbol{c}\boldsymbol{A}\boldsymbol{b}u(k-2) \\ &\quad + \boldsymbol{c}\boldsymbol{b}u(k-1) + du(k) \end{aligned} \tag{4.2}$$

Definition 4.1
The state equation in (4.1) is said to be *controllable* if for any initial state $\boldsymbol{x}(0)$ in the state space $\mathbf{\Sigma}$, and any state $\boldsymbol{x}_d$ in $\mathbf{\Sigma}$, there exist a finite step $k_o > 0$ and an input sequence that will transfer the initial state $\boldsymbol{x}(0)$ to the state $\boldsymbol{x}_d$ at step k_o. Otherwise, the state equation is said to be *uncontrollable*.

Definition 4.2
The state-space model $(\boldsymbol{A}, \boldsymbol{b}, \boldsymbol{c}, d)_n$ in (4.1) is said to be *observable* if for any initial state $\boldsymbol{x}(0)$ in the state space $\mathbf{\Sigma}$, there exist a finite step $k_o > 0$ such that the knowledge of the input and output over the finite interval $0 \le k \le k_o$ suffices to determine the initial state $\boldsymbol{x}(0)$ uniquely. Otherwise, the state-space model $(\boldsymbol{A}, \boldsymbol{b}, \boldsymbol{c}, d)_n$ is said to be *unobservable*.

Definition 4.3
The state-space model $(\boldsymbol{A}, \boldsymbol{b}, \boldsymbol{c}, d)_n$ in (4.1) is said to be *minimal* if it is controllable and observable.

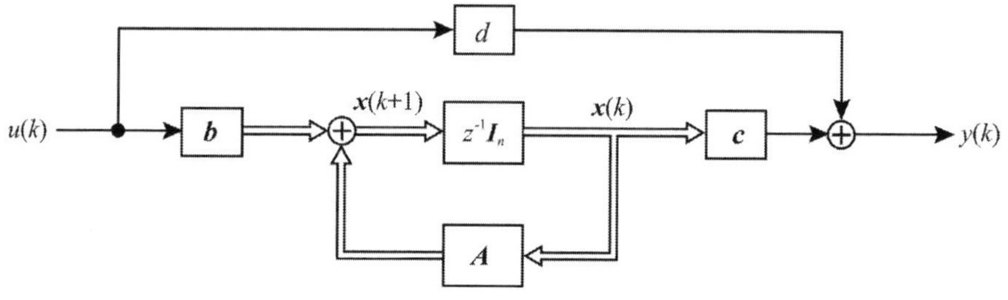

Figure 4.1 A state-space model.

From (4.2), it follows that

$$\boldsymbol{x}(n) - \boldsymbol{A}^n \boldsymbol{x}(0) = \begin{bmatrix} \boldsymbol{b} & \boldsymbol{A}\boldsymbol{b} & \cdots & \boldsymbol{A}^{n-1}\boldsymbol{b} \end{bmatrix} \begin{bmatrix} u(n-1) \\ u(n-2) \\ \vdots \\ u(0) \end{bmatrix}$$

$$\begin{bmatrix} y(0) \\ y(1) \\ \vdots \\ y(n-1) \end{bmatrix} = \begin{bmatrix} \boldsymbol{c} \\ \boldsymbol{c}\boldsymbol{A} \\ \vdots \\ \boldsymbol{c}\boldsymbol{A}^{n-1} \end{bmatrix} \boldsymbol{x}(0) + \begin{bmatrix} d & 0 & \cdots & 0 \\ \boldsymbol{c}\boldsymbol{b} & d & \ddots & \vdots \\ \vdots & \ddots & \ddots & 0 \\ \boldsymbol{c}\boldsymbol{A}^{n-2}\boldsymbol{b} & \cdots & \boldsymbol{c}\boldsymbol{b} & d \end{bmatrix} \begin{bmatrix} u(0) \\ u(1) \\ \vdots \\ u(n-1) \end{bmatrix} \tag{4.3}$$

By using (4.3) and applying the *Cayley-Hamilton theorem* described in Section 4.3.3, we can readily obtain the following two theorems.

Theorem 4.1
The state equation in (4.1) is controllable if and only if

$$\text{rank} \begin{bmatrix} \boldsymbol{b} \ \boldsymbol{A}\boldsymbol{b} & \cdots & \boldsymbol{A}^{n-1}\boldsymbol{b} \end{bmatrix} = n \tag{4.4}$$

Theorem 4.2
The state-space model $(\boldsymbol{A}, \boldsymbol{b}, \boldsymbol{c}, d)_n$ in (4.1) is observable if and only if

$$\text{rank} \begin{bmatrix} \boldsymbol{c} \\ \boldsymbol{c}\boldsymbol{A} \\ \vdots \\ \boldsymbol{c}\boldsymbol{A}^{n-1} \end{bmatrix} = n \tag{4.5}$$

The matrices $\boldsymbol{V}_n = \begin{bmatrix} \boldsymbol{b} \ \boldsymbol{A}\boldsymbol{b} \cdots \boldsymbol{A}^{n-1}\boldsymbol{b} \end{bmatrix}$ and $\boldsymbol{U}_n = \begin{bmatrix} \boldsymbol{c}^T \ (\boldsymbol{c}\boldsymbol{A})^T \cdots (\boldsymbol{c}\boldsymbol{A}^{n-1})^T \end{bmatrix}^T$ are called the *controllability matrix* and *observability matrix*, respectively.

Under the assumption that $\boldsymbol{A}^k$ converges to $\boldsymbol{0}$ as k goes to infinity (i.e., the state-space model $(\boldsymbol{A}, \boldsymbol{b}, \boldsymbol{c}, d)_n$ is stable), we define two symmetric matrices $\boldsymbol{K}_c$ and $\boldsymbol{W}_o$ as

$$\begin{aligned} \boldsymbol{K}_c &= \sum_{k=0}^{\infty} \boldsymbol{A}^k \boldsymbol{b}\boldsymbol{b}^T (\boldsymbol{A}^k)^T \\ \boldsymbol{W}_o &= \sum_{k=0}^{\infty} (\boldsymbol{A}^k)^T \boldsymbol{c}^T \boldsymbol{c} \boldsymbol{A}^k \end{aligned} \tag{4.6}$$

These marices $\boldsymbol{K}_c$ and $\boldsymbol{W}_o$ can be obtained by solving the Lyapunov equations

$$\begin{aligned}\boldsymbol{K}_c &= \boldsymbol{A}\boldsymbol{K}_c\boldsymbol{A}^T + \boldsymbol{b}\boldsymbol{b}^T \\ \boldsymbol{W}_o &= \boldsymbol{A}^T\boldsymbol{W}_o\boldsymbol{A} + \boldsymbol{c}^T\boldsymbol{c}\end{aligned} \tag{4.7}$$

Marices $\boldsymbol{K}_c$ and $\boldsymbol{W}_o$ are called the *controllability Grammian* and *the observability Grammian*, respectively. It can readily be shown that if the state-space model in (4.1) is controllable and observable, then $\boldsymbol{K}_c$ and $\boldsymbol{W}_o$ are positive definite.

4.3 Transfer Function

4.3.1 Impulse Response

By taking the z-transform of both sides in (4.1), we obtain

$$\begin{aligned}z[\boldsymbol{X}(z) - \boldsymbol{x}(0)] &= \boldsymbol{A}\boldsymbol{X}(z) + \boldsymbol{b}U(z) \\ Y(z) &= \boldsymbol{c}\boldsymbol{X}(z) + dU(z)\end{aligned} \tag{4.8}$$

where $\boldsymbol{X}(z)$, $U(z)$ and $Y(z)$ denote the z-transforms of state-variable vector $\boldsymbol{x}(k)$, input $u(k)$ and output $y(k)$, respectively. From (4.8), we have

$$\begin{aligned}\boldsymbol{X}(z) &= (z\boldsymbol{I}_n - \boldsymbol{A})^{-1}z\boldsymbol{x}(0) + (z\boldsymbol{I}_n - \boldsymbol{A})^{-1}\boldsymbol{b}U(z) \\ Y(z) &= \boldsymbol{c}(z\boldsymbol{I}_n - \boldsymbol{A})^{-1}z\boldsymbol{x}(0) + \left[\boldsymbol{c}(z\boldsymbol{I}_n - \boldsymbol{A})^{-1}\boldsymbol{b} + d\right]U(z)\end{aligned} \tag{4.9}$$

The first and second terms on the right-hand side of (4.9) represent the transient response and steady state characteristic, respectively. By setting $\boldsymbol{x}(0) = \boldsymbol{0}$, the transfer function $H(z)$ of the state-space model $(\boldsymbol{A}, \boldsymbol{b}, \boldsymbol{c}, d)_n$ in (4.1) is obtained as

$$H(z) = \frac{Y(z)}{U(z)} = \boldsymbol{c}(z\boldsymbol{I}_n - \boldsymbol{A})^{-1}\boldsymbol{b} + d \tag{4.10}$$

By noting that

$$\begin{aligned}(z\boldsymbol{I}_n - \boldsymbol{A})^{-1} &= \boldsymbol{I}z^{-1} + \boldsymbol{A}z^{-2} + \boldsymbol{A}^2z^{-3} + \cdots \\ &= \sum_{i=1}^{\infty} \boldsymbol{A}^{i-1}z^{-i}\end{aligned} \tag{4.11}$$

we can write (4.10) as

$$H(z) = d + \sum_{i=1}^{\infty} \boldsymbol{c}\boldsymbol{A}^{i-1}\boldsymbol{b}\, z^{-i} = \sum_{i=0}^{\infty} h_i\, z^{-i} \tag{4.12}$$

where

$$h_0 = d, \qquad h_i = \boldsymbol{c}\boldsymbol{A}^{i-1}\boldsymbol{b} \text{ for } i = 1, 2, 3, \cdots$$

Sequence $\{h_0, h_1, h_2, \cdots\}$ is called the *unit-pulse response* or *impulse response* of the state-space model $(\boldsymbol{A}, \boldsymbol{b}, \boldsymbol{c}, d)_n$ in (4.1).

4.3.2 Faddeev's Formula

To compute the transfer function in (4.10), we express the inverse of $(z\boldsymbol{I}_n - \boldsymbol{A})$ as

$$(z\boldsymbol{I}_n - \boldsymbol{A})^{-1} = \frac{\text{adj}(z\boldsymbol{I}_n - \boldsymbol{A})}{\det(z\boldsymbol{I}_n - \boldsymbol{A})} \tag{4.13a}$$

where $\det(z\boldsymbol{I}_n - \boldsymbol{A})$ and $\text{adj}(z\boldsymbol{I}_n - \boldsymbol{A})$ stand for the determinant and adjoint matrix of $(z\boldsymbol{I}_n - \boldsymbol{A})$, respectively, and the denominator and numerator in (4.13a) are given by

$$\begin{aligned} \det(z\boldsymbol{I}_n - \boldsymbol{A}) &= z^n + a_1 z^{n-1} + \cdots + a_{n-1}z + a_n \\ \text{adj}(z\boldsymbol{I}_n - \boldsymbol{A}) &= \boldsymbol{B}_0 z^{n-1} + \boldsymbol{B}_1 z^{n-2} + \cdots + \boldsymbol{B}_{n-2} z + \boldsymbol{B}_{n-1} \end{aligned} \tag{4.13b}$$

respectively. The scalars $a_1, a_2, \cdots, a_n$ and the matrices $\boldsymbol{B}_0, \boldsymbol{B}_1, \cdots, \boldsymbol{B}_{n-1}$ in (4.13b) can be computed from

$$\boldsymbol{B}_0 = \boldsymbol{I}_n, \qquad \boldsymbol{B}_i = \boldsymbol{A}\boldsymbol{B}_{i-1} + a_i \boldsymbol{I}_n \tag{4.14a}$$

$$a_i = -\frac{1}{i}\,\text{tr}[\boldsymbol{A}\boldsymbol{B}_{i-1}] \text{ for } i = 1, 2, \cdots, n \tag{4.14b}$$

where $\boldsymbol{B}_n = \boldsymbol{0}$ that can be used to check whether the program is correct or not. The equations in (4.14) are known as *Faddeev's formula.*

Proof

From (4.13a), it follows that

$$(z\boldsymbol{I}_n - \boldsymbol{A})\,\text{adj}(z\boldsymbol{I}_n - \boldsymbol{A}) = \det(z\boldsymbol{I}_n - \boldsymbol{A})\boldsymbol{I}_n \tag{4.15}$$

Substituting (4.13b) into (4.15) provides

$$(z\boldsymbol{I}_n - \boldsymbol{A})\Big(\sum_{i=0}^{n-1} \boldsymbol{B}_i z^{n-1-i}\Big) = \Big(\sum_{i=0}^{n} a_i z^{n-i}\Big)\boldsymbol{I}_n, \qquad a_0 = 1 \tag{4.16}$$

By making a comparison of coefficients between both sides of (4.16), we obtain

$$\begin{aligned} \boldsymbol{B}_0 &= \boldsymbol{I}_n \\ \boldsymbol{B}_i &= \boldsymbol{A}\boldsymbol{B}_{i-1} + a_i\boldsymbol{I}_n \ \text{ for } \ i = 1, 2, \cdots, n-1 \\ \boldsymbol{0} &= \boldsymbol{A}\boldsymbol{B}_{n-1} + a_n\boldsymbol{I}_n \end{aligned} \tag{4.17}$$

which is the same as (4.14a). Moreover, it can easily be verified that

$$\begin{aligned} \frac{\partial \det(z\boldsymbol{I}_n - \boldsymbol{A})}{\partial z} &= \sum_{i=1}^{n}\sum_{j=1}^{n} \frac{\partial \det(z\boldsymbol{I}_n - \boldsymbol{A})}{\partial\,(z\boldsymbol{I}_n - \boldsymbol{A})_{ij}} \cdot \frac{\partial\,(z\boldsymbol{I}_n - \boldsymbol{A})_{ij}}{\partial z} \\ &= \sum_{i=1}^{n}\sum_{j=1}^{n} \big[\text{adj}(z\boldsymbol{I}_n - \boldsymbol{A})\big]_{ji} \cdot \big(\boldsymbol{I}_n\big)_{ij} \\ &= \text{tr}\big[\text{adj}(z\boldsymbol{I}_n - \boldsymbol{A})\big] \end{aligned} \tag{4.18}$$

where $(\boldsymbol{A})_{ij}$ denotes the (i, j)th element of matrix $\boldsymbol{A}$. By substituting (4.13b) into (4.18), we obtain

$$\begin{aligned} &nz^{n-1} + (n-1)a_1 z^{n-2} + \cdots + a_{n-1} \\ &\quad = (\text{tr}\boldsymbol{B}_0)z^{n-1} + (\text{tr}\boldsymbol{B}_1)z^{n-2} + \cdots + \text{tr}\boldsymbol{B}_{n-1} \end{aligned} \tag{4.19}$$

By making a comparison of coefficients between both sides of (4.19), we have

$$a_i = \frac{1}{n-i}\text{tr}\boldsymbol{B}_i \ \text{ for } \ i = 1, 2, \cdots, n-1 \tag{4.20}$$

Finally, substituting $\boldsymbol{B}_i$ in (4.17) into (4.20) gives

$$\begin{aligned} a_i &= \frac{1}{n-i}\text{tr}[\boldsymbol{A}\boldsymbol{B}_{i-1} + a_i\boldsymbol{I}_n] \\ &= \frac{1}{n-i}\text{tr}[\boldsymbol{A}\boldsymbol{B}_{i-1}] + \frac{n}{n-i}a_i \end{aligned} \tag{4.21}$$

which readily yields (4.14b). This completes the proof of Faddeev's formula. ■

4.3.3 Cayley-Hamilton's Theorem

By setting the denominator polynomial of (4.13b) to zero, we obtain

$$\det(zI_n - \boldsymbol{A}) = z^n + a_1 z^{n-1} + \cdots + a_n = 0 \tag{4.22}$$

which is called the *characteristic equation*. From (4.14a), it follows that

$$\begin{aligned} \boldsymbol{B}_1 &= \boldsymbol{A}\boldsymbol{B}_0 + a_1\boldsymbol{I}_n = \boldsymbol{A} + a_1\boldsymbol{I}_n \\ \boldsymbol{B}_2 &= \boldsymbol{A}\boldsymbol{B}_1 + a_2\boldsymbol{I}_n = \boldsymbol{A}^2 + a_1\boldsymbol{A} + a_2\boldsymbol{I}_n \\ &\vdots \\ \boldsymbol{B}_n &= \boldsymbol{A}\boldsymbol{B}_{n-1} + a_n\boldsymbol{I}_n = \boldsymbol{A}^n + a_1\boldsymbol{A}^{n-1} + \cdots + a_n\boldsymbol{I}_n \end{aligned} \tag{4.23}$$

Since $\boldsymbol{B}_n = \boldsymbol{0}$, we readily obtain

$$\boldsymbol{A}^n + a_1\boldsymbol{A}^{n-1} + \cdots + a_n\boldsymbol{I}_n = \boldsymbol{0} \tag{4.24}$$

which is known as the *Cayley-Hamilton theorem.* ■

It is noted that the values of z satisfying (4.22) coincide with the eigenvalues of matrix $\boldsymbol{A}$, and they are called the *poles* of the transfer function $H(z)$.

4.4 Equivalent Systems

4.4.1 Equivalent Transformation

In this section, we introduce the concept of equivalent linear time-invariant causal dynamical systems. This concept is found to be useful in constructing canonical forms as well as developing input-normal, output-normal, and balanced state-space descriptions.

Definition 4.4

Let $\boldsymbol{T}$ be an $n \times n$ nonsingular real matrix, and let $\overline{\boldsymbol{x}}(k) = \boldsymbol{T}^{-1}\boldsymbol{x}(k)$. Then the state-space model $(\overline{\boldsymbol{A}}, \overline{\boldsymbol{b}}, \overline{\boldsymbol{c}}, d)_n$ described by

$$\begin{aligned} \overline{\boldsymbol{x}}(k+1) &= \overline{\boldsymbol{A}}\overline{\boldsymbol{x}}(k) + \overline{\boldsymbol{b}}u(k) \\ y(k) &= \overline{\boldsymbol{c}}\,\overline{\boldsymbol{x}}(k) + du(k) \end{aligned} \tag{4.25}$$

is said to be *equivalent* to the state-space model $(\boldsymbol{A}, \boldsymbol{b}, \boldsymbol{c}, d)_n$ in (4.1) and $\boldsymbol{T}$ is called the *equivalent transformation matrix, similarity transformation matrix* or *coordinate transformation matrix* where

$$\overline{A} = T^{-1}AT, \quad \overline{b} = T^{-1}b, \quad \overline{c} = cT$$

By taking the z-transform of both sides in (4.25), we obtain

$$\begin{aligned} z\left[\overline{X}(z) - \overline{x}(0)\right] &= \overline{A}\,\overline{X}(z) + \overline{b}U(z) \\ Y(z) &= \overline{c}\overline{X}(z) + dU(z) \end{aligned} \tag{4.26}$$

where $\overline{X}(z)$ is the z-transform of state-vaiable vector $\overline{x}(k)$. By setting $\overline{x}(0) = \mathbf{0}$, the transfer function $\overline{H}(z)$ of the state-space model $(\overline{A}, \overline{b}, \overline{c}, d)_n$ in (4.25) can be written as

$$\begin{aligned} \overline{H}(z) = \frac{Y(z)}{U(z)} &= \overline{c}(zI_n - \overline{A})^{-1}\overline{b} + d \\ &= c(zI_n - A)^{-1}b + d \\ &= H(z) \end{aligned} \tag{4.27}$$

In other words, the transfer function of a state-space model remains invariant under equivalent transformation. That is, $\overline{H}(z) = H(z)$ holds true for any equivalent transformation defined by $\overline{x}(k) = T^{-1}x(k)$.

4.4.2 Canonical Forms

1. Controllable Canonical Form

If the state-space model $(A, b, c, d)_n$ in (4.1) is controllable, then by applying an equivalent transformation $\overline{x}(k) = T_c^{-1}x(k)$ with $T_c = [b \;\; Ab \;\; \cdots \;\; A^{n-1}b]$, the model can be transformed into the following controllable canonical form:

$$\begin{aligned} \overline{x}(k+1) &= \begin{bmatrix} 0 & \cdots & 0 & -a_n \\ 1 & \ddots & \vdots & \vdots \\ \vdots & \ddots & 0 & -a_2 \\ 0 & \cdots & 1 & -a_1 \end{bmatrix} \overline{x}(k) + \begin{bmatrix} 1 \\ 0 \\ \vdots \\ 0 \end{bmatrix} u(k) \\ y(k) &= \begin{bmatrix} h_1 & h_2 & \cdots & h_n \end{bmatrix} \overline{x}(k) + du(k) \end{aligned} \tag{4.28}$$

By writing the Cayley-Hamilton theorem in (4.24) as

$$A^n b = -a_n b - a_{n-1}Ab - \cdots - a_1 A^{n-1}b \tag{4.29}$$

Equation (4.28) can easily be verified by

$$\begin{aligned}
\boldsymbol{A}\boldsymbol{T}_c &= \begin{bmatrix} \boldsymbol{A}\boldsymbol{b} & \boldsymbol{A}^2\boldsymbol{b} & \cdots & \boldsymbol{A}^n\boldsymbol{b} \end{bmatrix} \\
&= \boldsymbol{T}_c \begin{bmatrix} 0 & \cdots & 0 & -a_n \\ 1 & \ddots & \vdots & \vdots \\ \vdots & \ddots & 0 & -a_2 \\ 0 & \cdots & 1 & -a_1 \end{bmatrix}
\end{aligned} \tag{4.30}$$

$$\boldsymbol{b} = \boldsymbol{T}_c \begin{bmatrix} 1 \\ 0 \\ \vdots \\ 0 \end{bmatrix}, \qquad \boldsymbol{c}\boldsymbol{T}_c = \begin{bmatrix} h_1 & h_2 & \cdots & h_n \end{bmatrix}$$

where $h_i = \boldsymbol{c}\boldsymbol{A}^{i-1}\boldsymbol{b}$ for $i \geq 1$ in (4.12).

Moreover, if an equivalent transformation $\overline{\boldsymbol{x}}(k) = \boldsymbol{T}^{-1}\boldsymbol{x}(k)$ with

$$\boldsymbol{T} = \boldsymbol{T}_c \begin{bmatrix} a_{n-1} & \cdots & a_1 & 1 \\ \vdots & \iddots & 1 & 0 \\ a_1 & \iddots & \iddots & \vdots \\ 1 & 0 & \cdots & 0 \end{bmatrix} \tag{4.31}$$

is applied to the state-space model $(\boldsymbol{A}, \boldsymbol{b}, \boldsymbol{c}, d)_n$ in (4.1), the model can be transformed into the following controllable canonical form:

$$\begin{aligned}
\overline{\boldsymbol{x}}(k+1) &= \begin{bmatrix} 0 & 1 & \cdots & 0 \\ \vdots & \ddots & \ddots & \vdots \\ 0 & \cdots & 0 & 1 \\ -a_n & \cdots & -a_2 & -a_1 \end{bmatrix} \overline{\boldsymbol{x}}(k) + \begin{bmatrix} 0 \\ \vdots \\ 0 \\ 1 \end{bmatrix} u(k) \\
y(k) &= \begin{bmatrix} \sum_{i=1}^{n} a_{n-i}h_i & \sum_{i=1}^{n-1} a_{n-1-i}h_i & \cdots & h_1 \end{bmatrix} \overline{\boldsymbol{x}}(k) + du(k)
\end{aligned} \tag{4.32}$$

This can easily be shown by noticing the following:

$$
\begin{aligned}
&\boldsymbol{T}_c^{-1}\boldsymbol{A}\boldsymbol{T}_c \begin{bmatrix} a_{n-1} & \cdots & a_1 & 1 \\ \vdots & \iddots & 1 & 0 \\ a_1 & \iddots & \iddots & \vdots \\ 1 & 0 & \cdots & 0 \end{bmatrix} = \begin{bmatrix} -a_n & 0 & \cdots & 0 & 0 \\ 0 & a_{n-2} & \cdots & a_1 & 1 \\ \vdots & \vdots & \iddots & 1 & 0 \\ 0 & a_1 & \iddots & \iddots & \vdots \\ 0 & 1 & 0 & \cdots & 0 \end{bmatrix} \\
&= \begin{bmatrix} a_{n-1} & \cdots & a_1 & 1 \\ \vdots & \iddots & 1 & 0 \\ a_1 & \iddots & \iddots & \vdots \\ 1 & 0 & \cdots & 0 \end{bmatrix} \begin{bmatrix} 0 & 1 & \cdots & 0 \\ \vdots & \ddots & \ddots & \vdots \\ 0 & \cdots & 0 & 1 \\ -a_n & \cdots & -a_2 & -a_1 \end{bmatrix} \\
&\begin{bmatrix} a_{n-1} & \cdots & a_1 & 1 \\ \vdots & \iddots & 1 & 0 \\ a_1 & \iddots & \iddots & \vdots \\ 1 & 0 & \cdots & 0 \end{bmatrix} \begin{bmatrix} 0 \\ \vdots \\ 0 \\ 1 \end{bmatrix} = \begin{bmatrix} 1 \\ 0 \\ \vdots \\ 0 \end{bmatrix} = \boldsymbol{T}_c^{-1}\boldsymbol{b}
\end{aligned}
\tag{4.33}
$$

Moreover, we have

$$
\operatorname{adj}\left(z\boldsymbol{I}_n - \boldsymbol{T}^{-1}\boldsymbol{A}\boldsymbol{T}\right)\boldsymbol{T}^{-1}\boldsymbol{b} = \begin{bmatrix} 1 \\ z \\ \vdots \\ z^{n-1} \end{bmatrix} \tag{4.34}
$$

which leads the transfer function to

$$
\begin{aligned}
H(z) &= \frac{\boldsymbol{c}\boldsymbol{T}\left[1 \; z \; \cdots \; z^{n-1}\right]^T}{\det\left(z\boldsymbol{I}_n - \boldsymbol{T}^{-1}\boldsymbol{A}\boldsymbol{T}\right)} + d \\
&= \frac{b_1 z^{n-1} + b_2 z^{n-2} + \cdots + b_n}{z^n + a_1 z^{n-1} + \cdots + a_n} + d
\end{aligned}
\tag{4.35}
$$

where

$$
b_l = \sum_{i=1}^{l} a_{l-i} h_i \text{ with } a_0 = 1 \text{ for } l = 1, 2, \cdots, n
$$

2. Observable Canonical Form

If the state-space model $(\boldsymbol{A}, \boldsymbol{b}, \boldsymbol{c}, d)_n$ in (4.1) is observable, then by applying an equivalent transformation $\overline{\boldsymbol{x}}(k) = \boldsymbol{T}_o\boldsymbol{x}(k)$ with $\boldsymbol{T}_o = [\boldsymbol{c}^T\ (\boldsymbol{c}\boldsymbol{A})^T\ \cdots\ (\boldsymbol{c}\boldsymbol{A}^{n-1})^T]^T$, the model can be transformed into the following observable canonical form:

$$\overline{\boldsymbol{x}}(k+1) = \begin{bmatrix} 0 & 1 & \cdots & 0 \\ \vdots & \ddots & \ddots & \vdots \\ 0 & \cdots & 0 & 1 \\ -a_n & \cdots & -a_2 & -a_1 \end{bmatrix} \overline{\boldsymbol{x}}(k) + \begin{bmatrix} h_1 \\ h_2 \\ \vdots \\ h_n \end{bmatrix} u(k) \tag{4.36}$$

$$y(k) = \begin{bmatrix} 1 & 0 & \cdots & 0 \end{bmatrix} \overline{\boldsymbol{x}}(k) + du(k)$$

By writing the Cayley-Hamilton theorem in (4.24) as

$$\boldsymbol{c}\boldsymbol{A}^n = -a_n\boldsymbol{c} - a_{n-1}\boldsymbol{c}\boldsymbol{A} - \cdots - a_1\boldsymbol{c}\boldsymbol{A}^{n-1} \tag{4.37}$$

Equation (4.36) can easily be demonstrated by

$$\boldsymbol{T}_o\boldsymbol{A} = \begin{bmatrix} \boldsymbol{c}\boldsymbol{A} \\ \boldsymbol{c}\boldsymbol{A}^2 \\ \vdots \\ \boldsymbol{c}\boldsymbol{A}^n \end{bmatrix} = \begin{bmatrix} 0 & 1 & \cdots & 0 \\ \vdots & \ddots & \ddots & \vdots \\ 0 & \cdots & 0 & 1 \\ -a_n & \cdots & -a_2 & -a_1 \end{bmatrix} \boldsymbol{T}_o$$

$$\boldsymbol{T}_o\boldsymbol{b} = \begin{bmatrix} h_1 \\ h_2 \\ \vdots \\ h_n \end{bmatrix}, \qquad \boldsymbol{c} = \begin{bmatrix} 1 & 0 & \cdots & 0 \end{bmatrix} \boldsymbol{T}_o \tag{4.38}$$

where $h_i = \boldsymbol{c}\boldsymbol{A}^{i-1}\boldsymbol{b}$ for $i \geq 1$ in (4.12).

Moreover, if an equivalent transformation $\overline{\boldsymbol{x}}(k) = \boldsymbol{T}\boldsymbol{x}(k)$ with

$$\boldsymbol{T} = \begin{bmatrix} a_{n-1} & \cdots & a_1 & 1 \\ \vdots & \cdot^{\cdot^{\cdot}} & 1 & 0 \\ a_1 & \cdot^{\cdot^{\cdot}} & \cdot^{\cdot^{\cdot}} & \vdots \\ 1 & 0 & \cdots & 0 \end{bmatrix} \boldsymbol{T}_o \tag{4.39}$$

is applied to the state-space model $(\boldsymbol{A}, \boldsymbol{b}, \boldsymbol{c}, d)_n$ in (4.1), the model can be transformed into the following observable canonical form:

$$\overline{\boldsymbol{x}}(k+1) = \begin{bmatrix} 0 & \cdots & 0 & -a_n \\ 1 & \ddots & \vdots & \vdots \\ \vdots & \ddots & 0 & -a_2 \\ 0 & \cdots & 1 & -a_1 \end{bmatrix} \overline{\boldsymbol{x}}(k) + \begin{bmatrix} \sum_{i=1}^{n} a_{n-i} h_i \\ \sum_{i=1}^{n-1} a_{n-1-i} h_i \\ \vdots \\ h_1 \end{bmatrix} u(k) \tag{4.40}$$

$$y(k) = \begin{bmatrix} 0 & \cdots & 0 & 1 \end{bmatrix} \overline{\boldsymbol{x}}(k) + du(k)$$

This can easily be verified by noticing the following:

$$\begin{aligned} \begin{bmatrix} a_{n-1} & \cdots & a_1 & 1 \\ \vdots & \cdot^{\cdot^{\cdot}} & 1 & 0 \\ a_1 & \cdot^{\cdot^{\cdot}} & \cdot^{\cdot^{\cdot}} & \vdots \\ 1 & 0 & \cdots & 0 \end{bmatrix} \boldsymbol{T}_o \boldsymbol{A} \boldsymbol{T}_o^{-1} &= \begin{bmatrix} -a_n & 0 & \cdots & 0 & 0 \\ 0 & a_{n-2} & \cdots & a_1 & 1 \\ \vdots & \vdots & \cdot^{\cdot^{\cdot}} & 1 & 0 \\ 0 & a_1 & \cdot^{\cdot^{\cdot}} & \cdot^{\cdot^{\cdot}} & \vdots \\ 0 & 1 & 0 & \cdots & 0 \end{bmatrix} \\ &= \begin{bmatrix} 0 & \cdots & 0 & -a_n \\ 1 & \ddots & \vdots & \vdots \\ \vdots & \ddots & 0 & -a_2 \\ 0 & \cdots & 1 & -a_1 \end{bmatrix} \begin{bmatrix} a_{n-1} & \cdots & a_1 & 1 \\ \vdots & \cdot^{\cdot^{\cdot}} & 1 & 0 \\ a_1 & \cdot^{\cdot^{\cdot}} & \cdot^{\cdot^{\cdot}} & \vdots \\ 1 & 0 & \cdots & 0 \end{bmatrix} \\ \begin{bmatrix} 0 & \cdots & 0 & 1 \end{bmatrix} \begin{bmatrix} a_{n-1} & \cdots & a_1 & 1 \\ \vdots & \cdot^{\cdot^{\cdot}} & 1 & 0 \\ a_1 & \cdot^{\cdot^{\cdot}} & \cdot^{\cdot^{\cdot}} & \vdots \\ 1 & 0 & \cdots & 0 \end{bmatrix} &= \begin{bmatrix} 1 & 0 & \cdots & 0 \end{bmatrix} = \boldsymbol{c} \boldsymbol{T}_o^{-1} \end{aligned} \tag{4.41}$$

We conclude this section by introducing the concept of duality for two discrete-time dynamical linear systems. Two systems in (4.32) and (4.40) are said to be *dual*, since the controllability of system in (4.32) is equivalent to the observability of system in (4.40), and the observability of system in (4.32) is equivalent to the controllability of system in (4.40). Notice that their transfer functions are identical.

4.4.3 Balanced, Input-Normal, and Output-Normal State-Space Models

In this section, the state-space model $(\boldsymbol{A}, \boldsymbol{b}, \boldsymbol{c}, d)_n$ in (4.1) is assumed to be stable, controllable and observable. From (4.6), the controllability and observability Grammians $\overline{\boldsymbol{K}}_c$ and $\overline{\boldsymbol{W}}_o$ for the state-space model $(\overline{\boldsymbol{A}}, \overline{\boldsymbol{b}}, \overline{\boldsymbol{c}}, d)_n$ in (4.25) can be written as

$$\overline{\boldsymbol{K}}_c = \boldsymbol{T}^{-1}\boldsymbol{K}_c\boldsymbol{T}^{-T}, \qquad \overline{\boldsymbol{W}}_o = \boldsymbol{T}^T\boldsymbol{W}_o\boldsymbol{T} \tag{4.42}$$

which lead to

$$\overline{\boldsymbol{K}}_c\overline{\boldsymbol{W}}_o = \boldsymbol{T}^{-1}\boldsymbol{K}_c\boldsymbol{W}_o\boldsymbol{T} \tag{4.43}$$

Therefore, the eigenvalues of $\boldsymbol{K}_c\boldsymbol{W}_o$ are invariant under the algebraic equivalence. Since matrices $\boldsymbol{K}_c$ and $\boldsymbol{W}_o$ are symmetric and positive definite, it is obvious that the eigenvalues of $\boldsymbol{K}_c\boldsymbol{W}_o$ are all strictly positive. We thus denote the ith eigenvalue of $\boldsymbol{K}_c\boldsymbol{W}_o$ by σ_i^2 for $i = 1, 2, \cdots, n$ with the ordering $\sigma_1 \geq \sigma_2 \geq \cdots \geq \sigma_n > 0$ and define

$$\boldsymbol{\Sigma} = \text{diag}\{\sigma_1, \sigma_2, \cdots, \sigma_n\} \tag{4.44}$$

Definition 4.5

(1) A state-space model $(\overline{\boldsymbol{A}}, \overline{\boldsymbol{b}}, \overline{\boldsymbol{c}}, d)_n$ in (4.25) is said to be *balanced* if $\overline{\boldsymbol{K}}_c = \overline{\boldsymbol{W}}_o = \boldsymbol{\Sigma}$.
(2) The state-space model $(\overline{\boldsymbol{A}}, \overline{\boldsymbol{b}}, \overline{\boldsymbol{c}}, d)_n$ is said to be *input-normal* if $\overline{\boldsymbol{K}}_c = \boldsymbol{I}_n$ and $\overline{\boldsymbol{W}}_o = \boldsymbol{\Sigma}^2$.
(3) The state-space model $(\overline{\boldsymbol{A}}, \overline{\boldsymbol{b}}, \overline{\boldsymbol{c}}, d)_n$ is said to be *output-normal* if $\overline{\boldsymbol{K}}_c = \boldsymbol{\Sigma}^2$ and $\overline{\boldsymbol{W}}_o = \boldsymbol{I}_n$.

By applying the Cholesky decomposition to $\boldsymbol{K}_c$, we have

$$\boldsymbol{K}_c = \boldsymbol{L}\boldsymbol{L}^T \tag{4.45}$$

where $\boldsymbol{L}$ is an $n \times n$ lower triangular matrix. Let $\boldsymbol{S}$ and $\boldsymbol{\Sigma}$ be obtained by eigenvalue-eigenvector decomposition of $\boldsymbol{L}^T\boldsymbol{W}_o\boldsymbol{L}$ as

$$\boldsymbol{L}^T\boldsymbol{W}_o\boldsymbol{L} = \boldsymbol{S}\boldsymbol{\Sigma}^2\boldsymbol{S}^T \tag{4.46}$$

where $\boldsymbol{\Sigma}^2$ and $\boldsymbol{S}$ are $n \times n$ diagonal and orthogonal matrices composed of the eigenvalues and eigenvectors of $\boldsymbol{L}^T\boldsymbol{W}_o\boldsymbol{L}$, respectively, and $\boldsymbol{S}^T\boldsymbol{S} = \boldsymbol{I}_n$.

1. Balanced State-Space Model

Suppose an equivalent transformation $\overline{\boldsymbol{x}}(k) = \boldsymbol{T}^{-1}\boldsymbol{x}(k)$ with

$$\boldsymbol{T} = \boldsymbol{L}\boldsymbol{S}\boldsymbol{\Sigma}^{-\frac{1}{2}} \tag{4.47}$$

is applied, we obtain

$$\overline{\boldsymbol{K}}_c = \overline{\boldsymbol{W}}_o = \boldsymbol{\Sigma} \tag{4.48}$$

Hence an equivalent state-space model is balanced, and the balanced state-space model $(\overline{\boldsymbol{A}}, \overline{\boldsymbol{b}}, \overline{\boldsymbol{c}}, d)_n$ is characterized by

$$\begin{aligned} \overline{\boldsymbol{A}} &= \boldsymbol{\Sigma}^{\frac{1}{2}}\boldsymbol{S}^T\boldsymbol{L}^{-1}\boldsymbol{A}\boldsymbol{L}\boldsymbol{S}\boldsymbol{\Sigma}^{-\frac{1}{2}} \\ \overline{\boldsymbol{b}} &= \boldsymbol{\Sigma}^{\frac{1}{2}}\boldsymbol{S}^T\boldsymbol{L}^{-1}\boldsymbol{b}, \qquad \overline{\boldsymbol{c}} = \boldsymbol{c}\boldsymbol{L}\boldsymbol{S}\boldsymbol{\Sigma}^{-\frac{1}{2}} \end{aligned} \tag{4.49}$$

2. Input-Normal State-Space Model

Suppose an equivalent transformation $\overline{\boldsymbol{x}}(k) = \boldsymbol{T}^{-1}\boldsymbol{x}(k)$ with

$$\boldsymbol{T} = \boldsymbol{L}\boldsymbol{S} \tag{4.50}$$

is applied, we can derive

$$\overline{\boldsymbol{K}}_c = \boldsymbol{I}_n, \qquad \overline{\boldsymbol{W}}_o = \boldsymbol{\Sigma}^2 \tag{4.51}$$

Hence an equivalent state-space model is input-normal, and the input-normal state-space model $(\overline{\boldsymbol{A}}, \overline{\boldsymbol{b}}, \overline{\boldsymbol{c}}, d)_n$ is specified by

$$\begin{aligned} \overline{\boldsymbol{A}} &= \boldsymbol{S}^T\boldsymbol{L}^{-1}\boldsymbol{A}\boldsymbol{L}\boldsymbol{S} \\ \overline{\boldsymbol{b}} &= \boldsymbol{S}^T\boldsymbol{L}^{-1}\boldsymbol{b}, \qquad \overline{\boldsymbol{c}} = \boldsymbol{c}\boldsymbol{L}\boldsymbol{S} \end{aligned} \tag{4.52}$$

3. Output-Normal State-Space Model

Suppose an equivalent transformation $\overline{\boldsymbol{x}}(k) = \boldsymbol{T}^{-1}\boldsymbol{x}(k)$ with

$$\boldsymbol{T} = \boldsymbol{L}\boldsymbol{S}\boldsymbol{\Sigma}^{-1} \tag{4.53}$$

is applied, we have

$$\overline{\boldsymbol{K}}_c = \boldsymbol{\Sigma}^2, \qquad \overline{\boldsymbol{W}}_o = \boldsymbol{I}_n \tag{4.54}$$

Hence an equivalent state-space model is output-normal, and the output-normal state-space model $(\overline{\boldsymbol{A}}, \overline{\boldsymbol{b}}, \overline{\boldsymbol{c}}, d)_n$ is found to be

$$\begin{aligned} \overline{\boldsymbol{A}} &= \boldsymbol{\Sigma}\boldsymbol{S}^T\boldsymbol{L}^{-1}\boldsymbol{A}\boldsymbol{L}\boldsymbol{S}\boldsymbol{\Sigma}^{-1} \\ \overline{\boldsymbol{b}} &= \boldsymbol{\Sigma}\boldsymbol{S}^T\boldsymbol{L}^{-1}\boldsymbol{b}, \qquad \overline{\boldsymbol{c}} = \boldsymbol{c}\boldsymbol{L}\boldsymbol{S}\boldsymbol{\Sigma}^{-1} \end{aligned} \tag{4.55}$$

The Use of Balanced State-Space Model for Reduced-Order Approximation of a System

The balanced state-space model characterized by (4.49) is of particular interest. Suppose a state-space model $(\boldsymbol{A}, \boldsymbol{b}, \boldsymbol{c}, d)_n$ is balanced with $\boldsymbol{\Sigma}$ as its controllability and observability Grammian and we partition $\boldsymbol{\Sigma}$ as

$$\boldsymbol{\Sigma} = \begin{bmatrix} \boldsymbol{\Sigma}_1 & \boldsymbol{0} \\ \boldsymbol{0} & \boldsymbol{\Sigma}_2 \end{bmatrix} \tag{4.56}$$

where

$$\boldsymbol{\Sigma}_1 = \text{diag}\{\sigma_1, \sigma_2, \cdots, \sigma_r\}, \qquad \boldsymbol{\Sigma}_2 = \text{diag}\{\sigma_{r+1}, \cdots, \sigma_n\}$$

$$\sigma_1 \geq \sigma_2 \geq \cdots \geq \sigma_n, \qquad \sigma_r \gg \sigma_{r+1}, \qquad 0 < r < n$$

the corresponding partition of $(\boldsymbol{A}, \boldsymbol{b}, \boldsymbol{c}, d)_n$ becomes

$$\boldsymbol{A} = \begin{bmatrix} \boldsymbol{A}_{11} & \boldsymbol{A}_{12} \\ \boldsymbol{A}_{21} & \boldsymbol{A}_{22} \end{bmatrix}, \qquad \boldsymbol{b} = \begin{bmatrix} \boldsymbol{b}_1 \\ \boldsymbol{b}_2 \end{bmatrix}, \qquad \boldsymbol{c} = \begin{bmatrix} \boldsymbol{c}_1 & \boldsymbol{c}_2 \end{bmatrix} \tag{4.57}$$

Since the rth-order subsystem $(\boldsymbol{A}_{11}, \boldsymbol{b}_1, \boldsymbol{c}_1, d)_r$ is associated with the r largest eigenvalues of $\boldsymbol{K}_c\boldsymbol{W}_o$, in a sense it is the closest to the original system $(\boldsymbol{A}, \boldsymbol{b}, \boldsymbol{c}, d)_n$ among all rth-order subsystems. Therefore it is natural to take $(\boldsymbol{A}_{11}, \boldsymbol{b}_1, \boldsymbol{c}_1, d)_r$ to be a r-order approximation of $(\boldsymbol{A}, \boldsymbol{b}, \boldsymbol{c}, d)_n$. We remark that the reduced-order system $(\boldsymbol{A}_{11}, \boldsymbol{b}_1, \boldsymbol{c}_1, d)_r$ is always stable for all r in the range $[1, n-1]$ as long as the original system $(\boldsymbol{A}, \boldsymbol{b}, \boldsymbol{c}, d)_n$ is stable. The interested reader is referred to Section 8.5.3 for details.

4.5 Kalman's Canonical Structure Theorem

Given a state-space model $(\boldsymbol{A}, \boldsymbol{b}, \boldsymbol{c}, d)_n$ in (4.1), consider its transfer function in (4.12) in terms of the impulse response, namely $H(z) = d + \boldsymbol{c}\boldsymbol{b}z^{-1} + \boldsymbol{c}\boldsymbol{A}\boldsymbol{b}z^{-2} + \boldsymbol{c}\boldsymbol{A}^2\boldsymbol{b}z^{-3} + \cdots$. Since d involves only a direct path from the input to the output, it does not affect the procedure described below. Hence, for simplicity, d is assumed to be zero, and the state-space model $(\boldsymbol{A}, \boldsymbol{b}, \boldsymbol{c}, d)_n$ with $d = 0$ in (4.1) will be denoted by $(\boldsymbol{A}, \boldsymbol{b}, \boldsymbol{c})_n$ henceforth.

Definition 4.6

Given a state-space model $(\boldsymbol{A}, \boldsymbol{b}, \boldsymbol{c})_n$, the controllable state-space X^c is defined as

$$X^c = \{\boldsymbol{x} \in R^n | \, \boldsymbol{x} \in \text{Range}\,[\boldsymbol{V}_n]\} \tag{4.58}$$

where R and R^n denote the set of all real numbers and the set of all ordered n-tuples of real numbers, respectively, and $\boldsymbol{V}_n = \begin{bmatrix}\boldsymbol{b} \; \boldsymbol{A}\boldsymbol{b} \; \cdots \; \boldsymbol{A}^{n-1}\boldsymbol{b}\end{bmatrix}$ is the controllability matrix.

Definition 4.7
Given a state-space model $(\boldsymbol{A}, \boldsymbol{b}, \boldsymbol{c})_n$, the uncontrollable state-space X^u is defined as

$$X^u = \{\boldsymbol{x} \in R^n | \, \boldsymbol{x} \in \text{Null}\,[\boldsymbol{U}_n]\} \tag{4.59}$$

where $\boldsymbol{U}_n = \begin{bmatrix}\boldsymbol{c}^T \; (\boldsymbol{c}\boldsymbol{A})^T \; \cdots \; (\boldsymbol{c}\boldsymbol{A}^{n-1})^T\end{bmatrix}^T$ is the observability matrix.

As is shown in (4.4) and (4.5), the system $(\boldsymbol{A}, \boldsymbol{b}, \boldsymbol{c})_n$ is controllable (observable) if and only if rank $\boldsymbol{V}_n = n$ (rank $\boldsymbol{U}_n = n$). From Definitions 4.6 and 4.7, we obtain

$$X^c = \text{Range}\,[\boldsymbol{V}_n], \qquad X^u = \text{Null}[\boldsymbol{U}_n] \tag{4.60}$$

Lemma 4.1
The controllable state-space X^c is invariant under matrix transformation $\boldsymbol{A}$.

Proof
Using (4.60) and applying the Cayley-Hamilton theorem in (4.24) to the controllability matrix $\boldsymbol{U}_n$, it follows that

$$\boldsymbol{A}X^c = \text{Range}\,[\boldsymbol{A}\boldsymbol{V}_n] \subseteq \text{Range}\,[\boldsymbol{V}_{n+1}] = \text{Range}\,[\boldsymbol{V}_n] = X^c \tag{4.61}$$

Hence, the state-space X^c is invariant under matrix transformation $\boldsymbol{A}$. ■

Lemma 4.2
The unobservable state-space X^u is invariant under matrix transformation $\boldsymbol{A}$.

Proof
Making use of (4.60) and applying the Cayley-Hamilton theorem in (4.24) to the observability matrix $\boldsymbol{U}_n$, we have

$$\text{Null}\,[\boldsymbol{U}_n]\boldsymbol{A} \supseteq \text{Null}\,[\boldsymbol{U}_{n+1}] = \text{Null}\,[\boldsymbol{U}_n] = X^u \tag{4.62}$$

Thus, for any $\boldsymbol{x} \in X^u$, $\boldsymbol{A}\boldsymbol{x} \in X^u$ since $\boldsymbol{U}_n\boldsymbol{A}\boldsymbol{x} = \boldsymbol{0}$. Namely, the state-space X^u is invariant under matrix transformation $\boldsymbol{A}$. This completes the proof of Lemma 4.2. ■

In terms of the properties of controllability and observability, or more precisely, in terms of the subspaces X^c and X^u, the structure of state space R^n can be exposed as the direct sum of four subspaces X_1, X_2, X_3, and X_4:

$$\begin{aligned} X_1 &= X^c \cap X^u, & X^c &= X_1 \oplus X_2 \\ X^u &= X_1 \oplus X_3, & R^n &= X_1 \oplus X_2 \oplus X_3 \oplus X_4 \end{aligned} \tag{4.63}$$

where dimension $[X_i] = n_i$ for $i = 1, 2, 3, 4$ and $n_1 + n_2 + n_3 + n_4 = n$.

The theorem below follows on the basis of the above decomposition.

Theorem 4.3: *Kalman's Canonical Structure Theorem*
By performing an appropriate equivalent transformation $\overline{\boldsymbol{x}}(k) = \boldsymbol{T}^{-1}\boldsymbol{x}(k)$, the system $(\boldsymbol{A}, \boldsymbol{b}, \boldsymbol{c})_n$ can be transformed into an equivalent system $(\overline{\boldsymbol{A}}, \overline{\boldsymbol{b}}, \overline{\boldsymbol{c}})_n$ with the canonical structure

$$\overline{\boldsymbol{A}} = \begin{bmatrix} \boldsymbol{A}_{11} & \boldsymbol{A}_{12} & \boldsymbol{A}_{13} & \boldsymbol{A}_{14} \\ \boldsymbol{0} & \boldsymbol{A}_{22} & \boldsymbol{0} & \boldsymbol{A}_{24} \\ \boldsymbol{0} & \boldsymbol{0} & \boldsymbol{A}_{33} & \boldsymbol{A}_{34} \\ \boldsymbol{0} & \boldsymbol{0} & \boldsymbol{0} & \boldsymbol{A}_{44} \end{bmatrix}, \qquad \overline{\boldsymbol{b}} = \begin{bmatrix} \boldsymbol{b}_1 \\ \boldsymbol{b}_2 \\ \boldsymbol{0} \\ \boldsymbol{0} \end{bmatrix} \tag{4.64}$$
$$\overline{\boldsymbol{c}} = \begin{bmatrix} \boldsymbol{0} & \boldsymbol{c}_2 & \boldsymbol{0} & \boldsymbol{c}_4 \end{bmatrix}$$

where the sizes of submatrices are as follows: $\boldsymbol{A}_{ij}$ is $n_i \times n_j$ for $i, j = 1, 2, 3, 4$; $\boldsymbol{b}_i$ is $n_i \times 1$ for $i = 1, 2$ and $\boldsymbol{c}_i$ is $1 \times n_i$ for $i = 2, 4$.

Proof
Let $\boldsymbol{T}_i$ for $i = 1, 2, 3, 4$ be an $n \times n_i$ real matrix whose columns consist of the basis of the subspace X_i and define an equivalent transformation matrix $\boldsymbol{T}$ by

$$\boldsymbol{T} = \begin{bmatrix} \boldsymbol{T}_1 & \boldsymbol{T}_2 & \boldsymbol{T}_3 & \boldsymbol{T}_4 \end{bmatrix} \tag{4.65}$$

From Lemmas 4.1 and 4.2, Equations (4.63) and (4.65), and $\boldsymbol{b} \in X^c$, it follows that

$$\boldsymbol{AT} = \begin{bmatrix} \boldsymbol{AT}_1 & \boldsymbol{AT}_2 & \boldsymbol{AT}_3 & \boldsymbol{AT}_4 \end{bmatrix}$$
$$= \begin{bmatrix} \boldsymbol{T}_1 & \boldsymbol{T}_2 & \boldsymbol{T}_3 & \boldsymbol{T}_4 \end{bmatrix} \begin{bmatrix} \boldsymbol{A}_{11} & \boldsymbol{A}_{12} & \boldsymbol{A}_{13} & \boldsymbol{A}_{14} \\ 0 & \boldsymbol{A}_{22} & 0 & \boldsymbol{A}_{24} \\ 0 & 0 & \boldsymbol{A}_{33} & \boldsymbol{A}_{34} \\ 0 & 0 & 0 & \boldsymbol{A}_{44} \end{bmatrix} \quad (4.66)$$
$$\boldsymbol{b} = \begin{bmatrix} \boldsymbol{T}_1 & \boldsymbol{T}_2 & \boldsymbol{T}_3 & \boldsymbol{T}_4 \end{bmatrix} \begin{bmatrix} \boldsymbol{b}_1 \\ \boldsymbol{b}_2 \\ 0 \\ 0 \end{bmatrix}$$

which yields $\overline{\boldsymbol{A}} = \boldsymbol{T}^{-1}\boldsymbol{AT}$ and $\overline{\boldsymbol{b}} = \boldsymbol{T}^{-1}\boldsymbol{b}$ in (4.64). By virtue of (4.60), (4.63) and (4.65), we have

$$\boldsymbol{cT} = \begin{bmatrix} \boldsymbol{cT}_1 & \boldsymbol{cT}_2 & \boldsymbol{cT}_3 & \boldsymbol{cT}_4 \end{bmatrix} = \begin{bmatrix} \boldsymbol{0} & \boldsymbol{c}_2 & \boldsymbol{0} & \boldsymbol{c}_4 \end{bmatrix} \quad (4.67)$$

which leads to $\overline{\boldsymbol{c}} = \boldsymbol{cT}$ in (4.64). This completes the proof of Theorem 4.3. ■

The canonical decomposition is illustrated in Figure 4.2, where the subsystem $S_{co} = (\boldsymbol{A}_{22}, \boldsymbol{b}_2, \boldsymbol{c}_2)_{n_2}$ is controllable and observable, the subsystem $S_{c\bar{o}} = (\boldsymbol{A}_{11}, \boldsymbol{b}_1, \boldsymbol{0})_{n_1}$ is controllable and unobservable, the subsystem $S_{\bar{c}\bar{o}} = (\boldsymbol{A}_{33}, \boldsymbol{0}, \boldsymbol{0})_{n_3}$ is uncontrollable and unobservable, and the subsystem $S_{\bar{c}o} = (\boldsymbol{A}_{44}, \boldsymbol{0}, \boldsymbol{c}_4)_{n_4}$ is uncontrollable and observable.

Corollary 4.1

The impulse response as well as the transfer function of the system $(\boldsymbol{A}, \boldsymbol{b}, \boldsymbol{c})_n$ are the same as those of the subsystem $(\boldsymbol{A}_{22}, \boldsymbol{b}_2, \boldsymbol{c}_2)_{n_2}$.

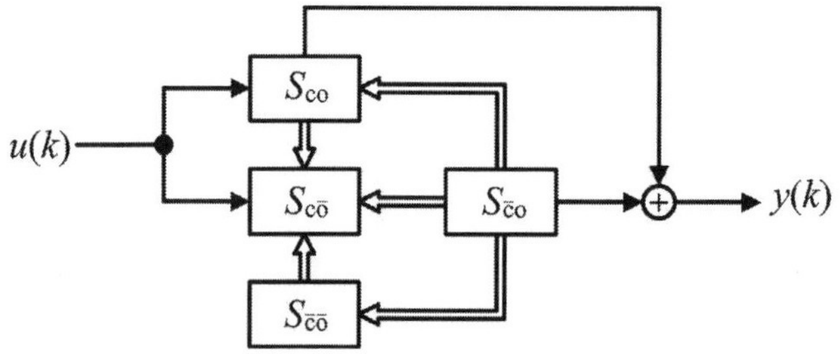

Figure 4.2 Canonical decomposition of a state-space model.

Proof
It is evident that $\boldsymbol{c}\boldsymbol{A}^{i-1}\boldsymbol{b} = \overline{\boldsymbol{c}}\,\overline{\boldsymbol{A}}^{i-1}\overline{\boldsymbol{b}} = \boldsymbol{c}_2\boldsymbol{A}_{22}^{i-1}\boldsymbol{b}_2$ holds for $i = 1, 2, 3, \cdots$ and hence $H(z) = \boldsymbol{c}(z\boldsymbol{I}_n - \boldsymbol{A})^{-1}\boldsymbol{b} = \boldsymbol{c}_2(z\boldsymbol{I}_{n_2} - \boldsymbol{A}_{22})^{-1}\boldsymbol{b}_2$. This completes the proof of Corollary 4.1. ■

It is noted that if a system $(\boldsymbol{A}, \boldsymbol{b}, \boldsymbol{c})_n$ is controllable and observable, this system is said to be *minimal.*

4.6 Hankel Matrix and Realization

4.6.1 Minimal Realization

Definition 4.8
Given a sequence of real numbers $\{h_i |\, i = 1, 2, 3, \cdots\}$, a triple $\{\boldsymbol{A}, \boldsymbol{b}, \boldsymbol{c}\}$ of constant matrices is said to be the *minimal realization* of an input-output map if $h_i = \boldsymbol{c}\boldsymbol{A}^{i-1}\boldsymbol{b}$ holds for $i = 1, 2, 3, \cdots$, and the size of $\boldsymbol{A}$ is minimal among all the realizations.

The problem of finding such a realization, if it does exist, is known as the *minimal realization problem.*

Definition 4.9
Given a sequence of real numbers $\{h_i |\, i = 1, 2, 3, \cdots\}$, the Hankel matrix $\boldsymbol{H}_{i,j}$ is defined as

$$\boldsymbol{H}_{i,j} = \begin{bmatrix} h_1 & h_2 & \cdots & h_j \\ h_2 & h_3 & \cdots & h_{j+1} \\ \vdots & \vdots & \ddots & \vdots \\ h_i & h_{i+1} & \cdots & h_{i+j-1} \end{bmatrix} \tag{4.68}$$

Definition 4.10
Let σ^K be a shift operator satisfying

$$\sigma^K \boldsymbol{H}_{i,j} = \begin{bmatrix} h_{1+K} & h_{2+K} & \cdots & h_{j+K} \\ h_{2+K} & h_{3+K} & \cdots & h_{j+1+K} \\ \vdots & \vdots & \ddots & \vdots \\ h_{i+K} & h_{i+1+K} & \cdots & h_{i+j-1+K} \end{bmatrix} \tag{4.69}$$

where matrix $\boldsymbol{H}_{i,j}$ is defined by (4.68).

Theorem 4.4
A sequence of real numbers $\{h_i|\, i = 1, 2, 3, \cdots\}$ has finite-dimensional realization if and only if there exist integers M and N such that

$$\operatorname{rank} \boldsymbol{H}_{M,N} = \operatorname{rank} \boldsymbol{H}_{M+i,N+j} \tag{4.70}$$

for all $i, j = 0, 1, 2, \cdots$.

Theorem 4.5
If a sequence of real numbers $\{h_i|\, i = 1, 2, 3, \cdots\}$ has finite-dimensional realization, and if integers M and N satisfy the condition in (4.70), then the minimal dimension n_o of realizing the sequence can be determined as

$$n_o = \operatorname{rank} \boldsymbol{H}_{M,N} \tag{4.71}$$

Given a sequence $\{h_i|\, i = 1, 2, 3, \cdots\}$ satisfying the condition in (4.70), the two algorithms described below compute a minimal realization $(\boldsymbol{A}, \boldsymbol{b}, \boldsymbol{c})_{n_o}$ such that $h_i = \boldsymbol{c}\boldsymbol{A}^{i-1}\boldsymbol{b}$ for $i = 1, 2, 3, \cdots$.

1. Direct Realization Algorithm

By applying maximum rank decomposition to the Hankel matrix $\boldsymbol{H}_{M,N}$, we obtain

$$\boldsymbol{H}_{M,N} = \boldsymbol{U}_M \boldsymbol{V}_N \tag{4.72}$$

where

$$\operatorname{rank} \boldsymbol{U}_M = \operatorname{rank} \boldsymbol{V}_N = n_o$$

and a minimal realization $(\boldsymbol{A}, \boldsymbol{b}, \boldsymbol{c})_{n_o}$ can be constructed as

$$\begin{aligned} \boldsymbol{A} &= (\boldsymbol{U}_M^T \boldsymbol{U}_M)^{-1} \boldsymbol{U}_M^T (\sigma \boldsymbol{H}_{M,N}) \boldsymbol{V}_N^T (\boldsymbol{V}_N \boldsymbol{V}_N^T)^{-1} \\ \boldsymbol{b} &= \text{first column of } \boldsymbol{V}_N \\ \boldsymbol{c} &= \text{first row of } \boldsymbol{U}_M \end{aligned} \tag{4.73}$$

2. Canonical Form Realization Algorithm

If the condition in (4.70) is satisfied, we can write

$$\operatorname{rank} \boldsymbol{H}_{M,n_o} = \operatorname{rank} \boldsymbol{H}_{M,n_o+1} = n_o \tag{4.74}$$

As a result, there exists a unique real vector $\boldsymbol{a} = [a_{n_o}, \cdots, a_2, a_1]^T$ satisfying

$$\boldsymbol{H}_{M,n_o}\boldsymbol{a} = \boldsymbol{\eta} \tag{4.75}$$

which leads to

$$\boldsymbol{a} = (\boldsymbol{H}_{M,n_o}^T \boldsymbol{H}_{M,n_o})^{-1} \boldsymbol{H}_{M,n_o}^T \boldsymbol{\eta} \tag{4.76}$$

where

$$\boldsymbol{\eta} = \text{the last column of } \boldsymbol{H}_{M,n_o+1}$$

Therefore, based on (4.28), a minimal realization $(\boldsymbol{A}, \boldsymbol{b}, \boldsymbol{c})_{n_o}$ can be found as

$$\boldsymbol{A} = \begin{bmatrix} 0 & \cdots & 0 & -a_{n_o} \\ 1 & \ddots & \vdots & \vdots \\ \vdots & \ddots & 0 & -a_2 \\ 0 & \cdots & 1 & -a_1 \end{bmatrix}, \quad \boldsymbol{b} = \begin{bmatrix} 1 \\ 0 \\ \vdots \\ 0 \end{bmatrix}, \quad \boldsymbol{c} = \begin{bmatrix} h_1 & h_2 & \cdots & h_{n_o} \end{bmatrix} \tag{4.77}$$

4.6.2 Minimal Partial Realization

Definition 4.11
A triple $\{\boldsymbol{A}, \boldsymbol{b}, \boldsymbol{c}\}$ of constant matrices is said to be the *partial realization* of an input-output map if $h_i = \boldsymbol{c}\boldsymbol{A}^{i-1}\boldsymbol{b}$ holds for $i = 1, 2, \cdots, N$.

Definition 4.12
A triple $\{\boldsymbol{A}, \boldsymbol{b}, \boldsymbol{c}\}$ of constant matrices is said to be the *minimal partial realization* of an input-output map if the size of $\boldsymbol{A}$ is minimal among all the partial realizations satisfying Definition 4.11.

The problem of finding a realization that satisfies Definition 4.12 will be called the *minimal partial realization problem.*

Theorem 4.5
If

$$\text{rank}\, \boldsymbol{H}_{\lambda,\,\mu} = \text{rank}\, \boldsymbol{H}_{\lambda,\,\mu+1} = \text{rank}\, \boldsymbol{H}_{\lambda+1,\,\mu} \tag{4.78}$$

holds for some positive integers λ and μ with $\lambda + \mu = N$, then finite sequence $\{h_i |\, i = 1, 2, \cdots, N\}$ admits the partial realization.

We now consider the minimal partial realization problem from a given finite sequence such that the rank condition in (4.78) is not satisfied.

Definition 4.13

Given a finite sequence of real numbers $\{h_i|\, i = 1, 2, \cdots, N\}$, the incomplete Hankel matrix $\boldsymbol{H}_{NN}$ is defined as

$$\boldsymbol{H}_{N,N} = \begin{bmatrix} h_1 & h_2 & \cdots & h_N \\ h_2 & \vdots & \therefore & * \\ \vdots & h_N & \therefore & \vdots \\ h_N & * & \cdots & * \end{bmatrix} \tag{4.79}$$

where the asterisks denote scalars which extend the given sequence without affecting the rank of the Hankel matrix.

It is well known [3] that a lower bound for the dimension of a state-space model realizing a given finite sequence is provided by the rank of the incomplete Hankel matrix. Since the matrix elements denoted by asterisks do not change the rank of the Hankel matrix, the rank of the incomplete Hankel matrix can be readily determined by testing its columns for dependency starting from the left and ignoring all scalars denoted by asterisks. Based on the observations made above, we have

Theorem 4.6

The rank of the incomplete Hankel matrix in (4.79), denoted by $n(N)$, can be computed as

$$\begin{aligned} n(N) &= \text{rank}\, \boldsymbol{H}_{N,1} + (\text{rank}\, \boldsymbol{H}_{N-1,2} - \text{rank}\, \boldsymbol{H}_{N-1,1}) \\ &\quad + \cdots + (\text{rank}\, \boldsymbol{H}_{1,N} - \text{rank}\, \boldsymbol{H}_{1,N-1}) \\ &= \sum_{i=1}^{N} \text{rank}\, \boldsymbol{H}_{N-i+1,i} - \sum_{i=1}^{N-1} \text{rank}\, \boldsymbol{H}_{N-i,i} \end{aligned} \tag{4.80}$$

Explicitly, the dimension $n(N)$ is the minimal dimension for the realizations of a given sequence $\{h_i|\, i = 1, 2, \cdots, N\}$. This dimension is non-decreasing regardless of the choice of those elements denoted by asterisks in the incomplete Hankel matrix in (4.79).

Described below is a concrete procedure for finding the elements denoted by asterisks such that the rank of the Hankel matrix so constructed remains to be equal to that of the incomplete Hankel matrix.

Step 1: Obtain an $n(N) \times 1$ real vector $\boldsymbol{\alpha} = [\alpha_{n(N)}, \cdots, \alpha_2, \alpha_1]^T$ satisfying

$$\begin{bmatrix} h_{n(N)+1} \\ h_{n(N)+1} \\ \vdots \\ h_N \end{bmatrix} = \begin{bmatrix} h_1 & h_2 & \cdots & h_{n(N)} \\ h_2 & h_3 & \cdots & h_{n(N)+1} \\ \vdots & \vdots & \ddots & \vdots \\ h_{N-n(N)} & h_{N-n(N)+1} & \cdots & h_{N-1} \end{bmatrix} \boldsymbol{\alpha} \tag{4.81}$$

Step 2: Generate the elements of the extended sequence in order using

$$h_{N+i} = \begin{bmatrix} h_{N-n(N)+i} & h_{N-n(N)+1+i} & \cdots & h_{N-1+i} \end{bmatrix} \boldsymbol{\alpha} \tag{4.82}$$

for $i = 1, 2, \cdots, n(N)$.

Note that although $\boldsymbol{\alpha}$ is not necessarily determined uniquely, its existence is ensured from (4.79) and (4.80). Hence

$$\text{rank}\, \boldsymbol{H}_{N,n(N)} = \text{rank}\, \boldsymbol{H}_{N+1,n(N)} = \text{rank}\, \boldsymbol{H}_{N,n(N)+1} = n(N) \tag{4.83}$$

which coincides with the condition in (4.78). Therefore, the minimal partial realization problem can be readily solved by applying a method similar to those studied in Section 4.6.1. For example, a simple minimal partial realization $(\boldsymbol{A}, \boldsymbol{b}, \boldsymbol{c})_{n(N)}$ can be found by the controllable canonical form

$$\boldsymbol{A} = \begin{bmatrix} 0 & \cdots & 0 & -\alpha_{n(N)} \\ 1 & \ddots & \vdots & \vdots \\ \vdots & \ddots & 0 & -\alpha_2 \\ 0 & \cdots & 1 & -\alpha_1 \end{bmatrix}, \quad \boldsymbol{b} = \begin{bmatrix} 1 \\ 0 \\ \vdots \\ 0 \end{bmatrix}, \quad \boldsymbol{c} = \begin{bmatrix} h_1 & h_2 & \cdots & h_{n(N)} \end{bmatrix} \tag{4.84}$$

4.6.3 Balanced Realization

Consider a sequence $\{h_i |\, i = 1, 2, 3, \cdots\}$ that satisfies

$$\sum_{i=1}^{\infty} |h_i| < \infty \tag{4.85}$$

Let $\boldsymbol{H}$ be the Hankel matrix $\boldsymbol{H}_{M,N}$ with sufficiently large integers M and N, and $\boldsymbol{H} = \boldsymbol{U}\boldsymbol{V}$ be a maximum rank decomposition of $\boldsymbol{H}$.

Lemma 4.3

(1) If $\boldsymbol{H}\boldsymbol{H}^T\boldsymbol{x} = \lambda\boldsymbol{x}$ with $\lambda \neq 0$ and $\boldsymbol{x} \neq \boldsymbol{0}$, then $(\boldsymbol{U}^T\boldsymbol{x})^T\boldsymbol{K}_c\boldsymbol{W}_o = \lambda(\boldsymbol{U}^T\boldsymbol{x})^T$.

(2) If $\boldsymbol{K}_c\boldsymbol{W}_o\boldsymbol{y} = \lambda\boldsymbol{y}$ with $\boldsymbol{y} \neq \boldsymbol{0}$, then $\boldsymbol{H}\boldsymbol{H}^T(\boldsymbol{U}\boldsymbol{y}) = \lambda(\boldsymbol{U}\boldsymbol{y})$.

Proof

If $\boldsymbol{H}\boldsymbol{H}^T\boldsymbol{x} = \lambda\boldsymbol{x}$ with $\lambda \neq 0$ and $\boldsymbol{x} \neq \boldsymbol{0}$, then $\boldsymbol{x}^T\boldsymbol{H}\boldsymbol{H}^T = \lambda\boldsymbol{x}^T$ which leads to $\boldsymbol{x}^T\boldsymbol{U}\boldsymbol{V}\boldsymbol{V}^T\boldsymbol{U}^T\boldsymbol{U} = \lambda\boldsymbol{x}^T\boldsymbol{U}$. Hence $(\boldsymbol{U}^T\boldsymbol{x})^T\boldsymbol{K}_c\boldsymbol{W}_o = \lambda(\boldsymbol{U}^T\boldsymbol{x})^T$ because $\boldsymbol{K}_c = \boldsymbol{V}\boldsymbol{V}^T$ and $\boldsymbol{W}_o = \boldsymbol{U}^T\boldsymbol{U}$. This means that $(\lambda, \boldsymbol{U}^T\boldsymbol{x})$ is an eigenvalue left-eigenvector pair for $\boldsymbol{K}_c\boldsymbol{W}_o$ because $\boldsymbol{U}^T\boldsymbol{x} \neq \boldsymbol{0}$. Conversely, if $\boldsymbol{K}_c\boldsymbol{W}_o\boldsymbol{y} = \lambda\boldsymbol{y}$ with $\boldsymbol{y} \neq \boldsymbol{0}$, then $\boldsymbol{U}\boldsymbol{V}\boldsymbol{V}^T\boldsymbol{U}^T\boldsymbol{U}\boldsymbol{y} = \lambda\boldsymbol{U}\boldsymbol{y}$, namely, $\boldsymbol{H}\boldsymbol{H}^T(\boldsymbol{U}\boldsymbol{y}) = \lambda(\boldsymbol{U}\boldsymbol{y})$. Hence, $\boldsymbol{x} = \boldsymbol{U}\boldsymbol{y} \neq \boldsymbol{0}$ is an eigenvector for $\boldsymbol{H}\boldsymbol{H}^T$. This completes the proof of the lemma. ■

By Lemma 4.3, nonzero eigenvalues of $\boldsymbol{H}\boldsymbol{H}^T$ coincide with those of $\boldsymbol{K}_c\boldsymbol{W}_o$. Hence $\boldsymbol{H}$ can be factorized using singular value decomposition (SVD) as

$$\boldsymbol{H} = \boldsymbol{U}_o\boldsymbol{\Sigma}\boldsymbol{V}_o^T \tag{4.86}$$

where $\boldsymbol{U}_o^T\boldsymbol{U}_o = \boldsymbol{V}_o^T\boldsymbol{V}_o = \boldsymbol{I}_n$ and $\boldsymbol{\Sigma}$ is defined by (4.44) (i.e., $\boldsymbol{\Sigma} = \mathrm{diag}\{\sigma_1, \sigma_2, \cdots, \sigma_n\}$ and $\sigma_1 \geq \sigma_2 \geq \cdots \geq \sigma_n > 0$). In this case, $\boldsymbol{H}^T\boldsymbol{H} = \boldsymbol{V}_o\boldsymbol{\Sigma}^2\boldsymbol{V}_o^T$ and $\boldsymbol{H}\boldsymbol{H}^T = \boldsymbol{U}_o\boldsymbol{\Sigma}^2\boldsymbol{U}_o^T$ hold.

As demonstrated below, several useful realizations can be deduced based on above analysis.

1. Balanced Realization

By letting $\boldsymbol{U} = \boldsymbol{U}_o\boldsymbol{\Sigma}^{\frac{1}{2}}$ and $\boldsymbol{V} = \boldsymbol{\Sigma}^{\frac{1}{2}}\boldsymbol{V}_o^T$ in $\boldsymbol{H} = \boldsymbol{U}\boldsymbol{V}$, we obtain $\boldsymbol{K}_c = \boldsymbol{W}_o = \boldsymbol{\Sigma}$ which gives a balanced realization $(\boldsymbol{A}, \boldsymbol{b}, \boldsymbol{c})_n$ as follows.

$$\begin{aligned} &\boldsymbol{A} = \boldsymbol{\Sigma}^{-\frac{1}{2}}\boldsymbol{U}_o^T\sigma\boldsymbol{H}\boldsymbol{V}_o\boldsymbol{\Sigma}^{-\frac{1}{2}} \\ &\boldsymbol{b} = \text{first column of } \boldsymbol{\Sigma}^{\frac{1}{2}}\boldsymbol{V}_o^T, \qquad \boldsymbol{c} = \text{first row of } \boldsymbol{U}_o\boldsymbol{\Sigma}^{\frac{1}{2}} \end{aligned} \tag{4.87}$$

2. Input-Normal Realization

By letting $\boldsymbol{U} = \boldsymbol{U}_o\boldsymbol{\Sigma}$ and $\boldsymbol{V} = \boldsymbol{V}_o^T$ in $\boldsymbol{H} = \boldsymbol{U}\boldsymbol{V}$, we obtain $\boldsymbol{K}_c = \boldsymbol{I}_n$ and $\boldsymbol{W}_o = \boldsymbol{\Sigma}^2$ which gives an input-normal realization $(\boldsymbol{A}, \boldsymbol{b}, \boldsymbol{c})_n$ as follows.

$$\begin{aligned} &\boldsymbol{A} = \boldsymbol{\Sigma}^{-1}\boldsymbol{U}_o^T\sigma\boldsymbol{H}\boldsymbol{V}_o \\ &\boldsymbol{b} = \text{first column of } \boldsymbol{V}_o^T, \qquad \boldsymbol{c} = \text{first row of } \boldsymbol{U}_o\boldsymbol{\Sigma} \end{aligned} \tag{4.88}$$

3. Output-Normal Realization
By letting $\boldsymbol{U} = \boldsymbol{U}_o$ and $\boldsymbol{V} = \boldsymbol{\Sigma}\boldsymbol{V}_o^T$ in $\boldsymbol{H} = \boldsymbol{U}\boldsymbol{V}$, we obtain $\boldsymbol{K}_c = \boldsymbol{\Sigma}^2$ and $\boldsymbol{W}_o = \boldsymbol{I}_n$ which gives an output-normal realization $(\boldsymbol{A}, \boldsymbol{b}, \boldsymbol{c})_n$ as follows.

$$
\begin{aligned}
&\boldsymbol{A} = \boldsymbol{U}_o^T \sigma \boldsymbol{H} \boldsymbol{V}_o \boldsymbol{\Sigma}^{-1} \\
&\boldsymbol{b} = \text{first column of } \boldsymbol{\Sigma}\boldsymbol{V}_o^T, \qquad \boldsymbol{c} = \text{first row of } \boldsymbol{U}_o
\end{aligned}
\tag{4.89}
$$

Obviously, these realizations correspond to the three normalized state-space models of Section 4.4.3.

4. Reduced-Order Approximation
Using SVD of the Hankel matrix and the balanced realization induced from it, an algorithm for reduced-order approximation can be deduced. Let SVD of $\boldsymbol{H}$ in (4.86) be partitioned as

$$
\boldsymbol{H} = \begin{bmatrix} \boldsymbol{U}_{o1} & \boldsymbol{U}_{o2} \end{bmatrix} \begin{bmatrix} \boldsymbol{\Sigma}_1 & \boldsymbol{0} \\ \boldsymbol{0} & \boldsymbol{\Sigma}_2 \end{bmatrix} \begin{bmatrix} \boldsymbol{V}_{o1}^T \\ \boldsymbol{V}_{o2} \end{bmatrix}
\tag{4.90}
$$

where

$$
\boldsymbol{\Sigma}_1 = \{\sigma_1, \sigma_2, \cdots, \sigma_r\}, \qquad \boldsymbol{\Sigma}_2 = \{\sigma_{r+1}, \cdots, \sigma_n\}, \qquad \sigma_r >> \sigma_{r+1}
$$

A good rth-order approximation $(\boldsymbol{A}_{11}, \boldsymbol{b}_1, \boldsymbol{c}_1)_r$ of (4.87) can be obtained as

$$
\begin{aligned}
&\boldsymbol{A}_{11} = \boldsymbol{\Sigma}_1^{-\frac{1}{2}} \boldsymbol{U}_{o1}^T \sigma \boldsymbol{H} \boldsymbol{V}_{o1} \boldsymbol{\Sigma}_1^{-\frac{1}{2}} \\
&\boldsymbol{b}_1 = \text{first column of } \boldsymbol{\Sigma}_1^{\frac{1}{2}} \boldsymbol{V}_{o1}^T, \qquad \boldsymbol{c}_1 = \text{first row of } \boldsymbol{U}_{o1} \boldsymbol{\Sigma}_1^{\frac{1}{2}}
\end{aligned}
\tag{4.91}
$$

It is known that $\boldsymbol{\Delta}_1 = \boldsymbol{U}_{o1}\boldsymbol{\Sigma}_1\boldsymbol{V}_{o1}^T$ minimizes $||\boldsymbol{H} - \boldsymbol{\Delta}_1||_s$ over all matrices $\boldsymbol{\Delta}_1$ of rank r and $||\boldsymbol{H} - \boldsymbol{\Delta}_1||_s = \sigma_{r+1}$ where $||\boldsymbol{A}||_s$ max eigenvalue of $\boldsymbol{A}^T\boldsymbol{A}$. Unfortunately, $\boldsymbol{\Delta}_1$ is not a Hankel matrix in general and hence, $\boldsymbol{\Delta}_1$ does not admit an exact realization. However, by employing an algorithm similar to (4.87), often times a good reduced-order approximation can be attained [7].

4.7 Discrete-Time Lossless Bounded-Real Lemma

The discrete-time lossless bounded-real lemma is a passivity property of discrete-time systems, which finds applications in network synthesis and stability analysis. To introduce this important property, consider a state-space model $(\boldsymbol{A}, \boldsymbol{b}, \boldsymbol{c}, d)_n$ described by (4.1), whose transfer function is given by (4.10). It is assumed that the state-space model $(\boldsymbol{A}, \boldsymbol{b}, \boldsymbol{c}, d)_n$ is asymptotically stable, controllable and observable, i.e., $(\boldsymbol{A}, \boldsymbol{b}, \boldsymbol{c}, d)_n$ is a minimal realization.

Definition 4.14

A transfer function $H(z)$ is said to be lossless bounded-real (LBR) if it is asymptotically stable and $|H(e^{j\omega})|^2 = 1$ holds for all ω.

An equivalent characterization of $|H(e^{j\omega})|^2 = 1$ for all ω is that for every finite-energy input sequence $u(k)$, the output sequence $y(k)$ of the system satisfies

$$\sum_{k=0}^{\infty} |y(k)|^2 = \sum_{k=0}^{\infty} |u(k)|^2 \tag{4.92}$$

where the initial state vector $\boldsymbol{x}(0)$ is assumed to be null. Note that the losslessness property is satisfied by all-pass filters.

Theorem 4.7: *Discrete-Time LBR Lemma*

$H(z)$ is lossless bounded-real (LBR) if and only if there exists a real positive-definite symmetric matrix $\boldsymbol{P}$ such that [12]

$$\boldsymbol{A}^T\boldsymbol{P}\boldsymbol{A} + \boldsymbol{c}^T\boldsymbol{c} = \boldsymbol{P} \tag{4.93a}$$

$$\boldsymbol{b}^T\boldsymbol{P}\boldsymbol{b} + d^T d = 1 \tag{4.93b}$$

$$\boldsymbol{A}^T\boldsymbol{P}\boldsymbol{b} + \boldsymbol{c}^T d = \boldsymbol{0} \tag{4.93c}$$

Proof: (Sufficiency) Assuming that the equations in (4.93) holds true, Equation (4.93a) can be written as

$$\begin{aligned}\boldsymbol{P} &= \boldsymbol{A}^T[\boldsymbol{A}^T\boldsymbol{P}\boldsymbol{A} + \boldsymbol{c}^T\boldsymbol{c}]\boldsymbol{A} + \boldsymbol{c}^T\boldsymbol{c} \\ &= (\boldsymbol{A}^T)^2\boldsymbol{P}\boldsymbol{A}^2 + (\boldsymbol{c}\boldsymbol{A})^T\boldsymbol{c}\boldsymbol{A} + \boldsymbol{c}^T\boldsymbol{c} \\ &= \cdots \\ &= (\boldsymbol{A}^T)^n\boldsymbol{P}\boldsymbol{A}^n + \boldsymbol{U}_n^T\boldsymbol{U}_n\end{aligned} \tag{4.94}$$

where $\boldsymbol{U}_n = [\boldsymbol{c}^T\ (\boldsymbol{c}\boldsymbol{A})^T\ \cdots\ (\boldsymbol{c}\boldsymbol{A}^{n-1})^T]^T$ is the observability matrix of the system in (4.1). Recall the standard Lyapunov stability theorem in [13], which states that matrix $\boldsymbol{B}$ has all eigenvalues in the open unit disk if and only if there exist two positive-definite symmetric matrices $\boldsymbol{V}$ and $\boldsymbol{W}$ for which $\boldsymbol{V} = \boldsymbol{B}^T\boldsymbol{V}\boldsymbol{B} + \boldsymbol{W}$. By taking $\boldsymbol{V} = \boldsymbol{P}$ and $\boldsymbol{W} = \boldsymbol{U}_n^T\boldsymbol{U}_n$, and noticing the observability of the system in (4.1) hence the nonsingularity of $\boldsymbol{U}_n$, we conclude that both $\boldsymbol{V}$ and $\boldsymbol{W}$ are positive-definite, hence (4.94) satisfies the condition in the standard Lyapunov stability theorem, and all eigenvalues of $\boldsymbol{A}^n$ must lie inside the open unit disk. Evidently, these eigenvalues are simply the nth powers of the eigenvalues of $\boldsymbol{A}$, and thus all eigenvalues of $\boldsymbol{A}$ must lie inside the open unit disk. Thus $H(z)$ is asymptotically stable.

Next, since $\boldsymbol{P} = \boldsymbol{P}^T > 0$, matrix $\boldsymbol{P}$ can be decomposed as $\boldsymbol{P} = \boldsymbol{T}^{-T}\boldsymbol{T}^{-1}$. Hence, (4.93) can be written as

$$\overline{\boldsymbol{A}}^T\overline{\boldsymbol{A}} + \overline{\boldsymbol{c}}^T\overline{\boldsymbol{c}} = \boldsymbol{I}_n \tag{4.95a}$$

$$\overline{\boldsymbol{b}}^T\overline{\boldsymbol{b}} + d^T d = 1 \tag{4.95b}$$

$$\overline{\boldsymbol{A}}^T\overline{\boldsymbol{b}} + \overline{\boldsymbol{c}}^T d = \boldsymbol{0} \tag{4.95c}$$

where

$$\overline{\boldsymbol{A}} = \boldsymbol{T}^{-1}\boldsymbol{A}\boldsymbol{T}, \qquad \overline{\boldsymbol{b}} = \boldsymbol{T}^{-1}\boldsymbol{b}, \qquad \overline{\boldsymbol{c}} = \boldsymbol{c}\boldsymbol{T}$$

We can now consider an equivalent state-space model of the system $H(z)$ given by

$$\begin{bmatrix} \overline{\boldsymbol{x}}(k+1) \\ y(k) \end{bmatrix} = \begin{bmatrix} \overline{\boldsymbol{A}} & \overline{\boldsymbol{b}} \\ \overline{\boldsymbol{c}} & d \end{bmatrix} \begin{bmatrix} \overline{\boldsymbol{x}}(k) \\ u(k) \end{bmatrix} \tag{4.96}$$

where $\overline{\boldsymbol{x}}(k) = \boldsymbol{T}^{-1}\boldsymbol{x}(k)$. Alternatively, by defining

$$\mathcal{R} = \begin{bmatrix} \overline{\boldsymbol{A}} & \overline{\boldsymbol{b}} \\ \overline{\boldsymbol{c}} & d \end{bmatrix} \tag{4.97}$$

Equation (4.95) can be expressed as

$$\mathcal{R}^T\mathcal{R} = \boldsymbol{I}_{n+1} \tag{4.98}$$

which means that $\mathcal{R}$ is an orthogonal matrix. As a result,

$$||\overline{\boldsymbol{x}}(k+1)||^2 + |y(k)|^2 = ||\overline{\boldsymbol{x}}(k)||^2 + |u(k)|^2 \tag{4.99}$$

holds for any nonnegative integer k, hence

$$\sum_{k=0}^{N} |y(k)|^2 = \sum_{k=0}^{N} |u(k)|^2 + ||\overline{\boldsymbol{x}}(0)||^2 - ||\overline{\boldsymbol{x}}(N+1)||^2 \tag{4.100}$$

is satisfied for every positive integer N. If we assume that $u(k) = 0$ for $k > N$, then

$$y(k) = \overline{\boldsymbol{c}}\,\overline{\boldsymbol{x}}(k) \ \text{ for } \ k > N \tag{4.101}$$

which, by virtue of (4.95), leads to

$$\begin{aligned} |y(k)|^2 &= \overline{\boldsymbol{x}}(k)^T \overline{\boldsymbol{c}}^T \overline{\boldsymbol{c}}\,\overline{\boldsymbol{x}}(k) \\ &= \overline{\boldsymbol{x}}(k)^T \left[\boldsymbol{I}_n - \overline{\boldsymbol{A}}^T \overline{\boldsymbol{A}}\right] \overline{\boldsymbol{x}}(k) \end{aligned} \tag{4.102}$$

Hence,

$$\begin{aligned} \sum_{k=N+1}^{\infty} |y(k)|^2 &= \sum_{k=N+1}^{\infty} \left[||\overline{\boldsymbol{x}}(k)||^2 - ||\overline{\boldsymbol{x}}(k+1)||^2\right] \\ &= ||\overline{\boldsymbol{x}}(N+1)||^2 \end{aligned} \tag{4.103}$$

Equations (4.100) and (4.103) result in

$$\sum_{k=0}^{\infty} |y(k)|^2 = \sum_{k=0}^{\infty} |u(k)|^2 + ||\overline{\boldsymbol{x}}(0)||^2 \tag{4.104}$$

for every finite-energy input that is identically zero for $k > N$, where N is an arbitrary finite positive integer. This reveals that $H(z)$ is LBR.

(Necessity) Assume that $H(z)$ is LBR. Let a minimal realization of $H(z)$ be given by $(\boldsymbol{A}, \boldsymbol{b}, \boldsymbol{c}, d)_n$. Since $H(z)$ is asymptotically stable, the matrix defined by

$$\boldsymbol{P} = \sum_{i=0}^{\infty} (\boldsymbol{A}^T)^i \boldsymbol{c}^T \boldsymbol{c} \boldsymbol{A}^i \tag{4.105}$$

is a symmetric positive-definite matrix and satisfies

$$\boldsymbol{P} = \boldsymbol{A}^T \boldsymbol{P} \boldsymbol{A} + \boldsymbol{c}^T \boldsymbol{c} \tag{4.106}$$

By decomposing matrix $\boldsymbol{P}$ in (4.105) as $\boldsymbol{P} = \boldsymbol{T}^{-T}\boldsymbol{T}^{-1}$, and defining a state-space model $(\overline{\boldsymbol{A}}, \overline{\boldsymbol{b}}, \overline{\boldsymbol{c}}, d)_n$ as in (4.96), we can derive

$$\boldsymbol{I}_n = \overline{\boldsymbol{A}}^T \overline{\boldsymbol{A}} + \overline{\boldsymbol{c}}^T \overline{\boldsymbol{c}} \tag{4.107}$$

from (4.106). Next, by LBR property of $H(z)$,

$$\sum_{k=0}^{\infty} |y(k)|^2 = \sum_{k=0}^{\infty} |u(k)|^2 \tag{4.108}$$

holds for any finite-energy input, under the assumption that the initial states are zero. In particular, suppose $u(k) = 0$ for $k > N$ where N is an arbitrary finite positive integer, then $y(k) = \overline{\boldsymbol{c}}\,\overline{\boldsymbol{x}}(k)$ holds for $k > N$, hence

$$\begin{aligned} |y(k)|^2 &= \overline{\boldsymbol{x}}(k)^T \left[\boldsymbol{I}_n - \overline{\boldsymbol{A}}^T\overline{\boldsymbol{A}}\right] \overline{\boldsymbol{x}}(k) \\ &= ||\overline{\boldsymbol{x}}(k)||^2 - ||\overline{\boldsymbol{x}}(k+1)||^2 \ \text{ for } \ k > N \end{aligned} \tag{4.109}$$

is obtained by (4.107). Thus, (4.108) can be written as

$$\sum_{k=0}^{N} |y(k)|^2 + ||\overline{\boldsymbol{x}}(N+1)||^2 = \sum_{k=0}^{N} |u(k)|^2 \tag{4.110}$$

because

$$\sum_{k=N+1}^{\infty} |y(k)|^2 = ||\overline{\boldsymbol{x}}(N+1)||^2 \tag{4.111}$$

By replacing N by $N+1$ and then subtracting, we obtain

$$\begin{bmatrix}\overline{\boldsymbol{x}}(N+1)^T & y(N)\end{bmatrix} \begin{bmatrix}\overline{\boldsymbol{x}}(N+1) \\ y(N)\end{bmatrix} = \begin{bmatrix}\overline{\boldsymbol{x}}(N)^T & u(N)\end{bmatrix} \begin{bmatrix}\overline{\boldsymbol{x}}(N) \\ u(N)\end{bmatrix} \tag{4.112}$$

for any finite positive integer N. This means that $\mathcal{R}$ defined by (4.97) is orthogonal, and we arrive at (4.93). This completes the proof of Theorem 4.7.

4.8 Summary

In this chapter, we have presented fundamental properties of the state-space description of linear causal dynamical systems, a method for deriving its transfer function from a given state-space model, the concepts of the equivalent transformation, the canonical structure decomposition, and methods for state-space realization using the Hankel matrix. Both the state-space description and the external (input-output) description of linear causal dynamical systems are of significance and useful in practice. However, when it comes to choosing the most appropriate description, the matter is often problem dependent. In addition, we have studied a passivity property of discrete-time systems, known as lossless bounded-real lemma, which finds applications in network synthesis and stability analysis.

References

[1] R. E. Kalman, "Mathematical description of linear systems," *J. SIAM Contr.*, Ser. A, vol. 1, no. 2, pp. 152–192, 1963.

[2] R. E. Kalman, P. L. Falb and M. A. Arbib, *Topics in Mathematical System Theory*, New York, McGraw-Hill, 1969.

[3] A. J. Tether, "Construction of minimal linear state-variable models from finite input-output data," *IEEEE Trans. Automat. Contr.*, vol. AC-15, no. 4, pp. 427–436, Aug. 1970.

[4] H. Kogo and T. Mita, *Introduction to System Control Theory*, Tokyo, Japan, Jikkyo Shuppan, 1979.

[5] T. Hinamoto, "Realizations of a state-space model from two-dimensional input-output map," *IEEE Trans. Circuits Syst.*, vol. CAS-27, no. 1, pp. 36–44, Jan. 1980.

[6] V. C. Klema and A. J. Laub, "The singular value decomposition: Its computation and some applications," *IEEEE Trans. Automat. Contr.*, vol. AC-25, no. 2, pp. 164–176, Apr. 1980.

[7] L. M. Silverman, "Optimal approximation of linear systems," in Proc, *Joint Automat. Contr. Conf.*, S. F., 1980, FA8-A.

[8] T. Hinamoto and F. W. Fairman, "Separable-denominator state-space realization of two-dimensional filters using a canonic form," *IEEE Trans. Acoust. Speech, Signal Process.*, vol. ASSP-29, no. 4, pp. 846–853, Aug, 1981.

[9] J. R. Sveinsson and F. W. Fairman, "Minimal balanced realization of transfer function matrices using Markov parameters," *IEEEE Trans. Automat. Contr.*, vol. AC-30, no. 10, pp. 1014–1016, Oct. 1985.

[10] T. Hinamoto, S. Maekawa, J. Shimonishi and A. N. Venetsanopoulos, "Balanced realization and model reduction of 3-D separable-denominator transfer functions," *Franklin Institute*, vol. 325, no. 2, pp. 207–219, 1988.

[11] M. S. Santina, A. R. Stubberud and G. H. Hostetter, *Digital Control System Design*, 2nd ed. Orlando, FL, Saunders College Publishing, Harcourt Brace College Publishers, 1994.

[12] P. P. Vaidyanathan, "The discrete-time bounded-real lemma in digital filtering," *IEEE Trans. Circuits Syst.*, vol. CAS-32, no. 9, pp. 918–924, Sep. 1985.

[13] R. E. Kalman and J. Bertram, "Control system design via the second method of Liapunov, part II, discrete time systems," *ASME J. Basic Engineering*, vol. 82, pp. 394–400, 1960.

5

FIR Digital Filter Design

5.1 Preview

Digital filters with finite sequence of the impulse response are called *FIR digital filters*, *nonrecursive digital filters* or *digital transversal filters*, where "*FIR*" is the acronym of terms, "*Finite Impulse Response*". A general FIR digital filter of order $N-1$ is described by

$$y(k) = \sum_{i=0}^{N-1} h_i u(k-i) \tag{5.1}$$

where $u(k)$ and $y(k)$ are scalar input and output, respectively, and h_i for $i = 0, 1, \cdots, N-1$ denote the impulse response. The transfer function of the FIR digital filter in (5.1) can be expressed as

$$H(z) = h_0 + h_1 z^{-1} + \cdots + h_{N-1} z^{-(N-1)} \tag{5.2}$$

A block diagram of an FIR digital filter in (5.1) is depicted in Figure 5.1.

FIR digital filters are the preferred filtering scheme in many DSP applications, mainly due to the advantages of the FIR digital filters as compared to their IIR counterparts, i.e.,

1. FIR digital filters are always stable.
2. Exact linear-phase response can easily be achieved by imposing either symmetric or antisymmetric condition on the FIR filter's coefficients.
3. FIR digital filters possess low output noise due to coefficient quantization and multiplication roundoff errors.
4. Effective methods for the design of a variety of FIR digital filters are available.

On the other hand, in the case of designing IIR digital filters, stability is always a concern. For IIR digital filters, exact linear-phase responses cannot

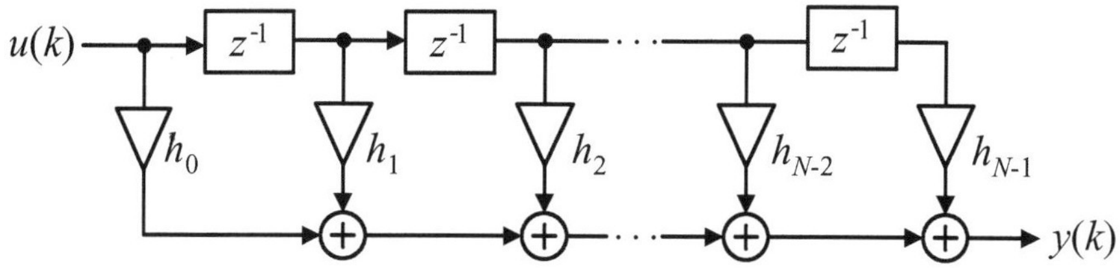

Figure 5.1 A block diagram of an FIR digital filter.

be realized in general, even in the passband. Coefficient sensitivity as well as output roundoff noise due to multiplications often become severe and therefore particular cares might be taken to deal with these problems.

The main disadvantage of FIR digital filters is that the order of an FIR digital filter is usually considerably higher than its IIR counterpart to meet the same design specification, especially when the transition bands are narrow. As a result, the implementation of an FIR digital filter with narrow transition bands is often costly.

5.2 Filter Classification

The frequency response of a digital filter can be described by

$$H(e^{j\omega}) = M(\omega)e^{j\theta(\omega)} \tag{5.3}$$

where $M(\omega)$ is the magnitude response of the filter, and $\theta(\omega)$ is the phase characteristic of the filter. The ideal magnitude responses can commonly be used to classify digital filters. Even if such digital filters are not realizable, they can be approximated in practice with some acceptable tolerance.

The magnitude responses of the four typical types of ideal digital filters are illustrated in Figure 5.2. For *lowpass filter* of Figure 5.2(a), the passband and the stopband are given by $0 \le \omega \le \omega_p$ and $\omega_p < \omega \le \pi$, respectively. For *highpass filter* of Figure 5.2(b), the stopband is given by $0 \le \omega < \omega_p$, while the passband is given by and $\omega_p \le \omega \le \pi$, respectively. For *bandpass filter* of Figure 5.2(c), the passband region is given by $\omega_{p1} \le \omega \le \omega_{p2}$, and the stopband regions are specified by $0 \le \omega < \omega_{p1}$ and $\omega_{p2} < \omega \le \pi$. Finally, for *bandstop filter* of Figure 5.2(d), the passband regions are given by $0 \le \omega \le \omega_{p1}$ and $\omega_{p2} \le \omega \le \pi$, while the stopband region is $\omega_{p1} < \omega < \omega_{p2}$. The frequencies ω_p, ω_{p1}, and ω_{p2} are called the *passband edges* of their respective filters. It is observed from the figure that an ideal filter

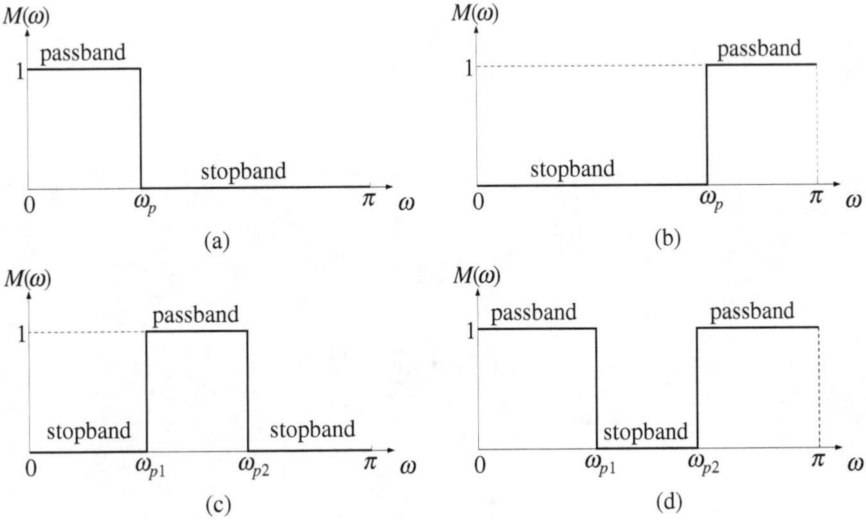

Figure 5.2 Four types of ideal filters. (a) Ideal lowpass filter. (b) Ideal highpass filter. (c) Ideal bandpass filter. (d) Ideal bandstop filter.

has the magnitude response equal to unity in the passband and zero in the stopband.

The specifications of magnitude responses with some acceptable tolerance for the four typical types of digital filters are shown in Figure 5.3. For the digital filters of Figure 5.3, the passband and the stopband allow to have some acceptable tolerance, respectively, while the transition band regions are free from any specifications.

For the design of a lowpass FIR digital filter, a few formulas exist for estimating the minimum value of filter length N directly from the digital filter specifications. Let ω_p and ω_s denote the normalized passband edge frequency and the normalized stopband edge frequency, respectively, and let δ_p and δ_s indicate the peak passband ripple and the peak stopband ripple, respectively.

Kaiser's Formula

A simple formula for estimating filter length N which meets the desired specifications is given by [3]

$$N \simeq \frac{-20\log_{10}(\sqrt{\delta_p\delta_s}) - 13}{14.6\Delta} \tag{5.4}$$

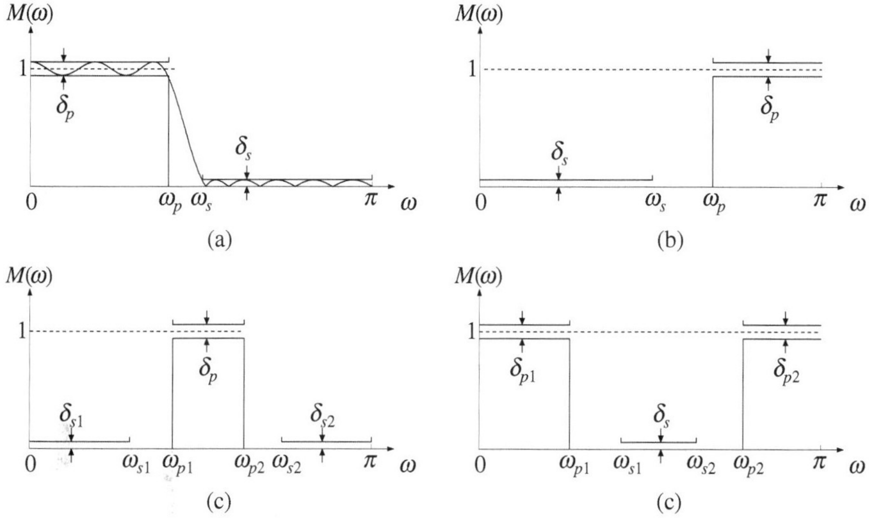

Figure 5.3 Typical magnitude response specifications. (a) Lowpass filter. (b) Highpass filter. (c) Bandpass filter. (d) Bandstop filter.

where Δ is the transition band width normalized by sampling frequency, namely,

$$\Delta = \frac{\omega_s - \omega_p}{2\pi}$$

Bellanger's Formula

Another simple formula for estimating filter length N to meet the desired specifications is given by [3]

$$N \simeq -\frac{2\log_{10}(10\delta_p\delta_s)}{3\Delta} - 1 \tag{5.5}$$

where Δ is defined in (5.4).

5.3 Linear-phase Filters

5.3.1 Frequency Transfer Function

The frequency transfer function of the filter in (5.2) can be obtained by setting $z = e^{j\omega}$ as

$$H(e^{j\omega}) = \sum_{i=0}^{N-1} h_i e^{-j\omega i} = |H(e^{j\omega})|e^{j\theta(\omega)} \tag{5.6}$$

where

$$|H(e^{j\omega})| = \sqrt{\left(\sum_{i=0}^{N-1} h_i \cos i\omega\right)^2 + \left(\sum_{i=0}^{N-1} h_i \sin i\omega\right)^2}$$

$$\theta(\omega) = -\tan^{-1}\left(\frac{\sum_{i=0}^{N-1} h_i \sin i\omega}{\sum_{i=0}^{N-1} h_i \cos i\omega}\right)$$

$|H(e^{j\omega})|$ and $\theta(\omega)$ in (5.6) are called the *amplitude response* and the *phase characteristic*, respectively. It is obvious that $|H(e^{j\omega})|$ is an even function, and $\theta(\omega)$ is an odd function. The phase delay and the group delay for the filter in (5.6) are defined as

$$\tau_p(\omega) = -\frac{\theta(\omega)}{\omega}, \qquad \tau_g(\omega) = -\frac{d\theta(\omega)}{d\omega} \tag{5.7}$$

respectively.

5.3.2 Symmetric Impulse Responses

For constant phase delay, $\theta(\omega)$ must be linear with respect to ω, that is,

$$\theta(\omega) = -\tau\omega \tag{5.8}$$

From (5.6) and (5.8), it follows that

$$\frac{\sum_{i=0}^{N-1} h_i \sin i\omega}{\sum_{i=0}^{N-1} h_i \cos i\omega} = \frac{\sin \tau\omega}{\cos \tau\omega} \tag{5.9}$$

which leads to

$$\sum_{i=0}^{N-1} h_i\big(\cos i\omega \sin \tau\omega - \sin i\omega \cos \tau\omega\big) = \sum_{i=0}^{N-1} h_i \sin(\tau\omega - i\omega) = 0 \tag{5.10}$$

The values of τ and h_i for $i = 0, 1, \cdots, N-1$ satisfying (5.10) are given by

$$\tau = \frac{N-1}{2}, \qquad h_i = h_{N-1-i} \text{ for } i = 0, 1, \cdots, N-1 \tag{5.11}$$

Hence, it is only necessary for the impulse response to be symmetric about the shifted origin $(N-1)/2$. Unlike IIR digital filters, FIR digital filters can have a linear phase over the entire baseband. The impulse responses which are symmetric about the shifted origin $(N-1)/2$ for odd N as well as even N are illustrated in Figure 5.4.

1. Symmetric Impulse Response of Even Order $N-1$ (Type 1)

Since $h_i = h_{N-1-i}$ for $i = 0, 1, \cdots, N-1$, we write (5.2) as

$$\begin{aligned} H(z) &= \sum_{i=0}^{\frac{N-1}{2}-1} h_i z^{-i} + h_{\frac{N-1}{2}} z^{-\frac{N-1}{2}} + \sum_{i=\frac{N-1}{2}+1}^{N-1} h_{N-1-i} z^{-i} \\ &= \sum_{i=0}^{\frac{N-1}{2}-1} h_i z^{-i} + h_{\frac{N-1}{2}} z^{-\frac{N-1}{2}} + \sum_{l=0}^{\frac{N-1}{2}-1} h_l z^{-(N-1-l)} \\ &= z^{-\frac{N-1}{2}} \left[h_{\frac{N-1}{2}} + \sum_{i=0}^{\frac{N-1}{2}-1} h_i \left(z^{\frac{N-1}{2}-i} + z^{-(\frac{N-1}{2}-i)} \right) \right] \end{aligned} \tag{5.12}$$

The frequency transfer function of the filter in (5.12) becomes

$$\begin{aligned} H(e^{j\omega}) &= \left[h_{\frac{N-1}{2}} + \sum_{i=0}^{\frac{N-1}{2}-1} 2h_i \cos\left(\frac{N-1}{2} - i \right) \omega \right] e^{-j\frac{N-1}{2}\omega} \\ &= \left[\sum_{k=0}^{\frac{N-1}{2}} c_k \cos(k\omega) \right] e^{-j\frac{N-1}{2}\omega} \end{aligned} \tag{5.13}$$

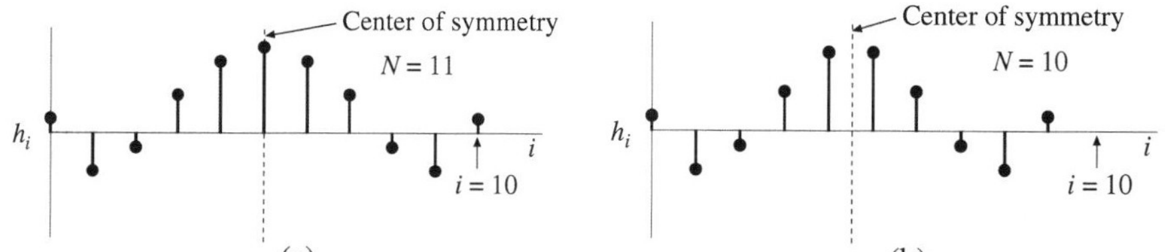

(a) (b)

Figure 5.4 Symmetric impulse responses. (a) N is odd. (b) N is even.

where

$$c_0 = h_{\frac{N-1}{2}}, \qquad c_k = 2h_{\frac{N-1}{2}-k} \text{ for } 1 \le k \le (N-1)/2$$

2. Symmetric Impulse Response of Odd Order $N-1$ (Type 2)

In this case, (5.2) is changed to

$$\begin{aligned} H(z) &= \sum_{i=0}^{\frac{N}{2}-1} h_i z^{-i} + \sum_{i=\frac{N}{2}}^{N-1} h_{N-1-i} z^{-i} \\ &= \sum_{i=0}^{\frac{N}{2}-1} h_i z^{-i} + \sum_{l=0}^{\frac{N}{2}-1} h_l z^{-(N-1-l)} \\ &= z^{-\frac{N-1}{2}} \left[\sum_{i=0}^{\frac{N}{2}-1} h_i \left(z^{\frac{N-1}{2}-i} + z^{-(\frac{N-1}{2}-i)} \right) \right] \end{aligned} \tag{5.14}$$

The frequency transfer function of the filter in (5.14) can be expressed as

$$\begin{aligned} H(e^{j\omega}) &= \left[\sum_{i=0}^{\frac{N}{2}-1} 2h_i \cos\left(\frac{N-1}{2} - i \right) \omega \right] e^{-j\frac{N-1}{2}\omega} \\ &= \left[\sum_{k=1}^{\frac{N}{2}} c_k \cos\left(k - \frac{1}{2} \right) \omega \right] e^{-j\frac{N-1}{2}\omega} \end{aligned} \tag{5.15}$$

where

$$c_k = 2h_{\frac{N}{2}-k} \text{ for } k = 1, 2, \cdots, \frac{N}{2}$$

Since $\cos(k-1/2)\pi = 0$ in (5.15), it follows that $H(e^{j\pi}) = 0$. This reveals that odd-order FIR filters with symmetric impulse response are not suitable for the design of highpass filters.

By defining

$$\begin{aligned} c_1 &= \frac{1}{2}\tilde{c}_1 + \tilde{c}_0, \qquad c_{\frac{N}{2}} = \frac{1}{2}\tilde{c}_{\frac{N}{2}-1} \\ c_k &= \frac{1}{2}(\tilde{c}_k + \tilde{c}_{k-1}) \text{ for } 2 \le k \le N/2-1 \end{aligned} \tag{5.16}$$

we obtain

$$
\begin{aligned}
\sum_{k=1}^{\frac{N}{2}} c_k \cos\Big(k-\frac{1}{2}\Big)\omega &= \tilde{c}_0 \cos\frac{1}{2}\omega + \frac{1}{2}\tilde{c}_1 \left[\cos\frac{1}{2}\omega + \cos\left(1+\frac{1}{2}\right)\omega\right] \\
&\quad + \frac{1}{2}\tilde{c}_2 \left[\cos\left(1+\frac{1}{2}\right)\omega\right. \\
&\quad \left. + \cos\left(2+\frac{1}{2}\right)\omega\right] + \cdots \\
&\quad + \frac{1}{2}\tilde{c}_{\frac{N}{2}-1} \left[\cos\left(\frac{N}{2}-2+\frac{1}{2}\right)\omega\right. \\
&\quad \left. + \cos\left(\frac{N}{2}-1+\frac{1}{2}\right)\omega\right] \\
&= \cos\frac{1}{2}\omega \sum_{k=0}^{\frac{N}{2}-1} \tilde{c}_k \cos(k\omega)
\end{aligned}
\tag{5.17}
$$

Substituting (5.17) into (5.15) yields

$$
H(e^{j\omega}) = \cos\Big(\frac{\omega}{2}\Big) \left[\sum_{k=0}^{\frac{N}{2}-1} \tilde{c}_k \cos(k\omega)\right] e^{-j\frac{N-1}{2}\omega} \tag{5.18}
$$

5.3.3 Antisymmetric Impulse Responses

In many applications, only the group delay needs to be constant. A phase response with constant group delay assumes the form

$$
\theta(\omega) = -\tau\omega + \theta_o, \qquad 0 \le \omega \le \pi \tag{5.19}
$$

Applying the arguments similar to those in (5.9) and (5.10), we arrive at

$$
\sum_{i=0}^{N-1} h_i \sin(\tau\omega - i\omega - \theta_o) = 0 \tag{5.20}
$$

In order to find a solution of (5.20), we set $\theta_o = \pm\pi/2$. Then, (5.20) is changed to

$$\sum_{i=0}^{N-1} h_i \cos(\tau\omega - i\omega) = 0 \tag{5.21}$$

The values of τ and h_i for $i = 0, 1, \cdots, N-1$ satisfying (5.21) are given by

$$\tau = \frac{N-1}{2}, \qquad h_i = -h_{N-1-i} \text{ for } i = 0, 1, \cdots, N-1 \tag{5.22}$$

Here, the impulse response is required to be antisymmetric about the shifted origin $(N-1)/2$ in which an FIR digital filter has a linear phase over the entire baseband. The impulse responses which are antisymmetric about the shifted origin $(N-1)/2$ for odd N as well as even N are illustrated in Figure 5.5.

3. Antisymmetric Impulse Response of Even Order $N-1$ (Type 3)

Since $h_i = -h_{N-1-i}$ for $i = 0, 1, \cdots, N-1$ and $h_{\frac{N-1}{2}} = 0$, we can write (5.2) as

$$\begin{aligned} H(z) &= \sum_{i=0}^{\frac{N-1}{2}-1} h_i z^{-i} - \sum_{i=\frac{N-1}{2}+1}^{N-1} h_{N-1-i} z^{-i} \\ &= \sum_{i=0}^{\frac{N-1}{2}-1} h_i z^{-i} - \sum_{l=0}^{\frac{N-1}{2}-1} h_l z^{-(N-1-l)} \\ &= z^{-\frac{N-1}{2}} \left[\sum_{i=0}^{\frac{N-1}{2}-1} h_i \left(z^{\frac{N-1}{2}-i} - z^{-(\frac{N-1}{2}-i)} \right) \right] \end{aligned} \tag{5.23}$$

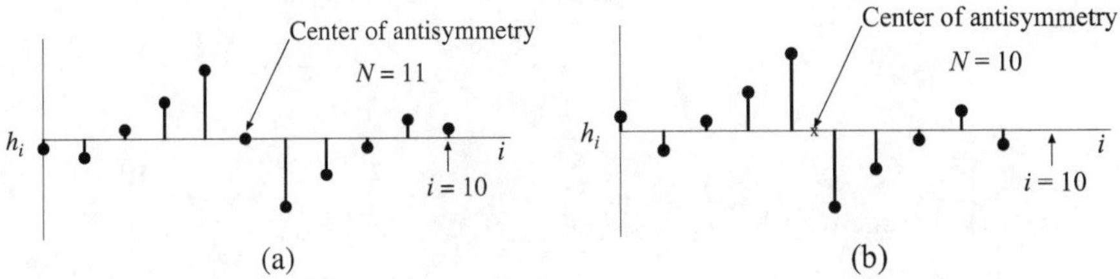

Figure 5.5 Antisymmetric impulse responses. (a) N is odd. (b) N is even.

The frequency transfer function of the filter in (5.23) becomes

$$\begin{aligned} H(e^{j\omega}) &= \left[\sum_{i=0}^{\frac{N-1}{2}-1} 2h_i \sin\left(\frac{N-1}{2} - i\right)\omega \right] je^{-j\frac{N-1}{2}\omega} \\ &= \left[\sum_{k=1}^{\frac{N-1}{2}} c_k \sin(k\omega) \right] e^{-j\theta(\omega)} \end{aligned} \tag{5.24}$$

where

$$c_k = 2h_{\frac{N-1}{2}-k} \text{ for } 1 \le k \le (N-1)/2,$$

$$\theta(\omega) = \begin{cases} \dfrac{\pi}{2} - \dfrac{N-1}{2}\omega & \text{for } \omega > 0 \\ -\dfrac{\pi}{2} - \dfrac{N-1}{2}\omega & \text{for } \omega < 0 \end{cases}$$

Since $\sin(k0) = \sin(k\pi) = 0$ in (5.24), it follows that $H(e^{j0}) = H(e^{j\pi}) = 0$. Therefore, even-order FIR filters with antisymmetric impulse response are inadequate for the design of lowpass and highpass filters.

By defining

$$\begin{aligned} c_1 &= \tilde{c}_0 - \frac{1}{2}\tilde{c}_2, \qquad c_{\frac{N-1}{2}} = \frac{1}{2}\tilde{c}_{\frac{N-3}{2}} \\ c_k &= \frac{1}{2}(\tilde{c}_{k-1} - \tilde{c}_{k+1}) \text{ for } 2 \le k \le (N-5)/2 \end{aligned} \tag{5.25}$$

we have

$$\begin{aligned} \sum_{k=1}^{\frac{N-1}{2}} c_k \sin k\omega &= \tilde{c}_0 \sin\omega + \frac{1}{2}\tilde{c}_1 \sin 2\omega + \frac{1}{2}\tilde{c}_2 \left[\sin 3\omega - \sin\omega\right] \\ &\quad + \frac{1}{2}\tilde{c}_3 \left[\sin 4\omega - \sin 2\omega\right] + \frac{1}{2}\tilde{c}_4 \left[\sin 5\omega - \sin 3\omega\right] \\ &\quad + \cdots + \tilde{c}_{\frac{N-3}{2}} \left[\sin\frac{N-1}{2}\omega - \sin\frac{N-5}{2}\omega\right] \\ &= \sin\omega \sum_{k=0}^{\frac{N-3}{2}} \tilde{c}_k \cos(k\omega) \end{aligned} \tag{5.26}$$

Substituting (5.26) into (5.24) gives

$$H(e^{j\omega}) = \sin\omega \left[\sum_{k=0}^{\frac{N-3}{2}} \tilde{c}_k \cos(k\omega) \right] e^{-j\theta(\omega)} \tag{5.27}$$

4. Antisymmetric Impulse Response of Odd Order $N-1$ (Type 4)
In this case, (5.2) is changed to

$$\begin{aligned} H(z) &= \sum_{i=0}^{\frac{N}{2}-1} h_i z^{-i} - \sum_{i=\frac{N}{2}}^{N-1} h_{N-1-i} z^{-i} \\ &= \sum_{i=0}^{\frac{N}{2}-1} h_i z^{-i} - \sum_{l=0}^{\frac{N}{2}-1} h_l z^{-(N-1-l)} \\ &= z^{-\frac{N-1}{2}} \left[\sum_{i=0}^{\frac{N}{2}-1} h_i \left(z^{\frac{N-1}{2}-i} - z^{-(\frac{N-1}{2}-i)} \right) \right] \end{aligned} \tag{5.28}$$

The frequency transfer function of the filter in (5.28) can be expressed as

$$\begin{aligned} H(e^{j\omega}) &= \left[\sum_{i=0}^{\frac{N}{2}-1} 2h_i \sin\left(\frac{N-1}{2} - i \right)\omega \right] je^{-j\frac{N-1}{2}\omega} \\ &= \left[\sum_{k=1}^{\frac{N}{2}} c_k \sin\left(k - \frac{1}{2} \right)\omega \right] e^{j\theta(\omega)} \end{aligned} \tag{5.29}$$

where

$$c_k = 2h_{\frac{N}{2}-k} \text{ for } 1 \le k \le N/2, \quad \theta(\omega) = \begin{cases} \dfrac{\pi}{2} - \dfrac{N-1}{2}\omega & \text{for } \omega > 0 \\ -\dfrac{\pi}{2} - \dfrac{N-1}{2}\omega & \text{for } \omega < 0 \end{cases}$$

Since $\sin(k-1/2)0 = 0$ in (5.29), it follows that $H(e^{j0}) = 0$. Hence odd-order FIR filters with antisymmetric impulse response are inadequate for the design of lowpass filters.

By defining

$$
\begin{aligned}
c_1 &= \tilde{c}_0 - \frac{1}{2}\tilde{c}_1, \qquad c_{\frac{N}{2}} = \frac{1}{2}\tilde{c}_{\frac{N}{2}-1} \\
c_k &= \frac{1}{2}(\tilde{c}_{k-1} - \tilde{c}_k) \text{ for } 2 \le k \le N/2 - 1
\end{aligned}
\tag{5.30}
$$

we can write

$$
\begin{aligned}
\sum_{k=1}^{\frac{N}{2}} c_k \sin\Big(k - \frac{1}{2}\Big)\omega &= \tilde{c}_0 \sin\frac{1}{2}\omega + \frac{1}{2}\tilde{c}_1 \left[\sin\left(1 + \frac{1}{2}\right)\omega - \sin\frac{1}{2}\omega\right] \\
&\quad + \frac{1}{2}\tilde{c}_2 \left[\sin\left(2 + \frac{1}{2}\right)\omega - \sin\left(1 + \frac{1}{2}\right)\omega\right] + \cdots \\
&\quad + \frac{1}{2}\tilde{c}_{\frac{N}{2}-1} \left[\sin\left(\frac{N}{2} - 1 + \frac{1}{2}\right)\omega \right.\\
&\quad \left. - \sin\left(\frac{N}{2} - 2 + \frac{1}{2}\right)\omega\right] \\
&= \sin\frac{1}{2}\omega \sum_{k=0}^{\frac{N}{2}-1} \tilde{c}_k \cos(k\omega)
\end{aligned}
\tag{5.31}
$$

By substituting (5.31) into (5.29), we obtain

$$
H(e^{j\omega}) = \sin\Big(\frac{\omega}{2}\Big) \left[\sum_{k=0}^{\frac{N}{2}-1} \tilde{c}_k \cos(k\omega)\right] e^{-j\theta(\omega)} \tag{5.32}
$$

5.4 Design Using Window Function

5.4.1 Fourier Series Expansion

Suppose that $H_d(e^{j\omega})$ is the desired frequency response. Since $H_d(e^{j\omega})$ a periodic function of ω with a period 2π, it can be represented by its Fourier series as

$$
H_d(e^{j\omega}) = \sum_{i=-\infty}^{\infty} h_i e^{-j\omega i} \tag{5.33}
$$

where the Fourier coefficients given by

$$h_i = \frac{1}{2\pi}\int_{-\pi}^{\pi} H_d(e^{j\omega})e^{j\omega i}d\omega, \qquad -\infty < i < \infty \tag{5.34}$$

correspond precisely to the impulse response samples. Substituting $e^{j\omega} = z$ into (5.33) yields

$$H_d(z) = \sum_{i=-\infty}^{\infty} h_i z^{-i} \tag{5.35}$$

Therefore, the transfer function $H_d(z)$ in (5.35) can be determined by computing its h_i's using (5.34). We remark, however, that the corresponding impulse response is of infinite length and noncausal.

In order to find a finite-duration impulse response sequence $\{h_i\}$ of length N, we have to truncate the impulse response sequence with finite terms, that is,

$$h_i = 0 \text{ for } |\,i\,| > \frac{N-1}{2} \tag{5.36}$$

where $N - 1$ is assumed to be even. Then, we obtain

$$H(z) = h_0 + \sum_{i=1}^{\frac{N-1}{2}} (h_{-i}\, z^i + h_i\, z^{-i}) \tag{5.37}$$

and a causal FIR digital filter can then be derived from (5.37) by setting $H_c(z) = z^{-\frac{N-1}{2}} H(z)$.

As an example, consider the design of an ideal lowpass filter with linear phase whose magnitude response and phase characteristic are shown in Figure 5.6. The frequency transfer function is given by

$$H_d(e^{j\omega}) = M(\omega)e^{j\theta(\omega)} = \begin{cases} 1 \cdot e^{-j\omega\tau} & \text{for } \ |\omega| \le \omega_c \\ 0 & \text{for } \ |\omega| > \omega_c \end{cases} \tag{5.38}$$

Using (5.34) yields

$$\begin{aligned} h_i &= \frac{1}{2\pi}\int_{-\omega_c}^{\omega_c} 1 \cdot e^{-j\omega\tau} e^{j\omega i} d\omega \\ &= \frac{1}{2\pi}\left[\frac{e^{j\omega(i-\tau)}}{j(i-\tau)}\right]_{-\omega_c}^{\omega_c} = \frac{\sin \omega_c(i-\tau)}{\pi(i-\tau)} \end{aligned} \tag{5.39}$$

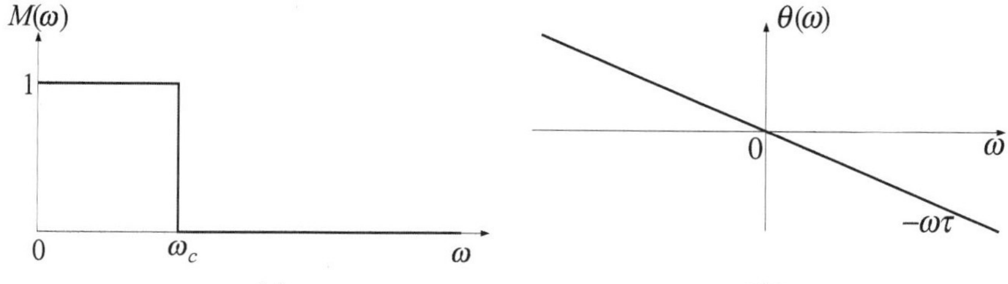

(a) (b)

Figure 5.6 Ideal lowpass filter characteristics. (a) Magnitude response. (b) Phase characteristic.

By applying the truncation in (5.36) to (5.39) and setting

$$H_c(z) = \sum_{i=-\frac{N-1}{2}}^{\frac{N-1}{2}} h_i z^{-(i+\frac{N-1}{2})} \tag{5.40}$$

we obtain a causal and feasible FIR digital filter of order $N-1$.

5.4.2 Window Functions

There exist many tapered windows in the literature, however an introduction of all these windows is beyond the scope of this text. Our discussion will be restricted to several well-known tapered windows of length N.

Let an infinite impulse response sequence $\{h_i| -\infty < i < \infty\}$ be converted into a finite impulse response sequence $\{w_i h_i| \, |i| \le (N-1)/2\}$ where $\{w_i| \, |i| \le (N-1)/2\}$ is said to be the *window function*. By applying the z-transform to the finite impulse response sequence $\{w_i h_i\}$, the frequency transfer function of the resulting causal FIR digital filter can be expressed as

$$H_c(e^{j\omega}) = \sum_{i=-\frac{N-1}{2}}^{\frac{N-1}{2}} w_i h_i \, e^{-j\omega(i+\frac{N-1}{2})} \tag{5.41}$$

1. Rectangular Window

The rectangular window is defined by

$$w_i = \begin{cases} 1 & \text{for} \quad |i| \le \frac{N-1}{2} \\ 0 & \text{for} \quad |i| > \frac{N-1}{2} \end{cases} \tag{5.42}$$

Since the Fourier series are truncated outside the N terms, the undesirable *Gibbs phenomenon*, which is known to be inherently associated with the Fourier series near the function's discontinuities, will occur. Various windows have been proposed to deal with this problem.

2. Bartlett Window

The Bartlett window is described by [3]

$$w_i = \begin{cases} 1 - \frac{2|i|}{N-1} & \text{for } |i| \le \frac{N-1}{2} \\ 0 & \text{for } |i| > \frac{N-1}{2} \end{cases} \tag{5.43}$$

3. Generalized Hamming Window

The generalized Hamming window is given by [8]

$$w_i = \begin{cases} \alpha + (1-\alpha)\cos(\frac{2\pi i}{N-1}) & \text{for } |i| \le \frac{N-1}{2} \\ 0 & \text{for } |i| > \frac{N-1}{2} \end{cases} \tag{5.44}$$

where $0 \le \alpha \le 1$. The window in (5.44) is called the *Hamming window* in case $\alpha = 0.54$ and the *Hanning window* in case $\alpha = 0.50$.

4. Blackman Window

The Blackman window is specified by [3]

$$w_i = \begin{cases} 0.42 + 0.5\cos(\frac{2\pi i}{N-1}) + 0.08\cos(\frac{4\pi i}{N-1}) & \text{for } |i| \le \frac{N-1}{2} \\ 0 & \text{for } |i| > \frac{N-1}{2} \end{cases} \tag{5.45}$$

Continuous profiles of these windows are depicted in Figure 5.7. For illustration purposes the amplitude responses of the FIR filter specified by (5.39) and (5.41) with $N = 31$ and window $\{w_i \,|\, -(N-1)/2 \le i \le (N-1)/2\}$ being each of the above windows are displayed in Figure 5.8.

5.4.3 Frequency Transformation

Suppose that an FIR lowpass filter with cutoff frequency ω_c has been designed and its transfer function is given by

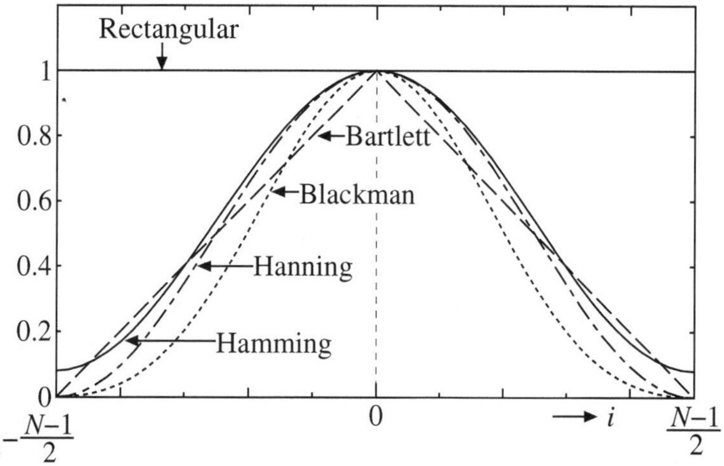

Figure 5.7 Plots of the fixed windows shown with solid lines for clearness.

$$H_{LP}(z) = \sum_{i=0}^{N-1} h_i^{LP} z^{-i} \tag{5.46}$$

where $\{h_i^{LP}\}$ denotes the impulse response of the filter.

1. FIR Highpass Filter

The impulse response of an FIR highpass filter can be expressed as

$$h_i^{HP} = (-1)^i h_i^{LP} \ \text{ for } i = 0, 1, \cdots, N-1 \tag{5.47}$$

The highpass characteristic of $\{h_i^{HP}\}$ can be verified by evaluating its frequency response in terms of that of the FIR lowpass filter:

$$\begin{aligned} H_{HP}(e^{j\omega}) &= \sum_{i=0}^{N-1} (-1)^i h_i^{LP} e^{-j\omega i} = \sum_{i=0}^{N-1} (e^{j\pi})^i h_i^{LP} e^{-j\omega i} \\ &= \sum_{i=0}^{N-1} h_i^{LP} e^{-j(\omega-\pi)i} = H_{LP}(e^{j(\omega-\pi)}) \end{aligned} \tag{5.48}$$

where the cutoff frequency is $\pi - \omega_c$.

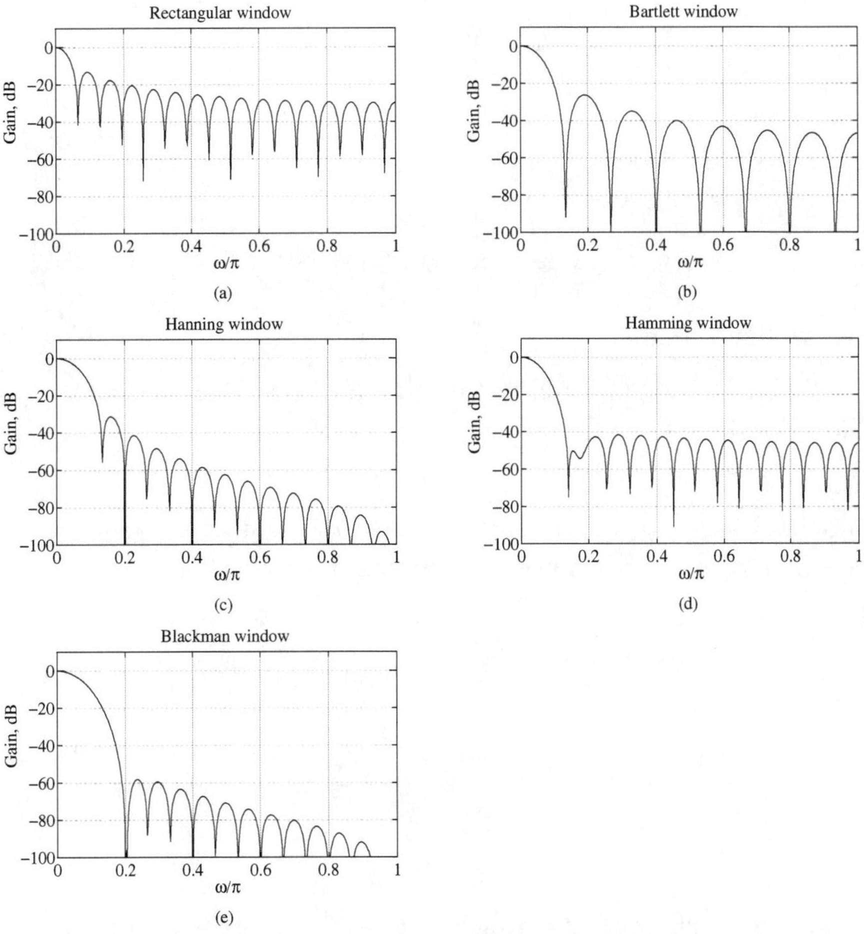

Figure 5.8 Gain responses of the fixed window functions.

2. FIR Bandpass Filter

The impulse response of an FIR bandpass filter can be written as

$$h_i^{BP} = (2\cos\omega_o i)h_i^{LP} \;\text{ for }\; i = 0, 1, \cdots, N-1 \tag{5.49}$$

where ω_o stands for the center frequency of the passband. The bandpass characteristic of $\{h_i^{BP}\}$ can be demonstrated by evaluating its frequency response in terms of that of the FIR lowpass filter:

$$\begin{aligned}H_{BP}(e^{j\omega}) &= \sum_{i=0}^{N-1} \left(e^{j\omega_o i} + e^{-j\omega_o i}\right) h_i^{LP} e^{-j\omega i} \\ &= \sum_{i=0}^{N-1} h_i^{LP} e^{-j(\omega-\omega_o)i} + \sum_{i=0}^{N-1} h_i^{LP} e^{-j(\omega+\omega_o)i} \\ &= H_{LP}\left(e^{j(\omega-\omega_o)}\right) + H_{LP}\left(e^{j(\omega+\omega_o)}\right)\end{aligned} \tag{5.50}$$

3. FIR Bandstop Filter

The impulse response of an FIR bandstop filter can be specified by

$$h_0^{BS} = 1 - h_0^{BP}, \qquad h_i^{BS} = -h_i^{BP} \text{ for } i = 1, 2, \cdots, N-1 \tag{5.51}$$

The bandstop characteristic of $\{h_i^{BS}\}$ can be verified by evaluating its frequency response in terms of that of the FIR bandpass filter:

$$\begin{aligned}H_{BS}(e^{j\omega}) &= 1 - h_0^{BP} - \sum_{i=1}^{N-1} h_i^{BP} e^{-j\omega i} \\ &= 1 - \sum_{i=0}^{N-1} h_i^{BP} e^{-j\omega i} \\ &= 1 - H_{BP}(e^{j\omega})\end{aligned} \tag{5.52}$$

The frequency transformation methods stated above are illustrated in Figure 5.9.

5.5 Least-Squares Design

5.5.1 Quadratic-Measure Minimization

In a typical design of lowpass filters, the desired frequency response of a lowpass filter is given by

$$D(e^{j\omega}) = \begin{cases} e^{-j\theta(\omega)} & \text{for } |\omega| \leq \omega_p \\ 0 & \text{for } |\omega| \geq \omega_s \end{cases} \tag{5.53}$$

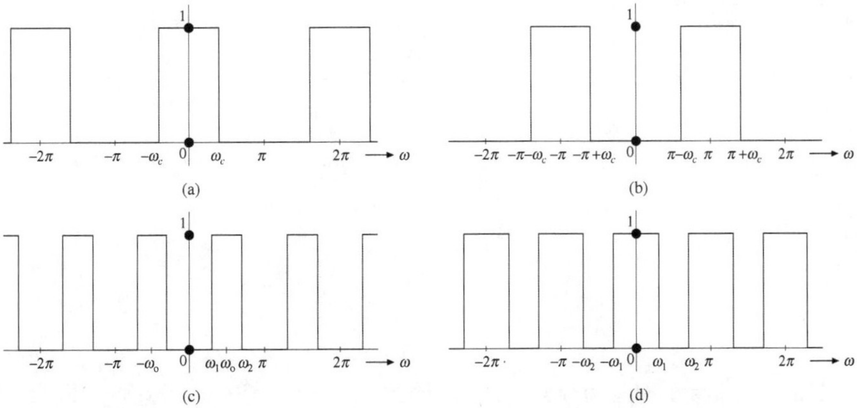

Figure 5.9 Magnitude responses of ideal filters. (a) Ideal lowpass filter. (b) Ideal highpass filter. (c) Ideal bandpass filter. (d) Ideal bandstop filter.

where ω_p and ω_s denote the passband edge and the stopband edge, respectively, and the characteristics of transition band $\omega_p < \omega < \omega_s$ are not specified.

To minimize a quadratic measure of the passband and stopband error in the frequency domain, the total error function is formulated as

$$\begin{aligned} E &= \alpha \int_0^{\omega_p} |D(e^{j\omega}) - H(e^{j\omega})|^2 d\omega + \beta \int_{\omega_s}^{\pi} |H(e^{j\omega})|^2 d\omega \\ &= \alpha E_p + \beta E_s \end{aligned} \tag{5.54}$$

where $H(e^{j\omega})$ is the frequency response of the filter to be designed, E_p and E_s are the passband and stopband errors, respectively, and α and β are positive weighting parameters which control the relative accuracies of approximation in the passband and stopband, respectively. In general, the larger the weighting parameter is, the better the performance of its corresponding band becomes. We remark that weighting parameters can also be functions of frequency ω.

For simplicity, we assume that $H(e^{j\omega})$ is specified by (5.13), the desired phase response is chosen as $\theta(\omega) = \frac{N-1}{2}\omega$ in (5.53), and α and β are chosen to be constant. Then (5.44) can be expressed as

$$\begin{aligned} E &= \alpha E_p + \beta E_s \\ &= \alpha \left(\boldsymbol{c}^T \boldsymbol{A} \boldsymbol{c} - 2\boldsymbol{c}^T \boldsymbol{p} + q \right) + \beta\, \boldsymbol{c}^T \boldsymbol{B} \boldsymbol{c} \end{aligned} \tag{5.55}$$

where

$$\boldsymbol{c} = \begin{pmatrix} c_0 & c_1 & \cdots & c_{\frac{N-1}{2}} \end{pmatrix}^T, \qquad \boldsymbol{\phi}(\omega) = \begin{pmatrix} 1 & \cos\omega & \cdots & \cos(\frac{N-1}{2}\omega) \end{pmatrix}^T$$

$$\boldsymbol{A} = \int_0^{\omega_p} \boldsymbol{\phi}(\omega)\boldsymbol{\phi}(\omega)^T d\omega, \qquad \boldsymbol{p} = \int_0^{\omega_p} \boldsymbol{\phi}(\omega) d\omega, \qquad q = \int_0^{\omega_p} d\omega$$

$$\boldsymbol{B} = \int_{\omega_s}^{\pi} \boldsymbol{\phi}(\omega)\boldsymbol{\phi}(\omega)^T d\omega, \qquad c_0 = h_{\frac{N-1}{2}}$$

$$c_k = 2\, h_{\frac{N-1}{2}-k} \ \text{ for } \ 1 \le k \le (N-1)/2$$

Here, $\boldsymbol{A}$ and $\boldsymbol{B}$ are real, symmetric and positive-definite matrices. By differentiating (5.55) with respect to vector $\boldsymbol{c}$ and setting the result to null, we obtain

$$\frac{dE}{d\boldsymbol{c}} = 2\,(\alpha \boldsymbol{A} + \beta \boldsymbol{B})\,\boldsymbol{c} - 2\alpha \boldsymbol{p} = \boldsymbol{0} \tag{5.56}$$

which leads to

$$\boldsymbol{c} = \left(\boldsymbol{A} + \frac{\beta}{\alpha}\boldsymbol{B}\right)^{-1} \boldsymbol{p} \tag{5.57}$$

This is the optimal least-squares solution which minimizes the quadratic measure in (5.55).

5.5.2 Eigenfilter Method

Notice that in (5.55) $E_s = \boldsymbol{c}^T \boldsymbol{B} \boldsymbol{c}$ is a quadratic form, but $E_p = \boldsymbol{c}^T \boldsymbol{A} \boldsymbol{c} - 2\boldsymbol{p}^T \boldsymbol{c} + q$ is not. To express E_p in a quadratic form, note that the zero-frequency response of the filter in (5.13) is given by

$$H(e^{j0}) = \begin{bmatrix} 1 & 1 & \cdots & 1 \end{bmatrix} \boldsymbol{c} = \boldsymbol{1}^T \boldsymbol{c} \tag{5.58}$$

where vector $\boldsymbol{1}$ is defined as $\boldsymbol{1} = [1\ 1\ \cdots\ 1]^T$. Therefore, the quantity $(\boldsymbol{1} - \boldsymbol{\phi}(\omega))^T \boldsymbol{c}$ represents the deviation of the frequency response $H(e^{j\omega})$ from the zero-frequency response $H(e^{j0})$. As a result, the error measure for the passband can be expressed as a quadratic form, that is,

$$E_p = \boldsymbol{c}^T \boldsymbol{D} \boldsymbol{c} \tag{5.59}$$

where

$$\boldsymbol{D} = \int_0^{\omega_p} \big(\boldsymbol{1} - \boldsymbol{\phi}(\omega)\big)\big(\boldsymbol{1} - \boldsymbol{\phi}(\omega)\big)^T d\omega$$

Hence (5.55) can be written as

$$E = \alpha E_p + \beta E_s = \boldsymbol{c}^T \left(\alpha \boldsymbol{D} + \beta \boldsymbol{B} \right) \boldsymbol{c} \tag{5.60}$$

which is a quadratic form. Since $\alpha\boldsymbol{D}+\beta\boldsymbol{B}$ is a symmetric and positive-definite matrix,

$$\lambda_{min} \leq \frac{\boldsymbol{c}^T \left(\alpha \boldsymbol{D} + \beta \boldsymbol{B} \right) \boldsymbol{c}}{||\boldsymbol{c}||^2} \leq \lambda_{max} \tag{5.61}$$

always holds for any $(N+1)/2 \times 1$ vector $\boldsymbol{c}$ where λ_{min} (λ_{max}) denotes the minimum (maximum) eigenvalue of $\alpha\boldsymbol{D} + \beta\boldsymbol{B}$.

Summarizing the optimal least-squares solution that minimizes E in (5.60) with respect to vector $\boldsymbol{c}$ is given by the eigenvector $\boldsymbol{c}_1$ of $\alpha\boldsymbol{D} + \beta\boldsymbol{B}$ corresponding to its minimum eigenvalue λ_{min}, the minimum value of E is described by $E_{min} = \lambda_{min} \, ||\boldsymbol{c}_1||^2$.

5.6 Analytical Approach

5.6.1 General FIR Filter Design

The frequency transfer function of the filter in (5.2) is given by

$$H(e^{j\omega}) = \sum_{i=0}^{N-1} h_i e^{-j\omega i} \tag{5.62}$$

Given a desired frequency response $D(e^{j\omega})$, designing an FIR filter amounts to obtaining a total of N independent parameters h_i's of (5.62) in such a way that a quadratic measure between the designed filter's magnitude response $H(e^{j\omega})$ and the desired magnitude response $D(e^{j\omega})$ is minimized.

Suppose the frequency grids in the range $0 \leq \omega < 2\pi$ are defined by

$$\omega_l = l \, \frac{2\pi}{M} \quad \text{for } l = 0, 1, \cdots, M-1 \tag{5.63}$$

then a quadratic error measure can be written as

$$\begin{aligned} E &= (\boldsymbol{W}\boldsymbol{h} - \boldsymbol{d})^H \, (\boldsymbol{W}\boldsymbol{h} - \boldsymbol{d}) \\ &= \boldsymbol{h}^T\boldsymbol{W}^H\boldsymbol{W}\boldsymbol{h} - 2\boldsymbol{h}^T \mathrm{Re}\left[\boldsymbol{W}^H\boldsymbol{d}\right] + \boldsymbol{d}^H\boldsymbol{d} \end{aligned} \tag{5.64}$$

where

$$\boldsymbol{W} = \begin{bmatrix} e^{-j0\frac{2\pi}{M}0} & e^{-j0\frac{2\pi}{M}1} & \cdots & e^{-j0\frac{2\pi}{M}(N-1)} \\ e^{-j1\frac{2\pi}{M}0} & e^{-j1\frac{2\pi}{M}1} & \cdots & e^{-j1\frac{2\pi}{M}(N-1)} \\ \vdots & \vdots & \ddots & \vdots \\ e^{-j(M-1)\frac{2\pi}{M}0} & e^{-j(M-1)\frac{2\pi}{M}1} & \cdots & e^{-j(M-1)\frac{2\pi}{M}(N-1)} \end{bmatrix}$$

$$\boldsymbol{h} = (h_0 \;\; h_1 \;\; \cdots \;\; h_{N-1})^T$$

$$\boldsymbol{d} = \left(D(e^{j0\frac{2\pi}{M}}) \;\; D(e^{j1\frac{2\pi}{M}}) \;\; \cdots \;\; D(e^{j(M-1)\frac{2\pi}{M}})\right)^T$$

$\boldsymbol{W}^H$ ($\boldsymbol{d}^H$) denotes the conjugate transpose of matrix $\boldsymbol{W}$ ($\boldsymbol{d}$), and Re[·] stands for the real part of [·]. By differentiating (5.64) with respect to vector $\boldsymbol{h}$ and setting the result to null, we obtain

$$\frac{dE}{d\boldsymbol{h}} = 2\boldsymbol{W}^H\boldsymbol{W}\boldsymbol{h} - 2\mathrm{Re}\left[\boldsymbol{W}^H\boldsymbol{d}\right] = \boldsymbol{0} \tag{5.65}$$

which leads to

$$\boldsymbol{h} = \frac{1}{M}\mathrm{Re}\left[\boldsymbol{W}^H\boldsymbol{d}\right] \tag{5.66}$$

where $\boldsymbol{W}^H\boldsymbol{W} = M\boldsymbol{I}_N$ because

$$(p, q)\text{th element of } \boldsymbol{W}^H\boldsymbol{W} = \sum_{k=0}^{M-1} e^{jk\frac{2\pi}{M}(p-q)} = \begin{cases} M & \text{for} \;\; p = q \\ 0 & \text{for} \;\; p \neq q \end{cases}$$

The expression in (5.66) provides the optimal least-squares solution which minimizes the quadratic measure in (5.64). Note that, due to the orthogonality of $\boldsymbol{W}$, the formula in (5.66) does not require matrix inversion.

5.6.2 Linear-Phase FIR Filter Design

The magnitude response of the filter in (5.13) is described by

$$M(\omega) = |H(e^{j\omega})| = \sum_{k=0}^{\frac{N-1}{2}} c_k \cos(k\omega) \tag{5.67}$$

Given a desired magnitude response $|D(e^{j\omega})|$, the problem of designing an FIR filter is to obtain a total of $(N+1)/2$ independent parameters c_k's of (5.67) that minimize a quadratic measure between the designed filter's magnitude response $M(\omega)$ and the desired magnitude response $|D(e^{j\omega})|$ over $0 \le \omega \le \pi$.

Suppose the frequency grids in the range $0 \le \omega \le \pi$ are defined by

$$\omega_l = l\frac{\pi}{M} \quad \text{for } l = 0, 1, \cdots, M \tag{5.68}$$

with $M > (N-1)/2$, then a quadratic error measure can be written as

$$\begin{aligned} E &= (\boldsymbol{V}\boldsymbol{c} - \boldsymbol{d})^T(\boldsymbol{V}\boldsymbol{c} - \boldsymbol{d}) \\ &= \boldsymbol{c}^T\boldsymbol{V}^T\boldsymbol{V}\boldsymbol{c} - 2\boldsymbol{c}^T\boldsymbol{V}^T\boldsymbol{d} + \boldsymbol{d}^T\boldsymbol{d} \end{aligned} \tag{5.69}$$

where

$$\boldsymbol{V} = \begin{bmatrix} \cos(0\frac{\pi}{M}0) & \cos(0\frac{\pi}{M}1) & \cdots & \cos(0\frac{\pi}{M}\frac{N-1}{2}) \\ \cos(1\frac{\pi}{M}0) & \cos(1\frac{\pi}{M}1) & \cdots & \cos(1\frac{\pi}{M}\frac{N-1}{2}) \\ \vdots & \vdots & \ddots & \vdots \\ \cos(M\frac{\pi}{M}0) & \cos(M\frac{\pi}{M}1) & \cdots & \cos(M\frac{\pi}{M}\frac{N-1}{2}) \end{bmatrix}$$

$$\boldsymbol{c} = \left(c_0 \;\; c_1 \;\; \cdots \;\; c_{\frac{N-1}{2}}\right)^T$$

$$\boldsymbol{d} = \left(|D(e^{j0\frac{\pi}{M}})| \;\; |D(e^{j1\frac{\pi}{M}})| \;\; \cdots \;\; |D(e^{jM\frac{\pi}{M}})| \right)^T$$

By differentiating (5.69) with respect to $\boldsymbol{c}$ and setting the result to null, we obtain

$$\frac{dE}{d\boldsymbol{c}} = 2\boldsymbol{V}^T\boldsymbol{V}\boldsymbol{c} - 2\boldsymbol{V}^T\boldsymbol{d} = \boldsymbol{0} \tag{5.70}$$

which leads to

$$\boldsymbol{c} = \left(\boldsymbol{V}^T\boldsymbol{V}\right)^{-1}\boldsymbol{V}^T\boldsymbol{d} \tag{5.71}$$

We now define a matrix $\boldsymbol{R} = [R_{ij}]$ for $i, j = 0, 1, \cdots, (N-1)/2$ as

$$\boldsymbol{R} = \boldsymbol{V}^T\boldsymbol{V} \tag{5.72}$$

Obviously, matrix $\boldsymbol{R}$ is symmetric whose (i, j)th element is given by

$$R_{ij} = \sum_{k=0}^{M} \cos\left(k\frac{\pi}{M}i\right)\cos\left(k\frac{\pi}{M}j\right) \tag{5.73}$$

In addition, because $M > (N-1)/2$, $\boldsymbol{R}$ is nonsingular, hence $\boldsymbol{R}^{-1}$ exists and is also symmetric. By substituting (5.72) into (5.71), we obtain

$$\boldsymbol{c} = \boldsymbol{R}^{-1}\boldsymbol{V}^T\boldsymbol{d} \tag{5.74}$$

Table 5.1 $\boldsymbol{R}^{-1} = [\lambda_{ij}]$ for $0 \le i, j \le N'$ and $N' < M$ with $N' = \frac{N-1}{2}$

i, j	λ_{ij} (N' odd)	λ_{ij} (N' even)
$i = j = 0$	$\frac{M + N' - 1}{M(M + N')}$	$\frac{M + N'}{M(M + N' + 1)}$
$i = 0$ and j even, or $j = 0$ and i even	$-\frac{2}{M(M + N')}$	$-\frac{2}{M(M + N' + 1)}$
$i = j \neq 0$ and i, j odd	$\frac{2(M + N' - 1)}{M(M + N' + 1)}$	$\frac{2(M + N' - 2)}{M(M + N')}$
$i = j \neq 0$ and i, j even	$\frac{2(M + N' - 2)}{M(M + N')}$	$\frac{2(M + N' - 1)}{M(M + N' + 1)}$
$i \neq j$ and i, j odd	$-\frac{4}{M(M + N' + 1)}$	$-\frac{4}{M(M + N')}$
$i \neq j$ and i, j even	$-\frac{4}{M(M + N')}$	$-\frac{4}{M(M + N' + 1)}$
$(i + j)$ odd	0	0

The elements of $\boldsymbol{R}^{-1} = \{\lambda_{ij} \,|\, 0 \le i, j \le (N-1)/2\}$ can be found in Table 5.1 [6].

5.7 Chebyshev Approximation

5.7.1 The Parks-McClellan Algorithm

From (5.13), the frequency response $H(e^{j\omega})$ of a causal linear-phase FIR filter of even order $N - 1$ is described by (Type 1)

$$H(e^{j\omega}) = M(\omega) e^{-j\frac{N-1}{2}\omega} \tag{5.75}$$

where

$$M(\omega) = \sum_{k=0}^{\frac{N-1}{2}} c_k \cos(k\omega)$$

Here, the amplitude response $M(\omega)$ of the filter is a real function of frequency ω. Given a desired amplitude response $D(\omega)$, the weighted error function is defined as

$$\varepsilon(\omega) = W(\omega)\left[M(\omega) - D(\omega)\right] \tag{5.76}$$

where $W(\omega)$ is a positive weighting function which controls the relative size of the peak error in the specified frequency band. The problem of designing an FIR digital filter, in this case, is to iteratively adjust the coefficients c_i's of the amplitude response $M(\omega)$ so that the peak absolute value of $\varepsilon(\omega)$ is minimized.

Suppose the minimum of the peak absolute value of $\varepsilon(\omega)$ in a band $\omega_a \leq \omega \leq \omega_b$ is ε_o, then the absolute value satisfies

$$|\varepsilon(\omega)| = |W(\omega)|\,|M(\omega) - D(\omega)| \leq \varepsilon_o \ \text{ for } \omega_a \leq \omega \leq \omega_b \tag{5.77}$$

Typically, the desired amplitude response is specified by

$$D(\omega) = \begin{cases} 1, & \text{in passband} \\ 0, & \text{in stopband} \end{cases} \tag{5.78}$$

and it is also required that the amplitude response $M(\omega)$ satisfies the above desired response with a ripple of $\pm\delta_p$ in the passband and a ripple of δ_s in the stopband. Hence, from (5.76) the weighting function can be chosen as either of

$$W(\omega) = \begin{cases} 1, & \text{in passband} \\ \delta_p/\delta_s, & \text{in stopband} \end{cases} \tag{5.79}$$

and

$$W(\omega) = \begin{cases} \delta_s/\delta_p, & \text{in passband} \\ 1, & \text{in stopband} \end{cases} \tag{5.80}$$

The optimization problem encountered here is to determine the coefficients c_i's of $M(\omega)$ in (5.75) that minimize the peak absolute value ε of the weighted approximation error $\varepsilon(\omega)$ of (5.76) over specified frequency bands R. As will be shown below, this problem can be solved by applying the alternation theorem from the theory of Chebyshev approximation [3].

5.7.2 Alternation Theorem

The alternation theorem [3] can be stated as follows: the amplitude response $M(\omega)$ in (5.75) obtained by minimizing the peak absolute value ε of $\varepsilon(\omega)$ in (5.76) is the optimal unique approximation of the desired amplitude response if and only if there are at least $(N+3)/2$ extremal frequencies $\omega_0, \omega_1, \cdots, \omega_{\frac{N+1}{2}}$ in a closed subset R of the frequency range $0 \leq \omega \leq \pi$ such that

$\omega_0 < \omega_1 < \cdots < \omega_{\frac{N-1}{2}} < \omega_{\frac{N+1}{2}}$ and $\varepsilon(\omega_i) = -\varepsilon(\omega_{i+1})$ with $|\varepsilon(\omega_i)| = \varepsilon$ for all i in the range $0 \leq i \leq (N+1)/2$.

When the approximation error $\varepsilon(\omega)$ for amplitude response $M(\omega)$ satisfies the condition of the above theorem, the peaks of $\varepsilon(\omega)$ occur at $\omega = \omega_i$ for $0 \leq i \leq (N+1)/2$ in which

$$\frac{d\varepsilon(\omega)}{d\omega} = 0 \tag{5.81}$$

Since $W(\omega)$ and $D(\omega)$ are piecewise constant in the passband and the stopband, from (5.76) it follows that

$$\left.\frac{d\varepsilon(\omega)}{d\omega}\right|_{\omega=\omega_i} = \left.\frac{dM(\omega)}{d\omega}\right|_{\omega=\omega_i} = 0 \tag{5.82}$$

which implies that the magnitude response $M(\omega)$ also has peaks at $\omega = \omega_i$. The Chebyshev polynomials of first kind is defined as

$$T_n(x) = \cos(n\omega) \;\text{ with }\; x = \cos(\omega) \tag{5.83}$$

hence

$$\begin{aligned}
&T_0(x) = 1 && \text{because} && \cos(0\,\omega) = 1\\
&T_1(x) = x && \text{because} && \cos(1\omega) = \cos(\omega)\\
&T_2(x) = 2x^2 - 1 && \text{because} && \cos(2\omega) = 2\cos^2(\omega) - 1\\
&T_3(x) = 4x^3 - 3x && \text{because} && \cos(3\omega) = 4\cos^3(\omega) - 3\cos(\omega)\\
&\;\;\vdots\\
&T_{n+1}(x) = 2xT_n(x) - T_{n-1}(x) && \text{for } n = 1, 2, \cdots && \text{in general}
\end{aligned} \tag{5.84}$$

The amplitude response $M(\omega)$ in (5.75) can be expressed as a power series in $\cos(\omega)$, i.e.,

$$M(\omega) = \sum_{k=0}^{\frac{N-1}{2}} \alpha_k \cos^k(\omega) \tag{5.85}$$

This equation is a polynomial of order $(N-1)/2$ in $\cos(\omega)$, hence $M(\omega)$ can have at most $(N-3)/2$ local minima inside the specified passband and stopband. Also, note that $|\varepsilon(\omega)|$ is a maximum at the band edges $\omega = \omega_p$ and $\omega = \omega_s$ and hence, $M(\omega)$ has extrema at these frequencies. Moreover, $M(\omega)$ may also have extrema at $\omega = 0$ and $\omega = \pi$. As a result, there exist at most $(N+3)/2$ extremal frequencies.

To obtain the optimal solution for the unknown c_k's and ε under the assumption that the $(N+3)/2$ extremal frequencies are known, we need to solve the set of $(N+3)/2$ equations

$$W(\omega_i)\left[M(\omega_i) - D(\omega_i)\right] = (-1)^i\varepsilon \quad \text{for } 0 \le i \le (N+1)/2 \tag{5.86}$$

which is equivalent to

$$\begin{bmatrix} 1 & \cos(\omega_0) & \cdots & \cos(\frac{N-1}{2}\omega_0) & \frac{-1}{W(\omega_0)} \\ 1 & \cos(\omega_1) & \cdots & \cos(\frac{N-1}{2}\omega_1) & \frac{1}{W(\omega_1)} \\ \vdots & \vdots & \ddots & \vdots & \vdots \\ 1 & \cos\left(\omega_{\frac{N-1}{2}}\right) & \cdots & \cos\left(\frac{N-1}{2}\omega_{\frac{N-1}{2}}\right) & \frac{(-1)^{\frac{N-3}{2}}}{W\left(\omega_{\frac{N-1}{2}}\right)} \\ 1 & \cos\left(\omega_{\frac{N+1}{2}}\right) & \cdots & \cos\left(\frac{N+1}{2}\omega_{\frac{N+1}{2}}\right) & \frac{(-1)^{\frac{N-1}{2}}}{W\left(\omega_{\frac{N+1}{2}}\right)} \end{bmatrix} \begin{bmatrix} c_0 \\ c_1 \\ \vdots \\ c_{\frac{N-1}{2}} \\ \varepsilon \end{bmatrix}$$

$$= \begin{bmatrix} D(\omega_0) \\ D(\omega_1) \\ \vdots \\ D\left(\omega_{\frac{N-1}{2}}\right) \\ D\left(\omega_{\frac{N+1}{2}}\right) \end{bmatrix} \tag{5.87}$$

The above simultaneous equations can be solved in principle for the unknown parameters provided that the locations of the $(N+3)/2$ extremal frequencies are known a priori. This problem is resolved by the *Remez exchange algorithm* outlined below.

The Remez Exchange Algorithm:

This algorithm is a very efficient iterative procedure for determining the locations of the extremal frequencies, and is composed of the following steps.

Step 1: Choose a set of initial values for the extremal frequencies, or use the values available from the termination of the previous iteration.

Step 2: Compute the value ε by solving (5.87).
Step 3: Compute the values of the amplitude response $M(\omega)$ at $\omega = \omega_i$ using

$$M(\omega_i) = \frac{(-1)^i \varepsilon}{W(\omega_i)} + D(\omega_i) \quad \text{for } 0 \le i \le (N+1)/2 \tag{5.88}$$

Step 4: Determine the polynomial $M(\omega)$ by interpolating the above values at the $(N+3)/2$ extremal frequencies using the Lagrange interpolation formula

$$M(\omega) = \sum_{k=0}^{\frac{N+1}{2}} M(\omega_k) P_k[\cos(\omega)] \tag{5.89}$$

where

$$P_k[\cos(\omega)] = \prod_{l=0,\, l \ne k}^{\frac{N+1}{2}} \frac{\cos(\omega) - \cos(\omega_l)}{\cos(\omega_k) - \cos(\omega_l)} \quad \text{for } 0 \le k \le (N+1)/2$$

Step 5: Compute the new weighted error function $\varepsilon(\omega)$ of (5.76) at a dense set S of frequencies where $S >> (N-1)/2$ and the transition band is excluded. Setting $S \simeq 16(N-1)$ is adequate in practice.
Step 6: Determine the $(N+3)/2$ new extremal frequencies from the values of $\varepsilon(\omega)$ evaluated at the dense set of frequencies.
Step 7: Stop if the peak values ε are approximately equal. Otherwise, go back to Step 2.

The above arguments can be applied to the other type of linear-phase FIR digital filtes (Type 2, Type 3, and Type 4) with the slight modifications of the algorithm [3].

5.8 Cascaded Lattice Realization of FIR Digital Filters

For the cascaded lattice realization, an $(N-1)$th-order FIR transfer function is assumed to be of the form

$$H_N(z) = h_0 + h_1 z^{-1} + \cdots + h_{N-1} z^{-(N-1)} \tag{5.90}$$

which is related to

$$\begin{aligned} \tilde{H}_N(z) &= h_{N-1} + h_{N-2} z^{-1} + \cdots + h_0 z^{-(N-1)} \\ &= z^{-(N-1)} H_N(z^{-1}) \end{aligned} \tag{5.91}$$

From (5.90) and (5.91), it follows that

$$\begin{aligned} h_{N-1}H_N(z) - h_0\tilde{H}_N(z) &= z^{-1}\left[h'_1 + h'_2 z^{-1} + \cdots + h'_{N-1}z^{-(N-2)}\right] \\ &= z^{-1}H_{N-1}(z) \\ h_{N-1}\tilde{H}_N(z) - h_0 H_N(z) &= h'_{N-1} + h'_{N-2}z^{-1} + \cdots + h'_1 z^{-(N-2)} \\ &= \tilde{H}_{N-1}(z) \end{aligned} \tag{5.92}$$

which is equivalent to

$$\begin{bmatrix} h_{N-1} & -h_0 \\ -h_0 & h_{N-1} \end{bmatrix} \begin{bmatrix} H_N(z) \\ \tilde{H}_N(z) \end{bmatrix} = \begin{bmatrix} z^{-1}H_{N-1}(z) \\ \tilde{H}_{N-1}(z) \end{bmatrix} \tag{5.93}$$

where

$$\begin{bmatrix} h'_1 \\ h'_2 \\ \vdots \\ h'_{N-1} \end{bmatrix} = \begin{bmatrix} h_1 h_{N-1} - h_0 h_{N-2} \\ h_2 h_{N-1} - h_0 h_{N-3} \\ \vdots \\ h_{N-1}^2 - h_0^2 \end{bmatrix}$$

Assuming that $h_{N=1} \neq \pm h_0$, (5.93) can be expressed as

$$\begin{bmatrix} H_N(z) \\ \tilde{H}_N(z) \end{bmatrix} = \Delta_0 \begin{bmatrix} 1 & k_0 \\ k_0 & 1 \end{bmatrix} \begin{bmatrix} z^{-1} & 0 \\ 0 & 1 \end{bmatrix} \begin{bmatrix} H_{N-1}(z) \\ \tilde{H}_{N-1}(z) \end{bmatrix} \tag{5.94}$$

where

$$k_0 = \frac{h_0}{h_{N-1}}, \qquad \Delta_0 = \frac{h_{N-1}}{(h_{N-1} + h_0)(h_{N-1} - h_0)}$$

A block diagram of the system in (5.94) is illustrated in Figure 5.10.

Similarly, from the transfer functions $H_{N-1}(z)$ and $\tilde{H}_{N-1}(z)$, we obtain

$$\begin{bmatrix} H_{N-1}(z) \\ \tilde{H}_{N-1}(z) \end{bmatrix} = \Delta_1 \begin{bmatrix} 1 & k_1 \\ k_1 & 1 \end{bmatrix} \begin{bmatrix} z^{-1} & 0 \\ 0 & 1 \end{bmatrix} \begin{bmatrix} H_{N-2}(z) \\ \tilde{H}_{N-2}(z) \end{bmatrix} \tag{5.95}$$

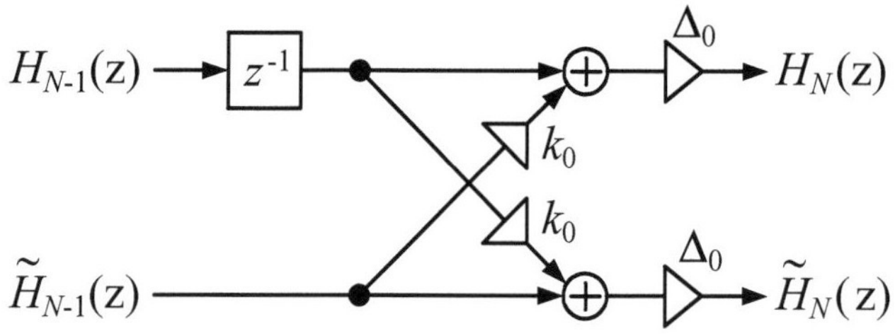

Figure 5.10 Normalized lattice structure of a section.

where

$$k_1 = \frac{h'_1}{h'_{N-1}}, \qquad \Delta_1 = \frac{h'_{N-1}}{(h'_{N-1} + h'_1)(h'_{N-1} - h'_1)}$$

Eventually, we arrive at

$$\begin{aligned}\begin{bmatrix} H_2(z) \\ \tilde{H}_2(z) \end{bmatrix} &= \Delta_{N-2} \begin{bmatrix} 1 & k_{N-2} \\ k_{N-2} & 1 \end{bmatrix} \begin{bmatrix} z^{-1} & 0 \\ 0 & 1 \end{bmatrix} \begin{bmatrix} H_1(z) \\ \tilde{H}_1(z) \end{bmatrix} \\ &= \Delta_{N-2}\Delta_{N-1} \begin{bmatrix} 1 & k_{N-2} \\ k_{N-2} & 1 \end{bmatrix} \begin{bmatrix} z^{-1} \\ 1 \end{bmatrix}\end{aligned} \tag{5.96}$$

where $H_1(z) = \tilde{H}_1(z) = \Delta_{N-1}$. By substituting (5.95) into (5.94), we obtain

$$\begin{bmatrix} H_N(z) \\ \tilde{H}_N(z) \end{bmatrix} = \Delta_0\Delta_1 \begin{bmatrix} 1 & k_0 \\ k_0 & 1 \end{bmatrix} \begin{bmatrix} z^{-1} & 0 \\ 0 & 1 \end{bmatrix} \begin{bmatrix} 1 & k_1 \\ k_1 & 1 \end{bmatrix} \begin{bmatrix} z^{-1} & 0 \\ 0 & 1 \end{bmatrix} \begin{bmatrix} H_{N-2}(z) \\ \tilde{H}_{N-2}(z) \end{bmatrix} \tag{5.97}$$

A block diagram of the system in (5.97) is shown in Figure 5.11. Moreover, by making use of (5.96) and (5.97), we have

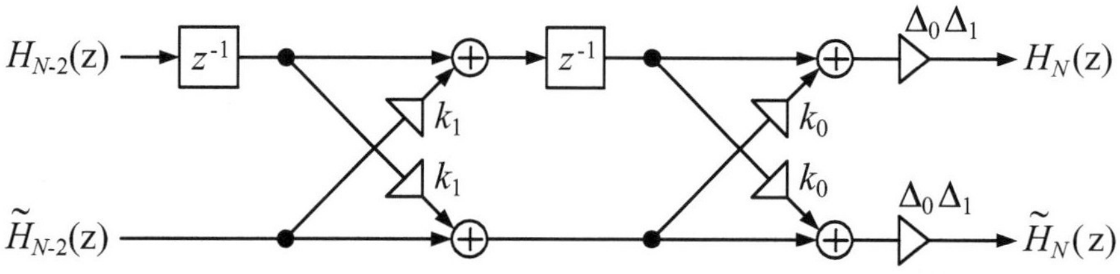

Figure 5.11 Normalized lattice structure of cascaded two sections.

$$\begin{bmatrix} H_N(z) \\ \tilde{H}_N(z) \end{bmatrix} = \Delta \begin{bmatrix} 1 & k_0 \\ k_0 & 1 \end{bmatrix} \begin{bmatrix} z^{-1} & 0 \\ 0 & 1 \end{bmatrix} \begin{bmatrix} 1 & k_1 \\ k_1 & 1 \end{bmatrix} \begin{bmatrix} z^{-1} & 0 \\ 0 & 1 \end{bmatrix} \cdots \begin{bmatrix} 1 & k_{N-2} \\ k_{N-2} & 1 \end{bmatrix} \begin{bmatrix} z^{-1} \\ 1 \end{bmatrix} \quad (5.98)$$

where $\Delta = \Delta_0 \Delta_1 \cdots \Delta_{N-1}$. A block diagram of the system in (5.98) is depicted in Figure 5.12.

Next, we consider a linear-phase FIR digital filter with even integer N, i.e., odd order $N-1$. In this case, since $h_i = \pm h_{N-1-i}$ holds for $i = 0, 1, \cdots, N-1$, we can write

$$\begin{aligned} H(z) = \sum_{i=0}^{N-1} h_i z^{-i} &= \sum_{i=0}^{\frac{N}{2}-1} h_i z^{-i} \pm \sum_{i=\frac{N}{2}}^{N-1} h_{N-1-i} z^{-i} \\ &= H_{\frac{N}{2}}(z) \pm z^{-\frac{N}{2}} \tilde{H}_{\frac{N}{2}}(z) \end{aligned} \quad (5.99)$$

where positive sign is for the symmetric case, while negative sign is for the antisymmetric case and

$$H_{\frac{N}{2}}(z) = h_0 + h_1 z^{-1} + \cdots + h_{\frac{N}{2}-1} z^{-(\frac{N}{2}-1)}$$

$$\tilde{H}_{\frac{N}{2}}(z) = h_{\frac{N}{2}-1} + h_{\frac{N}{2}-2} z^{-1} + \cdots + h_0 z^{-(\frac{N}{2}-1)}$$

A block diagram of the system in (5.99) is drawn in Figure 5.13 where the $(N/2-1)$th-order lattice structure is used and $\Delta = \Delta_0 \Delta_1 \cdots \Delta_{N/2-1}$.

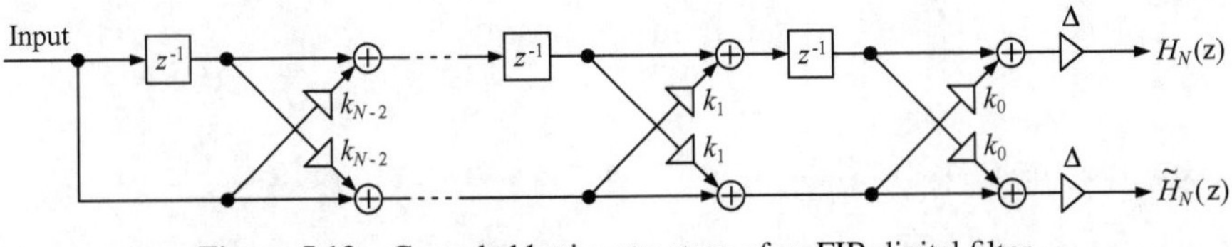

Figure 5.12 Cascaded lattice structure of an FIR digital filter.

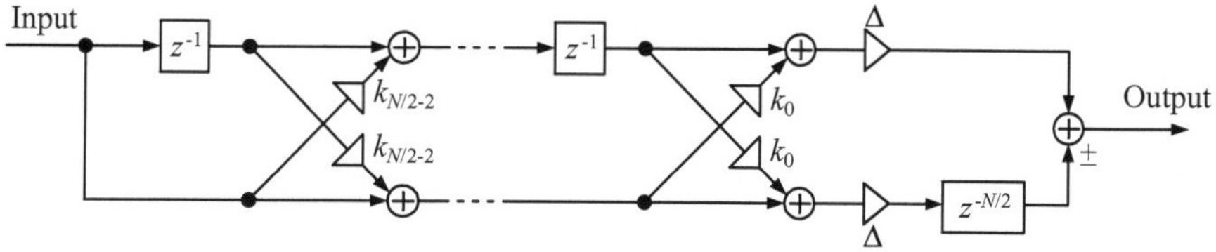

Figure 5.13 The lattice structure of linear-phase FIR digital filters.

Notice that since $h_0 = \pm h_{N-1}$, $h_{N-1}^2 - h_0^2 = 0$ holds, the $(N-1)$th-order lattice structure is not available in the linear-phase FIR digital filter.

5.9 Numerical Experiments

As a numerical example, Suppose that the desired frequency response of a lowpass digital filter is specified by

$$D(e^{j\omega}) = \begin{cases} e^{-j\frac{N-1}{2}\omega} & \text{for} \quad |\omega| \le \omega_p \\ 0 & \text{for} \quad |\omega| \ge \omega_s \end{cases}$$

where the passband edge and the stopband edge are $\omega_p = 0.3\pi$ and $\omega_s = 0.35\pi$, respectively, and the order of the filter is assumed to be $N-1 = 30$.

5.9.1 Least-Squares Design

5.9.1.1 Quadratic measure minimization

When the weighting parameters in (5.54) were chosen as $\alpha = \beta = 1/2$, $\mathbf{c} = (c_0, c_1, \cdots, c_{15})^T = (h_{15}, 2h_{14}, \cdots, 2h_0)^T$ was computed from (5.57) yielding

$$\begin{bmatrix} h_0 & h_1 & h_2 & h_3 \\ h_4 & h_5 & h_6 & h_7 \\ h_8 & h_9 & h_{10} & h_{11} \\ h_{12} & h_{13} & h_{14} & h_{15} \end{bmatrix} = 10^{-1} \begin{bmatrix} 0.058420 & 0.154451 & 0.109441 & -0.069545 \\ -0.226197 & -0.181544 & 0.078904 & 0.339349 \\ 0.311192 & -0.085985 & -0.565692 & -0.620779 \\ 0.090398 & 1.411959 & 2.704779 & 3.241437 \end{bmatrix}$$

The magnitude response of the resulting filter is shown in Figure 5.14.

5.9.1.2 Eigenfilter method

The same design problem was addressed using the eigenfilter method. In this case, the eigenvector $\mathbf{c}_1 = (c_0, c_1, \cdots, c_{15})^T = (h_{15}, 2h_{14}, \cdots, 2h_0)^T$

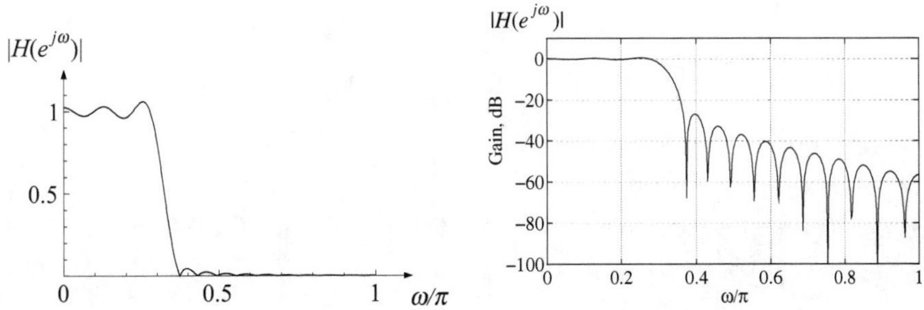

Figure 5.14 The magnitude response of the resulting filter.

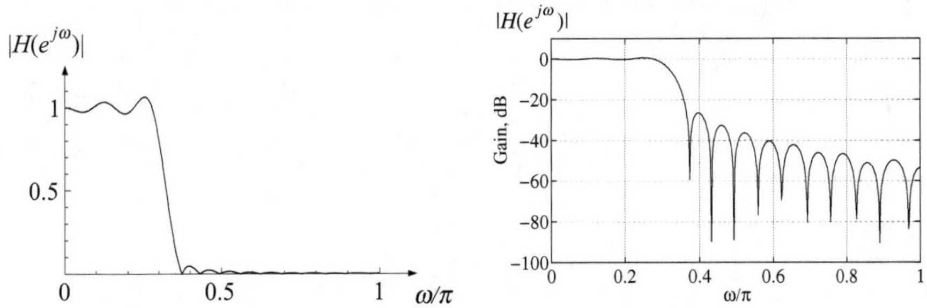

Figure 5.15 The magnitude response of the resulting filter.

corresponding to the minimum eigenvalue $\lambda_{min} = 2.392118 \times 10^{-3}$ which satisfies (5.61) was computed as

$$\kappa \begin{bmatrix} h_0 & h_1 & h_2 & h_3 \\ h_4 & h_5 & h_6 & h_7 \\ h_8 & h_9 & h_{10} & h_{11} \\ h_{12} & h_{13} & h_{14} & h_{15} \end{bmatrix} = 10^{-1} \begin{bmatrix} 0.048062 & 0.145306 & 0.101539 & -0.077514 \\ -0.235591 & -0.192292 & 0.068380 & 0.330571 \\ 0.303905 & -0.093823 & -0.575924 & -0.632817 \\ 0.079784 & 1.406269 & 2.704706 & 3.238881 \end{bmatrix}$$

where $\kappa = D(1)/\mathbf{1}^T \boldsymbol{c}_1$ so that $D(1) = H(1) = 1$. The magnitude response of the resulting filter is shown in Figure 5.15.

5.9.2 Analytical Approach

5.9.2.1 General FIR filter design

Consider designing a lowpass digital filter that approximates the desired frequency response

$$
D(e^{j\omega}) = \begin{cases} e^{-j\frac{N-1}{2}\omega} & \text{for} \quad 0 \le \omega \le 0.3\pi \\ (-\dfrac{20}{\pi}\omega + 7)\, e^{-j\frac{N-1}{2}\omega} & \text{for} \quad 0.3\pi < \omega < 0.35\pi \\ 0 & \text{for} \quad 0.35\pi \le \omega| \le 1.65\pi \\ (\dfrac{20}{\pi}\omega - 33)\, e^{-j\frac{N-1}{2}\omega} & \text{for} \quad 1.65\pi < \omega < 1.7\pi \\ e^{-j\frac{N-1}{2}\omega} & \text{for} \quad 1.7\pi \le \omega < 2\pi \end{cases}
$$

and $M = 200$ was chosen in (5.63). The coefficient vector $\boldsymbol{h} = (h_0, h_1, \cdots, h_{30})^T$ was computed from (5.66) as

$$
\begin{bmatrix} h_0 & h_1 & h_2 & h_3 \\ h_4 & h_5 & h_6 & h_7 \\ h_8 & h_9 & h_{10} & h_{11} \\ h_{12} & h_{13} & h_{14} & h_{15} \\ h_{16} & h_{17} & h_{18} & h_{19} \\ h_{20} & h_{21} & h_{22} & h_{23} \\ h_{24} & h_{25} & h_{26} & h_{27} \\ h_{28} & h_{29} & h_{30} & \end{bmatrix} = 10^{-1} \begin{bmatrix} 0.064876 & 0.184934 & 0.134657 & -0.071201 \\ -0.250137 & -0.204317 & 0.076366 & 0.355871 \\ 0.329948 & -0.080191 & -0.574339 & -0.634091 \\ 0.082541 & 1.412721 & 2.711472 & 3.250000 \\ 2.711472 & 1.412721 & 0.082541 & -0.634091 \\ -0.574339 & -0.080191 & 0.329948 & 0.355871 \\ 0.076366 & -0.204317 & -0.250137 & -0.071201 \\ 0.134657 & 0.184934 & 0.064876 & \end{bmatrix}
$$

The magnitude response of the resulting filter is drawn in Figure 5.16.

5.9.2.2 Linear-Phase FIR filter design

We now consider designing a lowpass digital filter that approximates the desired frequency response

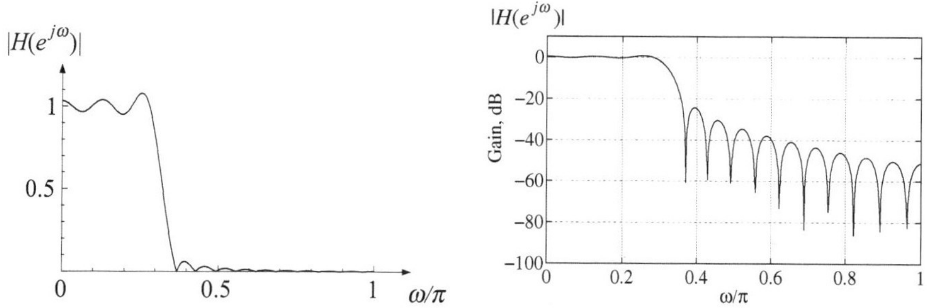

Figure 5.16 The magnitude response of the resulting filter.

$$D(e^{j\omega}) = \begin{cases} e^{-j\frac{N-1}{2}\omega} & \text{for} \quad 0 \le \omega \le 0.3\pi \\ (-\dfrac{20}{\pi}\omega + 7)\, e^{-j\frac{N-1}{2}\omega} & \text{for} \quad 0.3\pi < \omega < 0.35\pi \\ 0 & \text{for} \quad 0.35\pi \le \omega \le \pi \end{cases}$$

and $M = 100$ was chosen in (5.68). The vector $\boldsymbol{c} = (c_0, c_1, \cdots, c_{15})^T = (h_{15}, 2h_{14}, \cdots, 2h_0)^T$ was computed from (5.74) as

$$\begin{bmatrix} h_0 & h_1 & h_2 & h_3 \\ h_4 & h_5 & h_6 & h_7 \\ h_8 & h_9 & h_{10} & h_{11} \\ h_{12} & h_{13} & h_{14} & h_{15} \end{bmatrix} = 10^{-1} \begin{bmatrix} 0.063576 & 0.183391 & 0.133357 & -0.072744 \\ -0.251437 & -0.205860 & 0.075066 & 0.354328 \\ 0.328648 & -0.081734 & -0.575639 & -0.635634 \\ 0.081241 & 1.411178 & 2.710172 & 3.248457 \end{bmatrix}$$

The magnitude response of the resulting filter is shown in Figure 5.17.

5.9.3 Chebyshev Approximation

The FIR digital filter was described by (5.75) with $N = 31$. The set of frequencies S in Step 5 of the Remez exchange algorithm was chosen to be

$$S = \{\omega_i \in \Omega \,|\, 0 \le \omega_i \le \omega_p\} \cup \{\omega_i \in \Omega \,|\, \omega_s \le \omega_i \le \pi\}$$

where $\Omega = \{\omega_i = (\pi/500)i \,|\, i = 0, 1, \cdots, 500\}$. With $\omega_p = 0.3\pi$ and $\omega_s = 0.35\pi$, we have

$$\{\omega_i \in \Omega \,|\, 0 \le \omega_i \le \omega_p\} = \{\omega_i \,|\, i = 0, 1, \cdots, 150\}$$

$$\{\omega_i \in \Omega \,|\, \omega_s \le \omega_i \le \pi\} = \{\omega_i \,|\, i = 175, 176, \cdots, 500\}$$

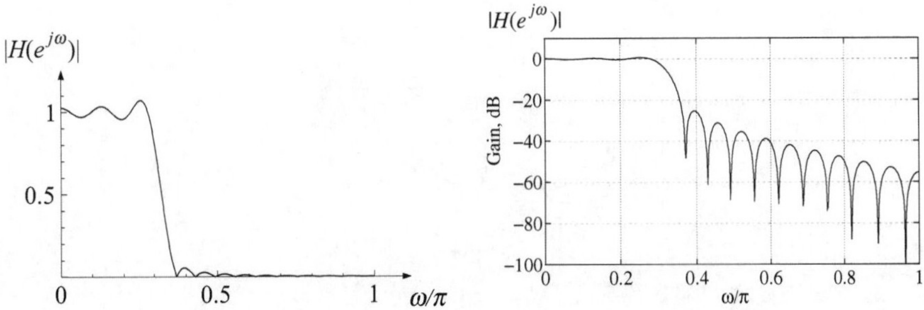

Figure 5.17 The magnitude response of the resulting filter.

We now denote the elements of set S as $S = \{\omega'_i \mid i = 0, 1, \cdots, 476\}$ where

$$\omega'_i = \begin{cases} (\pi/500)i & \text{for} \quad i = 0, 1, \cdots, 150 \\ (\pi/500)(i+24) & \text{for} \quad i = 151, 152, \cdots, 476 \end{cases}$$

In the Remez exchange algorithm, the initial values for the extremal frequencies were chosen to be

$$\omega'_{round\{(N_s-1)/16\}r} \text{ for } r = 0, 1, \cdots, 16$$

where $N_s = 477$ is the number of elements in set S and $(N+1)/2 = 16$. In other words, the extremal frequencies were expressed in terms of the ω_i's as

$$\begin{bmatrix} \omega_0 & \omega_{30} & \omega_{60} & \omega_{89} & \omega_{119} & \omega_{149} & \omega_{203} & \omega_{232} & \omega_{262} \\ \omega_{292} & \omega_{322} & \omega_{351} & \omega_{381} & \omega_{411} & \omega_{441} & \omega_{470} & \omega_{500} & \end{bmatrix}$$

Then vector $\boldsymbol{c} = (c_0, c_1, \cdots, c_{15})^T = (h_{15}, 2h_{14}, \cdots, 2h_0)^T$ was computed from (5.87) and we obtained

$$\begin{bmatrix} h_0 & h_1 & h_2 & h_3 \\ h_4 & h_5 & h_6 & h_7 \\ h_8 & h_9 & h_{10} & h_{11} \\ h_{12} & h_{13} & h_{14} & h_{15} \end{bmatrix} = 10^{-1} \begin{bmatrix} -0.034827 & 0.601336 & 0.158336 & -0.069666 \\ -0.248594 & -0.203219 & 0.075809 & 0.353959 \\ 0.328456 & -0.079881 & -0.572778 & -0.632930 \\ 0.082451 & 1.412009 & 2.711147 & 3.250054 \end{bmatrix}$$

The magnitude response of the resulting filter is shown in Figure 5.18.

5.9.4 Comparison of Algorithms' Performances

In order to compare their performances, the design results obtained above are summarized in Table 5.2 where

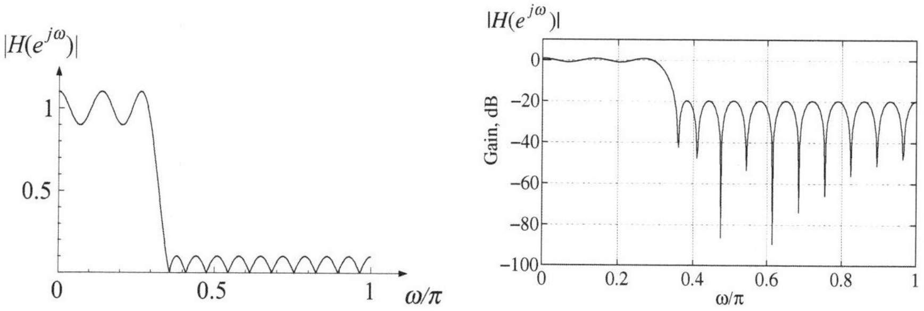

Figure 5.18 The magnitude response of the resulting filter.

Table 5.2 Performance comparisons among algorithms

Algorithms	ε_2	ε_∞	Max. Negative Ripple on Range $0 \le i \le 1000$
Quadratic Measure Minimization	5.170751	20.051857	–0.045462
Eigenfilter Method	5.246651	19.864566	–0.046884
General FIR Filter Design	5.349644	17.245433	–0.059975
Linear-Phase FIR Filter Design	5.345929	17.273949	–0.060041
Chebyshev Approximation	12.731666	10.142531	–0.101418

$$\varepsilon_2 = \frac{\sqrt{\sum_{i=0}^{1000} \left(|D(e^{j\omega_i})| - |H(e^{j\omega_i})|\right)^2}}{\sqrt{\sum_{i=0}^{1000} |D(e^{j\omega_i})|^2}} \times 100$$

$$\varepsilon_\infty = \frac{\max_{0 \le i \le 1000} \left| |D(e^{j\omega_i})| - |H(e^{j\omega_i})| \right|}{\max_{0 \le i \le 1000} |D(e^{j\omega_i})|} \times 100$$

where $\omega_i = \pi i/1000$ for $i = 0, 1, 2, \cdots, 1000$ and set $D(e^{j\omega_i}) = H(e^{j\omega_i}) = 0$ for ω_i in the transition band $0.3\pi < \omega_i < 0.35\pi$.

5.10 Summary

This chapter has shown that exact linear-phase responses can be achieved by imposing either symmetric or antisymmetric condition on the FIR filter's coefficients. Several window functions have been introduced with their application to FIR digital filter design. An approach for designing least squares linear-phase FIR digital filters that minimize a quadratic measure has been studied. An eigenfilter method for designing least squares linear-phase FIR digital filters has been presented in which an eigenvector corresponding to the minimum eigenvalue of a symmetric positive-definite matrix has been computed to obtain the optimal solution. A closed-form least square solution to the problem of analytically designing linear-phase FIR digital filters has been given. The Parks-McClellan algorithm based on the minimax optimality criterion has been reviewed. A method for realizing FIR digital filters by cascaded lattice forms has also been examined. Finally, performance comparisons among these algorithms have been performed through a numerical example.

References

[1] A. V. Oppenheim and R. W. Schafer, *Digital Signal Processing*, NJ: Prentice-Hall, 1975.

[2] A. Antoniou, *Digital Filters*, 2nd ed. NJ: McGraw-Hill, 1993.

[3] S. K. Mitra, *Digital Signal Processing*, 3rd ed. NJ: McGraw-Hill, 2006.

[4] S. Takahashi and M. Ikehara, *Digital Filters*, Tokyo, Japan, Baifukan, 1999.

[5] P. P. Vaidyanathan and T. Q. Nguyen, "Eigenfilters: A new approach to least-squares FIR filter design and applications including Nyquist filters," *IEEE Trans. Circuits Syst.*, vol. CAS-34, no. 1, pp. 11–23, Jan. 1987.

[6] M. O. Ahmad and J.-D. Wang, "An analytical least square solution to the design problem of two-dimensional FIR filters with quadrantally symmetric or antisymmetric frequency response," *IEEE Trans. Circuits Syst.*, vol. 36, no. 7, pp. 968–979, July 1989.

[7] T. W. Parks and J. H. McClellan, "Chebyshev approximation for non-recursive digital filters with linear phase," *IEEE Trans. Circuits Theory*, vol. CT-19, no. 2, pp. 189–194, Mar. 1972.

[8] T. Higuchi, *Fundamentals of Digital Signal Processing*, Tokyo, Japan, Shokodo, 1986.

[9] M. Hagiwara, *Digital Signal Processing*, Tokyo, Japan, Morikita Publishing, 2001.

6

Design Methods Using Analog Filter Theory

6.1 Preview

One of the approaches to the design of an IIR digital filter is to use analog filter theory in conjunction with bilinear transformation that maps frequencies in the analog domain to the digital domain. This indirect design method works well, especially for the design of standard IIR digital filters such as lowpass, highpass, bandpass, and bandstop filters. This chapter starts by a brief review of several design techniques for analog filters. These include designs based on Butterworth, Chebyshev, inverse-Chebyshev, and elliptic approximations as well as analog-filter approximations by transformations which transform normalized lowpass analog filters to denormalized lowpass, highpass, bandpass, and bandstop analog filters. The bilinear transformation method for the design of IIR digital filters is then studied in detail and illustrated by a design example.

6.2 Design Methods Using Analog Filter Theory

Standard IIR digital filters such as lowpass, highpass, bandpass, and bandstop filters can be designed through indirect methods in which a continuous-time transfer function satisfying certain specifications is obtained by a standard analog-filter approximation, and then a corresponding discrete-time transfer function is obtained by one of the following methods: invariant-impulse-response method and its variants, matched-z transformation method, and bilinear transformation method [1]. Loss function is a concept often involved in the study of analog filters. A loss function $L(-s^2)$ is related its corresponding transfer function $H(s)$ as

$$L(-s^2) = \frac{D(s)D(-s)}{N(s)N(-s)}, \qquad H(s) = \frac{N(s)}{D(s)}$$

where $N(s)$ and $D(s)$ are polynomials in s. To ensure the stability of the analog approximation, the poles of $H(s)$ (i.e. the zeros of polynomial $D(s)$) must lie strictly inside the left-half s-plane.

6.2.1 Lowpass Analog-Filter Approximations

6.2.1.1 Butterworth approximation

The transfer function of the nth-order normalized lowpass Butterworth filter assumes the form

$$H_N(s) = \frac{1}{\prod_{i=1}^{n}(s - p_i)}$$

where p_i for $i = 1, 2, \cdots, n$ are the left-half s-plane zeros of the corresponding loss function $L(-s^2)$, which are given by

$$L(-s^2) = 1 + (-s^2)^n = \prod_{k=1}^{2n}(s - s_k)$$

where

$$s_k = \begin{cases} e^{j(2k-1)\pi/2n} & \text{for even } n \\ e^{j(k-1)\pi/n} & \text{for odd } n \end{cases}$$

The term "normalized" refers to the constraint that at $s = j\omega$ with $\omega = 1$, $L(1) = 2$. As a result, the magnitude of the normalized Butterworth filter at $\omega = 1$ assumes the value $|H_N(j)| = 1/\sqrt{2} \simeq 0.707$, namely a 3 dB loss relative to the filter gain at $\omega = 0$.

6.2.1.2 Chebyshev approximation

The magnitude response of the lowpass Butterworth filter is a monotonically increasing function of ω. A more balanced characteristic may be achieved using the Chebyshev approximation where the magnitude response in passband oscillates between one and a less-than-one value $10^{-0.05A_p}$ which means an A_p dB loss in passband.

The normalized transfer function $H_N(s)$ of nth-order lowpass Chebyshev filter is given by [1]

$$H_N(s) = \frac{H_0}{D_0(s)\prod_{i=1}^{r}(s - p_i)(s - p_i^*)}$$

where

$$r = \begin{cases} (n-1)/2 & \text{for odd } n \\ n/2 & \text{for even } n \end{cases} \quad \text{and} \quad D_0(s) = \begin{cases} s - 1/p_0 & \text{for odd } n \\ 1 & \text{for even} \end{cases}$$

and constant H_0 and poles p_i are calculated for a given $A_p > 0$ (in dB) as follows:

$$\begin{aligned}
\varepsilon &= \left(10^{0.1A_p} - 1\right)^{1/2} \\
p_0 &= \sigma_{(n+1)/2} \quad \text{with } \sigma_{(n+1)/2} = -\sinh\left(\frac{1}{n}\sinh^{-1}\frac{1}{\varepsilon}\right) \\
p_i &= \sigma_i + j\omega_i \quad \text{for } i = 1, 2, \ldots, r \\
\sigma_i &= -\sinh\left(\frac{1}{n}\sinh^{-1}\frac{1}{\varepsilon}\right)\sin\frac{(2i-1)\pi}{2n} \\
\omega_i &= \cosh\left(\frac{1}{n}\sinh^{-1}\frac{1}{\varepsilon}\right)\cos\frac{(2i-1)\pi}{2n} \\
H_0 &= \begin{cases} -p_0 \prod_{i=1}^{r} |p_i|^2 & \text{for odd } n \\ 10^{-0.05A_p} \prod_{i=1}^{r} |p_i|^2 & \text{for even } n \end{cases}
\end{aligned}$$

6.2.1.3 Inverse-Chebyshev approximation

The inverse-Chebyshev filters are closely related to the Chebyshev filters, whose magnitude response is a monotonically decreasing function of ω in the passband and oscillates between zero and a prescribed minimum attenuation $\left(10^{0.1A_a} - 1\right)^{-1/2}$ (with $A_a > 0$ in dB) in the stopband.

The normalized transfer function of the nth-order lowpass inverse-Chebyshev filter is given by [1]

$$H_N(s) = \frac{H_0}{D_0(s)} \prod_{i=1}^{r} \frac{(s - 1/z_i)(s - 1/z_i^*)}{(s - 1/p_i)(s - 1/p_i^*)}$$

where

$$r = \begin{cases} (n-1)/2 & \text{for odd } n \\ n/2 & \text{for even } n \end{cases} \quad \text{and} \quad D_0(s) = \begin{cases} s - 1/p_0 & \text{for odd } n \\ 1 & \text{for even } n \end{cases}$$

and constant H_0, zeros z_i, and poles p_i are calculated for a given $A_a > 0$ (in dB) as follows:

$$\delta = \left(10^{0.1A_a} - 1\right)^{-1/2}$$

$$z_i = j\cos\tfrac{(2i-1)\pi}{2n} \quad \text{for } i = 1, 2, \cdots, r$$

$$p_0 = \sigma_{(n+1)/2} \quad \text{with } \sigma_{(n+1)/2} = -\sinh\left(\tfrac{1}{n}\sinh^{-1}\tfrac{1}{\delta}\right)$$

$$p_i = \sigma_i + j\omega_i \quad \text{for } i = 1, 2, \cdots, r$$

$$\sigma_i = -\sinh\left(\tfrac{1}{n}\sinh^{-1}\tfrac{1}{\delta}\right)\sin\tfrac{(2i-1)\pi}{2n}$$

$$\omega_i = \cosh\left(\tfrac{1}{n}\sinh^{-1}\tfrac{1}{\delta}\right)\cos\tfrac{(2i-1)\pi}{2n}$$

$$H_0 = \begin{cases} \frac{1}{-p_0}\prod_{i=1}^{r}\frac{|z_i|^2}{|p_i|^2} & \text{for odd } n \\ \prod_{i=1}^{r}\frac{|z_i|^2}{|p_i|^2} & \text{for even } n \end{cases}$$

6.2.1.4 Elliptic approximation

Elliptic filters are a class of analog filters that are more efficient than the Butterworth, Chebyshev and inverse-Chebyshev filters, in which the magnitude response oscillates between one and a maximum passband loss in passband and oscillates between zero and a minimum stopband attenuation in stopband.

Given a selectivity factor $k > 0$, a maximum passband loss of A_p dB and a minimum stopband attenuation of A_a dB, the transfer function of a normalized lowpass elliptic filter with passband edge $\omega_p = \sqrt{k}$ and stopband edge $\omega_a = 1/\sqrt{k}$ assumes the form

$$H_N(s) = \frac{H_0}{D_0(s)}\prod_{i=1}^{r}\frac{s^2 + a_{0i}}{s^2 + b_{1i}s + b_{0i}}$$

where

$$r = \begin{cases} (n-1)/2 & \text{for odd } n \\ n/2 & \text{for even } n \end{cases} \quad \text{and} \quad D_0(s) = \begin{cases} s + \sigma_0 & \text{for odd } n \\ 1 & \text{for even } n \end{cases}$$

and constant H_0 and transfer-function coefficients can be evaluated using the following formulas [1]:

$$k' = \sqrt{1-k^2}, \qquad q_0 = \frac{1}{2}\left(\frac{1-\sqrt{k'}}{1+\sqrt{k'}}\right)$$

$$q = q_0 + 2q_0^5 + 15q_0^9 + 150q_0^{13}$$

$$D = \frac{10^{0.1A_a} - 1}{10^{0.1A_p} - 1}, \qquad n \geq \frac{\log 16D}{\log(1/q)}$$

$$\Lambda = \frac{1}{2n} \ln \frac{10^{0.05A_p} + 1}{10^{0.05A_p} - 1}$$

$$\sigma_0 = \left| \frac{2q^{1/4} \sum_{m=0}^{\infty} (-1)^m q^{m(m+1)} \sinh\left[(2m+1)\Lambda\right]}{1 + 2\sum_{m=1}^{\infty} (-1)^m q^{m^2} \cosh 2m\Lambda} \right|$$

$$W = \sqrt{\left(1 + k\sigma_0^2\right)\left(1 + \frac{\sigma_0^2}{k}\right)}$$

$$\Omega_i = \frac{2q^{1/4} \sum_{m=0}^{\infty} (-1)^m q^{m(m+1)} \sin \frac{(2m+1)\pi\mu}{n}}{1 + 2\sum_{m=1}^{\infty} (-1)^m q^{m^2} \cos \frac{2m\pi\mu}{n}}$$

where

$$\mu = \begin{cases} i & \text{for odd } n \\ i - \dfrac{1}{2} & \text{for even } n \end{cases} \qquad i = 1, 2, \cdots, r$$

$$V_i = \sqrt{\left(1 - k\Omega_i^2\right)\left(1 - \frac{\Omega_i^2}{k}\right)}, \qquad a_{0i} = \frac{1}{\Omega_i^2}$$

$$b_{0i} = \frac{(\sigma_0 V_i)^2 + (\Omega_i W)^2}{\left(1 + \sigma_0^2 \Omega_i^2\right)^2}, \qquad b_{1i} = \frac{2\sigma_0 V_i}{1 + \sigma_0^2 \Omega_i^2}$$

$$H_0 = \begin{cases} \sigma_0 \displaystyle\prod_{i=1}^{r} \frac{b_{0i}}{a_{0i}} & \text{for odd } n \\ 10^{-0.05A_p} \displaystyle\prod_{i=1}^{r} \frac{b_{0i}}{a_{0i}} & \text{for even } n \end{cases}$$

The actual minimum stopband attenuation is given by

$$A_a = 10 \log \left(\frac{10^{0.1A_p} - 1}{16q^n} + 1 \right)$$

The series involved in calculating σ_0 and Ω_i converge rapidly, and 3 or 4 terms are sufficient for most designs.

6.2.2 Other Analog-Filter Approximations by Transformations

Denormalized lowpass, highpass, bandpass, and bandstop approximations can be deduced from normalized lowpass approximations using transformations of the form $s = f(\bar{s})$ [1]. In what follows, $H_N(s)$ denotes the transfer function of a normalized lowpass analog filter with stopband and passband edges ω_p and ω_a, respectively.

6.2.2.1 Lowpass-to-lowpass transformation

The transformation

$$s = \lambda \bar{s} \tag{6.1}$$

maps the ranges $[0, j\omega_p]$ and $[j\omega_a, j\infty)$ onto the ranges $[0, j\omega_p/\lambda]$ and $[j\omega_a/\lambda, j\infty)$, respectively. Hence

$$H_{\text{LP}}(\bar{s}) = H_N(s)\big|_{s=\lambda\bar{s}}$$

is a denormalized lowpass approximation with passband edge ω_p/λ and stopband edge ω_a/λ.

6.2.2.2 Lowpass-to-highpass transformation

The transformation

$$s = \frac{\lambda}{\bar{s}} \tag{6.2}$$

maps the ranges $[0,\ j\omega_p]$ and $[j\omega_a,\ j\infty)$ onto the ranges $-j\infty,\ -j\lambda/\omega_p]$ and $[-j\lambda/\omega_a,\ 0]$, respectively. Hence

$$H_{\text{HP}}(\bar{s}) = H_N(s)\big|_{s=\lambda/\bar{s}}$$

is a denormalized highpass approximation with stopband edge λ/ω_a and passband edge λ/ω_p.

6.2.2.3 Lowpass-to-bandpass transformation

A transformation that converts a normalized lowpass approximation $H_N(s)$ to a bandpass approximation is given by

$$s = \frac{1}{B}\left(\bar{s} + \frac{\omega_0^2}{\bar{s}}\right) \tag{6.3}$$

where B and ω_0 are constants. The passband and stopband edges of the transformed bandpass filter are given by

$$\bar{\omega}_{p1}, \bar{\omega}_{p2} = \mp\frac{\omega_p B}{2} + \sqrt{\omega_0^2 + \left(\frac{\omega_p B}{2}\right)^2}$$

$$\bar{\omega}_{a1}, \bar{\omega}_{a2} = \mp\frac{\omega_a B}{2} + \sqrt{\omega_0^2 + \left(\frac{\omega_a B}{2}\right)^2}$$

6.2.2.4 Lowpass-to-bandstop transformation

A transformation that converts a normalized lowpass approximation $H_N(s)$ to a bandstop approximation is given by

$$s = \frac{B\bar{s}}{\bar{s}^2 + \omega_0^2} \tag{6.4}$$

where B and ω_0 are constants. The passband and stopband edges of the transformed bandpass filter are given by

$$\bar{\omega}_{p1}, \bar{\omega}_{p2} = \mp\frac{B}{2\omega_p} + \sqrt{\omega_0^2 + \left(\frac{B}{2\omega_p}\right)^2}$$

$$\bar{\omega}_{a1}, \bar{\omega}_{a2} = \mp\frac{B}{2\omega_a} + \sqrt{\omega_0^2 + \left(\frac{B}{2\omega_a}\right)^2}$$

6.2.3 Design Methods Based on Analog Filter Theory

In order for an IIR digital filter to be realizable, the transfer function of an IIR filter must be a rational function of z with the degree of numerator polynomial equal to or less than that of the denominator polynomial, and its poles must lie within the unit circle of the z plane. These conditions are called the realizability constraints [1].

6.2.3.1 Invariant impulse-response method

Let $H_A(s) = N(s)/D(s)$ be an analog IIR filter whose impulse response is denoted by $h_A(t)$. The digital IIR filter designed by the invariant impulse-response method requires that the impulse response $h_D(k)$ of the digital filter exactly equals to equally spaced samples of the impulse response $h_A(t)$ [2], namely

$$h_D(k) = h_A(t)|_{t=kT} = h_A(kT) \tag{6.5}$$

The transfer function of the digital filter can be expressed in terms of its impulse response as

$$H_D(z) = \sum_{k=0}^{\infty} h_D(k) z^{-k} \tag{6.6}$$

and $H_A(s)$ can be expanded in terms of partial fractions as

$$H_A(s) = \sum_{i=1}^{n} \frac{K_i}{s - p_i} \tag{6.7}$$

where p_i for $i = 1, 2, \cdots, n$ are the poles of $H_A(s)$. Consequently, we have

$$h_A(t) = \sum_{i=1}^{n} K_i \, e^{p_i t}$$

and

$$h_A(kT) = \sum_{i=1}^{n} K_i \, e^{p_i kT} = \sum_{i=1}^{n} K_i \, \left(e^{p_i T}\right)^k$$

which in conjunction with (6.5) leads (6.6) to an IIR transfer function

$$H_D(z) = \sum_{k=0}^{\infty} \left[\sum_{i=1}^{n} K_i \, \left(e^{p_i T}\right)^k \right] z^{-n} = \sum_{i=1}^{n} \frac{K_i z}{z - e^{p_i T}} \tag{6.8}$$

It is noted that $H_D(z)$ in (6.8) also has n poles and that if the analog $H_A(s)$ is stable, i.e., all p_i have negative real parts, then the magnitudes of $e^{P_i T}$ are less than one, hence the digital $H_D(z)$ is also stable.

The design method may be summarized in three steps as follows:

1. Design a prototype analog filter with transfer function $H_A(s)$.
2. Obtain a partial fraction expansion of $H_A(s)$ as in (6.7).
3. Use the values of K_i and p_i obtained from Step 2 to construct a digital transfer function $H_D(z)$ using (6.8).

One may choose one of the methods in Sections 6.2.1 and 6.2.2 to implement Step 1. To illustrate Steps 2 and 3 of the above method, consider the transfer function of the sixth-order normalized lowpass Butterworth filter given by [1]

$$H_A(s) = \frac{1}{(s^2 + 0.517638s + 1)(s^2 + 1.414214s + 1)(s^2 + 1.931852s + 1)}$$

The partial fraction expansion of $H_A(s)$ is found to be

$$H_A(s) = \sum_{i=1}^{6} \frac{K_i}{s - p_i}$$

with

$$\begin{aligned} K_{1,2} &= -1.523603, & K_{3,4} &= 0.204124 \pm j0.353553 \\ K_{5,6} &= 1.319479 \mp j2.285412, & p_{1,2} &= -0.707107 \pm j0.707107 \\ p_{3,4} &= -0.258819 \pm j0.965926, & p_{5,6} &= -0.965926 \pm j0.258819 \end{aligned}$$

Now using (6.4) with $T = 1$, one obtains

$$H_D(z) = \frac{N_1(z)}{D_1(z)} + \frac{N_2(z)}{D_2(z)} + \frac{N_3(z)}{D_3(z)}$$

where

$$\begin{aligned} \frac{N_1(z)}{D_1(z)} &= \frac{-3.047201z^2 + 1.143354z}{z^2 - 0.749706z + 0.243117} \\ \frac{N_2(z)}{D_2(z)} &= \frac{0.408248z^2 - 0.628224z}{z^2 - 0.877962z + 0.595926} \\ \frac{N_3(z)}{D_3(z)} &= \frac{2.638959z^2 - 0.525732z}{z^2 - 0.735906z + 0.144880} \end{aligned}$$

The largest magnitude of the poles of $H_D(z)$ is 0.771963, hence the IIR digital filter is stable. The magnitude response of $H_D(z)$ is depicted in Figure 6.1.

6.2.3.2 Bilinear-transformation method

Basic Concepts and Properties

The bilinear-transformation method is a frequency-domain method to convert an analog prototype filter into a desired digital filter. In this process, passbands and stopbands in the analog filter are translated into the corresponding passbands and stopbands of the digital filter with the stability, passband ripple, and stopband attenuation preserved. As such, the bilinear transformation method has been one of the most important methods for the design of IIR digital filters.

The bilinear transformation is a mapping that is linear in the numerator as well as the denominator and is given by

$$s = \frac{1}{T}\left(\frac{z-1}{z+1}\right) \tag{6.9}$$

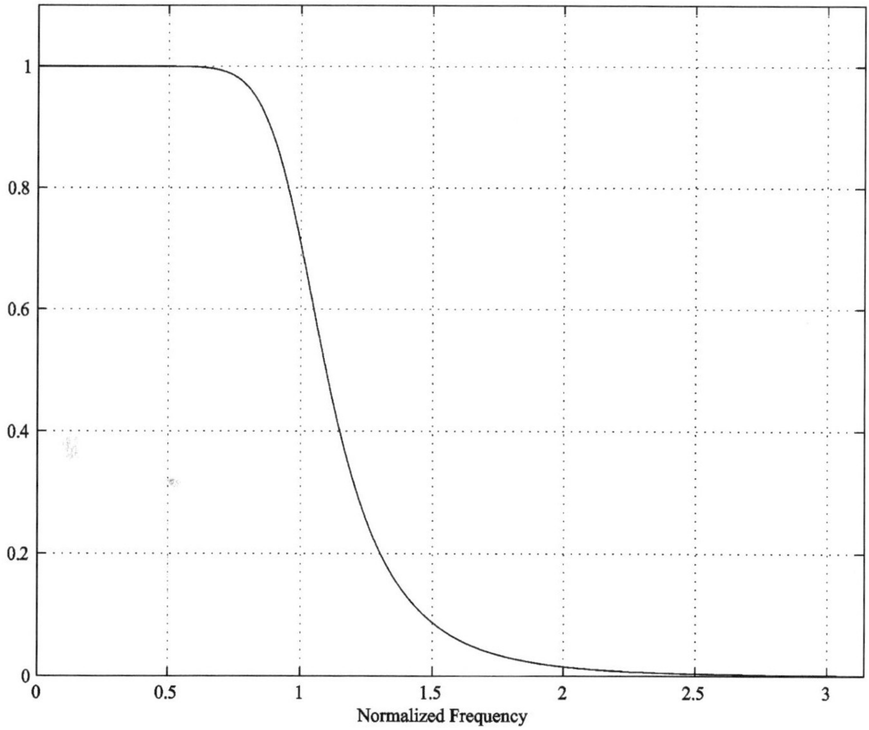

Figure 6.1 Magnitude response of the 6th-order IIR filter.

where T is the sampling interval in seconds. Application of the bilinear transformation to an analog transfer function $H_A(s)$ leads to a digital transfer function $H_D(z)$ with

$$H_D(z) = H_A(s)|_{s=\frac{1}{T}\left(\frac{z-1}{z+1}\right)}$$

The time-domain response of the digital filter so obtained is approximately the same as that of the prototype analog filter, and the two time-domain responses get closer as T gets smaller [1]. From (6.9), it follows that

$$z = \frac{T/2 + s}{T/2 - s}$$

which with $s = \sigma + j\omega$ and $z = re^{j\theta}$ gives

$$r = \left[\frac{\left(\frac{2}{T} + \sigma\right)^2 + \omega^2}{\left(\frac{2}{T} - \sigma\right)^2 + \omega^2}\right]^{1/2} \tag{6.10a}$$

and

$$\theta = \tan^{-1}\frac{\omega}{T/2+\sigma} + \tan^{-1}\frac{\omega}{T/2-\sigma} \tag{6.10b}$$

From (6.10a) and (6.10b), it is concluded that the bilinear transformation maps the open right-half s plane onto the exterior to the unit circle of the z plane, j axis of the s plane onto the unit circle of the z plane, and open left-half s plane onto the interior of the unit circle of the z plane. In addition, if in (6.9) we let $s = j\omega$ and $z = e^{j\Omega T}$, then we obtain a frequency interpretation of the bilinear transformation that

$$H_D(e^{j\Omega T}) = H_A(j\omega) \quad \text{if} \quad \omega = \frac{2}{T}\tan\frac{\Omega T}{2} \tag{6.11}$$

Note that $\omega \simeq \Omega$ if $\Omega \leq 0.3/T$. In other words, the digital filter has the same frequency response as the prototype analog filter as long as $\Omega \leq 0.3/T$. For high frequencies the relation between ω and Ω becomes nonlinear as can be seen from (6.11). The distortion introduced by the bilinear transformation in the frequency scale of the digital filter is known as the warping effect [1]. As far as the amplitude response is concerned, the warping effect can be eliminated by prewarping the analog filter as follows. Suppose $\omega_1, \omega_2, \cdots, \omega_L$ are the passband and stopband edges of the analog filter. In order for the digital filter to have passband and stopband edges $\Omega_1, \Omega_2, \cdots, \Omega_L$, the analog filter should be prewarped before application of the bilinear transformation so that its passband and stopband edges $\omega_1, \omega_2, \cdots, \omega_L$ satisfy

$$\omega_i = \frac{2}{T}\tan\frac{\Omega_i T}{2} \quad \text{for } i = 1, 2, \cdots, L \tag{6.12}$$

Lowpass, highpass, bandpass, and bandstop digital IIR filters satisfying prescribed specifications can be designed by first transforming a normalized analog lowpass transfer function $H_N(s)$ into a denormalized lowpass, highpass, bandpass, and bandstop transfer function (see Section 6.2.3.2):

$$H_X(\bar{s}) = H_N(s)|_{s=f_X(\bar{s})} \tag{6.13}$$

and then applying the bilinear transformation to $H_X(\bar{s})$ to obtain a digital transfer function

$$H_D(z) = H_X(\bar{s})|_{\bar{s}=\frac{2}{T}\left(\frac{z-1}{z+1}\right)} \tag{6.14}$$

It is of interest to note that the second step of the design can be carried out with ease when the transfer function of the analog filter is given in terms of its poles and zeros as [3]

$$H_A(s) = H_0 \frac{\prod_{i=1}^{m} (s - z_i^{(a)})}{\prod_{i=1}^{n} (s - p_i^{(a)})} \tag{6.15}$$

The application of bilinear transformation to (6.15) yields

$$H_D(z) = A(z+1)^{n-m} \frac{\prod_{i=1}^{m} (z - z_i)}{\prod_{i=1}^{n} (z - p_i)} \tag{6.16a}$$

where

$$z_i = \frac{1 + \frac{T}{2} z_i^{(a)}}{1 - \frac{T}{2} z_i^{(a)}}, \quad p_i = \frac{1 + \frac{T}{2} p_i^{(a)}}{1 - \frac{T}{2} p_i^{(a)}} \tag{6.16b}$$

and

$$A = \frac{\prod_{i=1}^{m} \left(\frac{2}{T} - z_i^{(a)}\right)}{\prod_{i=1}^{n} \left(\frac{2}{T} - p_i^{(a)}\right)} \tag{6.16c}$$

Design Procedure

To describe the design procedure in detail, let the design specifications of a digital IIR filter be given in terms of passband edge Ω_p (Ω_{p1} and Ω_{p2} for bandpass and bandstop filters) (rad/s), stopband edge Ω_a (Ω_{a1} and Ω_{a2} for bandpass and bandstop filters) (rad/s), sampling frequency ω_s (rad/s), maximum passband loss A_p (dB), and minimum stopband attenuation A_a (dB). The sampling period T is evaluated as $T = 2\pi/\omega_s$.

1. For lowpass (LP) and highpass (HP) filters, compute parameter K_0 as

$$K_0 = \frac{\tan(\Omega_p T/2)}{\tan(\Omega_a T/2)} \tag{6.17}$$

For bandpass (BP) and bandstop (BS) filters, compute

$$K_A = \tan \frac{\Omega_{p2}T}{2} - \tan \frac{\Omega_{p1}T}{2}, \qquad K_B = \tan \frac{\Omega_{p1}T}{2} \tan \frac{\Omega_{p2}T}{2}$$

$$K_C = \tan \frac{\Omega_{a1}T}{2} \tan \frac{\Omega_{a2}T}{2}, \qquad K_1 = \frac{K_A \tan(\Omega_{a1}T/2)}{K_B - \tan^2(\Omega_{a1}T/2)}$$

$$K_2 = \frac{K_A \tan(\Omega_{a2}T/2)}{\tan^2 \Omega_{a2}T/2} - K_B$$

2. Determine n and ω_p. For elliptic filters, also determine k.
   - For Butterworth filters, first compute K as

$$K = \begin{cases} K_0 & \text{for LP} \\ 1/K_0 & \text{for HP} \\ K_1 \text{ (if } K_C \geq K_B) \text{ or } K_2 \text{ (if } K_C < K_B) & \text{for BP} \\ 1/K_2 \text{ (if } K_C \geq K_B) \text{ or } 1/K_1 \text{ (if } K_C < K_B) & \text{for BS} \end{cases} \tag{6.18a}$$

then compute

$$D = \frac{10^{0.1A_a} - 1}{10^{0.1A_p} - 1}, \qquad n \geq \frac{\log D}{2\log(1/K)}, \qquad \omega_p = \left(10^{0.1A_p} - 1\right)^{1/2n} \tag{6.18b}$$

   - For Chebyshev filters, first compute K using (6.18a), then compute

$$D = \frac{10^{0.1A_a} - 1}{10^{0.1A_p} - 1}, \qquad n \geq \frac{\cosh^{-1}\sqrt{D}}{\cosh^{-1}(1/K)}, \qquad \omega_p = 1 \tag{6.19}$$

   - For elliptic filters, compute k using

$$k = \begin{cases} K_0 & \text{for LP} \\ 1/K_0 & \text{for HP} \\ K_1 \text{ (if } K_C \geq K_B) \text{ or } K_2 \text{ (if } K_C < K_B) & \text{for BP} \\ 1/K_2 \text{ (if } K_C \geq K_B) \text{ or } 1/K_1 \text{ (if } K_C < K_B) & \text{for BS} \end{cases} \tag{6.20a}$$

then compute ω_p using

$$\omega_p = \begin{cases} \sqrt{K_0} & \text{for LP} \\ 1/\sqrt{K_0} & \text{for HP} \\ \sqrt{K_1} \text{ (if } K_C \geq K_B) \text{ or } \sqrt{K_2} \text{ (if } K_C < K_B) & \text{for BP} \\ 1/\sqrt{K_2} \text{ (if } K_C \geq K_B) \text{ or } 1/\sqrt{K_1} \text{ (if } K_C < K_B) & \text{for BS} \end{cases} \tag{6.20b}$$

and finally compute

$$\begin{aligned} k' &= \sqrt{1 - k^2}, \qquad q_0 = \tfrac{1}{2}\left(\tfrac{1-\sqrt{k'}}{1+\sqrt{k'}}\right) \\ q &= q_0 + 2q_0^5 + 15q_0^9 + 150q_0^{13} \\ D &= \frac{10^{0.1A_a} - 1}{10^{0.1A_p} - 1}, \qquad n \geq \frac{\log 16D}{\log(1/q)} \end{aligned} \tag{6.20c}$$

3. Determine λ for LP and HP, or B and ω_0 for BP and BS.
   For LP and HP, compute λ as

$$\lambda = \begin{cases} \dfrac{\omega_p T}{2\tan(\Omega_p T/2)} & \text{for LP} \\ \dfrac{2\omega_p \tan(\Omega_p T/2)}{T} & \text{for HP} \end{cases} \tag{6.21}$$

   For BP, compute

$$\omega_0 = 2\sqrt{K_B}/T, \qquad B = 2K_A/T\omega_p \tag{6.22}$$

   For BS, compute

$$\omega_0 = 2\sqrt{K_B}/T, \qquad B = 2K_A\omega_p/T \tag{6.23}$$

4. Form the normalized lowpass transfer function $H_N(s)$, see Section 6.2.1.4.
5. Apply the analog-filter transformation in (6.13), see Section 6.2.2.
6. Apply the bilinear transformation in (6.14).

To illustrate the bilinear-transformation method, consider designing a digital elliptic bandpass filter with specifications

$$\begin{aligned} A_p &= 1 \text{ dB}, & A_a &= 42 \text{ dB} \\ \Omega_{p1} &= 1000 \text{ rad/s}, & \Omega_{p2} &= 1300 \text{ rad/s} \\ \Omega_{a1} &= 900 \text{ rad/s}, & \Omega_{a2} &= 140 \text{ rad/s} \\ \omega_s &= 6000 \text{ rad/s} \end{aligned}$$

Following the above design steps, we compute

$$\begin{aligned} k &= 0.609806, & \omega_p &= 0.780900, & k' &= 0.792551 \\ q_0 &= 0.029030, & q &= 0.029030, & D &= 6.120655 \times 10^4 \\ n &= 4, & \omega_0 &= 1.305887 \times 10^3, & B &= 5.684665 \times 10^2 \end{aligned}$$

By applying the LP-to-BP transformation followed by the bilinear transformation, an eighth-order transfer function is obtained as

$$H_D(z) = H_0 \prod_{j=1}^{2} \frac{z^4 + a_{3j}z^3 + a_{2j}z^3 + a_{1j}z^3 + a_{0j}}{z^4 + b_{3j}z^3 + b_{2j}z^3 + b_{1j}z^3 + b_{0j}}$$

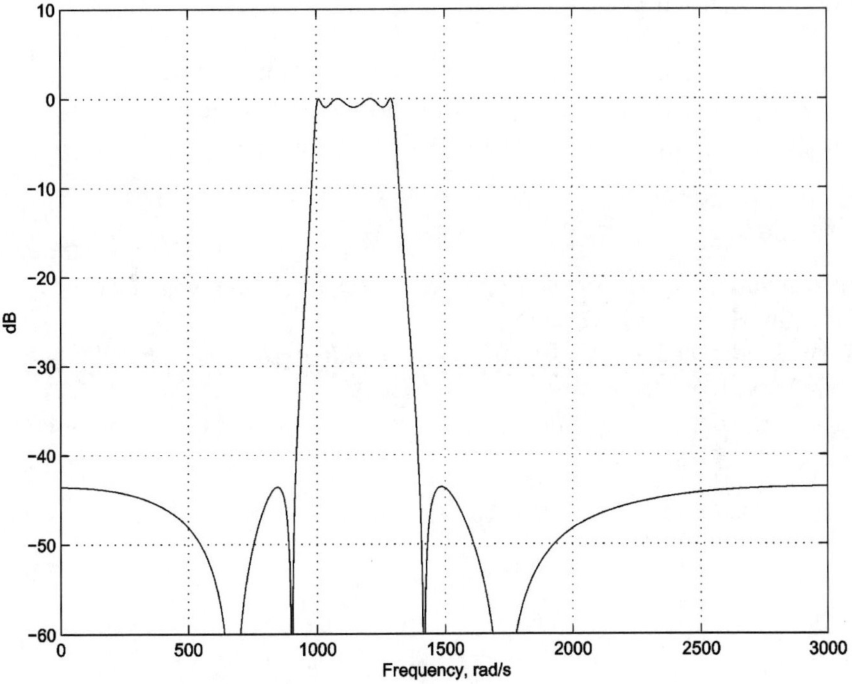

Figure 6.2 Magnitude response of the 8th-order IIR filter.

where $H_0 = 8.272767 \times 10^{-3}$ and

$$\begin{bmatrix} a_{01} & a_{02} \\ a_{11} & a_{12} \\ a_{21} & a_{22} \\ a_{31} & a_{32} \end{bmatrix} = \begin{bmatrix} 1 & 1 \\ -1.051264 & -1.348091 \\ 1.278797 & 2.204574 \\ -1.051264 & -1.348091 \end{bmatrix}$$

$$\begin{bmatrix} b_{01} & b_{02} \\ b_{11} & b_{12} \\ b_{21} & b_{22} \\ b_{31} & b_{32} \end{bmatrix} = \begin{bmatrix} 0.797075 & 0.933566 \\ -1.219252 & -1.344800 \\ 2.235302 & 2.335923 \\ -1.366509 & -1.393009 \end{bmatrix}$$

The magnitude response of the digital IIR filter is depicted in Figure 6.2.

6.3 Summary

This chapter has shown that the design of IIR filters can be achieved using analog filter theory in conjunction with bilinear transformation. Several methods for the design of analog filters have been reviewed. The bilinear transform is then introduced and utilized as the key technical component for the design of stable IIR digital filters. A design procedure for the design of lowpass, highpass, bandpass, and bandstop IIR digital filters using this technique is presented. The design procedure is illustrated through an example of an eighth-order digital elliptic bandpass filter.

References

[1] A. Antoniou, *Digital Signal Processing: Signals, Systems, and Filters*, McGraw-Hill, New York, 2006.
[2] T. W. Parks and C. S. Burrus, *Digital Filter Design*, John Wiley, New York, 1987.
[3] L. B. Jackson, *Digital Filters and Signal Processing*, 3rd ed., Kluwer Academic, Boston, 1996.

7

Design Methods in the Frequency Domain

7.1 Preview

This chapter presents an alternative approach to the design of IIR digital filters by applying optimization methods [2] where an objective function is formulated in the frequency domain based on an error between certain desired and actual magnitude and/or phase responses. Specifically, we present four methods for the design of stable IIR filters, which are based on mean squared error minimization, equal-ripple minimization of squared magnitude error, weighted least squares subject to stability, and minimization of maximum error subject to stability, respectively. In addition, two Remez exchange type of techniques for designing an all-pass digital filter to approximate a desired phase response are also examined.

7.2 Design Methods in the Frequency Domain

Many methods for the design of IIR digital filters in the frequency domain have been developed since early 1970s. Below we describe several of them with representative design ideas and techniques.

7.2.1 Minimum Mean Squared Error Design

The method presented below is a slightly modified algorithm proposed by Steiglitz [1]. Let the IIR filter be expressed in terms of K second-order cascaded sections as

$$H(z,\boldsymbol{x}) = A \prod_{k=1}^{K} \frac{1 + a_k z^{-1} + b_k z^{-2}}{1 + c_k z^{-1} + d_k z^{-2}} \tag{7.1}$$

where

$$\boldsymbol{x} = \left[\ a_1 \quad b_1 \quad c_1 \quad d_1 \quad \cdots \quad a_K \quad b_K \quad c_K \quad d_K \quad A\ \right]^T$$

is a vector of $4K + 1$ design variables. Let $H_d(\omega)$ be the desired frequency response and $\Omega_L = \{\omega_i | \, i = 1, 2, \cdots, L\}$ be a discrete set of frequencies at which the error of the actual magnitude response $\left|H(e^{j\omega}, \boldsymbol{x})\right|$ approximating a desired magnitude response $|H_d(\omega)|$ is evaluated in a mean squared manner:

$$J_2(\boldsymbol{x}) = \sum_{i=1}^{L} \left[\left|H(e^{j\omega_i}, \boldsymbol{x})\right| - |H_d(\omega_i)| \right]^2 \tag{7.2}$$

The design problem is formulated as the problem of finding a vector $\boldsymbol{x}^*$ that minimizes the mean-squared type of objective function $J_2(\boldsymbol{x})$ in (7.2). Many optimization techniques suitable for the problem at hand exist [2]. These techniques require the gradient of the objective function. By (7.2), the gradient of $J_2(\boldsymbol{x})$ is given by

$$\nabla J_2(\boldsymbol{x}) = 2 \sum_{i=1}^{L} \nabla \left(\left|H(e^{j\omega_i}, \boldsymbol{x})\right| \right) \left[\left|H(e^{j\omega_i}, \boldsymbol{x})\right| - |H_d(\omega_i)| \right] \tag{7.3}$$

with

$$\nabla \left(\left|H(e^{j\omega_i}, \boldsymbol{x})\right| \right) = \begin{bmatrix} \partial \left|H(e^{j\omega_i}, \boldsymbol{x})\right| \Big/ \partial a_1 \\ \partial \left|H(e^{j\omega_i}, \boldsymbol{x})\right| \Big/ \partial b_1 \\ \partial \left|H(e^{j\omega_i}, \boldsymbol{x})\right| \Big/ \partial c_1 \\ \partial \left|H(e^{j\omega_i}, \boldsymbol{x})\right| \Big/ \partial d_1 \\ \vdots \\ \partial \left|H(e^{j\omega_i}, \boldsymbol{x})\right| \Big/ \partial A \end{bmatrix}$$

where

$$\frac{\partial \left|H(e^{j\omega_i}, \boldsymbol{x})\right|}{\partial a_k} = \left|H(e^{j\omega_i}, \boldsymbol{x})\right| \frac{(1 + a_k \cos\omega_i + b_k \cos 2\omega_i)\cos\omega_i + (a_k \sin\omega_i + b_k \sin 2\omega_i)\sin\omega_i}{(1 + a_k \cos\omega_i + b_k \cos 2\omega_i)^2 + (a_k \sin\omega_i + b_k \sin 2\omega_i)^2}$$

$$\frac{\partial \left|H(e^{j\omega_i}, \boldsymbol{x})\right|}{\partial b_k} = \left|H(e^{j\omega_i}, \boldsymbol{x})\right| \frac{(1 + a_k \cos\omega_i + b_k \cos 2\omega_i)\cos 2\omega_i + (a_k \sin\omega_i + b_k \sin 2\omega_i)\sin 2\omega_i}{(1 + a_k \cos\omega_i + b_k \cos 2\omega_i)^2 + (a_k \sin\omega_i + b_k \sin 2\omega_i)^2}$$

$$\frac{\partial \left|H(e^{j\omega_i}, \boldsymbol{x})\right|}{\partial c_k} = - \left|H(e^{j\omega_i}, x)\right|$$
$$\frac{(1 + c_k \cos\omega_i + d_k \cos 2\omega_i)\cos\omega_i + (c_k \sin\omega_i + d_k \sin 2\omega_i)\sin\omega_i}{(1 + c_k \cos\omega_i + d_k \cos 2\omega_i)^2 + (c_k \sin\omega_i + d_k \sin 2\omega_i)^2}$$
$$\frac{\partial \left|H(e^{j\omega_i}, \boldsymbol{x})\right|}{\partial d_k} = - \left|H(e^{j\omega_i}, \boldsymbol{x})\right|$$
$$\frac{(1 + c_k \cos\omega_i + d_k \cos 2\omega_i)\cos 2\omega_i + (c_k \sin\omega_i + d_k \sin 2\omega_i)\sin 2\omega_i}{(1 + c_k \cos\omega_i + d_k \cos 2\omega_i)^2 + (c_k \sin\omega_i + d_k \sin 2\omega_i)^2}$$
$$\frac{\partial \left|H(e^{j\omega_i}, \boldsymbol{x})\right|}{\partial A} = \frac{\left|H(e^{j\omega_i}, \boldsymbol{x})\right|}{A}$$

A quasi-Newton algorithm [2] can be applied to the design problem where its kth iteration updates the design variable $\boldsymbol{x}_k$ to $\boldsymbol{x}_{k+1}$ as

$$\boldsymbol{x}_{k+1} = \boldsymbol{x}_k + \alpha_k \boldsymbol{d}_k$$

where

$$\boldsymbol{d}_k = -\boldsymbol{S}_k \nabla J_2(\boldsymbol{x}_k), \qquad \alpha_k = \arg\min_{\alpha} J_2(\boldsymbol{x}_k + \alpha \boldsymbol{d}_k)$$

$$\boldsymbol{S}_{k+1} = \boldsymbol{S}_k + \left(1 + \frac{\boldsymbol{\gamma}_k^T \boldsymbol{S}_k \boldsymbol{\gamma}_k}{\boldsymbol{\gamma}_k^T \boldsymbol{\delta}_k}\right) \frac{\delta_k \delta_k^T}{\boldsymbol{\gamma}_k^T \boldsymbol{\delta}_k} - \frac{\boldsymbol{\delta}_k \boldsymbol{\gamma}_k^T \boldsymbol{S}_k + \boldsymbol{S}_k \boldsymbol{\gamma}_k \boldsymbol{\delta}_k^T}{\boldsymbol{\gamma}_k^T \boldsymbol{\delta}_k}, \qquad \boldsymbol{S}_0 = \boldsymbol{I}_{4K+1}$$

$$\boldsymbol{\delta}_k = \boldsymbol{x}_{k+1} - \boldsymbol{x}_k, \qquad \boldsymbol{\gamma}_k = \nabla J_2(\boldsymbol{x}_{k+1}) - \nabla J_2(\boldsymbol{x}_k)$$

The step size α_k is calculated by minimizing the single-variable function $J_2(\boldsymbol{x}_k + \alpha \boldsymbol{d}_k)$. The inexact line search technique initiated by Fletcher is often found effective to perform this step [2]. The quasi-Newton algorithm starts with an initial point $\boldsymbol{x}_0$ that is associated with a stable but trivial IIR filter, and the iterations continue until $|J_2(\boldsymbol{x}_{k+1}) - J_2(\boldsymbol{x}_k)|$ is less than a prescribed tolerance ε.

Since the above optimization is carried out without constraints on filter stability, it is possible that at convergence some poles are outside the unit circle. Suppose the ith second-order section of the IIR filter, i.e.

$$\frac{1 + a_i z^{-1} + b_i z^{-2}}{1 + c_i z^{-1} + d_i z^{-2}} \tag{7.4}$$

is found to be unstable. Denoting its two poles by p_{i1} and p_{i2}, we can write the denominator as

$$1 + c_i z^{-1} + d_i z^{-2} = \left(1 - p_{i1} z^{-1}\right)\left(1 - p_{i2} z^{-1}\right)$$

If only one of the poles, say p_{i1}, is an unstable pole, then replacing the filter section in (7.4) with

$$\frac{1 + a_i z^{-1} + b_i z^{-2}}{-p_{i1}\left(1 - p_{i1}^{-1} z^{-1}\right)\left(1 - p_{i2} z^{-1}\right)}$$

stabilizes the filter section without changing its magnitude response. Similarly, if both poles are unstable, then replacing the section in (7.4) with

$$\frac{1 + a_i z^{-1} + b_i z^{-2}}{p_{i1} p_{i2}\left(1 - p_{i1}^{-1} z^{-1}\right)\left(1 - p_{i2}^{-1} z^{-1}\right)}$$

stabilizes the section.

As an example, the above algorithm is applied to the design of a sixth-order lowpass IIR digital filter with normalized passband edge $\omega_p = 0.5\pi$ and stopband edge $\omega_a = 0.575\pi$. The total number of frequency grid points in the passband and stopband is set to L=70 and the convergence tolerance is set to $\varepsilon = 10^{-6}$. The initial point is a 13-component vector $\boldsymbol{x}_0$ corresponding to a 7-tap averaging FIR filter whose impulse response is $\{1/7, 1/7, \cdots, 1/7\}$. It took the algorithm 201 iterations to converge and the magnitude response of the IIR obtained is depicted in Figure 7.1.

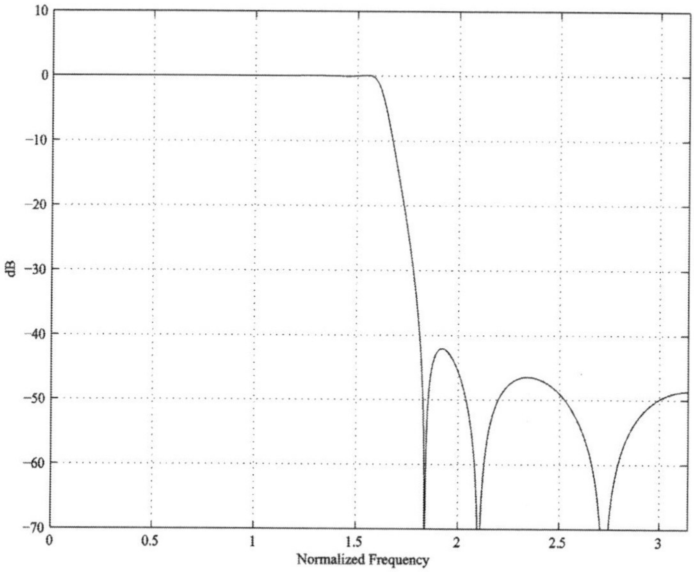

Figure 7.1 Magnitude response of the 6th-order lowpass IIR filter.

7.2.2 An Equiripple Design by Linear Programming

The design technique describe below is based on the algorithm proposed in [3], see also [4]. Let the transfer function of the IIR filter be given by

$$H(z) = \frac{A(z)}{B(z)} = \frac{\sum\limits_{i=0}^{M} a_i z^{-i}}{\sum\limits_{i=0}^{N} b_i z^{-i}} \quad \text{with } b_0 = 1 \tag{7.5}$$

The squared magnitude response of the filter can be expressed as

$$\begin{aligned} |H(z)|^2 &= H(z)H(z^{-1}) = \frac{A(z)A(z^{-1})}{B(z)B(z^{-1})} \\ &= \frac{\left(\sum\limits_{i=0}^{M} a_i z^{-i}\right)\left(\sum\limits_{i=0}^{M} a_i z^{i}\right)}{\left(\sum\limits_{i=0}^{N} b_i z^{-i}\right)\left(\sum\limits_{i=0}^{N} b_i z^{i}\right)} = \frac{\sum\limits_{i=-M}^{M} c_i z^{-i}}{\sum\limits_{i=-N}^{N} d_i z^{-i}} \end{aligned}$$

with $c_i = c_{-i}$ and $d_i = d_{-i}$. Hence the magnitude-squared function of the filter is given by

$$\left|H(e^{j\omega})\right|^2 = \frac{C(\omega)}{D(\omega)} = \frac{c_0 + 2\sum\limits_{i=1}^{M} c_i \cos(i\omega)}{d_0 + 2\sum\limits_{i=1}^{N} d_i \cos(i\omega)} \tag{7.6}$$

Let $F(\omega)$ be the desired magnitude-squared function and $\varepsilon(\omega)$ be a tolerance function on the approximation error, i.e.,

$$\left|\frac{C(\omega)}{D(\omega)} - F(\omega)\right| \leq \varepsilon(\omega)$$

It follows that

$$\begin{aligned} C(\omega) - [F(\omega) - \varepsilon(\omega)]\, D(\omega) \geq 0 \\ -C(\omega) + [F(\omega) + \varepsilon(\omega)]\, D(\omega) \geq 0 \end{aligned} \tag{7.7}$$

In addition, (7.6) implies that

$$\begin{aligned} C(\omega) \geq 0 \\ D(\omega) \geq 0 \end{aligned} \tag{7.8}$$

which are held in the entire baseband. By adding an auxiliary variable η to the right-hand side of each constraint in (7.7) and (7.8), one seeks for a solution of the optimization problem

$$\begin{aligned} &\text{minimize} \quad \eta \\ &\text{subject to:} \quad C(\omega) - [F(\omega) - \varepsilon(\omega)]\, D(\omega) + \eta \geq 0 \\ &\qquad -C(\omega) + [F(\omega) + \varepsilon(\omega)]\, D(\omega) + \eta \geq 0 \\ &\qquad C(\omega) + \eta \geq 0 \\ &\qquad D(\omega) + \eta \geq 0 \end{aligned} \tag{7.9}$$

where $F(\omega)$ and $\varepsilon(\omega)$ are given as design specifications with frequency ω varying over a dense but finite set of grids in the frequency bands of interest, and the unknowns are coefficients $\{c_i | \, i = 0, 1, \cdots, M\}$, $\{d_i | \, i = 0, 1, \cdots, N\}$, and η. Since the objective function and all constraints are linearly dependent on the design parameters, (7.9) is a linear programming (LP) problem which can be solved efficiently [2].

Concerning the solution of problem (7.9) for given $F(\omega)$ and $\varepsilon(\omega)$, there are three possibilities:

(i) Problem (7.9) has a solution with $\eta = 0$. This means there exists a solution satisfying all constraints in (7.7) and (7.8), and hence a stable IIR filter $H(z)$ satisfying the design specifications can be obtained by spectrum factorization [5] of the optimized $C(\omega)/D(\omega)$.

(ii) Problem (7.9) has a solution with $\eta < 0$. This means that there exist IIR filters with design specifications more restrictive than the current error tolerance. Such designs can be obtained by imposing a more demanding $\varepsilon(\omega)$ such that the solution of (7.9) yields $\eta = 0$. Once this is achieved, a stable IIR filter $H(z)$ satisfying the design specifications can be obtained by spectrum factorization of the optimized $C(\omega)/D(\omega)$.

(iii) Problem (7.9) has a solution with $\eta > 0$. This means that IIR filters with current specifications do not exist. In order to produce a design, a less demanding $\varepsilon(\omega)$ should be used so that the solution of (7.9) yields $\eta = 0$. Once this is achieved, a stable IIR filter is obtained.

A transfer function $H(z)$ satisfying the design specifications can be obtained by spectrum factorization of the optimized $C(\omega)/D(\omega)$.

As an example, the algorithm was applied to design a fourth-order IIR lowpass filter with normalized passband edge $\omega_p = 0.4\pi$ and stopband edge $\omega_a = 0.5\pi$. Suppose it is required that the magnitude responses of the IIR filter in the passband and stopband vary in the range $[1-\delta,\ 1+\delta]$ and $[0,\ \delta]$ respectively, then the passband and stopband squared magnitude responses, $C(\omega)/D(\omega)$, will vary in the range $[1+\delta^2-2\delta,\ 1+\delta^2+2\delta]$ and $[0,\ \delta^2]$ respectively. Hence the desired magnitude-squared function $F(\omega)$ and tolerance function $\varepsilon(\omega)$ are given by

$$F(\omega) = \begin{cases} 1+\delta^2 & \text{for } \omega \in [0,\ \omega_p] \\ \delta^2/2 & \text{for } \omega \in [\omega_a,\ \pi] \end{cases}$$

and

$$\varepsilon(\omega) = \begin{cases} 2\delta & \text{for } \omega \in [0,\ \omega_p] \\ \delta^2/2 & \text{for } \omega \in [\omega_a,\ \pi] \end{cases}$$

In the algorithm implementation, a total of 150 frequency grids uniformly placed in the passband and stopband were used for the first two constraints in (7.9), while 120 frequency grid uniformly placed on $[0,\ \pi]$ were used for the last two constraints in (7.9). By trial and error, it was found that with $\delta = 0.02645$, the LP problem in (7.9) has a solution with $\eta = -7.38 \times 10^{-5}$ which is practically zero. The largest magnitude of the poles of the IIR filter obtained from the LP solution was found to be 0.8961, and the magnitude response of the filter is depicted in Figure 7.2.

7.2.3 Weighted Least-Squares Design with Stability Constraints

Stable IIR digital filters that optimally approximate arbitrary magnitude as well as phase responses in weighted least-squares sense can be designed using convex quadratic programming (QP). The method described below is similar in spirit to that reported in [6].

Let the IIR filter be expressed in terms of K second-order cascaded sections as

$$H(z, \mathbf{x}) = A \prod_{k=1}^{K} \frac{1 + a_k z^{-1} + b_k z^{-2}}{1 + c_k z^{-1} + d_k z^{-2}} \tag{7.10}$$

where

$$\mathbf{x} = \begin{bmatrix} a_1 & b_1 & c_1 & d_1 & \cdots & a_K & b_K & c_K & d_K & A \end{bmatrix}^T$$

is a vector of $4K+1$ design variables. Let $H_d(\omega)$ be the desired frequency response and Ω be frequency region of interest at which the error of the actual

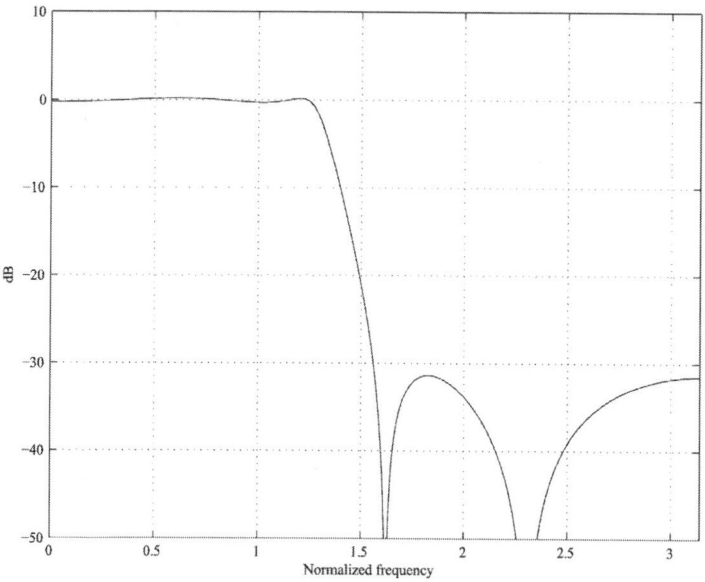

Figure 7.2 Magnitude response of the 4th-order lowpass IIR filter.

frequency response $H(e^{j\omega}, \boldsymbol{x})$ approximating a desired response $H_d(\omega)$ is evaluated in a weighted least-squares manner:

$$E_2(\boldsymbol{x}) = \int_\Omega W(\omega) \left| H(e^{j\omega}, \boldsymbol{x}) - H_d(\omega) \right|^2 d\omega \tag{7.11}$$

where $W(\omega) \geq 0$ is a weighting function defined over Ω.

The design is accomplished in an iterative manner. Suppose one has a reasonable initial point $\boldsymbol{x}_0$ to start, in the kth iteration one can write

$$H(e^{j\omega}, \boldsymbol{x}_k + \boldsymbol{\delta}) \simeq H(e^{j\omega}, \boldsymbol{x}_k) + g_k^T(\omega)\boldsymbol{\delta} \tag{7.12}$$

provided that $\|\boldsymbol{\delta}\|$ is small, where $\boldsymbol{g}_k(\omega)$ is the gradient of $H(e^{j\omega}, \boldsymbol{x})$ at $\boldsymbol{x}_k$. The optimal updating vector $\boldsymbol{\delta}_k$ is obtained by minimizing

$$\begin{aligned} E_2(\boldsymbol{x}_k + \boldsymbol{\delta}) &= \int_\Omega W(\omega) \left| H(e^{j\omega}, \boldsymbol{x}_k + \boldsymbol{\delta}) - H_d(\omega) \right|^2 d\omega \\ &\simeq \int_\Omega W(\omega) \left| H(e^{j\omega}, \boldsymbol{x}_k) + \boldsymbol{g}_k^T(\omega)\boldsymbol{\delta} - H_d(\omega) \right|^2 d\omega \\ &= \boldsymbol{\delta}^{\mathrm{T}} \boldsymbol{Q}_k \boldsymbol{\delta} + 2\boldsymbol{\delta}^{\mathrm{T}} \boldsymbol{q}_k + \kappa \end{aligned}$$

where

$$\boldsymbol{Q}_k = \int_\Omega W(\omega)\boldsymbol{g}_k(\omega)\boldsymbol{g}_k^H(\omega)\, d\omega$$

$$\boldsymbol{q}_k = Re\left\{\int_\Omega W(\omega)\left[\bar{H}(e^{j\omega}, x_k) - \bar{H}_d(\omega)\right]\boldsymbol{g}_k(\omega)\, d\omega\right\}$$

$$\kappa = \int_\Omega W(\omega)\left|H(e^{j\omega}, \boldsymbol{x}_k) - H_d(\omega)\right|^2 d\omega$$

subject to two constraints: the filter is stable and δ is small in magnitude. In order to implement the iteration, we use (7.10) to evaluate the gradient $\boldsymbol{g}_k(\omega)$ as

$$\boldsymbol{g}_k(\omega) = \begin{bmatrix} \partial H(e^{j\omega}, \boldsymbol{x}_k)\Big/\partial a_1 \\ \partial H(e^{j\omega}, \boldsymbol{x}_k)\Big/\partial b_1 \\ \partial H(e^{j\omega}, \boldsymbol{x}_k)\Big/\partial c_1 \\ \partial H(e^{j\omega}, \boldsymbol{x}_k)\Big/\partial d_1 \\ \vdots \\ \partial H(e^{j\omega}, \boldsymbol{x}_k)\Big/\partial A \end{bmatrix} \tag{7.13a}$$

with

$$\begin{aligned} \frac{\partial H(e^{j\omega}, \boldsymbol{x}_k)}{\partial a_i} &= \frac{e^{-j\omega}H(e^{j\omega}, \boldsymbol{x}_k)}{1 + a_i e^{-j\omega} + b_i e^{-j2\omega}} \\ \frac{\partial H(e^{j\omega}, \boldsymbol{x}_k)}{\partial b_i} &= \frac{e^{-j2\omega}H(e^{j\omega}, \boldsymbol{x}_k)}{1 + a_i e^{-j\omega} + b_i e^{-j2\omega}} \\ \frac{\partial H(e^{j\omega}, \boldsymbol{x}_k)}{\partial c_i} &= -\frac{e^{-j\omega}H(e^{j\omega}, \boldsymbol{x}_k)}{1 + c_i e^{-j\omega} + d_i e^{-j2\omega}} \\ \frac{\partial H(e^{j\omega}, \boldsymbol{x}_k)}{\partial d_i} &= -\frac{e^{-j2\omega}H(e^{j\omega}, \boldsymbol{x}_k)}{1 + c_i e^{-j\omega} + d_i e^{-j2\omega}} \\ \frac{\partial H(e^{j\omega}, \boldsymbol{x}_k)}{\partial A} &= \frac{H(e^{j\omega}, \boldsymbol{x}_k)}{A} \end{aligned} \tag{7.13b}$$

for $i = 1, 2, \cdots, K$. The desired frequency response is typically specified in terms of desired magnitude response $M(\omega)$ and phase response $\theta(\omega)$, hence one can write

$$H_d(\omega) = M(\omega)e^{j\theta(\omega)} = M(\omega)\cos\theta(\omega) + jM(\omega)\sin\theta(\omega) \tag{7.14}$$

To ensure the filter's stability, the denominator parameters of each 2nd-order section must lie inside the stability triangle (see Section 3.2.6 of Chapter 3), namely,

$$d_i < 1, \ \ c_i + d_i > -1, \ \ c_i - d_i < 1$$

These constraints can be expressed as $\boldsymbol{E}\boldsymbol{u}_i + \boldsymbol{e} > \boldsymbol{0}$ with

$$\boldsymbol{E} = \begin{bmatrix} 1 & 1 \\ -1 & 1 \\ 0 & -1 \end{bmatrix}, \quad \boldsymbol{u}_i = \begin{bmatrix} c_i \\ d_i \end{bmatrix}, \quad \boldsymbol{e} = \begin{bmatrix} 1 \\ 1 \\ 1 \end{bmatrix}$$

To prevent the poles from being too close to the boundary of the stability region, one may require that $\boldsymbol{E}\boldsymbol{u}_i + (1-\tau)\boldsymbol{e} \geq \boldsymbol{0}$ with a small $\tau > 0$. Let $\boldsymbol{I}_i$ be a selection matrix such that $\boldsymbol{u}_i = \boldsymbol{I}_i\boldsymbol{x}$, then the stability constraints are expressed as $\boldsymbol{E}_i\boldsymbol{x} + (1-\tau)\boldsymbol{e} \geq \boldsymbol{0}$ where $\boldsymbol{E}_i = \boldsymbol{E}\boldsymbol{I}_i$ for $i = 1, 2, \cdots, K$. The kth iteration of the design is now performed by solving the convex quadratic programming (QP) problem

$$\begin{aligned} &\text{minimize } \boldsymbol{\delta}^{\mathrm{T}}\boldsymbol{Q}_k\boldsymbol{\delta} + 2\boldsymbol{\delta}^{\mathrm{T}}\boldsymbol{q}_k + \kappa \\ &\text{subject to: } \boldsymbol{E}_i(\boldsymbol{x}_k + \boldsymbol{\delta}) + (1-\tau)\boldsymbol{e} \geq \boldsymbol{0} \quad \text{for } i = 1, 2, \cdots, K \\ &|(\boldsymbol{\delta})_j| \leq \beta \quad \text{for } j = 1, 2, \cdots, 4K+1 \end{aligned} \tag{7.15}$$

where $\beta > 0$ is a small bound to ensure $\|\boldsymbol{\delta}\|$ is small. Having obtained a solution $\boldsymbol{\delta}_k$ of (7.15), the design vector is updated to $\boldsymbol{x}_{k+1} = \boldsymbol{x}_k + \boldsymbol{\delta}_k$, and the iteration continues until $\|\boldsymbol{x}_{k+1} - \boldsymbol{x}_k\|$ is less than a prescribed convergence tolerance ε.

As an example, the algorithm was applied to design a fourteenth-order IIR lowpass filter with normalized passband edge $\omega_p = 0.4\pi$ and stopband edge $\omega_a = 0.45\pi$. The desired frequency response includes a linear phase response in the passband with group delay being 15.5. In design implementation, a total of 90 uniformly placed frequency grids were used to evaluate $\boldsymbol{Q}_k$ and $\boldsymbol{q}_k$ in (7.15), the weighting function, stability margin, and bound for increment of design vector were set to $W(\omega) \equiv 1$, $\tau = 0.12$, and $\beta = 0.03$, respectively. With tolerance $\varepsilon = 0.03$, it took the algorithm 62 iterations to converge and the magnitude and passband phase response of the IIR filter obtained are depicted in Figure 7.3(a) and (b), respectively. The largest magnitude of the poles was found to be 0.9381 that ensures the filter's stability.

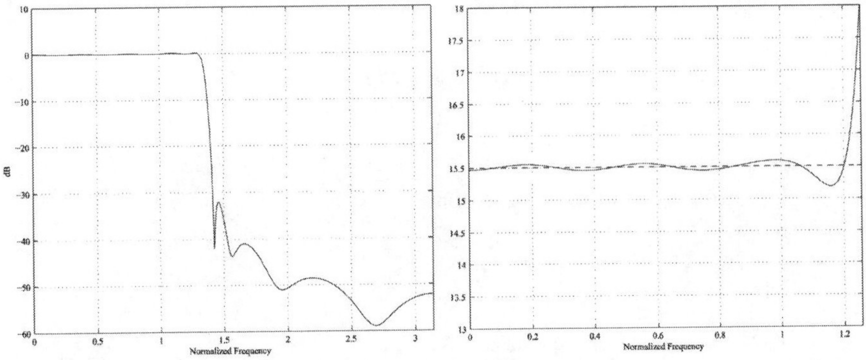

Figure 7.3 The frequency characteristics of the IIR filter. (a) Magnitude response (left side). (b) Passband phase response (right side).

7.2.4 Minimax Design with Stability Constraints

Convex programming techniques are also useful in designing stable IIR filters that optimally approximate arbitrary magnitude and phase responses in minimax sense in that maximum approximation error in the frequency domain is minimized. The design method described below is based on the work reported in [6].

Let the IIR filter be expressed in terms of K second-order cascaded sections as given by (7.10), $H_d(\omega)$ be the desired frequency response, and Ω be frequency region of interest at which the error of the actual frequency response $H(e^{j\omega}, \boldsymbol{x})$ approximating a desired response $H_d(\omega)$ is evaluated in a the following manner:

$$E_\infty(\boldsymbol{x}) = \underset{\omega\in\Omega}{\text{maximize}} \; W(\omega)\left|H(e^{j\omega}, \boldsymbol{x}) - H_d(\omega)\right| \qquad (7.16)$$

where $W(\omega) \geq 0$ is a weighting function defined over Ω. The design is accomplished in an iterative manner. Suppose one has a reasonable initial point $\boldsymbol{x}_0$ to start, in the kth iteration one can write

$$H(e^{j\omega}, \boldsymbol{x}_k + \boldsymbol{\delta}) \simeq H(e^{j\omega}, \boldsymbol{x}_k) + \boldsymbol{g}_k^T(\omega)\boldsymbol{\delta} \qquad (7.17)$$

provided that $\|\boldsymbol{\delta}\|$ is small, where $\boldsymbol{g}_k(\omega)$ is the gradient of $H(e^{j\omega}, \boldsymbol{x})$ at $\boldsymbol{x}_k$ and can be evaluated using (7.13a) and (7.13b). The optimal updating vector δ_k is obtained by minimizing

$$
\begin{aligned}
E_\infty(\boldsymbol{x}+\boldsymbol{\delta}) &= \underset{\omega\in\Omega}{\text{maximize}}\;\; W(\omega)\left|H(e^{j\omega},\boldsymbol{x}+\boldsymbol{\delta}) - H_d(\omega)\right| \\
&\simeq \underset{\omega\in\Omega}{\text{maximize}}\;\; W(\omega)\left|H(e^{j\omega},\boldsymbol{x}_k) + \boldsymbol{g}_k^T(\omega)\boldsymbol{\delta} - H_d(\omega)\right|
\end{aligned} \tag{7.18}
$$

By introducing an upper bound η of $W(\omega)\left|H(e^{j\omega},\boldsymbol{x}_k) + \boldsymbol{g}_k^T(\omega)\boldsymbol{\delta} - H_d(\omega)\right|$ for $\omega \in \Omega$, minimizing the approximate $E_\infty(\boldsymbol{x}+\boldsymbol{\delta})$ in (7.18) can be expressed as

$$
\begin{aligned}
&\text{minimize} \quad \eta \\
&\text{subject to: } W(\omega)\left|\boldsymbol{g}_k^T(\omega)\boldsymbol{\delta} + D_k(\omega)\right| \le \eta \quad \omega\in\Omega
\end{aligned}
$$

where $D_k(\omega) = H(e^{j\omega},\boldsymbol{x}_k) - H_d(\omega)$ in a known quantity in the kth iteration. To complete the design formulation, additional constraints concerning the stability of the filter and smallness of the increment vector δ are also imposed (see (7.15)) and the kth iteration of the algorithm is carried out by solving the optimization problem

$$
\begin{aligned}
&\text{minimize} \quad \eta \\
&\text{subject to: } W(\omega)\left|\boldsymbol{g}_k^T(\omega)\boldsymbol{\delta} + D_k(\omega)\right| \le \eta \quad \omega\in\Omega \\
&\boldsymbol{E}_i(\boldsymbol{x}_k+\boldsymbol{\delta}) + (1-\tau)\boldsymbol{e} \ge \boldsymbol{0} \quad \text{for } i = 1,2,\cdots,K \\
&|\boldsymbol{\delta}_j| \le \beta \quad \text{for } j = 1,2,\cdots,4K+1
\end{aligned} \tag{7.19}
$$

where scalar bound η and increment vector $\boldsymbol{\delta}$ are variables, the last two sets of constraints are linear, while the first set of constraints are of second-order cone type [2]. To see this, only a set of dense discrete frequencies $\Omega_d = \{\omega_i | \, i = 1,2,\cdots,L\} \subset \Omega$ is considered, and for each $\omega_i \in \Omega_d$ the first constraint in (7.19) is written as

$$
\left\| \begin{bmatrix} W(\omega_i)\boldsymbol{g}_{rk}^T(\omega_i) \\ W(\omega_i)\boldsymbol{g}_{jk}^T(\omega_i) \end{bmatrix} \boldsymbol{\delta} + \begin{bmatrix} W(\omega_i)D_{rk}(\omega_i) \\ W(\omega_i)D_{jk}(\omega_i) \end{bmatrix} \right\| \le \eta \quad \text{for } i = 1,\ 2,\ \cdots,\ L \tag{7.20}
$$

where $\boldsymbol{g}_{rk}(\omega_i)$ and $\boldsymbol{g}_{jk}(\omega_i)$ are the real and imaginary parts of $\boldsymbol{g}_k(\omega_i)$, respectively, and $D_{rk}(\omega_i)$ and $D_{jk}(\omega_i)$ are the real and imaginary parts of $D_k(\omega)$, respectively. Replacing the first set of constraints in (7.19) with those in (7.20), one obtains a standard second-order cone programming (SOCP)problem as

$$\begin{aligned}
&\text{minimize} \quad \eta \\
&\text{subject to:} \quad \left\| \begin{bmatrix} W(\omega_i)\boldsymbol{g}_{rk}^T(\omega_i) \\ W(\omega_i)\boldsymbol{g}_{jk}^T(\omega_i) \end{bmatrix} \boldsymbol{\delta} + \begin{bmatrix} W(\omega_i)D_{rk}(\omega_i) \\ W(\omega_i)D_{jk}(\omega_i) \end{bmatrix} \right\| \leq \eta \\
&\text{for } i = 1, 2, \cdots, L \\
&\boldsymbol{E}_i(\boldsymbol{x}_k + \boldsymbol{\delta}) + (1-\tau)\boldsymbol{e} \geq \boldsymbol{0} \quad \text{for } i = 1, 2, \cdots, K \\
&|(\boldsymbol{\delta})_j| \leq \beta \quad \text{for } j = 1, 2, \cdots, 4K+1
\end{aligned} \tag{7.21}$$

which can be solved efficiently [2]. Having obtained a solution δ_k of (7.21), the design vector is updated to $\boldsymbol{x}_{k+1} = \boldsymbol{x}_k + \boldsymbol{\delta}_k$, and the iteration continues until $\|\boldsymbol{x}_{k+1} - \boldsymbol{x}_k\|$ is less than a prescribed convergence tolerance ε.

As an example, the algorithm was applied to design a twelfth-order IIR lowpass filter with normalized passband edge $\omega_p = 0.5\pi$ and stopband edge $\omega_a = 0.6\pi$. The desired frequency response includes a linear phase response in the passband with group delay being 12. In design implementation, a total of $L = 80$ uniformly placed frequency grids were used in (7.21), the weighting function, stability margin, and bound for increment of design vector were set to $W(\omega) = 1$, $\tau = 0.12$, and $\beta = 0.03$, respectively. With tolerance $\varepsilon = 0.05$, it took the algorithm 49 iterations to converge and the magnitude and passband phase response of the IIR filter obtained are depicted in Figure 7.4(a) and (b), respectively. The largest magnitude of the poles was found to be 0.9381 that ensures the filter's stability.

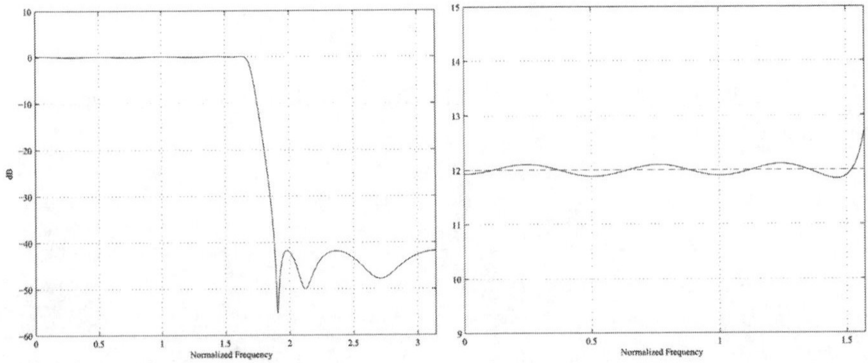

Figure 7.4 The frequency characteristics of the IIR filter. (a) Magnitude response (left side). (b) Passband phase response (right side).

7.3 Design of All-Pass Digital Filters

7.3.1 Design of All-Pass Filters Based on Frequency Response Error

The transfer function of an nth-order all-pass digital filter is described by

$$H(z) = \frac{a_n + a_{n-1}z^{-1} + \cdots + a_1 z^{-(n-1)} + z^{-n}}{1 + a_1 z^{-1} + \cdots + a_{n-1}z^{-(n-1)} + a_n z^{-n}}$$
$$= z^{-n}\frac{\sum_{k=0}^{n} a_k z^k}{\sum_{k=0}^{n} a_k z^{-k}}, \qquad a_0 = 1 \tag{7.22}$$

whose frequency response is given by

$$H(e^{j\omega}) = e^{-jn\omega}\frac{\sum_{k=0}^{n} a_k e^{jk\omega}}{\sum_{k=0}^{n} a_k e^{-jk\omega}} = e^{j\theta(\omega)} \tag{7.23}$$

where

$$\theta(\omega) = -n\omega + 2\tan^{-1}\left(\frac{\sum_{k=0}^{n} a_k \sin k\omega}{\sum_{k=0}^{n} a_k \cos k\omega}\right)$$

Let $d(\omega)$ be a desired phase characteristic. We now define a complex error function, called *frequency response error*, as

$$E(e^{j\omega}) = \frac{1}{2}\left[H(e^{j\omega}) - H_d(e^{j\omega})\right]$$
$$= \frac{1}{2}\left[e^{j\theta(\omega)} - e^{jd(\omega)}\right] \tag{7.24}$$

where $H_d(e^{j\omega}) = e^{jd(\omega)}$. By substituting (7.23) into (7.24), we obtain

$$
\begin{aligned}
E(e^{j\omega}) &= \frac{1}{2}\left[e^{-jn\omega}\frac{\sum_{k=0}^{n} a_k e^{jk\omega}}{\sum_{k=0}^{n} a_k e^{-jk\omega}} - e^{jd(\omega)}\right] \\
&= \frac{1}{2}\, e^{-j\frac{n\omega-d(\omega)}{2}} \cdot \frac{\sum_{k=0}^{n} a_k \left(e^{j(k\omega-\frac{n\omega+d(\omega)}{2})} - e^{-j(k\omega-\frac{n\omega+d(\omega)}{2})}\right)}{\sum_{k=0}^{n} a_k e^{-jk\omega}} \\
&= j e^{-j\frac{n\omega-d(\omega)}{2}} \cdot \frac{\sum_{k=0}^{n} a_k \sin\left(k\omega - \frac{n\omega+d(\omega)}{2}\right)}{\sum_{k=0}^{n} a_k e^{-jk\omega}}
\end{aligned}
\tag{7.25}
$$

From (7.25), the amplitude $\hat{E}(e^{j\omega})$ of the frequency response error in (7.24) is found to be

$$
\hat{E}(e^{j\omega}) = \frac{\sum_{k=0}^{n} a_k \sin\left(k\omega - \frac{n\omega+d(\omega)}{2}\right)}{\left|\sum_{k=0}^{n} a_k e^{-jk\omega}\right|} \tag{7.26}
$$

The Remez algorithm (see Section 5.7) can be applied to the numerator in (7.26) to minimize the amplitude error in (7.26) iteratively subject to some linearized constraints as detailed below. We first choose $n+1$ initial values of extremal frequencies ω_i for $i = 0, 1, \cdots, n$ over a frequency domain of interest appropriately. Then we design an all-pass digital filter so that $\hat{E}(e^{j\omega})$ alternates in equiripple with respect to value 0 at $n+1$ extremal frequencies over a frequency domain of interest. Hence, (7.26) is written as

$$
\hat{E}(e^{j\omega_i}) = (-1)^i \delta \tag{7.27}
$$

where δ an initial amplitude error. By substituting (7.26) into (7.27) and multiplying the both sides by $|A(e^{j\omega_i})|$, we obtain

$$\sum_{k=0}^{n} a_k \sin\Big(k\omega_i - \frac{n\omega_i + d(\omega_i)}{2}\Big) = (-1)^i \delta \, |A(e^{j\omega_i})| \;\text{ for } i = 0, 1, \cdots, n \tag{7.28}$$

where

$$\begin{aligned}|A(e^{j\omega})| &= \Big|\sum_{k=0}^{n} a_k e^{-jk\omega}\Big| \\ &= \sqrt{\Big(\sum_{k=0}^{n} a_k \cos k\omega\Big)^2 + \Big(\sum_{k=0}^{n} a_k \sin k\omega\Big)^2}\end{aligned}$$

Equation (7.28) can be expressed in matrix form as

$$\begin{bmatrix} \phi(0,0) & \phi(0,1) & \cdots & \phi(0,n) \\ \phi(1,0) & \phi(1,1) & \cdots & \phi(1,n) \\ \vdots & \vdots & \ddots & \vdots \\ \phi(n,0) & \phi(n,1) & \cdots & \phi(n,n) \end{bmatrix} \begin{bmatrix} a_0 \\ a_1 \\ \vdots \\ a_n \end{bmatrix} = \begin{bmatrix} \delta \, |A(e^{j\omega_0})| \\ (-1)\delta \, |A(e^{j\omega_1})| \\ \vdots \\ (-1)^n \delta \, |A(e^{j\omega_n})| \end{bmatrix} \tag{7.29}$$

where

$$\phi(i,k) = \sin\Big(k\omega_i - \frac{n\omega_i + d(\omega_i)}{2}\Big)$$

The above simultaneous equations can be solved for the filter coefficients $\{a_i | \, i = 0, 1, \cdots, n\}$ provided that the locations of the $n+1$ extremal frequencies are known a priori. However, the resulting $\hat{E}(e^{j\omega})$ is not always equiripple over a frequency domain of interest. To make $\hat{E}(e^{j\omega})$ equiripple, the $n+1$ locations ω_i for $i = 0, 1, \cdots, n$ are adjusted iteratively over a frequency domain of interest by employing the Remez exchange algorithm which is a very efficient iterative procedure for determining the locations of the extremal frequencies. Step-by-step algorithmic details are given below, where new extremal frequencies of $\hat{E}(e^{j\omega})$, say ω_i', replace previous extremal frequencies $\{\omega_i | \, i = 0, 1, \cdots, n\}$, and the simultaneous equations in (7.29) are solved again, and this process continues until

$$\max_{0 \le i \le n} \{|\omega_i' - \omega_i|\} < \varepsilon \tag{7.30}$$

is satisfied where ε is a prescribed tolerance.

The Remez Exchange Algorithm

Step 1: Set the order n of an all-pass filter and desired phase characteristic $d(\omega)$.

Step 2: Select the initial values of extremal frequencies $\{\omega_i | \, i = 0, 1, \cdots, n\}$ which are equally spaced over a frequency domain of interest.

Step 3: Set $|A(e^{j\omega_i})| = 1$ for $i = 0, 1, \cdots, n$.

Step 4: Compute the coefficients of an all-pass filter $\{a_k | \, k = 0, 1, \cdots, n\}$ by solving (7.29), then multiply the coefficients found by a scaling constant to normalize the coefficient a_0 to unity.

Step 5: Find the new extremal frequencies $\{\omega_i' | \, i = 0, 1, \cdots, n\}$ from the values of $|E(e^{j\omega})|$ evaluated at the dense set of frequencies.

Step 6: If (7.30) is satisfied, go to *Step 8*. Otherwise, replace ω_i by ω_i' for $i = 0, 1, \cdots, n$ and set

$$\delta = \frac{1}{n+1} \sum_{i=0}^{n} |\hat{E}(e^{j\omega_i'})|$$

Step 7: Compute $|A(e^{j\omega_i})|$ for $i = 0, 1, \cdots, n$ with the coefficients $\{a_k | \, k = 0, 1, \cdots, n\}$ obtained at the previous pass, and go back to *Step 4*.

Step 8: Obtain the transfer function $H(z)$ in (7.22).

7.3.2 Design of All-Pass Filters Based on Phase Characteristic Error

Let the phase characteristic error be defined by

$$e(\omega) = \theta(\omega) - d(\omega) \tag{7.31}$$

where $\theta(\omega)$ and $d(\omega)$ denote the actual and desired phase characteristics, respectively. It follows from (7.31) and (7.23) that

$$\begin{aligned} e^{je(\omega)} &= e^{j[\theta(\omega) - d(\omega)]} = H(e^{j\omega}) e^{-jd(\omega)} \\ &= e^{-j[n\omega + d(\omega)]} \cdot \frac{\sum_{k=0}^{n} a_k e^{jk\omega}}{\sum_{k=0}^{n} a_k e^{-jk\omega}} = \frac{\sum_{k=0}^{n} a_k e^{j\varphi_k(\omega)}}{\sum_{k=0}^{n} a_k e^{-j\varphi_k(\omega)}} \end{aligned} \tag{7.32}$$

where

$$\varphi_k(\omega) = k\omega - \frac{n\omega + d(\omega)}{2}$$

From (7.32), the phase characteristic error can be expressed as

$$e(\omega) = 2\tan^{-1}\left(\frac{\displaystyle\sum_{k=0}^{n} a_k \sin\varphi_k(\omega)}{\displaystyle\sum_{k=0}^{n} a_k \cos\varphi_k(\omega)}\right) \tag{7.33}$$

We now apply a Remez algorithm to (7.33) so that the sign of $e(\omega)$ alternates at $n+1$ extremal frequencies. We first select $n+1$ extremal frequencies $\{\omega_i | \, i = 0, 1, \cdots, n\}$ over a frequency domain of interest appropriately and define

$$e(\omega_i) = (-1)^i \delta \;\text{ for }\; i = 0, 1, \cdots, n \tag{7.34}$$

where δ stands for the phase characteristic error. By substituting (7.33) into (7.34), we arrive at

$$\sum_{k=0}^{n} a_k \sin\varphi_k(\omega_i) = \tan\frac{\delta}{2} \sum_{k=0}^{n} a_k \, (-1)^i \cos\varphi_k(\omega_i) \qquad i = 0, 1, \cdots, n \tag{7.35}$$

which can be expressed in matrix form as

$$\boldsymbol{P}\boldsymbol{a} = \tan\frac{\delta}{2}\boldsymbol{Q}\boldsymbol{a} \tag{7.36}$$

where

$$\boldsymbol{P} = \begin{bmatrix} \sin\varphi_0(\omega_0) & \sin\varphi_1(\omega_0) & \cdots & \sin\varphi_n(\omega_0) \\ \sin\varphi_0(\omega_1) & \sin\varphi_1(\omega_1) & \cdots & \sin\varphi_n(\omega_1) \\ \vdots & \vdots & \ddots & \vdots \\ \sin\varphi_0(\omega_n) & \sin\varphi_1(\omega_n) & \cdots & \sin\varphi_n(\omega_n) \end{bmatrix}, \qquad \boldsymbol{a} = \begin{bmatrix} 1 \\ a_1 \\ \vdots \\ a_n \end{bmatrix}$$

$$\boldsymbol{Q} = \begin{bmatrix} \cos\varphi_0(\omega_0) & \cos\varphi_1(\omega_0) & \cdots & \cos\varphi_n(\omega_0) \\ (-1)\cos\varphi_0(\omega_1) & (-1)\cos\varphi_1(\omega_1) & \cdots & (-1)\cos\varphi_n(\omega_1) \\ \vdots & \vdots & \ddots & \vdots \\ (-1)^n\cos\varphi_0(\omega_n) & (-1)^n\cos\varphi_1(\omega_n) & \cdots & (-1)^n\cos\varphi_n(\omega_n) \end{bmatrix}$$

Equation (7.36) can be written as

$$P^{-1}Qa = \lambda a \tag{7.37}$$

where

$$\lambda = \left(\tan\frac{\delta}{2}\right)^{-1}$$

Suppose the matrix $P^{-1}Q$ has the eigenvalues $\lambda_0, \lambda_1, \cdots, \lambda_n$ and corresponding eigenvectors $a_0, a_1, \cdots, a_n$ where $|\lambda_0| > |\lambda_1| > \cdots > |\lambda_n|$, then the eigenvector a_0 corresponding to the eigenvalue λ_0 is theoretically the optimal solution, because the phase error δ is related to λ by

$$\delta = 2\tan^{-1}\frac{1}{\lambda} \tag{7.38}$$

However, it was noted in [9] that the all-pass filter with the coefficient vector a_0 is not always stable. Moreover, it often occurs in this case that the Remez iterative algorithm either does not converge or converges to an unstable filter. Under these circumstances, an improved algorithm which takes filter stability into account is proposed in [9]. The algorithm is outlined below.

Step 1: Set the order n of an all-pass filter and desired phase characteristic $d(\omega)$.
Step 2: Select the initial values of extremal frequencies $\{\omega_i |\, i = 0, 1, \cdots, n\}$ which are equally spaced over a frequency domain of interest.
Step 3: Compute the eigenvalues $\lambda_0, \lambda_1, \cdots, \lambda_n$ and corresponding eigenvectors $a_0, a_1, \cdots, a_n$ of $P^{-1}Q$ where $|\lambda_0| > |\lambda_1| > \cdots > |\lambda_n|$.
Step 4: Set $l = 0$.
Step 5: Check whether the filter with a_l corresponding to λ_l is stable or not. If it is stable and the sign of $e(\omega_i)$ alternates at $n+1$ extremal frequencies, go to the next step. Otherwise, add l to 1 and go back to the top of this step.
Step 6: Find the new extremal frequencies $\{\omega_i' |\, i = 0, 1, \cdots, n\}$ from the values of $e(\omega)$ evaluated at the dense set of frequencies.
Step 7: If (7.30) is satisfied, go to the next step. Otherwise, replace ω_i by ω_i' for $i = 0, 1, \cdots, n$ and go back to *Step 3*.
Step 8: Obtain the transfer function $H(z)$ in (7.22).

7.3.3 A Numerical Example

As an example, the two algorithms studied above were applied to design an 8th-order all-pass digital filter where the desired phase characteristic is specified by

$$d(\omega) = \begin{cases} -7\omega & \text{for } |\omega| \le 0.5\pi \\ -7\omega - \pi & \text{for } |\omega| \ge 0.6\pi \end{cases}$$

The extremal frequencies were searched over discrete frequency points $\omega_i = \pi i/1000$ for $i = 0, 1, \cdots, 1000$. Nine initial values of extremal frequencies were chosen as $0.05\pi, 0.15\pi, 0.25\pi, 0.35\pi, 0.45\pi, 0.65\pi, 0.75\pi, 0.85\pi, 0.95\pi$ over a frequency domain of interest $\Omega_1 \bigcup \Omega_2$ where $\Omega_1 = \{\omega |\, 0 < \omega \le 0.5\pi\}$ and $\Omega_2 = \{\omega |\, 0.6 \le \omega < \pi\}$.

By applying the design method using frequency response error with initial amplitude of the frequency response error $E(e^{j\omega})$ being $\delta = 0.1$, it took the Remez exchange algorithm 5 iterations to converge to

$$\begin{bmatrix} a_0 & a_1 & a_2 \\ a_3 & a_4 & a_5 \\ a_6 & a_7 & a_8 \end{bmatrix} = \begin{bmatrix} 1.00000000 & 0.15520071 & 0.48184349 \\ -0.07326981 & -0.09658253 & 0.03963666 \\ 0.03849011 & -0.06678163 & -0.02050755 \end{bmatrix}$$

The maximum phase error and maximum magnitude of the frequency response error were found to be

$$7.93757266 \times 10^{-2} \text{ and } 3.96774452 \times 10^{-2}$$

respectively. The resulting phase and phase error characteristics are depicted in Figure 7.5, and the amplitude characteristic of the frequency response error is shown in Figure 7.6.

By employing the design method using phase characteristic error, it took the improved algorithm 5 iterations to converge to

$$\begin{bmatrix} a_0 & a_1 & a_2 \\ a_3 & a_4 & a_5 \\ a_6 & a_7 & a_8 \end{bmatrix} = \begin{bmatrix} 1.00000000 & 0.15519789 & 0.48184637 \\ -0.07326919 & -0.09658404 & 0.03963450 \\ 0.03849024 & -0.06677818 & -0.02050865 \end{bmatrix}$$

where all a_i's were scaled to normalize a_0 to unity. The maximum phase error was found to be $7.93644811 \times 10^{-2}$.

The resulting phase and phase error characteristics are depicted in Figure 7.7.

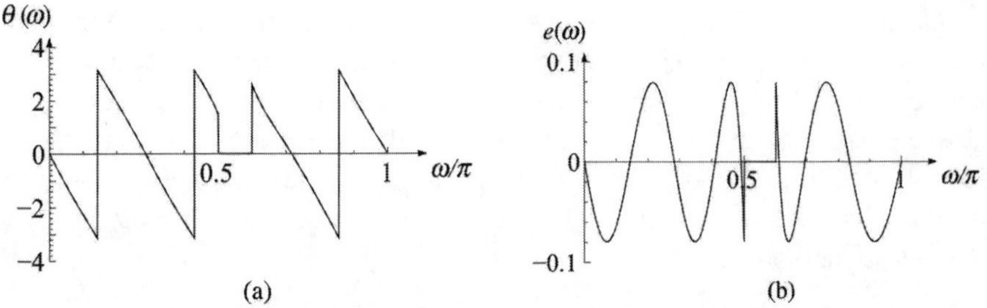

Figure 7.5 All-pass filter designed by using frequency response error. (a) Phase characteristic. (b) Phase error characteristic.

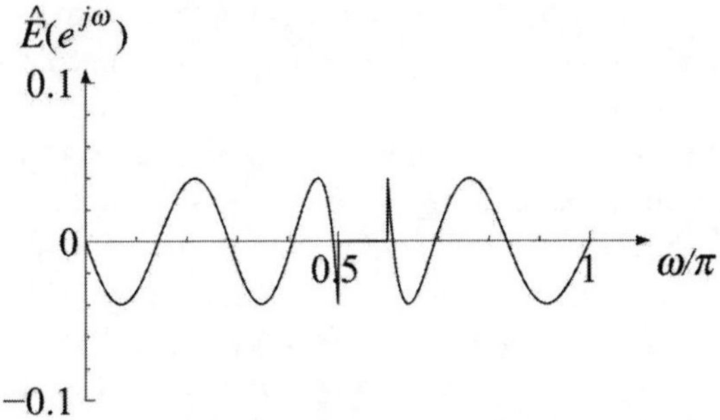

Figure 7.6 Amplitude characteristic of the frequency response error.

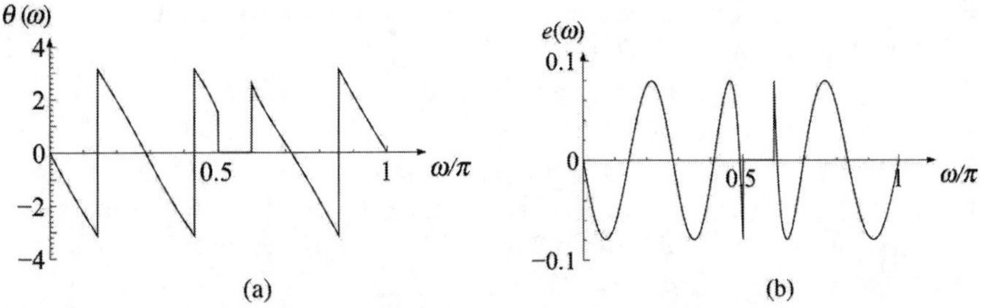

Figure 7.7 All-pass filter designed with phase characteristic error. (a) Phase characteristic. (b) Phase error characteristic.

7.4 Summary

This chapter has shown that the design of stable IIR filters can be achieved by applying optimization methods. In particular, an unconstrained optimization method based on a quasi-Newton algorithm has been applied to minimize mean squared error; linear programming has been applied to minimize a squared magnitude error for an equal-ripple design; convex quadratic programming has been used in weighted least squares design of stable IIR filters; and second-order cone programming has been used in minimax design of stable IIR filters. Based on the notion of frequency-response error and phase characteristic error, respectively, two Remez exchange type of algorithms for the design of all-pass digital filters have been presented. Design examples have been presented to illustrate these methods.

References

[1] K. Steiglitz, "Computer-aided design of recursive digital filters," *IEEE Trans. Audio and Electroacoust.*, vol. AU-18, no. 2, pp. 123–129, June 1970.

[2] A. Antoniou and W.-S. Lu, *Practical Optimization: Algorithms and Engineering Applications*, Springer, New York, 2007.

[3] L. R. Rabiner, N. Y. Graham, and H. D. Helms, "Linear programming design of IIR digital filters with arbitrary magnitude function," *IEEE Trans. Acoust., Speech, Signal Process.*, vol. ASSP-22, no. 2, pp. 117–123, Apr. 1974.

[4] L. R. Rabiner and R.-B. Gold, *Theory and Application of Digital Signal Processing*, Prentice-Hall, Englewood Cliffs, NJ., 1975.

[5] R. A. Roberts and C. T. Mullis, *Digital Signal Processing*, Addison-Wesley, Reading, MA, 1987.

[6] W.-S. Lu and T. Hinamoto, "Optimal design of IIR digital filters with robust stability using conic-quadratic-programming updates," *IEEE Trans. Signal Process.*, vol. 51, no. 6, pp. 1581–1592, June 2003.

[7] M. Ikehara, M. Funaishi and H. Kuroda, "Design of digital all-pass networks using Remez algorithm," *IEICE Trans. Fundamentals of Electronics, Communications and Computer Sciences*, vol. J74-A, no. 7, pp. 974–979, July 1991. (in Japanese)

[8] M. Ikehara, M. Funaishi and H. Kuroda, "Design of complex all-pass networks using Remez algorithm," *IEEE Trans. Circuits Syst. II*, vol. 39, no. 8, pp. 549–556, Aug. 1992.

[9] Y. Toguri and M. Ikehara, "A design method of all-pass networks based on the eigen filter method with consideration of the stability," *IEICE Trans. Fundamentals*, vol. E78-A, no. 7, pp. 885–889, July 1995.

[10] T. Q. Nguyen, T. I. Laakso and R. D. Koilpillai, "Eigenfilter approach for the design of allpass filters approximating a given phase response," *IEEE Trans. Signal Process.*, vol. 42, no. 9, pp. 2257–2263, Sep. 1994.

8

Design Methods in the Time Domain

8.1 Preview

The problem of designing an IIR digital filter involves determination of the coefficients a_i's and b_i's of a rational transfer function of the form

$$H(z) = \frac{N(z)}{D(z)} = \frac{b_0 + b_1 z^{-1} + \cdots + b_m z^{-m}}{1 + a_1 z^{-1} + \cdots + a_n z^{-n}} \tag{8.1}$$

There are two different approaches for the design of IIR digital filters. One approach is carried out in the *frequency domain*, which consists of minimizing some measure of the difference between the frequency response of the filter $H(e^{j\omega})$ and a desired frequency response $F(e^{j\omega})$. The other approach is carried out in the *time domain*, which consists of minimizing some measure of the difference between the impulse response of the filter

$$h_i = \frac{1}{2\pi j} \oint_C H(z) z^i \frac{dz}{z} \tag{8.2}$$

where C is a counterclockwise contour that encircles the origin, and a desired impulse response f_i in a direct way. Typically this direct minimization leads to a nonlinear problem. IIR filter design problems in the time domain can be mainly divided into two classes: least-squares approximation problem and modified least squares problem.

A *least-squares approximation problem* can be stated as follows. Given an impulse response sequence $\{f_0, f_1, f_2, \cdots\}$, find an IIR digital filter of the form in (8.1) which minimizes

$$\begin{aligned} ||\boldsymbol{f} - \boldsymbol{h}||^2 &= \sum_{i=0}^{\infty} (f_i - h_i)^2 \\ &= \frac{1}{2\pi} \int_{-\pi}^{\pi} |F(e^{j\omega}) - H(e^{j\omega})|^2 d\omega \end{aligned} \tag{8.3}$$

At a glance (8.3) provides a natural choice of error measure. However, it is ill-behaved in other respects. First, the problem of minimizing (8.3) with respect to unknown coefficients $\{a_1, a_2, \cdots, a_n, b_0, b_1, \cdots, b_m\}$ in (8.1) is highly nonquadratic, and is indeed a sophisticated nonlinear programming problem. Second, (8.3) requires the entire impulse response sequence. The problems in the literature have generally been specific to a particular input-output record or have considered only a truncated version of the impulse response sequence. A *modified least squares problem* consists of modifying the approximation problem whose cost function is (8.3) by considering instead a cost function which is quadratic in the coefficients a_i's and b_i's of the IIR digital filter in (8.1). In the modified problem one seeks the coefficients which minimize the quadratic form

$$J(\boldsymbol{a}, \boldsymbol{b}) = \frac{1}{2\pi} \int_{-\pi}^{\pi} |F(e^{j\omega})D(e^{j\omega}) - N(e^{j\omega})|^2 d\omega \tag{8.4}$$

where $\boldsymbol{a} = (a_1, a_2, \cdots, a_n)^T$ and $\boldsymbol{b} = (b_0, b_1, \cdots, b_m)^T$. The integral in (8.4) differs from that in (8.3) by the inclusion of $|D(e^{j\omega})|^2$ in the integrand. The difference between (8.3) and (8.4) is illustrated in Figure 8.1.

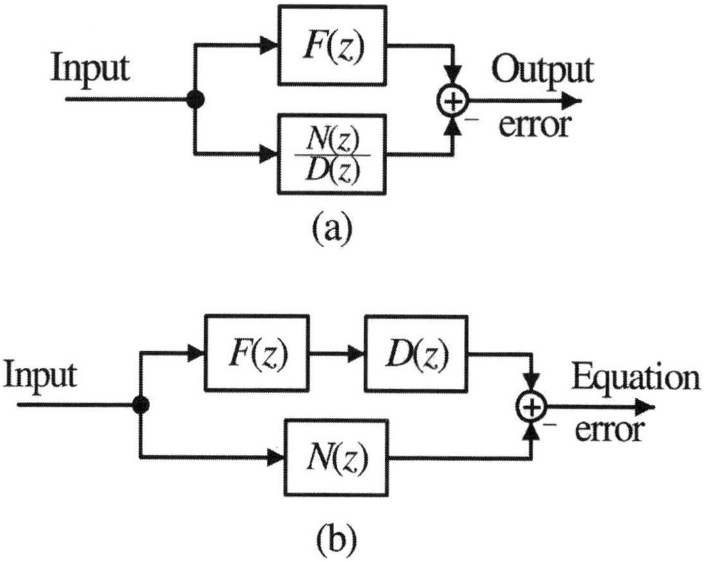

Figure 8.1 Time-domain IIR filter design. (a) Output error for least-squares approximation problem using (8.3). (b) Equation error for modified least-squares problem minimizing (8.4).

Another option for solving the least-squares approximation problem over the finite interval of the actual and desired impulse responses relies on a state-space approach where either the Hankel matrix or the controllability and observability Grammians are utilized to approximate a given finite impulse response sequence by a state-space model (or equivalently, an IIR digital filter).

In this chapter, several typical techniques for designing IIR digital filters in the time domain will be addressed concisely.

8.2 Design Based on Extended Pade's Approximation

Suppose that the desired transfer function $F(z)$ with a finite sequence $\{f_i | \, i = 0, 1, \cdots, N\}$ of the impulse response is given by

$$F(z) = f_0 + f_1 z^{-1} + f_2 z^{-2} + \cdots + f_N z^{-N} \tag{8.5}$$

We seek to design an IIR digital filter whose transfer function $H(z)$ is described by

$$\begin{aligned} H(z) &= \frac{b_0 + b_1 z^{-1} + \cdots + b_m z^{-m}}{1 + a_1 z^{-1} + \cdots + a_n z^{-n}} \\ &= \sum_{i=0}^{\infty} h_i z^{-i} \end{aligned} \tag{8.6}$$

The problem being considered here is to find the coefficients $a_1, a_2, \cdots, a_n$ and $b_0, b_1, \cdots, b_m$ of the transfer function $H(z)$ such that the actual impulse response h_i's approximates the desired impulse response f_i's in a certain sense over the finite interval $0 \le i \le N$ provided that $m + n \le N$.

From (8.6), it follows that

$$\begin{aligned} b_k &= \sum_{i=0}^{k} h_{k-i} a_i \quad \text{for } 0 \le k \le m \\ 0 &= \sum_{i=0}^{\min\{k,n\}} h_{k-i} a_i \quad \text{for } m < k \le N \end{aligned} \tag{8.7}$$

which is equivalent to

$$\begin{bmatrix} b_0 \\ b_1 \\ \vdots \\ b_m \end{bmatrix} = \begin{bmatrix} h_0 & 0 & \cdots & 0 \\ h_1 & h_0 & \ddots & 0 \\ \vdots & \vdots & \ddots & \vdots \\ h_m & h_{m-1} & \cdots & h_{m-n} \end{bmatrix} \begin{bmatrix} 1 \\ a_1 \\ \vdots \\ a_n \end{bmatrix}$$

$$\begin{bmatrix} 0 \\ 0 \\ \vdots \\ 0 \end{bmatrix} = \begin{bmatrix} h_{m+1} & h_m & \cdots & h_{m-n+1} \\ h_{m+2} & h_{m+1} & \cdots & h_{m-n+2} \\ \vdots & \vdots & \ddots & \vdots \\ h_N & h_{N-1} & \cdots & h_{N-n} \end{bmatrix} \begin{bmatrix} 1 \\ a_1 \\ \vdots \\ a_n \end{bmatrix} \tag{8.8}$$

where $h_k = 0$ for $k < 0$ and $a_0 = 1$.

8.2.1 A Direct Procedure

By replacing the unknown impulse response h_i's by the desired one f_i's in (8.8) and defining an $(N - m) \times 1$ error vector $\boldsymbol{e}_1$, (8.8) is changed to

$$\begin{bmatrix} \boldsymbol{b} \\ \boldsymbol{e}_1 \end{bmatrix} = \begin{bmatrix} \boldsymbol{F}_1 \\ \boldsymbol{F}_2 \end{bmatrix} \begin{bmatrix} 1 \\ \boldsymbol{a} \end{bmatrix} \tag{8.9}$$

where

$$\boldsymbol{a} = \begin{bmatrix} a_1 & a_2 & \cdots & a_n \end{bmatrix}^T, \qquad \boldsymbol{b} = \begin{bmatrix} b_0 & b_1 & \cdots & b_m \end{bmatrix}^T$$

$$\boldsymbol{F}_1 = \begin{bmatrix} f_0 & 0 & \cdots & 0 \\ f_1 & f_0 & \cdots & 0 \\ \vdots & \vdots & \ddots & \vdots \\ f_m & f_{m-1} & \cdots & f_{m-n} \end{bmatrix}, \quad \boldsymbol{F}_2 = \begin{bmatrix} f_{m+1} & f_m & \cdots & f_{m-n+1} \\ f_{m+2} & f_{m+1} & \cdots & f_{m-n+2} \\ \vdots & \vdots & \ddots & \vdots \\ f_N & f_{N-1} & \cdots & f_{N-n} \end{bmatrix}$$

with $f_k = 0$ for $k < 0$. A quadratic measure $J(\boldsymbol{a})$ is now defined as

$$J(\boldsymbol{a}) = \boldsymbol{e}_1^T \boldsymbol{e}_1 = (\boldsymbol{F}_3 \boldsymbol{a} - \hat{\boldsymbol{f}})^T (\boldsymbol{F}_3 \boldsymbol{a} - \hat{\boldsymbol{f}}) \tag{8.10}$$

where $\hat{\boldsymbol{f}}$ and $\boldsymbol{F}_3$ are $(N - m) \times 1$ and $(N - m) \times n$ matrices defined by

$$\boldsymbol{F}_3 = \begin{bmatrix} f_m & f_{m-1} & \cdots & f_{m-n+1} \\ f_{m+1} & f_m & \cdots & f_{m-n+2} \\ \vdots & \vdots & \ddots & \vdots \\ f_{N-1} & f_{N-2} & \cdots & f_{N-n} \end{bmatrix}, \qquad \hat{\boldsymbol{f}} = - \begin{bmatrix} f_{m+1} \\ f_{m+2} \\ \vdots \\ f_N \end{bmatrix}$$

respectively. By solving the equation

$$\frac{\partial J(\boldsymbol{a})}{\partial \boldsymbol{a}} = 2\boldsymbol{F}_3^T\left(\boldsymbol{F}_3\boldsymbol{a} - \hat{\boldsymbol{f}}\right) = \boldsymbol{0} \tag{8.11}$$

the denominator coefficient vector $\boldsymbol{a}$ which minimizes $J(\boldsymbol{a})$ in (8.10) can be obtained as

$$\boldsymbol{a} = \left(\boldsymbol{F}_3^T\boldsymbol{F}_3\right)^{-1}\boldsymbol{F}_3^T\hat{\boldsymbol{f}} \tag{8.12}$$

provided that rank $\boldsymbol{F}_3 = n$. Once the coefficient vector $\boldsymbol{a}$ is found, the numerator coefficient vector $\boldsymbol{b}$ can be readily derived from the upper portion of (8.9) as

$$\boldsymbol{b} = \boldsymbol{F}_1\begin{bmatrix}1\\ \boldsymbol{a}\end{bmatrix} \tag{8.13}$$

8.2.2 A Modified Procedure

Instead of an error vector $\boldsymbol{e}_1$ in (8.9), a more general $(N+1)\times 1$ error vector $\boldsymbol{e}_2$ is introduced as

$$\begin{bmatrix}\boldsymbol{b}\\ \boldsymbol{0}\end{bmatrix} + \boldsymbol{e}_2 = \begin{bmatrix}\boldsymbol{F}_1\\ \boldsymbol{F}_2\end{bmatrix}\begin{bmatrix}1\\ \boldsymbol{a}\end{bmatrix} \tag{8.14}$$

Equation (8.14) can be rearranged in the form

$$\begin{bmatrix}\boldsymbol{b}\\ \boldsymbol{0}\end{bmatrix} + \boldsymbol{e}_2 = \boldsymbol{A}_0\begin{bmatrix}f_0\\ \boldsymbol{f}\end{bmatrix} \tag{8.15}$$

where

$$\boldsymbol{A}_0 = \begin{bmatrix}a_0 & 0 & \cdots & \cdots & 0 & \cdots & 0\\ a_1 & a_0 & \ddots & & \vdots & & \vdots\\ \vdots & \ddots & \ddots & \ddots & \vdots & & \vdots\\ a_n & \cdots & a_1 & a_0 & 0 & \cdots & 0\\ 0 & a_n & \cdots & a_1 & a_0 & \ddots & \vdots\\ \vdots & \ddots & \ddots & & \ddots & \ddots & 0\\ 0 & \cdots & 0 & a_n & \cdots & a_1 & a_0\end{bmatrix}$$

Moreover, an $N \times 1$ error vector between the desired impulse response f_i's and the actual one h_i's over a finite interval $1 \le i \le N$ can be defined as

$$\boldsymbol{e} = \boldsymbol{f} - \boldsymbol{h} \tag{8.16}$$

where

$$\boldsymbol{f} = [f_1\ f_2\ \cdots\ f_N]^T, \qquad \boldsymbol{h} = [h_1\ h_2\ \cdots\ h_N]^T$$

Replacing the desired vector $\boldsymbol{f}$ by the actual vector $\boldsymbol{h}$ in (8.15) yields

$$\begin{bmatrix} \boldsymbol{b} \\ \boldsymbol{0} \end{bmatrix} = \boldsymbol{A}_0 \begin{bmatrix} h_0 \\ \boldsymbol{h} \end{bmatrix} = \boldsymbol{A}_0 \left(\begin{bmatrix} f_0 \\ \boldsymbol{f} \end{bmatrix} - \begin{bmatrix} e_0 \\ \boldsymbol{e} \end{bmatrix} \right) \tag{8.17}$$

where $e_0 = f_0 - h_0$. Since rank $\boldsymbol{A}_0 = N + 1$, matrix $\boldsymbol{A}_0$ is nonsingular. Hence (8.17) can be written as

$$\begin{bmatrix} f_0 \\ \boldsymbol{f} \end{bmatrix} - \begin{bmatrix} e_0 \\ \boldsymbol{e} \end{bmatrix} = [\boldsymbol{D}_1\ \ \boldsymbol{D}_2] \begin{bmatrix} \boldsymbol{b} \\ \boldsymbol{0} \end{bmatrix} = \boldsymbol{D}_1 \boldsymbol{b} \tag{8.18}$$

where $\boldsymbol{A}_0^{-1} = [\boldsymbol{D}_1\ \ \boldsymbol{D}_2]$.

The denominator coefficient vector $\boldsymbol{a}$ is determined by (8.12). The problem considered here is to obtain the numerator coefficient vector $\boldsymbol{b}$ that minimizes

$$I(\boldsymbol{b}) = \left[e_0\ \ \boldsymbol{e}^T \right] \boldsymbol{W} \begin{bmatrix} e_0 \\ \boldsymbol{e} \end{bmatrix} = \left(\begin{bmatrix} f_0 \\ \boldsymbol{f} \end{bmatrix} - \boldsymbol{D}_1 \boldsymbol{b} \right)^T \boldsymbol{W} \left(\begin{bmatrix} f_0 \\ \boldsymbol{f} \end{bmatrix} - \boldsymbol{D}_1 \boldsymbol{b} \right) \tag{8.19}$$

where $\boldsymbol{W}$ is an $(N + 1) \times (N + 1)$ symmetric positive-definite weighting matrix specified by the designer. Differentiating (8.19) with respect to vector $\boldsymbol{b}$ and setting it to null yields

$$\frac{\partial I(\boldsymbol{b})}{\partial \boldsymbol{b}} = 2 \left(\boldsymbol{D}_1^T \boldsymbol{W} \boldsymbol{D}_1 \boldsymbol{b} - \boldsymbol{D}_1^T \boldsymbol{W} \begin{bmatrix} f_0 \\ \boldsymbol{f} \end{bmatrix} \right) = \boldsymbol{0} \tag{8.20}$$

which leads to

$$\boldsymbol{b} = (\boldsymbol{D}_1^T \boldsymbol{W} \boldsymbol{D}_1)^{-1} \boldsymbol{D}_1^T \boldsymbol{W} \begin{bmatrix} f_0 \\ \boldsymbol{f} \end{bmatrix} \tag{8.21}$$

This is the numerator coefficient vector $\boldsymbol{b}$ that minimizes $I(\boldsymbol{b})$ in (8.19).

8.3 Design Using Second-Order Information

8.3.1 A Filter Design Method

Given an infinite sequence $\{f_i | \, i = 0, 1, \cdots\}$ of a desired impulse response in terms of the transfer function

$$F(z) = f_0 + f_1 z^{-1} + f_2 z^{-2} + \cdots + f_k z^{-k} + \cdots \tag{8.22}$$

the problem being considered here is to find the coefficients $a_1, a_2, \cdots, a_n$ and $b_0, b_1, \cdots, b_m$ of a transfer function of the form

$$H(z) = \frac{N(z)}{D(z)} = \frac{b_0 + b_1 z^{-1} + \cdots + b_m z^{-m}}{1 + a_1 z^{-1} + \cdots + a_n z^{-n}} \tag{8.23}$$

so as to minimize a quadratic measure defined by

$$\varepsilon = \frac{1}{2\pi} \int_{-\pi}^{\pi} |F(e^{j\omega})D(e^{j\omega}) - N(e^{j\omega})|^2 d\omega \tag{8.24}$$

Applying Parseval's formula to (8.24) yields

$$\varepsilon = \sum_{k=0}^{m} b_k^2 - 2\sum_{k=0}^{m} b_k \sum_{j=0}^{n} a_j f_{k-j} + \sum_{i=0}^{n}\sum_{j=0}^{n} a_i a_j r_{|i-j|} \tag{8.25}$$

where

$$r_k = \sum_{i=0}^{\infty} f_i f_{k+i}, \qquad a_0 = 1$$

Here, f_i and r_k are called the *first-order information* and the *second-order information* for the filter in (8.22), respectively. Differentiating (8.25) with respect to coefficients b_k's and setting the results to zero leads to

$$b_k = \sum_{j=0}^{n} a_j f_{k-j} \text{ for } k = 0, 1, \cdots, m \tag{8.26}$$

which is equivalent to

$$\begin{bmatrix} b_0 \\ b_1 \\ \vdots \\ b_m \end{bmatrix} = \begin{bmatrix} 1 & 0 & \cdots & 0 \\ a_1 & 1 & \cdots & 0 \\ \vdots & \ddots & \ddots & \vdots \\ a_m & \cdots & a_1 & 1 \end{bmatrix} \begin{bmatrix} f_0 \\ f_1 \\ \vdots \\ f_m \end{bmatrix} \tag{8.27}$$

Substituting (8.26) into (8.25) yields

$$\begin{aligned} \varepsilon &= \sum_{i=0}^{n}\sum_{j=0}^{n} a_i a_j r_{|i-j|} - \sum_{k=0}^{m}\Big(\sum_{j=0}^{n} a_j f_{k-j}\Big)^2 \\ &= [1 \ a_1 \ \cdots \ a_n]\, \boldsymbol{K}(m,n) \begin{bmatrix} 1 \\ a_1 \\ \vdots \\ a_n \end{bmatrix} \end{aligned} \tag{8.28a}$$

where $\boldsymbol{K}(m,n)$ is an $(n+1)\times(n+1)$ symmetric matrix denoted by

$$\boldsymbol{K}(m,n) = \begin{bmatrix} K_{00}(m,n) & K_{01}(m,n) & \cdots & K_{0n}(m,n) \\ K_{10}(m,n) & K_{11}(m,n) & \cdots & K_{1n}(m,n) \\ \vdots & \vdots & \ddots & \vdots \\ K_{n0}(m,n) & K_{n1}(m,n) & \cdots & K_{nn}(m,n) \end{bmatrix} \tag{8.28b}$$

whose (i,j)th component is given by

$$\begin{aligned} K_{ij}(m,n) &= r_{|i-j|} - \sum_{k=0}^{m} f_{k-i} f_{k-j} \\ &= r_{|i-j|} - \sum_{k=\max\{i,j\}}^{m} f_{k-i} f_{k-j} \quad \text{for } i,j = 0,1,\cdots,n \end{aligned} \tag{8.28c}$$

with $f_k = 0$ for $k < 0$.

Lemma 8.1
Let $\boldsymbol{K}$ be a positive semidefinite symmetric matrix, and $\boldsymbol{\psi}$ be a given vector. The solution $\boldsymbol{x}^*$ of the problem

$$\min_{\boldsymbol{x}} \ \boldsymbol{x}^T \boldsymbol{K} \boldsymbol{x} \ \text{subject to} \ \boldsymbol{x}^T \boldsymbol{\psi} = 1 \tag{8.29}$$

satisfies $\boldsymbol{K}\boldsymbol{x}^* = \alpha\boldsymbol{\psi}$ where α is the minimum value of problem (8.29).

Proof
We define the Lagrange function

$$J(\boldsymbol{x},\lambda) = \frac{1}{2}\boldsymbol{x}^T \boldsymbol{K}\boldsymbol{x} - \lambda(\boldsymbol{\psi} - 1) \tag{8.30}$$

where λ is a Lagrange multiplier, and compute the gradients

$$\begin{aligned} \begin{bmatrix} \dfrac{\partial J(\boldsymbol{x},\lambda)}{\partial \boldsymbol{x}} \\ \dfrac{\partial J(\boldsymbol{x},\lambda)}{\partial \lambda} \end{bmatrix} &= \begin{bmatrix} \boldsymbol{K}\boldsymbol{x} - \lambda\boldsymbol{\psi} \\ -(\boldsymbol{x}^T\boldsymbol{\psi} - 1) \end{bmatrix} \\ &= \begin{bmatrix} \boldsymbol{K} & -\boldsymbol{\psi} \\ -\boldsymbol{\psi}^T & 1/\lambda \end{bmatrix} \begin{bmatrix} \boldsymbol{x} \\ \lambda \end{bmatrix} \end{aligned} \tag{8.31}$$

By setting (8.31) to null, we obtain

$$\begin{bmatrix} \boldsymbol{K} & -\boldsymbol{\psi} \\ -\boldsymbol{\psi}^T & 1/\lambda \end{bmatrix} \begin{bmatrix} \boldsymbol{x}^* \\ \lambda^* \end{bmatrix} = \boldsymbol{0} \tag{8.32}$$

where $\boldsymbol{x}^*$ and λ^* denote the optimal values of $\boldsymbol{x}$ and λ, respectively. From (8.32), it follows that

$$\boldsymbol{K}\boldsymbol{x}^* = \lambda^*\boldsymbol{\psi}, \qquad \boldsymbol{x}^{*T}\boldsymbol{K}\boldsymbol{x}^* = \lambda^*\boldsymbol{x}^{*T}\boldsymbol{\psi} = \lambda^* = \alpha \tag{8.33}$$

which leads to

$$\boldsymbol{K}\boldsymbol{x}^* = \alpha\boldsymbol{\psi} \tag{8.34}$$

This completes the proof of Lemma 8.1. ■

By choosing $\boldsymbol{x} = [1\ a_1\ \cdots\ a_n]^T$ and $\boldsymbol{\psi} = [1\ 0\ \cdots\ 0]^T$ in Lemma 8.1, the optimal coefficients $a_1, a_2, \cdots, a_n$ minimizing (8.28a) must satisfy

$$\boldsymbol{K}(m,n) \begin{bmatrix} 1 \\ a_1 \\ \vdots \\ a_n \end{bmatrix} = \alpha_{mn} \begin{bmatrix} 1 \\ 0 \\ \vdots \\ 0 \end{bmatrix} \tag{8.35}$$

where α_{mn} is the nonnegative minimum value of (8.28a). Hence the optimal coefficient vector $\boldsymbol{a} = [a_1, a_2, \cdots, a_n]^T$ minimizing (8.28a) can be obtained from second to last equations in (8.35) as

$$\boldsymbol{a} = -\boldsymbol{K}_o^{-1}\boldsymbol{g} \tag{8.36}$$

where

$$\boldsymbol{K}_o = \begin{bmatrix} K_{11}(m,n) & K_{12}(m,n) & \cdots & K_{1n}(m,n) \\ K_{21}(m,n) & K_{22}(m,n) & \cdots & K_{2n}(m,n) \\ \vdots & \vdots & \ddots & \vdots \\ K_{n1}(m,n) & K_{n2}(m,n) & \cdots & K_{nn}(m,n) \end{bmatrix}, \quad \boldsymbol{g} = \begin{bmatrix} K_{10}(m,n) \\ K_{20}(m,n) \\ \vdots \\ K_{n0}(m,n) \end{bmatrix}$$

It is noted that (8.36) is the optimal solution for a given order n which minimizes ε in (8.28a) with respect to vector $\boldsymbol{a}$, because ε is a convex quadratic function whose minimizer can be found by taking the gradient of ε with respect to vector $\boldsymbol{a}$ and setting it to null. Moreover, (8.35) plays an important role later in stability issue as well as derivation of an efficient algorithm for solving (8.35).

Once the denominator coefficient vector $\boldsymbol{a}$ is found, the numerator coefficients $b_0, b_1, \cdots, b_m$ can be readily determined from (8.27).

8.3.2 Stability

We now address the stability of IIR digital filters designed by the method in Section 8.3.1.

Lemma 8.2 (*Lyapunov Stability Theorem*)
If $(\boldsymbol{A}, \boldsymbol{b})$ is a controllable pair, i.e., the matrix $\boldsymbol{G} = [\boldsymbol{b}, \boldsymbol{A}\boldsymbol{b}, \cdots, \boldsymbol{A}^{n-1}\boldsymbol{b}]$ is nonsingular, and there exist symmetric matrices $\boldsymbol{K}$ and $\boldsymbol{L}$ with $\boldsymbol{K}$ positive definite and $\boldsymbol{L}$ positive semidefinite such that

$$\boldsymbol{K} = \boldsymbol{A}\boldsymbol{K}\boldsymbol{A}^T + \alpha \boldsymbol{b}\boldsymbol{b}^T + \boldsymbol{L} \tag{8.37}$$

for some $\alpha > 0$, then the eigenvalues of $\boldsymbol{A}$ all lie in the open unit disk, where $\boldsymbol{A}$ and $\boldsymbol{b}$ are $n \times n$ and $n \times 1$ real matrices, respectively.

Proof
Equation (8.37) can be written as

$$\begin{aligned}
\boldsymbol{K} &= \boldsymbol{A}\big[\boldsymbol{A}\boldsymbol{K}\boldsymbol{A}^T + \alpha \boldsymbol{b}\boldsymbol{b}^T + \boldsymbol{L}\big]\boldsymbol{A}^T + \alpha \boldsymbol{b}\boldsymbol{b}^T + \boldsymbol{L} \\
&= \boldsymbol{A}^2\boldsymbol{K}(\boldsymbol{A}^T)^2 + \alpha \boldsymbol{A}\boldsymbol{b}\boldsymbol{b}^T\boldsymbol{A}^T + \boldsymbol{A}\boldsymbol{L}\boldsymbol{A}^T + \alpha \boldsymbol{b}\boldsymbol{b}^T + \boldsymbol{L} \\
&= \cdots \\
&= \boldsymbol{A}^n\boldsymbol{K}(\boldsymbol{A}^T)^n + \alpha \boldsymbol{G}\boldsymbol{G}^T + \boldsymbol{L} + \boldsymbol{A}\boldsymbol{L}\boldsymbol{A}^T + \cdots + \boldsymbol{A}^{n-1}\boldsymbol{L}(\boldsymbol{A}^T)^{n-1}
\end{aligned} \tag{8.38}$$

Recall the standard Lyapunov stability theorem in [6], which states that the matrix $\boldsymbol{B}$ has all eigenvalues in the open unit disk if and only if there exist two positive definite symmetric matrices $\boldsymbol{V}$ and $\boldsymbol{W}$ for which $\boldsymbol{V} = \boldsymbol{B}^T\boldsymbol{V}\boldsymbol{B} + \boldsymbol{W}$.

By taking $\boldsymbol{V} = \boldsymbol{K}$ and $\boldsymbol{W} = \alpha \boldsymbol{G}\boldsymbol{G}^T + \boldsymbol{L} + \boldsymbol{A}\boldsymbol{L}\boldsymbol{A}^T + \cdots + \boldsymbol{A}^{n-1}\boldsymbol{L}(\boldsymbol{A}^T)^{n-1}$, and noticing that $\boldsymbol{G}$ is nonsingular, we conclude that both $\boldsymbol{V}$ and $\boldsymbol{W}$ are positive definite, hence (8.38) satisfies the condition in the standard Lyapunov stability theorem, and the eigenvalues of $(\boldsymbol{A}^T)^n$ must all lie in the open unit disk. Evidently, these eigenvalues are simply the nth powers of the eigenvalues of $\boldsymbol{A}$, and thus the eigenvalues of $\boldsymbol{A}$ must all lie in the open unit disk. This completes the proof of the lemma. ■

Theorem 8.1
Suppose the coefficients of denominator $D(z)$ in (8.23) minimize (8.28a), the transfer function $H(z) = N(z)/D(z)$ is a stable filter.

Proof
Without loss of generality, we assume that the coordinates for the state space are chosen so that

$$\boldsymbol{A} = \begin{bmatrix} -a_1 & -a_2 & \cdots & -a_n \\ & & & 0 \\ & \boldsymbol{I}_{n-1} & & \vdots \\ & & & 0 \end{bmatrix}, \qquad \boldsymbol{b} = \begin{bmatrix} 1 \\ 0 \\ \vdots \\ 0 \end{bmatrix} \tag{8.39}$$

where $(\boldsymbol{A}, \boldsymbol{b})$ is a controllable pair. Form (8.28c), it follows that

$$K_{ij}(m-1, n-1) = K_{i+1,j+1}(m, n) \text{ for } i, j = 0, 1, \cdots, n-1 \tag{8.40}$$

which is equivalent to

$$\boldsymbol{K}(m-1, n-1) = \begin{bmatrix} K_{11}(m,n) & K_{12}(m,n) & \cdots & K_{1n}(m,n) \\ K_{21}(m,n) & K_{22}(m,n) & \cdots & K_{2n}(m,n) \\ \vdots & \vdots & \ddots & \vdots \\ K_{n1}(m,n) & K_{n2}(m,n) & \cdots & K_{nn}(m,n) \end{bmatrix} \tag{8.41}$$

where

$$\boldsymbol{K}(m-1, n-1) = \begin{bmatrix} K_{00}(m-1,n-1) & K_{01}(m-1,n-1) & \cdots & K_{0,n-1}(m-1,n-1) \\ K_{10}(m-1,n-1) & K_{11}(m-1,n-1) & \cdots & K_{1,n-1}(m-1,n-1) \\ \vdots & \vdots & \ddots & \vdots \\ K_{n-1,0}(m-1,n-1) & K_{n-1,1}(m-1,n-1) & \cdots & K_{n-1,n-1}(m-1,n-1) \end{bmatrix}$$

By post-multiplying (8.41) by $\boldsymbol{A}^T$ in (8.39) and using (8.35), we obtain

$$\boldsymbol{K}(m-1, n-1)\boldsymbol{A}^T = \begin{bmatrix} K_{10}(m,n) & K_{11}(m,n) & \cdots & K_{1,n-1}(m,n) \\ K_{20}(m,n) & K_{21}(m,n) & \cdots & K_{2,n-1}(m,n) \\ \vdots & \vdots & \ddots & \vdots \\ K_{n0}(m,n) & K_{n1}(m,n) & \cdots & K_{n,n-1}(m,n) \end{bmatrix} \tag{8.42}$$

Moreover, pre-multiplying (8.42) by $\boldsymbol{A}$ in (8.39) and using (8.35) produces

$$\boldsymbol{A}\boldsymbol{K}(m-1,n-1)\boldsymbol{A}^T = \begin{bmatrix} K_{00}(m,n) & K_{01}(m,n) & \cdots & K_{0,n-1}(m,n) \\ K_{10}(m,n) & K_{11}(m,n) & \cdots & K_{1,n-1}(m,n) \\ \vdots & \vdots & \ddots & \vdots \\ K_{n-1,0}(m,n) & K_{n-1,1}(m,n) & \cdots & K_{n-1,n-1}(m,n) \end{bmatrix} - \alpha_{mn}\boldsymbol{b}\boldsymbol{b}^T \tag{8.43}$$

By virtue of (8.28c), we have

$$K_{ij}(m,n) = K_{i,j}(m-1,n-1) - f_{m-i}f_{m-j} \quad \text{for} \quad i,j = 0,1,\cdots,n-1 \tag{8.44}$$

which leads to

$$\begin{bmatrix} K_{00}(m,n) & K_{01}(m,n) & \cdots & K_{0,n-1}(m,n) \\ K_{10}(m,n) & K_{11}(m,n) & \cdots & K_{1,n-1}(m,n) \\ \vdots & \vdots & \ddots & \vdots \\ K_{n-1,0}(m,n) & K_{n-1,1}(m,n) & \cdots & K_{n-1,n-1}(m,n) \end{bmatrix} = \boldsymbol{K}(m-1,n-1) - \boldsymbol{L} \tag{8.45}$$

where

$$\boldsymbol{L} = \begin{bmatrix} f_m \\ f_{m-1} \\ \vdots \\ f_{m-n+1} \end{bmatrix} \begin{bmatrix} f_m & f_{m-1} & \cdots & f_{m-n+1} \end{bmatrix}$$

By substituting (8.45) into (8.43), we obtain

$$\boldsymbol{K}(m-1,n-1) = \boldsymbol{A}\boldsymbol{K}(m-1,n-1)\boldsymbol{A}^T + \alpha_{mn}\boldsymbol{b}\boldsymbol{b}^T + \boldsymbol{L} \tag{8.46}$$

Clearly, the symmetric matrix $\boldsymbol{K}(m,n)$ is positive definite provided that $\alpha_{mn} > 0$. Since the symmetric matrix $\boldsymbol{K}(m-1,n-1)$ is the lower right $n \times n$ portion of $\boldsymbol{K}(m,n)$, it must also be positive definite. In addition, $(\boldsymbol{A},\boldsymbol{b})$ is a controllable pair and the symmetric matrix $\boldsymbol{L}$ is positive semidefinite. Thus, based on Lemma 8.2, (8.46) guarantees that all the eigenvalues of $\boldsymbol{A}$ exist inside the unit disk. It is noted that the eigenvalues of the companion matrix $\boldsymbol{A}$ in (8.39) are the roots of the polynomial $D(z) = \det(z\boldsymbol{I}_n - \boldsymbol{A}) = 0$. This completes the proof of the theorem. ■

8.3.3 An Efficient Algorithm for Solving (8.35)

We now present an algorithm that computes solutions to (8.35) efficiently. The algorithm provides not only an optimal solution but also a means of determining filter order n that falls in a reasonable range. In addition, unlike (8.36), the optimal vector $\boldsymbol{a}$ is calculated without matrix inversion. We shall focus on the case $m = n$, then treat the case $m \neq n$ using a variant of the algorithm derived for the case of $m = n$. For simplicity, let $\boldsymbol{K}(n) = \boldsymbol{K}(n, n)$ and $\alpha_n = \alpha_{nn}$. In this case, (8.28b) and (8.28c) can be expressed as

$$\boldsymbol{K}(n) = \begin{bmatrix} K_{00}(n) & K_{01}(n) & \cdots & K_{0n}(n) \\ K_{10}(n) & K_{11}(n) & \cdots & K_{1n}(n) \\ \vdots & \vdots & \ddots & \vdots \\ K_{n0}(n) & K_{n1}(n) & \cdots & K_{nn}(n) \end{bmatrix}$$

$$K_{ij}(n) = r_{|i-j|} - \sum_{k=\max\{i,j\}}^{n} f_{k-i} f_{k-j} \quad \text{for } i, j = 0, 1, \cdots, n \tag{8.47}$$

respectively. From (8.47), it follows that

$$\begin{aligned} K_{ij}(n+1) &= K_{ij}(n) - f_{n+1-i} f_{n+1-j} \\ K_{i+1,j+1}(n+1) &= K_{ij}(n) \quad \text{for } i, j = 0, 1, \cdots, n \end{aligned} \tag{8.48}$$

which are equivalent to

$$\begin{bmatrix} K_{00}(n+1) & K_{01}(n+1) & \cdots & K_{0n}(n+1) \\ K_{10}(n+1) & K_{11}(n+1) & \cdots & K_{1n}(n+1) \\ \vdots & \vdots & \ddots & \vdots \\ K_{n0}(n+1) & K_{n1}(n+1) & \cdots & K_{nn}(n+1) \end{bmatrix}$$

$$= \boldsymbol{K}(n) - \begin{bmatrix} f_{n+1} \\ f_n \\ \vdots \\ f_1 \end{bmatrix} \begin{bmatrix} f_{n+1} & f_n & \cdots & f_1 \end{bmatrix}$$

$$\begin{bmatrix} K_{11}(n+1) & K_{12}(n+1) & \cdots & K_{1,n+1}(n+1) \\ K_{21}(n+1) & K_{22}(n+1) & \cdots & K_{2,n+1}(n+1) \\ \vdots & \vdots & \ddots & \vdots \\ K_{n+1,1}(n+1) & K_{n+1,2}(n+1) & \cdots & K_{n+1,n+1}(n+1) \end{bmatrix} = \boldsymbol{K}(n) \tag{8.49}$$

respectively. Hence

$$\boldsymbol{K}(n+1) = \begin{bmatrix} & & & r_{n+1} \\ & \boldsymbol{K}(n) & & r_n \\ & & & \vdots \\ r_{n+1} & r_n & \cdots & r_0 \end{bmatrix} - \begin{bmatrix} f_{n+1} \\ f_n \\ \vdots \\ f_0 \end{bmatrix} \begin{bmatrix} f_{n+1} & f_n & \cdots & f_0 \end{bmatrix}$$
$$= \begin{bmatrix} d_{n0} & \boldsymbol{d}(n)^T \\ \boldsymbol{d}(n) & \boldsymbol{K}(n) \end{bmatrix} \tag{8.50}$$

where

$$\boldsymbol{d}(n) = \begin{bmatrix} d_{n1} & d_{n2} & \cdots & d_{n,n+1} \end{bmatrix}^T$$
$$d_{nl} = r_l - \sum_{k=l}^{n+1} f_k f_{k-l} \quad \text{for } l = 0, 1, \cdots, n+1$$

We now define three vectors

$$\boldsymbol{a}(n) = \begin{bmatrix} a_0(n) \\ a_1(n) \\ \vdots \\ a_n(n) \end{bmatrix}, \qquad \boldsymbol{p}(n) = \begin{bmatrix} p_0(n) \\ p_1(n) \\ \vdots \\ p_n(n) \end{bmatrix}, \qquad \boldsymbol{q}(n) = \begin{bmatrix} q_0(n) \\ q_1(n) \\ \vdots \\ q_n(n) \end{bmatrix} \tag{8.51}$$

such that

$$\boldsymbol{K}(n)\boldsymbol{a}(n) = \begin{bmatrix} \alpha_n \\ 0 \\ \vdots \\ 0 \end{bmatrix}, \qquad \boldsymbol{K}(n)\boldsymbol{p}(n) = \begin{bmatrix} 0 \\ \vdots \\ 0 \\ 1 \end{bmatrix}, \qquad \boldsymbol{K}(n)\boldsymbol{q}(n) = \begin{bmatrix} f_n \\ f_{n-1} \\ \vdots \\ f_0 \end{bmatrix} \tag{8.52}$$

where $a_0(n) = 1$. Using (8.50), it can be proved that

$$\boldsymbol{a}(n+1) = \begin{bmatrix} \boldsymbol{a}(n) \\ 0 \end{bmatrix} - \begin{bmatrix} 0 \\ \boldsymbol{p}(n) \end{bmatrix} \beta_n - \begin{bmatrix} 0 \\ \boldsymbol{q}(n) \end{bmatrix} \gamma_n$$

$$\boldsymbol{p}(n+1) = \begin{bmatrix} 0 \\ \boldsymbol{p}(n) \end{bmatrix} - \frac{1}{\alpha_{n+1}} \boldsymbol{a}(n+1)\tilde{\theta}_n$$

$$\boldsymbol{q}(n+1) = \begin{bmatrix} 0 \\ \boldsymbol{q}(n) \end{bmatrix} - \frac{1}{\alpha_{n+1}} \boldsymbol{a}(n+1)\tilde{\phi}_n$$
$$\alpha_{n+1} = \alpha_n - \tilde{\theta}_n \beta_n - \tilde{\phi}_n \gamma_n \tag{8.53}$$

yield

$$\boldsymbol{K}(n+1)\boldsymbol{a}(n+1) = \begin{bmatrix} \alpha_{n+1} \\ 0 \\ \vdots \\ 0 \end{bmatrix}$$

$$\boldsymbol{K}(n+1)\boldsymbol{p}(n+1) = \begin{bmatrix} 0 \\ \vdots \\ 0 \\ 1 \end{bmatrix}, \qquad \boldsymbol{K}(n+1)\boldsymbol{q}(n+1) = \begin{bmatrix} f_{n+1} \\ f_n \\ \vdots \\ f_0 \end{bmatrix} \tag{8.54}$$

where

$$\beta_n = \begin{bmatrix} r_{n+1} & r_n & \cdots & r_1 \end{bmatrix} \boldsymbol{a}(n)$$
$$\gamma_n = -\begin{bmatrix} f_{n+1} & f_n & \cdots & f_1 \end{bmatrix} \boldsymbol{a}(n)$$
$$\tilde{\theta}_n = \boldsymbol{d}(n)^T \boldsymbol{p}(n), \qquad \tilde{\phi}_n = \boldsymbol{d}(n)^T \boldsymbol{q}(n) - f_{n+1}$$

The numbers β_n and γ_n are the errors in the predicted values of r_{n+1} and f_{n+1}, respectively, based on the nth approximation. If both of these numbers vanish then the $(n+1)$st approximation is equivalent to the nth one.

To examine the intermediate variables in (8.53), we define

$$\boldsymbol{D}(n) = \begin{bmatrix} p_n(n) & q_n(n) \\ q_n(n) & \delta_n \end{bmatrix}$$
$$\delta_n = 1 + \begin{bmatrix} f_n & f_{n-1} & \cdots & f_0 \end{bmatrix} \boldsymbol{q}(n) \tag{8.55}$$
$$\begin{bmatrix} \theta_n \\ \phi_n \end{bmatrix} = \boldsymbol{D}(n) \begin{bmatrix} \beta_n \\ \gamma_n \end{bmatrix}$$

By noting that

$$q_n(n) = \boldsymbol{q}(n)^T \boldsymbol{K}(n) \boldsymbol{p}(n)$$
$$= \begin{bmatrix} f_n & f_{n-1} & \cdots & f_0 \end{bmatrix} \boldsymbol{p}(n) \tag{8.56}$$

it is easy to show that both $\tilde{\theta}_n - \theta_n$ and $\tilde{\phi}_n - \phi_n$ exclude the variables r_{n+1} and f_{n+1}, and thus are independent of r_{n+1} and f_{n+1}. Moreover, by using (8.52) and (8.56), matrix $\boldsymbol{D}(n)$ in (8.55) can be expressed as

$$\begin{aligned}\boldsymbol{D}(n) &= \begin{bmatrix} 0 & 0 \\ 0 & 1 \end{bmatrix} + \begin{bmatrix} 0 & \cdots & 0 & 1 \\ f_n & \cdots & f_1 & f_0 \end{bmatrix} \begin{bmatrix} \boldsymbol{p}(n) & \boldsymbol{q}(n) \end{bmatrix} \\ &= \begin{bmatrix} 0 & 0 \\ 0 & 1 \end{bmatrix} + \begin{bmatrix} 0 & \cdots & 0 & 1 \\ f_n & \cdots & f_1 & f_0 \end{bmatrix} \boldsymbol{K}(n)^{-1} \begin{bmatrix} 0 & f_n \\ \vdots & \vdots \\ 0 & f_1 \\ 1 & f_0 \end{bmatrix}\end{aligned} \tag{8.57}$$

hence $\boldsymbol{D}(n)$ is positive definite so long as $\alpha_n > 0$, i.e., $\det \boldsymbol{K}(n) \neq 0$.

Alternatively, from (8.55) α_{n+1} in (8.53) can be written as

$$\begin{aligned}\alpha_{n+1} &= \alpha_n - \begin{bmatrix} \beta_n & \gamma_n \end{bmatrix} \begin{bmatrix} \theta_n \\ \phi_n \end{bmatrix} - \begin{bmatrix} \beta_n & \gamma_n \end{bmatrix} \begin{bmatrix} \tilde{\theta}_n - \theta_n \\ \tilde{\phi}_n - \phi_n \end{bmatrix} \\ &= \alpha_n - \begin{bmatrix} \beta_n & \gamma_n \end{bmatrix} \boldsymbol{D}(n) \begin{bmatrix} \beta_n \\ \gamma_n \end{bmatrix} - \begin{bmatrix} \beta_n & \gamma_n \end{bmatrix} \begin{bmatrix} \tilde{\theta}_n - \theta_n \\ \tilde{\phi}_n - \phi_n \end{bmatrix}\end{aligned} \tag{8.58}$$

We now show that $\tilde{\theta}_n = \theta_n$ and $\tilde{\phi}_n = \phi_n$ by the method of "proof by contradiction". Suppose either $\tilde{\theta}_n \neq \theta_n$ or $\tilde{\phi}_n \neq \phi_n$ holds, then from (8.54) appropriate r_{n+1} and f_{n+1} could be chosen such that

$$\begin{bmatrix} \beta_n \\ \gamma_n \end{bmatrix} = -\frac{1}{2} \boldsymbol{D}(n)^{-1} \begin{bmatrix} \tilde{\theta}_n - \theta_n \\ \tilde{\phi}_n - \phi_n \end{bmatrix} \tag{8.59}$$

which would imply that

$$\alpha_{n+1} - \alpha_n = \begin{bmatrix} \beta_n & \gamma_n \end{bmatrix} \boldsymbol{D}(n) \begin{bmatrix} \beta_n \\ \gamma_n \end{bmatrix} > 0, \quad \text{or equivalently } \alpha_{n+1} > \alpha_n \tag{8.60}$$

that contradicts with the fact that $\alpha_{n+1} \leq \alpha_n$ must holds for consistent values of r_{n+1} and f_{n+1} which satisfy a certain Cayley-Hamilton's Theorem indirectly. Hence it can be concluded that $\tilde{\theta}_n = \theta_n$ and $\tilde{\phi}_n = \phi_n$.

From (8.53) and (8.54), it follows that

$$
\begin{aligned}
\phi_n &= -q_0(n+1)\alpha_{n+1} \\
&= -\boldsymbol{q}(n+1)^T \begin{bmatrix} \alpha_{n+1} & 0 & \cdots & 0 \end{bmatrix}^T \\
&= -\boldsymbol{q}(n+1)^T \boldsymbol{K}(n+1)\boldsymbol{a}(n+1) \\
&= -\begin{bmatrix} f_{n+1} & f_n & \cdots & f_0 \end{bmatrix} \boldsymbol{a}(n+1)
\end{aligned}
\tag{8.61}
$$

By virtue of (8.55), (8.53) and (8.61), we obtain

$$
\begin{aligned}
\delta_{n+1} &= 1 + \begin{bmatrix} f_{n+1} & f_n & \cdots & f_0 \end{bmatrix} \boldsymbol{q}(n+1) \\
&= 1 + \begin{bmatrix} f_{n+1} & f_n & \cdots & f_0 \end{bmatrix} \left\{ \begin{bmatrix} 0 \\ \boldsymbol{q}(n) \end{bmatrix} - \frac{1}{\alpha_{n+1}} \boldsymbol{a}(n+1)\phi_n \right\} \\
&= \delta_n + \frac{\phi_n^2}{\alpha_{n+1}}
\end{aligned}
\tag{8.62}
$$

An efficient algorithm for solving (8.35) for the case of $m = n$ can now be summarized as follows.

Initialization:

$$
\begin{aligned}
\boldsymbol{a}(0) &= a_0(0) = 1 \\
\alpha_0 &= r_0 - f_0^2 \\
\boldsymbol{p}(0) &= p_0(0) = 1/\alpha_0 \\
\boldsymbol{q}(0) &= q_0(0) = f_0/\alpha_0 \\
\delta_0 &= 1 + f_0^2/\alpha_0
\end{aligned}
\tag{8.63}
$$

Recursion:

$$
\begin{aligned}
\beta_n &= \begin{bmatrix} r_{n+1} & r_n & \cdots & r_1 \end{bmatrix} \boldsymbol{a}(n) \\
\gamma_n &= -\begin{bmatrix} f_{n+1} & f_n & \cdots & f_1 \end{bmatrix} \boldsymbol{a}(n) \\
\begin{bmatrix} \theta_n \\ \phi_n \end{bmatrix} &= \begin{bmatrix} p_n(n) & q_n(n) \\ q_n(n) & \delta_n \end{bmatrix} \begin{bmatrix} \beta_n \\ \gamma_n \end{bmatrix} \\
\alpha_{n+1} &= \alpha_n - \begin{bmatrix} \beta_n & \gamma_n \end{bmatrix} \begin{bmatrix} \theta_n \\ \phi_n \end{bmatrix}
\end{aligned}
$$

$$
\begin{aligned}
\delta_{n+1} &= \delta_n + \frac{\phi_n^2}{\alpha_{n+1}} \\
\boldsymbol{a}(n+1) &= \begin{bmatrix} \boldsymbol{a}(n) \\ 0 \end{bmatrix} - \beta_n \begin{bmatrix} 0 \\ \boldsymbol{p}(n) \end{bmatrix} - \gamma_n \begin{bmatrix} 0 \\ \boldsymbol{q}(n) \end{bmatrix} \\
\boldsymbol{p}(n+1) &= \begin{bmatrix} 0 \\ \boldsymbol{p}(n) \end{bmatrix} - \frac{\theta_n}{\alpha_{n+1}} \boldsymbol{a}(n+1) \\
\boldsymbol{q}(n+1) &= \begin{bmatrix} 0 \\ \boldsymbol{q}(n) \end{bmatrix} - \frac{\phi_n}{\alpha_{n+1}} \boldsymbol{a}(n+1)
\end{aligned} \tag{8.64}
$$

where

$$
\boldsymbol{a}(n) = \begin{bmatrix} a_0(n) \\ a_1(n) \\ \vdots \\ a_n(n) \end{bmatrix}, \qquad \boldsymbol{p}(n) = \begin{bmatrix} p_0(n) \\ p_1(n) \\ \vdots \\ p_n(n) \end{bmatrix}, \qquad \boldsymbol{q}(n) = \begin{bmatrix} q_0(n) \\ q_1(n) \\ \vdots \\ q_n(n) \end{bmatrix}
$$

This process continues until

$$
|\alpha_{n+1} - \alpha_n| < \varepsilon \tag{8.65}
$$

is satisfied where $\varepsilon > 0$ is a prescribed tolerance. If the recursion is terminated at step $n+1$, we set $\boldsymbol{a} = \boldsymbol{a}(n+1)$ and claim it to be a solution.

Once the denominator coefficient vector $\boldsymbol{a}$ is found, the numerator coefficients $b_0, b_1, \cdots, b_m$ can be readily determined from (8.27).

It is known [4] that the recursive algorithm shown in (8.63)–(8.65) can be applied to the case where $m \neq n$ by performing a certain number of left or right shifts to the original impulse response.

8.4 Least-Squares Design

The time-domain design of IIR digital filters involves obtaining the coefficients of an nth-order transfer function of the form

$$
\begin{aligned}
H(z, \boldsymbol{a}, \boldsymbol{b}) &= \frac{b_0 + b_1 z^{-1} + \cdots + b_n z^{-n}}{1 + a_1 z^{-1} + \cdots + a_n z^{-n}} \\
&= h_0 + \sum_{k=1}^{\infty} h_k(\boldsymbol{a}, \boldsymbol{b}) z^{-k}
\end{aligned} \tag{8.66}
$$

where

$$\boldsymbol{a} = [a_1\, a_2\, \cdots\, a_n]^T, \qquad \boldsymbol{b} = [b_0\, b_1\, \cdots\, b_n]^T, \qquad h_0 = b_0$$

With the N terms of the impulse response in (8.66), an $N \times 1$ vector is defined as

$$\boldsymbol{h}(\boldsymbol{a}, \boldsymbol{b}) = \begin{bmatrix} h_1(\boldsymbol{a}, \boldsymbol{b}) \\ h_2(\boldsymbol{a}, \boldsymbol{b}) \\ \vdots \\ h_N(\boldsymbol{a}, \boldsymbol{b}) \end{bmatrix} \tag{8.67}$$

An error vector between the desired impulse response f_i's and the actual one $h_i(\boldsymbol{a}, \boldsymbol{b})$'s is then defined by

$$\boldsymbol{e}(\boldsymbol{a}, \boldsymbol{b}) = \boldsymbol{f} - \boldsymbol{h}(\boldsymbol{a}, \boldsymbol{b}) \tag{8.68}$$

where $\boldsymbol{f}$ is defined in (8.16), and h_0 is chosen as $h_0 = f_0$. We seek the coefficient vectors $\boldsymbol{a}$ and $\boldsymbol{b}$ of the transfer function $H(z, \boldsymbol{a}, \boldsymbol{b})$ which minimize

$$J(\boldsymbol{a}, \boldsymbol{b}) = \boldsymbol{e}(\boldsymbol{a}, \boldsymbol{b})^T \boldsymbol{e}(\boldsymbol{a}, \boldsymbol{b}) = ||\boldsymbol{e}(\boldsymbol{a}, \boldsymbol{b})||^2 \tag{8.69}$$

provided that $2n \le N$.

To this end, we write the two equations in (8.8) as

$$\begin{bmatrix} \boldsymbol{b} \\ \boldsymbol{0} \end{bmatrix} = \begin{bmatrix} \boldsymbol{H}_1 \\ \boldsymbol{H}_2 \end{bmatrix} \begin{bmatrix} 1 \\ \boldsymbol{a} \end{bmatrix} \tag{8.70}$$

where it is assumed that $m = n$ and

$$\boldsymbol{H}_1 = \begin{bmatrix} h_0 & 0 & \cdots & 0 \\ h_1 & h_0 & \ddots & \vdots \\ \vdots & \vdots & \ddots & 0 \\ h_n & h_{n-1} & \cdots & h_0 \end{bmatrix}, \qquad \boldsymbol{H}_2 = \begin{bmatrix} h_{n+1} & h_n & \cdots & h_1 \\ h_{n+2} & h_{n+1} & \cdots & h_2 \\ \vdots & \vdots & \ddots & \vdots \\ h_N & h_{N-1} & \cdots & h_{N-n} \end{bmatrix}$$

with h_i's replacing terms $h_i(\boldsymbol{a}, \boldsymbol{b})$.

Hence, if the vectors $\boldsymbol{a}$ and $\boldsymbol{h}(\boldsymbol{a}, \boldsymbol{b})$ are known, the vector $\boldsymbol{b}$ can be determined directly from the upper portion of (8.70), i.e.,

$$\boldsymbol{b} = \boldsymbol{H}_1 \begin{bmatrix} 1 \\ \boldsymbol{a} \end{bmatrix}$$

The lower partition of (8.70) can be used to find a linear estimate of vector $\boldsymbol{a}$. Replacing the unknown terms of $\boldsymbol{h}(\boldsymbol{a}, \boldsymbol{b})$ in matrix $\boldsymbol{H}_2$ by the given terms of $\boldsymbol{f}$ produces a linear equation error as

$$\boldsymbol{d}(\boldsymbol{a}) = \boldsymbol{F}_2 \begin{bmatrix} 1 \\ \boldsymbol{a} \end{bmatrix} = \boldsymbol{G}\boldsymbol{a} - \boldsymbol{g} \tag{8.71}$$

where

$$\boldsymbol{F}_2 = \begin{bmatrix} f_{n+1} & f_n & \cdots & f_1 \\ f_{n+2} & f_{n+1} & \cdots & f_2 \\ \vdots & \vdots & \ddots & \vdots \\ f_N & f_{N-1} & \cdots & f_{N-n} \end{bmatrix}$$

$$\boldsymbol{G} = \begin{bmatrix} f_n & f_{n-1} & \cdots & f_1 \\ f_{n+1} & f_n & \cdots & f_2 \\ \vdots & \vdots & \ddots & \vdots \\ f_{N-1} & f_{N-2} & \cdots & f_{N-n} \end{bmatrix}, \qquad \boldsymbol{g} = -\begin{bmatrix} f_{n+1} \\ f_{n+2} \\ \vdots \\ f_N \end{bmatrix}$$

Differentiating $||\boldsymbol{d}(\boldsymbol{a})||^2 = \boldsymbol{d}(\boldsymbol{a})^T\boldsymbol{d}(\boldsymbol{a})$ with respect to vector $\boldsymbol{a}$ and setting it to null yields

$$\frac{\partial||\boldsymbol{d}(\boldsymbol{a})||^2}{\partial\boldsymbol{a}} = 2\left(\boldsymbol{G}^T\boldsymbol{G}\boldsymbol{a} - \boldsymbol{G}^T\boldsymbol{g}\right) = \boldsymbol{0} \tag{8.72}$$

which leads to

$$\boldsymbol{a} = (\boldsymbol{G}^T\boldsymbol{G})^{-1}\boldsymbol{G}^T\boldsymbol{g} \tag{8.73}$$

This can be used below as an initial estimate $\boldsymbol{a}^{(0)}$ in the iterative algorithm to obtain the suboptimal value of vector $\boldsymbol{a}$.

Now, (8.71) can be rearranged in the form

$$\boldsymbol{d}(\boldsymbol{a}) = \boldsymbol{A}(\boldsymbol{a})\boldsymbol{f} \tag{8.74}$$

where $\boldsymbol{A}(\boldsymbol{a})$ is an $(N-n) \times N$ matrix defined by

$$\boldsymbol{A}(\boldsymbol{a}) = \begin{bmatrix} a_n & \cdots & a_1 & 1 & 0 & \cdots & 0 \\ 0 & a_n & \cdots & a_1 & 1 & \ddots & \vdots \\ \vdots & \ddots & \ddots & & \ddots & \ddots & 0 \\ 0 & \cdots & 0 & a_n & \cdots & a_1 & 1 \end{bmatrix}$$

By substituting vector $\boldsymbol{f}$ in (8.68) into (8.74), we obtain

$$\boldsymbol{d}(\boldsymbol{a}) = \boldsymbol{A}(\boldsymbol{a})\boldsymbol{h}(\boldsymbol{a}, \boldsymbol{b}) + \boldsymbol{A}(\boldsymbol{a})\boldsymbol{e}(\boldsymbol{a}, \boldsymbol{b}) \tag{8.75}$$

Note that the lower partition of (8.70) can be rearranged in the form

$$\boldsymbol{H}_2 \begin{bmatrix} 1 \\ \boldsymbol{a} \end{bmatrix} = \boldsymbol{A}(\boldsymbol{a})\boldsymbol{h}(\boldsymbol{a}, \boldsymbol{b}) = \boldsymbol{0} \tag{8.76}$$

Hence (8.75) is reduced to

$$\boldsymbol{d}(\boldsymbol{a}) = \boldsymbol{A}(\boldsymbol{a})\boldsymbol{e}(\boldsymbol{a}, \boldsymbol{b}) \tag{8.77}$$

The expression in (8.77) relates the linear equation error $\boldsymbol{d}(\boldsymbol{a})$ to the error $\boldsymbol{e}(\boldsymbol{a}, \boldsymbol{b})$.

We now proceed to develop an iterative design procedure. We begin by writing an inverse relation of (8.77) between the errors as

$$\boldsymbol{e}(\boldsymbol{a}, \boldsymbol{b}^*) = \boldsymbol{W}(\boldsymbol{a})\boldsymbol{d}(\boldsymbol{a}) \tag{8.78}$$

where $\boldsymbol{e}(\boldsymbol{a}, \boldsymbol{b}^*)$ is the error vector corresponding to the optimal numerator coefficients $\boldsymbol{b}^*$ for a given $\boldsymbol{a}$, and the $N \times n$ matrix $\boldsymbol{W}(\boldsymbol{a})$ is a function of $\boldsymbol{a}$.

The projection theorem in [8] states that *for a given* $\boldsymbol{a}$*, the error* $\boldsymbol{e}(\boldsymbol{a}, \boldsymbol{b}^*)$ *corresponding to an optimal* $\boldsymbol{b}^*$ *is orthogonal to* $\boldsymbol{h}(\boldsymbol{a}, \boldsymbol{b})$. The expression in (8.76) means that the column vectors of $\boldsymbol{A}(\boldsymbol{a})^T$ are orthogonal to $\boldsymbol{h}(\boldsymbol{a}, \boldsymbol{b})$. Also, they are linearly independent and span the entire N-dimensional space. Hence $\boldsymbol{e}(\boldsymbol{a}, \boldsymbol{b}^*)$ can be expressed as a linear combination of the columns of $\boldsymbol{A}(\boldsymbol{a})^T$, that is,

$$\boldsymbol{e}(\boldsymbol{a}, \boldsymbol{b}^*) = \boldsymbol{A}(\boldsymbol{a})^T \gamma(\boldsymbol{a}) \tag{8.79}$$

where $\gamma(\boldsymbol{a})$ is an $(N - n) \times 1$ vector. Since (8.77) holds for all $\boldsymbol{e}(\boldsymbol{a}, \boldsymbol{b})$ which includes $\boldsymbol{e}(\boldsymbol{a}, \boldsymbol{b}^*)$, substituting (8.79) into (8.77) gives

$$\boldsymbol{d}(\boldsymbol{a}) = \boldsymbol{A}(\boldsymbol{a})\boldsymbol{A}(\boldsymbol{a})^T \gamma(\boldsymbol{a}) \tag{8.80}$$

Evidently, matrix $\boldsymbol{A}(\boldsymbol{a})\boldsymbol{A}(\boldsymbol{a})^T$ is nonsingular. Thus, from (8.80) it follows that

$$\gamma(\boldsymbol{a}) = \left[\boldsymbol{A}(\boldsymbol{a})\boldsymbol{A}(\boldsymbol{a})^T\right]^{-1} \boldsymbol{d}(\boldsymbol{a}) \tag{8.81}$$

Substituting (8.81) into (8.79) and comparing it with (8.78), $\boldsymbol{W}(\boldsymbol{a})$ can be expressed as

$$\boldsymbol{W}(\boldsymbol{a}) = \boldsymbol{A}(\boldsymbol{a})^T \left[\boldsymbol{A}(\boldsymbol{a})\boldsymbol{A}(\boldsymbol{a})^T\right]^{-1} \tag{8.82}$$

As a result, the problem of minimizing (8.69) is now converted into

$$\min_{\boldsymbol{a},\boldsymbol{b}} ||\boldsymbol{e}(\boldsymbol{a},\boldsymbol{b})||^2 = \min_{\boldsymbol{a}} ||\boldsymbol{W}(\boldsymbol{a})\boldsymbol{d}(\boldsymbol{a})||^2 \tag{8.83}$$

where $\boldsymbol{d}(\boldsymbol{a})$ is the linear equation error given by (8.71). By letting

$$\boldsymbol{W}(\boldsymbol{a}^{(k-1)})\boldsymbol{d}(\boldsymbol{a}^{(k)}) = \boldsymbol{W}(\boldsymbol{a}^{(k-1)})(\boldsymbol{G}\boldsymbol{a}^{(k)} - \boldsymbol{g}) \tag{8.84}$$

we compute

$$\frac{\partial||\boldsymbol{W}(\boldsymbol{a}^{(k-1)})\boldsymbol{d}(\boldsymbol{a}^{(k)})||^2}{\partial\boldsymbol{a}^{(k)}} = 2\,\boldsymbol{G}^T\boldsymbol{W}(\boldsymbol{a}^{(k-1)})^T\boldsymbol{W}(\boldsymbol{a}^{(k-1)})\big(\boldsymbol{G}\boldsymbol{a}^{(k)} - \boldsymbol{g}\big) \tag{8.85}$$

Setting (8.85) to null, we obtain

$$\boldsymbol{a}^{(k)} = \big[\boldsymbol{G}^T\boldsymbol{W}(\boldsymbol{a}^{(k-1)})^T\boldsymbol{W}(\boldsymbol{a}^{(k-1)})\boldsymbol{G}\big]^{-1}\boldsymbol{G}^T\boldsymbol{W}(\boldsymbol{a}^{(k-1)})^T\boldsymbol{W}(\boldsymbol{a}^{(k-1)})\boldsymbol{g} \tag{8.86}$$

This iteration process continues until

$$\Big|\,||\boldsymbol{W}(\boldsymbol{a}^{(k-2)})\boldsymbol{d}(\boldsymbol{a}^{(k-1)})||^2 - ||\boldsymbol{W}(\boldsymbol{a}^{(k-1)})\boldsymbol{d}(\boldsymbol{a}^{(k)})||^2\Big| < \varepsilon \tag{8.87}$$

is satisfied where $\varepsilon > 0$ is a prescribed tolerance. When the above iteration algorithm using (8.86) is complete, the resulting $\boldsymbol{a}^{(k)}$ is deemed to be the suboptimal solution.

At this point, the second phase of the algorithm begins in order to identify a stationary point of $||\boldsymbol{e}(\boldsymbol{a},\boldsymbol{b})||$. By (8.83), it follows that

$$||\boldsymbol{e}(\boldsymbol{a},\boldsymbol{b})||^2 = ||\boldsymbol{W}(\boldsymbol{a})\boldsymbol{d}(\boldsymbol{a})||^2 = \boldsymbol{d}(\boldsymbol{a})^T\boldsymbol{W}(\boldsymbol{a})^T\boldsymbol{W}(\boldsymbol{a})\boldsymbol{d}(\boldsymbol{a}) \tag{8.88}$$

which leads to

$$\begin{aligned}\frac{\partial||\boldsymbol{e}(\boldsymbol{a},\boldsymbol{b})||^2}{\partial a_i} &= 2\Big[\frac{\partial\boldsymbol{d}(\boldsymbol{a})^T}{\partial a_i}\boldsymbol{W}(\boldsymbol{a})^T + \Big(\frac{\partial\boldsymbol{W}(\boldsymbol{a})}{\partial a_i}\boldsymbol{d}(\boldsymbol{a})\Big)^T\Big]\boldsymbol{W}(\boldsymbol{a})\boldsymbol{d}(\boldsymbol{a}) \\ &= 2\big[\boldsymbol{e}_i^T\boldsymbol{G}^T\boldsymbol{W}(\boldsymbol{a})^T + \boldsymbol{l}_i(\boldsymbol{a})^T\big]\boldsymbol{W}(\boldsymbol{a})(\boldsymbol{G}\boldsymbol{a} - \boldsymbol{g})\end{aligned} \tag{8.89}$$

where a_i and $\boldsymbol{e}_i$ denote the ith element of $n \times 1$ vector $\boldsymbol{a}$ and the ith column of $n \times n$ identity matrix $\boldsymbol{I}_n$, respectively, and

$$\boldsymbol{l}_i(\boldsymbol{a}) = \frac{\partial\boldsymbol{W}(\boldsymbol{a})}{\partial a_i}\boldsymbol{d}(\boldsymbol{a})$$

Using (8.82), we have

$$\boldsymbol{W}(\boldsymbol{a})\boldsymbol{A}(\boldsymbol{a})\boldsymbol{A}(\boldsymbol{a})^T = \boldsymbol{A}(\boldsymbol{a})^T \tag{8.90}$$

which leads to

$$\begin{aligned}\frac{\partial \boldsymbol{W}(\boldsymbol{a})}{\partial a_i} = &\Big\{\frac{\partial \boldsymbol{A}(\boldsymbol{a})^T}{\partial a_i} - \boldsymbol{W}(\boldsymbol{a})\frac{\partial \boldsymbol{A}(\boldsymbol{a})}{\partial a_i}\boldsymbol{A}(\boldsymbol{a})^T - \boldsymbol{W}(\boldsymbol{a})\boldsymbol{A}(\boldsymbol{a})\frac{\partial \boldsymbol{A}(\boldsymbol{a})^T}{\partial a_i}\Big\}\\ &\cdot\big[\boldsymbol{A}(\boldsymbol{a})\boldsymbol{A}(\boldsymbol{a})^T\big]^{-1}\end{aligned} \tag{8.91}$$

where $\partial \boldsymbol{A}(\boldsymbol{a})/\partial a_i$ is simply the matrix $\boldsymbol{A}(\boldsymbol{a})$ with unity's replacing each and every a_i and setting the rest of the components to zero. The derivatives involved in computing $\boldsymbol{l}_i(\boldsymbol{a})$ in (8.89) are calculated in the same way.

It is clear from (8.89) that

$$\begin{aligned}\frac{\partial ||\boldsymbol{e}(\boldsymbol{a},\boldsymbol{b})||^2}{\partial \boldsymbol{a}} &= 2\big[\boldsymbol{G}^T\boldsymbol{W}(\boldsymbol{a})^T + \boldsymbol{L}(\boldsymbol{a})^T\big]\boldsymbol{W}(\boldsymbol{a})(\boldsymbol{G}\boldsymbol{a} - \boldsymbol{g})\\ &\triangleq 2\boldsymbol{U}(\boldsymbol{a})(\boldsymbol{G}\boldsymbol{a} - \boldsymbol{g})\end{aligned} \tag{8.92}$$

where

$$\begin{aligned}\boldsymbol{L}(\boldsymbol{a}) &= \big[\boldsymbol{l}_1(\boldsymbol{a}), \boldsymbol{l}_2(\boldsymbol{a}), \cdots, \boldsymbol{l}_n(\boldsymbol{a})\big]\\ \boldsymbol{U}(\boldsymbol{a}) &= \big[\boldsymbol{G}^T\boldsymbol{W}(\boldsymbol{a})^T + \boldsymbol{L}(\boldsymbol{a})^T\big]\boldsymbol{W}(\boldsymbol{a})\end{aligned}$$

By letting

$$\boldsymbol{U}(\boldsymbol{a}^{(k-1)})(\boldsymbol{G}\boldsymbol{a}^{(k)} - \boldsymbol{g}) = \boldsymbol{0} \tag{8.93}$$

we obtain

$$\boldsymbol{a}^{(k)} = \big[\boldsymbol{U}(\boldsymbol{a}^{(k-1)})\boldsymbol{G}\big]^{-1}\boldsymbol{U}(\boldsymbol{a}^{(k-1)})\boldsymbol{g} \tag{8.94}$$

This iteration process continues until

$$||\boldsymbol{a}^{(k)} - \boldsymbol{a}^{(k-1)}||^2 < \varepsilon \tag{8.95}$$

is satisfied for a prescribed tolerance $\varepsilon > 0$. As the iterate $\boldsymbol{a}_k$ converges to a vector $\boldsymbol{a}$, the gradient of $||\boldsymbol{e}(\boldsymbol{a},\boldsymbol{b})||$ with respect to $\boldsymbol{a}$ and $\boldsymbol{b}$ is expected to be practically zero. A step-by-step summary of the algorithm is given below.

Given a set of FIR filter coefficients $f_0, f_1, \cdots, f_N$

Sept 1: Find an initial estimate $\boldsymbol{a}^{(0)}$ using (8.73).
Sept 2: Continue the iteration process in (8.86) until (8.87) is satisfied.
Sept 3: Continue the iteration process in (8.94) until (8.95) is satisfied.
Sept 4: Determine $\boldsymbol{e}(\boldsymbol{a}^*, \boldsymbol{b}^*)$ from (8.78).
Sept 5: Compute $\boldsymbol{h}(\boldsymbol{a}^*, \boldsymbol{b}^*)$ from (8.68).
Sept 6: Obtain the optimal $\boldsymbol{b}^*$ from the upper portion of (8.70).

8.5 Design Using State-Space Models

8.5.1 Balanced Model Reduction

An Nth-order FIR digital filter is written in the form

$$F(z) = f_0 + F_1(z) \tag{8.96}$$

where

$$F_1(z) = f_1 z^{-1} + f_2 z^{-2} + \cdots + f_N z^{-N}$$

The FIR digital filter in (8.96) can be represented by a state-space model $(\boldsymbol{A}, \boldsymbol{b}, \boldsymbol{c}, d)_N$ as

$$\begin{aligned} \boldsymbol{x}(k+1) &= \boldsymbol{A}\boldsymbol{x}(k) + \boldsymbol{b}u(k) \\ y(k) &= \boldsymbol{c}\boldsymbol{x}(k) + du(k) \end{aligned} \tag{8.97}$$

where $\boldsymbol{x}(k)$ is an $N \times 1$ state-variable vector, $u(k)$ is a scalar input, $y(k)$ is a scalar output, and

$$\boldsymbol{A} = \begin{bmatrix} 0 & \cdots & 0 & 0 \\ 1 & \ddots & \vdots & \vdots \\ \vdots & \ddots & 0 & 0 \\ 0 & \cdots & 1 & 0 \end{bmatrix}, \qquad \boldsymbol{b} = \begin{bmatrix} 1 \\ 0 \\ \vdots \\ 0 \end{bmatrix}$$

$$\boldsymbol{c} = \begin{bmatrix} f_1 & f_2 & \cdots & f_N \end{bmatrix}, \qquad d = f_0$$

The Hankel matrix of the filter in (8.96) is written in the form

$$\boldsymbol{H}_{N,N} = \begin{bmatrix} f_1 & f_2 & \cdots & f_N \\ f_2 & \vdots & \cdot^{\cdot^{\cdot}} & 0 \\ \vdots & f_N & \cdot^{\cdot^{\cdot}} & \vdots \\ f_N & 0 & \cdots & 0 \end{bmatrix} \tag{8.98}$$

which is equivalent to

$$\boldsymbol{H}_{N,N} = \boldsymbol{U}_N \boldsymbol{V}_N \tag{8.99}$$

where

$$\boldsymbol{V}_N = \begin{bmatrix} \boldsymbol{b} \ \boldsymbol{A}\boldsymbol{b} \ \cdots \ \boldsymbol{A}^{N-1}\boldsymbol{b} \end{bmatrix} \quad \text{and} \quad \boldsymbol{U}_N = \begin{bmatrix} \boldsymbol{c} \\ \boldsymbol{c}\boldsymbol{A} \\ \vdots \\ \boldsymbol{c}\boldsymbol{A}^{N-1} \end{bmatrix}$$

are called the *controllability matrix* and the *observability matrix*, respectively. The Hankel matrix in (8.98) is symmetric, and can be factorized using eigenvalue-eigenvector decomposition as

$$\boldsymbol{H}_{N,N} = \boldsymbol{P}\boldsymbol{\Sigma}\boldsymbol{P}^T \tag{8.100}$$

where $\boldsymbol{\Sigma}$ and $\boldsymbol{P}$ are $N \times N$ diagonal and orthogonal matrices consisting of the eigenvalues and eigenvectors of $\boldsymbol{H}_{N,N}$, respectively, and $\boldsymbol{P}\boldsymbol{P}^T = \boldsymbol{I}_N$.

Lemma 8.3
For the system $(\boldsymbol{A}, \boldsymbol{b}, \boldsymbol{c}, d)_N$ in (8.97), the controllability matrix $\boldsymbol{V}_N$ and the controllability Grammian $\boldsymbol{K}_c$ are unit matrices, i.e.,

$$\boldsymbol{V}_N = \boldsymbol{K}_c = \boldsymbol{I}_N \tag{8.101}$$

and the coordinate transformation defined by

$$\boldsymbol{T} = \boldsymbol{P}\boldsymbol{\Sigma}^{-\frac{1}{2}} \tag{8.102}$$

will lead to a balanced realization of the system.

Proof
By simply substituting $\boldsymbol{A}$ and $\boldsymbol{b}$ in (8.97) into $\boldsymbol{V}_N$ in (8.99) and recalling that the controllability Grammian is defined by

$$\boldsymbol{K}_c = \sum_{i=0}^{\infty} \boldsymbol{A}^i \boldsymbol{b}\boldsymbol{b}^T (\boldsymbol{A}^i)^T$$

we arrive at (8.101). From (8.99)-(8.101), it follows that

$$\boldsymbol{H}_{N,N} = \boldsymbol{U}_N \boldsymbol{V}_N = \boldsymbol{U}_N = \boldsymbol{P}\boldsymbol{\Sigma}\boldsymbol{P}^T \tag{8.103}$$

Also recalling that observability Grammian is defined by

$$\boldsymbol{W}_o = \sum_{i=0}^{\infty} (\boldsymbol{A}^i)^T \boldsymbol{c}^T \boldsymbol{c} \boldsymbol{A}^i$$

and using (8.103), we obtain

$$\boldsymbol{W}_o = \boldsymbol{U}_N^T \boldsymbol{U}_N = \boldsymbol{P}\boldsymbol{\Sigma}^2\boldsymbol{P}^T \tag{8.104}$$

Hence, since $\overline{\boldsymbol{K}}_c = \boldsymbol{T}^{-1}\boldsymbol{K}_c\boldsymbol{T}^{-T}$ and $\overline{\boldsymbol{W}}_o = \boldsymbol{T}^T\boldsymbol{W}_o\boldsymbol{T}$, (8.102) produces

$$\overline{\boldsymbol{K}}_c = \overline{\boldsymbol{W}}_o = \boldsymbol{\Sigma} \tag{8.105}$$

This means that if the coordinate transformation is specified by (8.102), the resulting equivalent system $(\overline{\boldsymbol{A}}, \overline{\boldsymbol{b}}, \overline{\boldsymbol{c}}, d)_N$ is balanced where

$$\overline{\boldsymbol{A}} = \boldsymbol{T}^{-1}\boldsymbol{A}\boldsymbol{T}, \qquad \overline{\boldsymbol{b}} = \boldsymbol{T}^{-1}\boldsymbol{b}, \qquad \overline{\boldsymbol{c}} = \boldsymbol{c}\boldsymbol{T}$$

This completes the proof of Lemma 8.3. ■

Now, assume that

$$\boldsymbol{\Sigma} = \text{diag}\{\sigma_1, \sigma_2, \cdots, \sigma_N\} \text{ and } \sigma_n \gg \sigma_{n+1} \tag{8.106}$$

where $1 \le n < N$. The following theorem is helpful to obtain a reduced-order subsystem which approximates the whole system in a certain sense.

Theorem 8.2

Suppose the Hankel matrix $\boldsymbol{H}_{N,N}$ of an Nth-order FIR filter $(\boldsymbol{A}, \boldsymbol{b}, \boldsymbol{c}, d)_N$ in (8.97) is factorized as (8.100), an nth-order reduced balanced system is equivalent to the subsystem $(\boldsymbol{A}_{11}, \boldsymbol{b}_1, \boldsymbol{c}_1, d)_n$ where

$$\boldsymbol{A}_{11} = \boldsymbol{P}_1^T\boldsymbol{A}\boldsymbol{P}_1, \qquad \boldsymbol{b}_1 = \boldsymbol{P}_1^T\boldsymbol{b}, \qquad \boldsymbol{c}_1 = \boldsymbol{c}\boldsymbol{P}_1$$

and $\boldsymbol{P}_1$ is an $N \times n$ matrix obtained from the partition

$$\boldsymbol{P} = \begin{bmatrix}\boldsymbol{P}_1 & \boldsymbol{P}_2\end{bmatrix}$$

Proof

If $\boldsymbol{\Sigma}$ is partitioned as

$$\boldsymbol{\Sigma} = \begin{bmatrix}\boldsymbol{\Sigma}_1 & \boldsymbol{0} \\ \boldsymbol{0} & \boldsymbol{\Sigma}_4\end{bmatrix} \tag{8.107}$$

where

$$\boldsymbol{\Sigma}_1 = \text{diag}\{\sigma_1, \sigma_2, \cdots, \sigma_n\}, \qquad \boldsymbol{\Sigma}_4 = \text{diag}\{\sigma_{n+1}, \sigma_{n+2}, \cdots, \sigma_N\}$$

then use of (8.102) yields balanced realization $(\overline{\boldsymbol{A}}, \overline{\boldsymbol{b}}, \overline{\boldsymbol{c}})_N$ as

$$\begin{aligned}\overline{\boldsymbol{A}} &= \begin{bmatrix}\boldsymbol{\Sigma}_1^{\frac{1}{2}}\boldsymbol{P}_1^T\boldsymbol{A}\boldsymbol{P}_1\boldsymbol{\Sigma}_1^{-\frac{1}{2}} & \boldsymbol{\Sigma}_1^{\frac{1}{2}}\boldsymbol{P}_1^T\boldsymbol{A}\boldsymbol{P}_2\boldsymbol{\Sigma}_4^{-\frac{1}{2}} \\ \boldsymbol{\Sigma}_4^{\frac{1}{2}}\boldsymbol{P}_2^T\boldsymbol{A}\boldsymbol{P}_1\boldsymbol{\Sigma}_1^{-\frac{1}{2}} & \boldsymbol{\Sigma}_4^{\frac{1}{2}}\boldsymbol{P}_2^T\boldsymbol{A}\boldsymbol{P}_2\boldsymbol{\Sigma}_4^{-\frac{1}{2}}\end{bmatrix} \\ \overline{\boldsymbol{b}} &= \begin{bmatrix}\boldsymbol{\Sigma}_1^{\frac{1}{2}}\boldsymbol{P}_1^T\boldsymbol{b} \\ \boldsymbol{\Sigma}_4^{\frac{1}{2}}\boldsymbol{P}_2^T\boldsymbol{b}\end{bmatrix}, \qquad \overline{\boldsymbol{c}} = \begin{bmatrix}\boldsymbol{c}\boldsymbol{P}_1\boldsymbol{\Sigma}_1^{-\frac{1}{2}} & \boldsymbol{c}\boldsymbol{P}_2\boldsymbol{\Sigma}_4^{-\frac{1}{2}}\end{bmatrix}\end{aligned} \tag{8.108}$$

whose nth-order subsystem $(\overline{\boldsymbol{A}}_{11}, \overline{\boldsymbol{b}}_1, \overline{\boldsymbol{c}}_1)_n$ is specified by

$$\begin{aligned} \overline{\boldsymbol{A}}_{11} &= \boldsymbol{\Sigma}_1^{\frac{1}{2}} \boldsymbol{P}_1^T \boldsymbol{A} \boldsymbol{P}_1 \boldsymbol{\Sigma}_1^{-\frac{1}{2}} \\ \overline{\boldsymbol{b}}_1 &= \boldsymbol{\Sigma}_1^{\frac{1}{2}} \boldsymbol{P}_1^T \boldsymbol{b}, \qquad \overline{\boldsymbol{c}}_1 = \boldsymbol{c} \boldsymbol{P}_1 \boldsymbol{\Sigma}_1^{-\frac{1}{2}} \end{aligned} \tag{8.109}$$

The transfer function of the nth-order subsystem is described by

$$\begin{aligned} H(z) &= \overline{\boldsymbol{c}}_1 (z\boldsymbol{I}_n - \overline{\boldsymbol{A}}_{11})^{-1} \overline{\boldsymbol{b}}_1 + d \\ &= \boldsymbol{c}_1 (z\boldsymbol{I}_n - \boldsymbol{A}_{11})^{-1} \boldsymbol{b}_1 + d \end{aligned} \tag{8.110}$$

This completes the proof of the theorem. ■

Remark 8.1

Due to the special structure of matrices $\boldsymbol{A}$, $\boldsymbol{b}$, and $\boldsymbol{c}$ in (8.97), it can be shown that

$$\begin{aligned} \boldsymbol{A}_{11} &= \boldsymbol{P}(2:N, 1:n)^T \boldsymbol{P}(1:N-1, 1:n) \\ \boldsymbol{b}_1 &= \boldsymbol{P}(1, 1:n)^T, \qquad \boldsymbol{c}_1 = \boldsymbol{c}\boldsymbol{P}(1:N, 1:n) \end{aligned} \tag{8.111}$$

where $\boldsymbol{P}(i:j, k:m)$ denotes an extraction of matrix $\boldsymbol{P}$'s rows from i to j and its columns from k to m.

A step-by-step summary of the algorithm is given below.

Given a set of FIR digital filter coefficients $f_0, f_1, \cdots, f_N$

Sept 1: Construct the Hankel matrix $\boldsymbol{H}_{N,N}$ in (8.98).

Sept 2: Decompose the Hankel matrix $\boldsymbol{H}_{N,N}$ to obtain $\boldsymbol{\Sigma}$ and $\boldsymbol{P}$ as shown in (8.100).

Sept 3: Choose a derired order n of approximation according to the magnitudes of the elements of $\boldsymbol{\Sigma}$ as shown in (8.106).

Sept 4: Calculate matrices $\boldsymbol{A}_{11}$, $\boldsymbol{b}_1$, and $\boldsymbol{c}_1$ using (8.111).

Sept 5: Convert the state-space parameters $\boldsymbol{A}_{11}$, $\boldsymbol{b}_1$, and $\boldsymbol{c}_1$ into the transfer function form $H(z)$ where $d = f_0$ (see Faddeev's formula in Section 4.3.2).

8.5.2 Stability and Minimality

The controllability and observability Grammians of the balanced realization, which are both equal to $\boldsymbol{\Sigma}$, are given by the unique positive-definite solution to the Lyapunov equations

$$\overline{\boldsymbol{A}} \boldsymbol{\Sigma} \overline{\boldsymbol{A}}^T - \boldsymbol{\Sigma} = -\overline{\boldsymbol{b}}\, \overline{\boldsymbol{b}}^T \tag{8.112a}$$

$$\overline{\boldsymbol{A}}^T \boldsymbol{\Sigma} \overline{\boldsymbol{A}} - \boldsymbol{\Sigma} = -\overline{\boldsymbol{c}}^T \overline{\boldsymbol{c}} \tag{8.112b}$$

Theorem 8.3
For the balanced realization $(\overline{\boldsymbol{A}}, \overline{\boldsymbol{b}}, \overline{\boldsymbol{c}})_N$ in (8.108), we have

$$||\overline{\boldsymbol{A}}|| \le 1 \tag{8.113}$$

where $||\overline{\boldsymbol{A}}||$ denotes the spectral norm of matrix $\overline{\boldsymbol{A}}$. If $\boldsymbol{\Sigma}$ has distinct diagonal entries, then strict inequality holds in (8.113).

Proof
By multiplying (8.112a) from the left by $\overline{\boldsymbol{A}}^T$ and from the right by $\overline{\boldsymbol{A}}$ and adding the result to (8.112b), we obtain

$$\overline{\boldsymbol{A}}^T \overline{\boldsymbol{A}} \boldsymbol{\Sigma} \overline{\boldsymbol{A}}^T \overline{\boldsymbol{A}} - \boldsymbol{\Sigma} = -(\overline{\boldsymbol{A}}^T \overline{\boldsymbol{b}}\, \overline{\boldsymbol{b}}^T \overline{\boldsymbol{A}} + \overline{\boldsymbol{c}}^T \overline{\boldsymbol{c}}) \tag{8.114}$$

Let λ be an eigenvalue of $\overline{\boldsymbol{A}}^T \overline{\boldsymbol{A}}$ and let $\boldsymbol{v}$ be the corresponding eigenvector, i.e., $\overline{\boldsymbol{A}}^T \overline{\boldsymbol{A}} \boldsymbol{v} = \lambda \boldsymbol{v}$. Multiplying (8.114) from the left by $\boldsymbol{v}^H$ and from the right by $\boldsymbol{v}$ yields

$$(|\lambda|^2 - 1)\boldsymbol{v}^H \boldsymbol{\Sigma} \boldsymbol{v} = -(\boldsymbol{v}^H \overline{\boldsymbol{A}}^T \overline{\boldsymbol{b}}\, \overline{\boldsymbol{b}}^T \overline{\boldsymbol{A}} \boldsymbol{v} + \boldsymbol{v}^H \overline{\boldsymbol{c}}^T \overline{\boldsymbol{c}} \boldsymbol{v}) \le 0 \tag{8.115}$$

Since $\boldsymbol{v}^H \boldsymbol{\Sigma} \boldsymbol{v} > 0$, it follows that

$$|\lambda|^2 \le 1 \tag{8.116}$$

Since λ is an arbitrary eigenvalue of $\overline{\boldsymbol{A}}^T \overline{\boldsymbol{A}}$, the spectral norm of $\overline{\boldsymbol{A}}$ is defined as

$$||\overline{\boldsymbol{A}}|| = \sqrt{\lambda_{max}(\overline{\boldsymbol{A}}^T \overline{\boldsymbol{A}})} \tag{8.117}$$

and by virtue of (8.116), we arrive at (8.113).

Assume that $||\overline{\boldsymbol{A}}|| = 1$. Then, from (8.117) it follows that $\lambda = 1$ is an eigenvalue of $\overline{\boldsymbol{A}}^T \overline{\boldsymbol{A}}$. Hence matrix $\overline{\boldsymbol{A}}^T \overline{\boldsymbol{A}} - \boldsymbol{I}_N$ is singular. Letting $\boldsymbol{V}$ be a basis matrix for the right nullspace of $\overline{\boldsymbol{A}}^T \overline{\boldsymbol{A}} - \boldsymbol{I}_N$, it follows that

$$(\overline{\boldsymbol{A}}^T \overline{\boldsymbol{A}} - \boldsymbol{I}_N)\boldsymbol{V} = \boldsymbol{0} \tag{8.118}$$

By multiplying (8.114) from the left by $\boldsymbol{V}^H$ and from the right by $\boldsymbol{V}$ and then using (8.118), we arrive at

$$\overline{\boldsymbol{b}}^T \overline{\boldsymbol{A}} \boldsymbol{V} = \boldsymbol{0}, \qquad \overline{\boldsymbol{c}} \boldsymbol{V} = \boldsymbol{0} \tag{8.119}$$

Multiplying (8.114) from the right by $\boldsymbol{V}$ and employing (8.118) and (8.119) gives

$$(\overline{\boldsymbol{A}}^T\overline{\boldsymbol{A}} - \boldsymbol{I}_N)\boldsymbol{\Sigma}\boldsymbol{V} = \boldsymbol{0} \tag{8.120}$$

This means that $\boldsymbol{\Sigma}\boldsymbol{V}$ is in the right nullspace of $\overline{\boldsymbol{A}}^T\overline{\boldsymbol{A}} - \boldsymbol{I}_N$ and there exists a nonsingular matrix $\widetilde{\boldsymbol{\Sigma}}$ such that

$$\boldsymbol{\Sigma}\boldsymbol{V} = \boldsymbol{V}\widetilde{\boldsymbol{\Sigma}} \tag{8.121}$$

Since $\widetilde{\boldsymbol{\Sigma}}$ is the restriction of $\boldsymbol{\Sigma}$ to the space spanned by $\boldsymbol{V}$, it is possible to choose $\boldsymbol{V}$ such that $\widetilde{\boldsymbol{\Sigma}}$ is diagonal. By multiplying (8.112a) from the right by $\overline{\boldsymbol{A}}\boldsymbol{V}$ and using (8.118), (8.119) and (8.121), we obtain

$$\boldsymbol{\Sigma}\overline{\boldsymbol{A}}\boldsymbol{V} = \overline{\boldsymbol{A}}\boldsymbol{V}\widetilde{\boldsymbol{\Sigma}} \tag{8.122}$$

Let $\tilde{\sigma}$ be any diagonal entry of $\widetilde{\boldsymbol{\Sigma}}$ and let $\tilde{\boldsymbol{v}}$ be the corresponding column of $\boldsymbol{V}$. Then (8.121) and (8.122) provide

$$\boldsymbol{\Sigma}\tilde{\boldsymbol{v}} = \tilde{\sigma}\,\tilde{\boldsymbol{v}}, \qquad \boldsymbol{\Sigma}\overline{\boldsymbol{A}}\tilde{\boldsymbol{v}} = \tilde{\sigma}\overline{\boldsymbol{A}}\tilde{\boldsymbol{v}} \tag{8.123}$$

Therefore, both $\tilde{\boldsymbol{v}}$ and $\overline{\boldsymbol{A}}\tilde{\boldsymbol{v}}$ are eigenvectors of $\boldsymbol{\Sigma}$ corresponding to the eigenvalue $\tilde{\sigma}$. If $\boldsymbol{\Sigma}$ has distinct eigenvalues, then $\tilde{\boldsymbol{v}}$ and $\overline{\boldsymbol{A}}\tilde{\boldsymbol{v}}$ must be parallel, i.e., there exists an α such that

$$\overline{\boldsymbol{A}}\tilde{\boldsymbol{v}} = \alpha\tilde{\boldsymbol{v}} \tag{8.124}$$

From (8.119), it follows that

$$\overline{\boldsymbol{c}}\tilde{\boldsymbol{v}} = 0 \tag{8.125}$$

It is obvious that (8.124) and (8.125) contradict the assumption that the system in (8.97) is observable, since (8.98) and (8.99) hold. Hence $||\overline{\boldsymbol{A}}||$ cannot be equal to unity when $\boldsymbol{\Sigma}$ has distinct diagonal entries. This completes the proof of the theorem. ■

Equation (8.108) shows that the balanced realization $(\overline{\boldsymbol{A}}, \overline{\boldsymbol{b}}, \overline{\boldsymbol{c}})_N$ can be partitioned as

$$\overline{\boldsymbol{A}} = \begin{bmatrix} \overline{\boldsymbol{A}}_{11} & \overline{\boldsymbol{A}}_{12} \\ \overline{\boldsymbol{A}}_{21} & \overline{\boldsymbol{A}}_{22} \end{bmatrix}, \qquad \overline{\boldsymbol{b}} = \begin{bmatrix} \overline{\boldsymbol{b}}_1 \\ \overline{\boldsymbol{b}}_2 \end{bmatrix}, \qquad \overline{\boldsymbol{c}} = \begin{bmatrix} \overline{\boldsymbol{c}}_1 & \overline{\boldsymbol{c}}_2 \end{bmatrix} \tag{8.126}$$

when the controllability and observability Grammians are partitioned as $\boldsymbol{\Sigma} = \boldsymbol{\Sigma}_1 \oplus \boldsymbol{\Sigma}_4$. In this case, the Lyapunov equations in (8.112a) and (8.112b) can be written as

$$\begin{bmatrix} \overline{A}_{11} & \overline{A}_{12} \\ \overline{A}_{21} & \overline{A}_{22} \end{bmatrix} \begin{bmatrix} \Sigma_1 & 0 \\ 0 & \Sigma_4 \end{bmatrix} \begin{bmatrix} \overline{A}_{11} & \overline{A}_{12} \\ \overline{A}_{21} & \overline{A}_{22} \end{bmatrix}^T - \begin{bmatrix} \Sigma_1 & 0 \\ 0 & \Sigma_4 \end{bmatrix} = - \begin{bmatrix} \overline{b}_1 \\ \overline{b}_2 \end{bmatrix} \begin{bmatrix} \overline{b}_1^T & \overline{b}_2^T \end{bmatrix} \tag{8.127a}$$

$$\begin{bmatrix} \overline{A}_{11} & \overline{A}_{12} \\ \overline{A}_{21} & \overline{A}_{22} \end{bmatrix}^T \begin{bmatrix} \Sigma_1 & 0 \\ 0 & \Sigma_4 \end{bmatrix} \begin{bmatrix} \overline{A}_{11} & \overline{A}_{12} \\ \overline{A}_{21} & \overline{A}_{22} \end{bmatrix} - \begin{bmatrix} \Sigma_1 & 0 \\ 0 & \Sigma_4 \end{bmatrix} = - \begin{bmatrix} \overline{c}_1^T \\ \overline{c}_2^T \end{bmatrix} \begin{bmatrix} \overline{c}_1 & \overline{c}_2 \end{bmatrix} \tag{8.127b}$$

respectively.

Theorem 8.4

Suppose a system $(\overline{A}, \overline{b}, \overline{c})_N$ in (8.108) is asymptotically stable and either the controllability Grammian or the observability Grammian is nonsingular and diagonal, every subsystem is asymptotically stable.

Proof

Assume that the controllability Grammian is nonsingular, diagonal, and equal to Σ and that the system is partitioned as in (8.126). Then it will be shown that the subsystem $(\overline{A}_{11}, \overline{b}_1, \overline{c}_1)_n$ is asymptotically stable.

The upper left equation in (8.127a) becomes

$$\overline{A}_{11}\Sigma_1\overline{A}_{11}^T + \overline{A}_{12}\Sigma_4\overline{A}_{12}^T - \Sigma_1 = -\overline{b}_1\overline{b}_1^T \tag{8.128}$$

Let λ be an eigenvalue of $\overline{A}_{11}^T$ and let v be the corresponding eigenvector, i.e., $\overline{A}_{11}^T v = \lambda v$. Multiplying (8.128) from the left by v^H and from the right by v yields

$$(|\lambda|^2 - 1)v^H\Sigma_1 v = -(v^H\overline{A}_{12}\Sigma_4\overline{A}_{12}^T v + v^H\overline{b}_1\overline{b}_1^T v) \leq 0 \tag{8.129}$$

Since $v^H\Sigma_1 v > 0$, it follows that

$$|\lambda| \leq 1 \tag{8.130}$$

Suppose $|\lambda| = 1$, since Σ_4 is positive definite, it follows from (8.129) that

$$v^H\overline{A}_{12} = \mathbf{0}, \qquad v^H\overline{b}_1 = 0 \tag{8.131}$$

Hence

$$\begin{bmatrix} v^H & 0 \end{bmatrix} \begin{bmatrix} \overline{A}_{11} & \overline{A}_{12} \\ \overline{A}_{21} & \overline{A}_{22} \end{bmatrix} = \lambda^* \begin{bmatrix} v^H & 0 \end{bmatrix}, \qquad \begin{bmatrix} v^H & 0 \end{bmatrix} \begin{bmatrix} \overline{b}_1 \\ \overline{b}_2 \end{bmatrix} = 0 \tag{8.132}$$

These contradict the assumption that the system in (8.126) is controllable. Therefore, $|\lambda| \neq 1$ which means that $|\lambda| < 1$, i.e., the subsystem $(\overline{\boldsymbol{A}}_{11}, \overline{\boldsymbol{b}}_1, \overline{\boldsymbol{c}}_1)_n$ is asymptotically stable. This completes the proof of the theorem. ■

Theorem 8.4 states that if the system is balanced, then every subsystem is asymptotically stable. Assume that $\sigma_{min}(\boldsymbol{\Sigma}_1)$ is the smallest eigenvalue of $\boldsymbol{\Sigma}_1$ and $\sigma_{max}(\boldsymbol{\Sigma}_4)$ is the largest eigenvalue of $\boldsymbol{\Sigma}_4$.

Theorem 8.5

Let the partitioning in (8.126) be performed so that $\sigma_{min}(\boldsymbol{\Sigma}_1) > \sigma_{max}(\boldsymbol{\Sigma}_4)$. The subsystem $(\overline{\boldsymbol{A}}_{11}, \overline{\boldsymbol{b}}_1, \overline{\boldsymbol{c}}_1)_n$ is then controllable and observable.

Proof

The upper left equation in (8.127b) becomes

$$\overline{\boldsymbol{A}}_{11}^T \boldsymbol{\Sigma}_1 \overline{\boldsymbol{A}}_{11} + \overline{\boldsymbol{A}}_{21}^T \boldsymbol{\Sigma}_4 \overline{\boldsymbol{A}}_{21} - \boldsymbol{\Sigma}_1 = -\overline{\boldsymbol{c}}_1^T \overline{\boldsymbol{c}}_1 \tag{8.133}$$

Assume that the subsystem $(\overline{\boldsymbol{A}}_{11}, \overline{\boldsymbol{b}}_1, \overline{\boldsymbol{c}}_1)_n$ is not observable. Then there exists an eigenvalue λ of $\overline{\boldsymbol{A}}_{11}$ with corresponding eigenvector $\boldsymbol{v}$ such that

$$\overline{\boldsymbol{A}}_{11}\boldsymbol{v} = \lambda \boldsymbol{v}, \qquad \overline{\boldsymbol{c}}_1 \boldsymbol{v} = 0, \qquad ||\boldsymbol{v}|| = 1 \tag{8.134}$$

By multiplying (8.133) from the left by $\boldsymbol{v}^H$ and from the right by $\boldsymbol{v}$, we obtain

$$(1 - |\lambda|^2)\boldsymbol{v}^H \boldsymbol{\Sigma}_1 \boldsymbol{v} = \boldsymbol{v}^H \overline{\boldsymbol{A}}_{21}^T \boldsymbol{\Sigma}_4 \overline{\boldsymbol{A}}_{21} \boldsymbol{v} \tag{8.135}$$

Noting that

$$\boldsymbol{v}^H \boldsymbol{\Sigma}_1 \boldsymbol{v} \geq \sigma_{min}(\boldsymbol{\Sigma}_1), \qquad \boldsymbol{v}^H \overline{\boldsymbol{A}}_{21}^T \boldsymbol{\Sigma}_4 \overline{\boldsymbol{A}}_{21} \boldsymbol{v} \leq ||\overline{\boldsymbol{A}}_{21}\boldsymbol{v}||^2 \sigma_{max}(\boldsymbol{\Sigma}_4) \tag{8.136}$$

it follows from (8.135) that

$$(1 - |\lambda|^2)\, \sigma_{min}(\boldsymbol{\Sigma}_1) \leq ||\overline{\boldsymbol{A}}_{21}\boldsymbol{v}||^2 \sigma_{max}(\boldsymbol{\Sigma}_4) \tag{8.137}$$

From Theorem 8.3 it follows that

$$\left|\left| \overline{\boldsymbol{A}} \begin{bmatrix} \boldsymbol{v} \\ \boldsymbol{0} \end{bmatrix} \right|\right|^2 \leq 1 \iff ||\overline{\boldsymbol{A}}_{11}\boldsymbol{v}||^2 + ||\overline{\boldsymbol{A}}_{21}\boldsymbol{v}||^2 \leq 1 \iff ||\overline{\boldsymbol{A}}_{21}\boldsymbol{v}||^2 \leq 1 - |\lambda|^2 \tag{8.138}$$

By applying (8.138) to (8.137), we have

$$(1 - |\lambda|^2)\, \sigma_{min}(\boldsymbol{\Sigma}_1) \leq (1 - |\lambda|^2)\, \sigma_{max}(\boldsymbol{\Sigma}_4) \tag{8.139}$$

From Theorem 8.4, it is clear that the subsystem $(\overline{\boldsymbol{A}}_{11}, \overline{\boldsymbol{b}}_1, \overline{\boldsymbol{c}}_1)_n$ is asymptotically stable, i.e., $1 - |\lambda|^2 > 0$. Hence

$$\sigma_{min}(\boldsymbol{\Sigma}_1) \leq \sigma_{max}(\boldsymbol{\Sigma}_4) \tag{8.140}$$

This contradicts with the assumption of the theorem. Therefore, the subsystem is observable. Similarly, it can be shown that the subsystem is controllable. This completes the proof of the theorem. ■

8.6 Numerical Experiments

As an example, consider the problem of approximating an impulse response of the "Gaussian filter"

$$f_i = 0.256322 \exp\{-0.103203(i-4)^2\}$$

by an IIR digital filter over a finite interval $0 \leq i \leq 20$. The impulse response and the magnitude response of the Gaussian filter, i.e., the 20th-order FIR digital filter are depicted in Figure 8.2.

8.6.1 Design Based on Extended Pade's Approximation

A. A Direct Procedure

Using (8.12) and (8.13), the denominator coefficient vector $\boldsymbol{a}$ and the numerator coefficient vector $\boldsymbol{b}$ of a 3rd-order IIR digital filter were found to be

$$\boldsymbol{a} = [-2.084581,\ 1.657505,\ -0.499733]^T$$

$$\boldsymbol{b} = [0.049165,\ -0.001237,\ 0.040055,\ 0.020835]^T$$

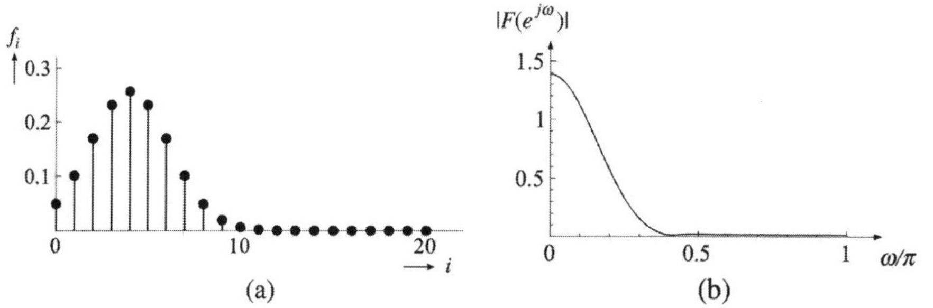

Figure 8.2 The Gaussian filter. (a) Its impulse response. (b) Its magnitude response.

respectively. The poles were given by

$$\lambda = 0.651023 \pm j0.463440 \; (|\lambda| = 0.799130), \qquad \lambda = 0.782534$$

hence the resulting filter is stable. Magnitude response of the 3rd-order IIR digital filter designed by a direct procedure is depicted in Figure 8.3.

B. A Modified Procedure

Using (8.12) and (8.21) with $\boldsymbol{W} = \boldsymbol{I}_{21}$, the denominator coefficient vector $\boldsymbol{a}$ and the numerator coefficient vector $\boldsymbol{b}$ of a 3rd-order IIR digital filter were found to be

$$\boldsymbol{a} = [-2.084581,\ 1.657505,\ -0.499733]^T$$

$$\boldsymbol{b} = [0.049165,\ -0.001355,\ 0.040022,\ 0.021245]^T$$

respectively, and the poles were naturally the same as those of the above filter. Magnitude response of the 3rd-order IIR digital filter designed by a modified procedure is depicted in Figure 8.4.

8.6.2 Design Using Second-Order Information

A. The Use of Solution (8.36) for Given Order $n = 3$

Using (8.36) and (8.27), the denominator coefficient vector $\boldsymbol{a}$ and the numerator coefficient vector $\boldsymbol{b}$ of a 3rd-order IIR digital filter were found to be

$$\boldsymbol{a} = [-2.084581,\ 1.657505,\ -0.499733]^T$$

$$\boldsymbol{b} = [0.049165,\ -0.001237,\ 0.040055,\ 0.020835]^T$$

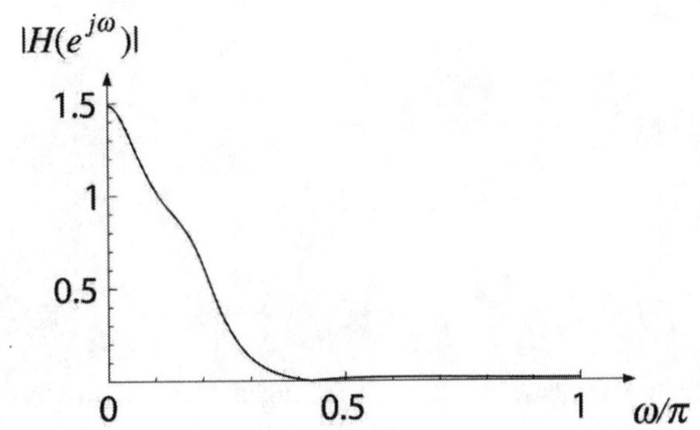

Figure 8.3 Magnitude response of a 3rd-order IIR digital filter designed by a direct procedure.

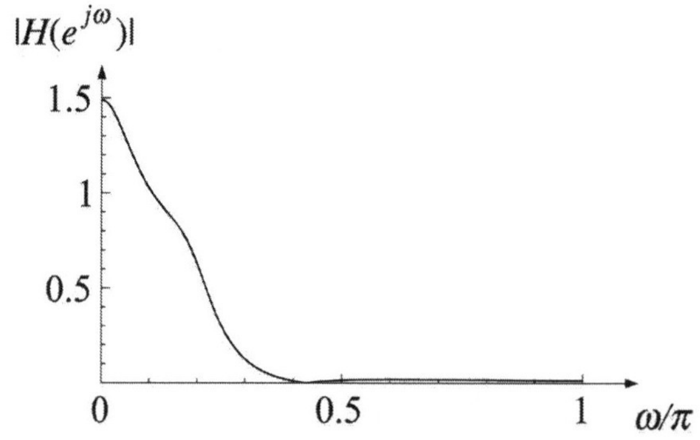

Figure 8.4 Magnitude response of a 3rd-order IIR filter designed by a modified procedure.

respectively. The poles were computed as

$$\lambda = 0.651023 \pm j0.463440 \; (|\lambda| = 0.799130), \qquad \lambda = 0.782534$$

hence the resulting filter is stable. Magnitude response of the 3rd-order IIR digital filter designed by using (8.36) and (8.27) is depicted in Figure 8.5.

B. The Use of Efficient Algorithm (8.63)–(8.65) for Solving (8.35)

By choosing $\varepsilon = 10^{-3}$ in (8.65) and using (8.63)–(8.65), it took the algorithm 4 iterations to converge to

$$\boldsymbol{a}(4) = \begin{bmatrix} -2.243631, & 2.191723, & -1.098524, & 0.237046 \end{bmatrix}^T$$

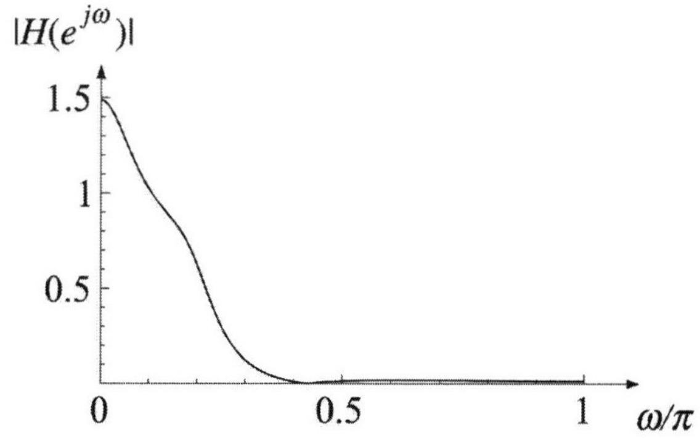

Figure 8.5 Magnitude response of a 3rd-order IIR digital filter designed by (8.36).

and the poles were found to be

$$\lambda = 0.459785 \pm j0.538302 \ (|\lambda| = 0.707934)$$

$$\lambda = 0.662031 \pm j0.186280 \ (|\lambda| = 0.687739)$$

This shows that the resulting filter is stable. The numerator coefficient vector $\boldsymbol{b} = [b_0, b_1, b_2, b_3, b_4]^T$ were then derived from (8.27) as

$$\boldsymbol{b} = \begin{bmatrix} 0.049165, & -0.009057, & 0.050216, & 0.018506, & 0.009832 \end{bmatrix}^T$$

Magnitude response of the 4th-order IIR digital filter designed by using (8.63)–(8.65) and (8.27) is depicted in Figure 8.6.

Detailed numerical results obtained by applying (8.63)–(8.65) are summarized in Table 8.1.

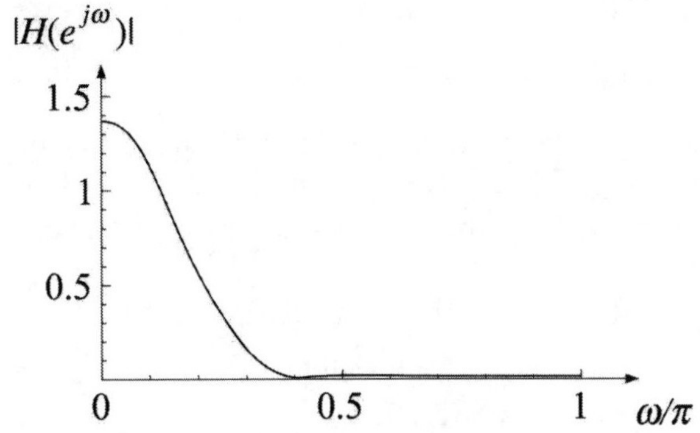

Figure 8.6 Magnitude response of a 4th-order IIR digital filter designed by (8.63)–(8.65) and (8.27).

Table 8.1 Convergence of the efficient algorithm using 2nd-order information

$n+1$	α_{n+1}	$\lvert\alpha_{n+1} - \alpha_n\rvert$
1	$209.627568 \times 10^{-4}$	$232.522911 \times 10^{-3}$
2	18.893975×10^{-4}	19.073359×10^{-3}
3	1.095412×10^{-4}	1.779856×10^{-3}
4	0.035643×10^{-4}	0.105977×10^{-3}
5	0.000589×10^{-4}	0.003506×10^{-3}
6	0.000004×10^{-4}	0.000059×10^{-3}

8.6.3 Least-Squares Design

Let a 3rd-order IIR digital filter be designed. An initial estimate was derived from (8.73) as

$$\boldsymbol{a}^{(0)} = [-2.084581,\ 1.657505,\ -0.499733]^T$$

By choosing $\varepsilon = 10^{-8}$ in (8.87), it took the algorithm in (8.86) 6 iterations to converge to

$$\boldsymbol{a}^{(6)} = [-1.813163,\ 1.234429,\ -0.313226]^T$$

By choosing $\varepsilon = 10^{-8}$ in (8.95) and continuing with the second phase, it took the algorithm in (8.94) 7 iterations to converge to

$$\boldsymbol{a}^{*} = [-1.811521,\ 1.231832,\ -0.312083]^T$$

The poles were calculated as

$$\lambda = 0.573478 \pm j0.375137\ (|\lambda| = 0.685277), \qquad \lambda = 0.664565$$

hence the resulting filter is stable. After determining the error vector $\boldsymbol{e}(\boldsymbol{a}^*, \boldsymbol{b}^*)$ and actual impulse response vector $\boldsymbol{h}(\boldsymbol{a}^*, \boldsymbol{b}^*)$ from (8.78) and (8.68), respectively, (8.70) was used to obtain the optimal numerator coefficient vector $\boldsymbol{b}^*$ as

$$\boldsymbol{b}^{*} = [0.049165,\ 0.013344,\ 0.039997,\ 0.048732]^T$$

Magnitude response of the 3rd-order IIR digital filter designed by a least-squares method is depicted in Figure 8.7.

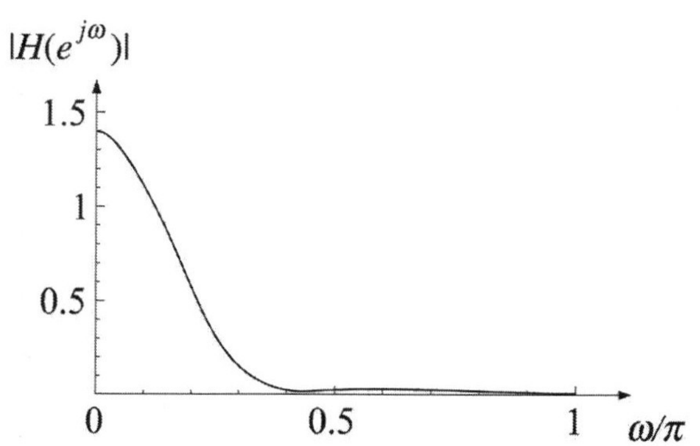

Figure 8.7 Magnitude response of a 3rd-order IIR digital filter designed by a least-squares method.

8.6.4 Design Using State-Space Model (Balanced Model Reduction)

By eigenvalue-eigenvector decomposition of the Hankel matrix $\boldsymbol{H}_{20,20}$ in (8.98), the eigenvalues of $\boldsymbol{H}_{20,20}$ in (8.106) were found to be

$$\begin{bmatrix} \sigma_1 & \sigma_2 \\ \sigma_3 & \sigma_4 \\ \sigma_5 & \sigma_6 \\ \sigma_7 & \sigma_8 \\ \sigma_9 & \sigma_{10} \end{bmatrix} = \begin{bmatrix} 0.95351959 & -0.33240485 \\ 0.07664440 & -0.01342993 \\ 0.00184360 & -0.00019436 \\ 0.00001498 & -0.00000079 \\ 0.00000003 & -0.00000000 \end{bmatrix}$$

Using (8.111), the reduced-order state-space model $(\boldsymbol{A}_{11}, \boldsymbol{b}_1, \boldsymbol{c}_1, d)_3$ was found to be

$$\boldsymbol{A}_{11} = \begin{bmatrix} 0.863908 & 0.160275 & -0.011029 \\ -0.459756 & 0.594112 & 0.174003 \\ -0.137212 & -0.754645 & 0.356728 \end{bmatrix}, \quad \boldsymbol{b}_1 = \begin{bmatrix} 0.477335 \\ 0.636658 \\ 0.509285 \end{bmatrix}$$

$$\boldsymbol{c}_1 = \begin{bmatrix} 0.455148 & -0.211628 & 0.039034 \end{bmatrix}, \qquad d = 0.049165$$

and the denominator coefficient vector $\boldsymbol{a}$ and the numerator coefficient vector $\boldsymbol{b}$ were found to be

$$\boldsymbol{a} = [-1.814749,\ 1.236860,\ -0.314268]^T$$

$$\boldsymbol{b} = [0.049165,\ 0.013180,\ 0.039977,\ 0.048437]^T$$

respectively. The poles were given by

$$\lambda = 0.574196 \pm j0.376723 \ (|\lambda| = 0.686747), \qquad \lambda = 0.666357$$

hence the resulting filter is stable. Magnitude response of the 3rd-order IIR digital filter designed by balanced model reduction is depicted in Figure 8.8.

8.6.5 Comparison of Algorithms' Performances

The performance of the design algorithms addressed in this chapter as applied to the above design example is summarized in Table 8.2 where

$$\varepsilon_2 = \frac{\sqrt{(f_0 - h_0)^2 + (f_1 - h_1)^2 + \cdots + (f_{20} - h_{20})^2}}{\sqrt{f_0^2 + f_1^2 + \cdots + f_{20}^2}} \times 100$$

$$\varepsilon_\infty = \frac{\max\limits_{0 \le i \le 20} |f_i - h_i|}{\max\limits_{0 \le i \le 20} |f_i|} \times 100$$

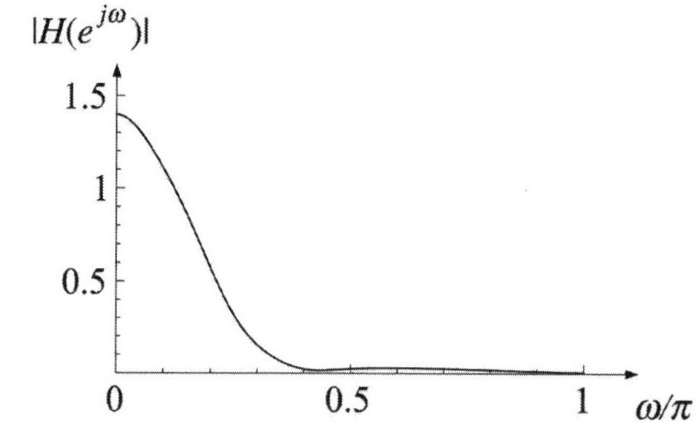

Figure 8.8 Magnitude response of a 3rd-order IIR digital filter designed by balanced model reduction.

Table 8.2 Performance comparison among algorithms

Algorithms	Order	ε_2	ε_∞	Max. Negative Ripple for $0 \leq i \leq 20$
Direct Procedure	$n = 3$	8.696952	8.820447	–0.000447
	$n = 4$	1.455143	1.552236	–0.003978
Modified Procedure	$n = 3$	8.691246	8.830924	–0.000440
	$n = 4$	1.455106	1.552335	–0.003978
2nd-Order Information	$n = 3$	8.696952	8.820447	–0.000447
	$n = 4$	1.455143	1.552236	–0.003978
Least-Squares Design	$n = 3$	2.501358	2.163644	–0.001453
	$n = 4$	0.353882	0.318031	–0.000500
Balanced Model Reduction	$n = 3$	2.502378	2.136857	–0.001435
	$n = 4$	0.353895	0.318568	–0.000503

8.7 Summary

In the techniques based on extended Pade's approximation, the problem of designing IIR digital filters to approximate a desired impulse response over a specified interval has been studied. Two design procedures that require only linear calculations have been illustrated for the approximation of IIR digital filters. In the filter design using second-order information, mixed first- and second-order information in the form of a finite portion of the impulse response and autocorrelation sequences has been applied to provide an efficient algorithm for designing IIR digital filters. In the least-squares design, a method for obtaining the coefficients of an nth-order IIR digital filter, which gives the optimal least-squares approximation to a desired impulse response over a finite interval, has been presented. In the filter design using state-space model, an algorithm, which is based on balanced model reduction, for the approximation of FIR digital filters by IIR digital filters has been examined. Finally, numerical experiments have been performed to compare their performances among the filter design techniques in the time domain.

References

[1] M. S. Bertran, "Approximation of digital filters in one and two dimensions," *IEEE Trans. Acoust., Speech, Signal Process.*, vol. ASSP-23, no. 5, pp. 438–443, Oct. 1975.

[2] C. S. Burrus and T. W. Parks, "Time domain design of recursive digital filters," *IEEE Trans. Audio Electroacoust.*, vol. AU-18, no. 2, pp. 137–141, June 1970.

[3] R. Hastings-James and S. K. Mehra, "Extensions of the Pade-approximant technique for the design of recursive digital filters," *IEEE Trans. Acoust. Speech, Signal Process.*, vol. ASSP-25, no. 6, pp. 501–509, Dec. 1977.

[4] C. T. Mullis and R. A. Roberts, "The use of second-order information in the approximation of discrete-time linear systems," *IEEE Trans. Acoust. Speech, Signal Process.*, vol. ASSP-24, no. 3, pp. 226–238, June. 1976.

[5] T. Hinamoto and S. Maekawa, "Separable-denominator 2-D rational approximation via 1-D based algorithm," *IEEE Trans. Circuits Syst.*, vol. CAS-32, no. 10, pp. 989–999, Oct. 1985.

[6] R. E. Kalman and J. Bertram, "Control system design via the second method of Liapunov, part II, discrete time systems," *ASME J. Basic Engineering*, vol. 82, pp. 394–400, 1960.

[7] A. G. Evans and R. Fischl, "Optimal least squares time-domain synthesis of recursive digital filters," *IEEE Trans. Audio Electroacoust.*, vol. AU-21, no. 1, pp. 61–65, Feb. 1973.

[8] D. G. Luenberger, *Optimization by Vector Space Method.* New York: Wiley, 1969.

[9] B. Beliczynski, I. Kale and G. D. Cain, "Approximation of FIR by IIR digital filters: An algorithm based on balanced model reduction," *IEEE Trans. Signal Process.*, vol. 40, no. 3, pp. 532–542, Mar. 1992.

[10] L. Pernebo and L. M. Silverman, "Model reduction via balanced state space representations," *IEEE Trans. Autom. Control*, vol. AC-27, no. 2, pp. 382–387, Apr. 1982.

9

Design of Interpolated and FRM FIR Digital Filters

9.1 Preview

Interpolated FIR (IFIR) filters [1, 2] and frequency-response-masking (FRM) FIR filters [3] are well-known classes of computationally efficient digital filters because the total number of multipliers and adders required to implement these filters are considerably less than those required by their conventional counterparts [4]. There has been a great deal of work in the literature following the aforementioned original development, see for example [5–15] and the references therein. Because of the importance of these filters, it is naturally desirable to develop a simple and unifying design method that is applicable to both of the filter classes. In this chapter, we propose such a design technique in that the subfilters involved are jointly optimized in the minimax sense. The core of the proposed design approach is the convex-concave procedure (CCP) [16–18] that allows the use of efficient convex optimization to deal with nonconvex design problems where the objective functions assume the form of difference of two convex functions.

Our focus is on the design of original IFIR and single stage FRM filters so as for the reader to sense this general design strategy in a simple and transparent manner. We also explain why the CCP is well-suited that simultaneously promotes sparsity of filter coefficients for the designs at hand. In addition, we present a variant of the CCP-based technique for the design of FRM filters with improved implementation efficiency.

9.2 Basics of IFIR and FRM Filters and CCP

9.2.1 Interpolated FIR Filters

IFIR filters are introduced in [1] and further investigated in [2]. As shown in Figure 9.1, an IFIR filter is composed of a cascade of two FIR filters, whose transfer function assumes the form

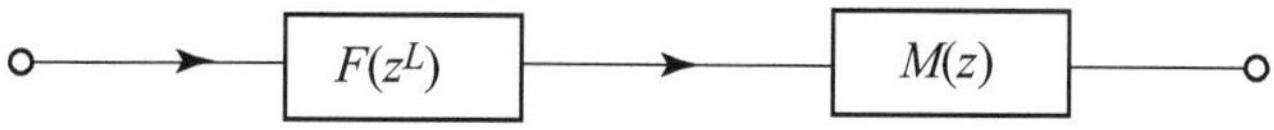

Figure 9.1 An IFIR filter.

$$H(z) = F(z^L)M(z) \tag{9.1}$$

where $L > 0$ is an integer that determines the degree of the filter's sparsity, hence its computational efficiency.

Suppose the parent transfer function $F(z)$ represents a lowpass filter with normalized passband $[0,\ \omega_p]$, transition band $[\omega_p,\ \omega_a]$, and stopband $[\omega_a, 1]$. Then $F(z^L)$ is a periodic filter because the baseband of the frequency response $F(e^{jL\omega})$ is reduced to $(1/L)$-th of the baseband of $F(z)$. This is illustrated in Figure 9.2 where the lowpass filter $F(z)$ possesses a passband [0, 0.2] and a transition band [0.2, 0.4]. With $L = 4$, Figure 9.2b depicts the magnitude response of $F(z^L)$. As expected, the first passband and transition band of $F(z^L)$ are reduced to $[0,\ 0.2]/L = [0,\ 0.05]$ and $[0.2\ 0.4]/L = [0.05\ 0.1]$, respectively. Consequently, the passband of $F(z^L)$ is considerably narrower than that of $F(z)$ and its magnitude roll-off is much sharper.

Clearly, for $F(z^L)$ to be of use in constructing a lowpass filter with narrow passband and sharp roll-off, the undesired passbands of the periodic filter $F(z^L)$ must be suppressed. As can be seen in Figure 9.1, this is done by connection $F(z^{-L})$ in cascade with a lowpass $M(z)$ which is known as interpolator. As an illustration, Figure 9.3(a) shows the magnitude response of a linear-phase lowpass filter $M(z)$ whose passband is [0, 0.05] and Figure 9.3(b) displays the magnitude response of the IFIR filter $H(z) = F(z^L)M(z)$.

9.2.2 Frequency-Response-Masking Filters

Another class of FIR filters with very narrow transition bands, known as FRM filters, is proposed in [3] and has since been a subject of study [4–15]. As illustrated in Figure 9.4, a single stage FRM filter has a connected parallel structure where the linear phase periodic filter $F(z^L)$ and its delay complementary periodic filers, $z^{-L(N-1)/2} - F(z^L)$, are cascaded with masking filters $M_a(z)$ and $M_c(z)$, respectively, so as to produce an FIR filter with sharper transition bands.

The transfer function of the FRM filter in Figure 9.4 is given by

$$H(z) = F(z^L)M_a(z) + [z^{-L(N-1)/2} - F(z^L)]M_c(z) \tag{9.2}$$

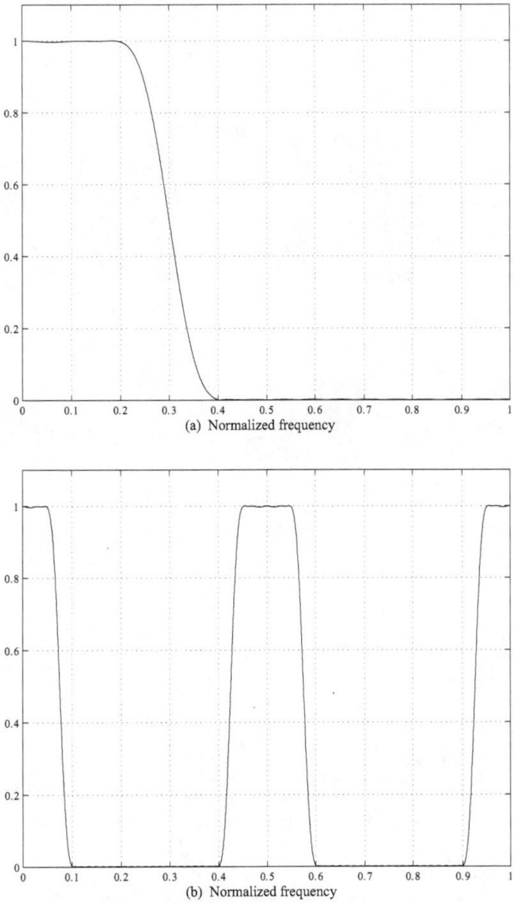

Figure 9.2 Magnitude response of (a) $F(z)$ and (b) $F(z^L)$ with $L = 4$.

where N denotes the length of $F(z)$ and is assumed to be an odd integer throughout. If we denote the complementary filter of $F(z)$ by $G(z) = z^{-(N-1)/2} - F(z)$, then (9.2) becomes

$$H(z) = F(z^L)M_a(z) + G(z^L)M_c(z) \tag{9.3}$$

Hence an FRM filter is essentially a two-channel filter bank with an interpolated FIR filter in each channel. Because $\{F(z), G(z)\}$ is a complementary pair, their passbands are complementary to each other. Consequently, by connecting these IFIR filters in parallel with adequately chosen interpolators $M_a(z)$ and $M_c(z)$, lowpass (as well as other standard types) FRM filers with

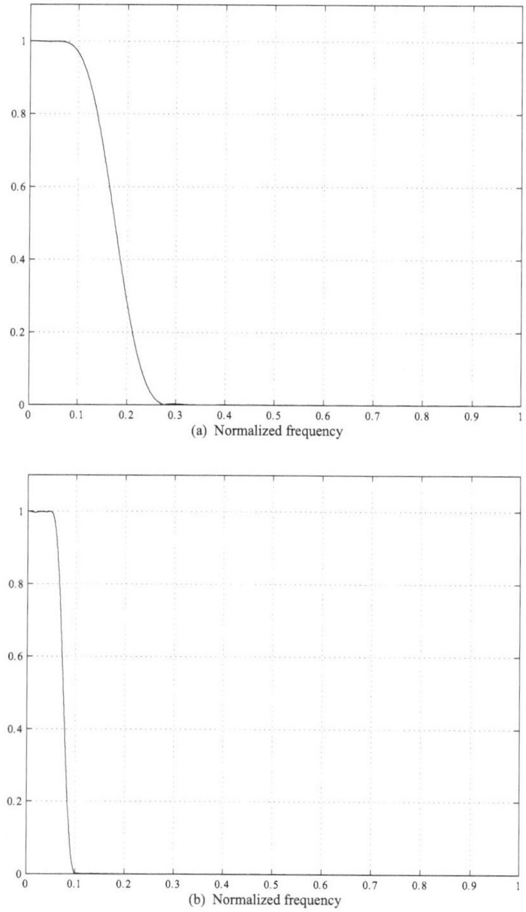

Figure 9.3 Magnitude response of (a) $M(z)$ and (b) $H(z) = F(z^L)M(z)$.

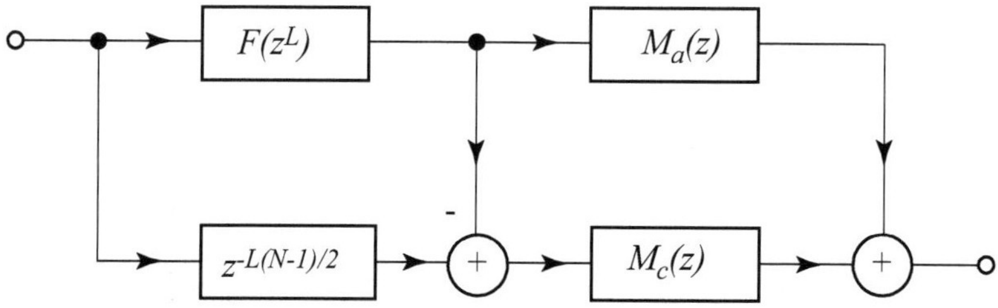

Figure 9.4 A single-stage FRM filter.

arbitrary passband can be designed. It is this feature that distinguishes FRM filters from IFIR filters which are limited to narrow-band frequency responses.

9.2.3 Convex-Concave Procedure (CCP)

As will become clear shortly, jointly optimizing all sub-filters involved in an IFIR or FRM filter is not a convex problem. The CCP is a heuristic method of convexifying nonconvex problems. It is known that a function $f(\boldsymbol{x})$ with continuous second-order derivatives can be expressed as a difference of two convex functions [23]. If the function in question has a bounded Hessian, such an expression, namely $f(\boldsymbol{x}) = u(\boldsymbol{x}) - v(\boldsymbol{x})$, can be explicitly constructed [19]. We are thus motivated to consider a nonconvex problem of the form

$$\text{minimize} \;\; f(\boldsymbol{x}) = u_0(\boldsymbol{x}) - v_0(\boldsymbol{x}) \tag{9.4a}$$

$$\text{subject to} \;\; u_j(\boldsymbol{x}) \leq v_j(\boldsymbol{x}) \;\; \text{for} \;\; j = 1, 2, \cdots, q \tag{9.4b}$$

where $u_j(\boldsymbol{x})$ and $v_j(\boldsymbol{x})$ for $j = 0, 1, \cdots, q$ are convex. The CCP is an iterative procedure for solving (9.4). In the kth iteration where iterate $\boldsymbol{x}_k$ is known, CCP performs two steps:

(i) convexification of the objective function and constraints at $\boldsymbol{x}_k$ by replacing each $v_j(\boldsymbol{x})$ by its affine approximation

$$\hat{v}_j(\boldsymbol{x}, \boldsymbol{x}_k) = v_j(\boldsymbol{x}_k) + \nabla v_j(\boldsymbol{x}_k)^T(\boldsymbol{x} - \boldsymbol{x}_k)$$

where ∇ denotes the gradient operator.
(ii) Solve the convex problem

$$\text{minimize} \;\; \hat{f}(\boldsymbol{x}) = u_0(\boldsymbol{x}) - \hat{v}_0(\boldsymbol{x}) \tag{9.5a}$$

$$\text{subject to} \;\; u_j(\boldsymbol{x}) \leq \hat{v}_j(\boldsymbol{x}) \;\; \text{for} \;\; j = 1, 2, \cdots, q \tag{9.5b}$$

One can start CCP by solving (9.5) with an initial $\boldsymbol{x}_0$ that is feasible for the original problem (9.4). This means that $u_j(\boldsymbol{x}_0) - v_j(\boldsymbol{x}_0) \leq 0$ for $1 \leq j \leq q$ which in conjunction with the convexity of $v_j(\boldsymbol{x})$ implies that

$$u_j(\boldsymbol{x}_0) - \hat{v}_j(\boldsymbol{x}_0, \boldsymbol{x}_0) = u_j(\boldsymbol{x}_0) - v_j(\boldsymbol{x}_0) \leq 0$$

hence $\boldsymbol{x}_0$ is also a feasible point for the convex problem in (9.5). Also note that if $\boldsymbol{x}_{k+1}$ is produced by solving (9.5) in the kth iteration, then $\boldsymbol{x}_{k+1}$ is also feasible for the original problem in (9.4) because the convexity of $v_j(\boldsymbol{x})$ implies that $v_j(\boldsymbol{x}_{k+1}) \geq \hat{v}_j(\boldsymbol{x}_{k+1}, \boldsymbol{x}_k)$ which leads to

$$u_j(\boldsymbol{x}_{k+1}) - v_j(\boldsymbol{x}_{k+1}) \leq u_j(\boldsymbol{x}_{k+1}) - \hat{v}_j(\boldsymbol{x}_{k+1}, \boldsymbol{x}_k) \leq 0$$

Another desirable property of CCP is that it is a descent method [20], namely the original objective function, $u_0(\boldsymbol{x}) - v_0(\boldsymbol{x})$, decreases monotonically, at the iterates $\{\boldsymbol{x}_k\}$ generated by solving convex problem (9.5). To see this, note that

$$\begin{aligned} f_k &= u_0(\boldsymbol{x}_k) - v_0(\boldsymbol{x}_k) = u_0(\boldsymbol{x}_k) - \hat{v}_0(\boldsymbol{x}_k, \boldsymbol{x}_k) \\ &\geq u_0(\boldsymbol{x}_{k+1}) - \hat{v}_0(\boldsymbol{x}_{k+1}, \boldsymbol{x}_k) \end{aligned} \tag{9.6}$$

Since $v_0(\boldsymbol{x})$ is convex, we have $v_0(\boldsymbol{x}_{k+1}) \geq \hat{v}_0(\boldsymbol{x}_{k+1}, \boldsymbol{x}_k)$, hence

$$u_0(\boldsymbol{x}_{k+1}) - \hat{v}_0(\boldsymbol{x}_{k+1}, \boldsymbol{x}_k) \geq u_0(\boldsymbol{x}_{k+1}) - v_0(\boldsymbol{x}_{k+1}) = f_{k+1}$$

which in conjunction with (9.6) gives $f_{k+1} \leq f_k$. It follows that if the objective is bounded from below, then $\{f_k\}$ converges to a finite limit. As expected, the convergence of CCP-produced iterates $\{\boldsymbol{x}_k\}$ cannot be assured in general because it deals with noncovex problems as a heuristic method after all. Nevertheless, it can be shown that under mild conditions $\{\boldsymbol{x}_k\}$ converges to critical points of the original problem [21]. The CCP may be terminated in several ways, including by a certain number of iterations or by monitoring if the difference in objective function between two consecutive iterations is less than a given convergence tolerance.

9.3 Minimax Design of IFIR Filters

9.3.1 Problem Formulation

Following (9.1), let the transfer functions of the parent filter and interpolator of an IFIR filter be given by

$$F(z) = \sum_{n=0}^{N-1} f_n z^{-n}, \qquad M(z) = \sum_{n=0}^{N_i-1} m_n z^{-n} \tag{9.7}$$

respectively, and the frequency response of an IFIR filter is given by

$$H(e^{j\omega}) = F(e^{jL\omega})M(e^{j\omega}) \tag{9.8}$$

Assume both $F(z)$ and $M(z)$ are of linear phase response, the zero-phase frequency response of the IFIR filter is given by

$$H_0(\boldsymbol{a}_f, \boldsymbol{a}_m, \omega) = [\boldsymbol{a}_f^T \boldsymbol{t}_f(L\omega)] \cdot [\boldsymbol{a}_m^T \boldsymbol{t}_m(\omega)] \tag{9.9}$$

where $\boldsymbol{a}_f$ and $\boldsymbol{a}_m$ are coefficient vectors determined by the impulse responses of $F(z)$ and $M(z)$, respectively, $\boldsymbol{t}_f(\omega)$ and $\boldsymbol{t}_m(\omega)$ are vectors with trigonometric components determined by the filter lengths and types (e.g. odd or even length and symmetrical or antisymmetrical filter coefficients). In the case of both $F(z)$ and $M(z)$ being of odd length, for example, we have

$$\boldsymbol{a}_f = \begin{bmatrix} f_{(N-1)/2} \\ 2f_{(N+1)/2} \\ \vdots \\ 2f_{N-1} \end{bmatrix}, \qquad \boldsymbol{a}_m = \begin{bmatrix} m_{(N_i-1)/2} \\ 2m_{(N_i+1)/2} \\ \vdots \\ 2m_{N_i-1} \end{bmatrix}$$

$$\boldsymbol{t}_f(\omega) = \begin{bmatrix} 1 \\ \cos\omega \\ \vdots \\ \cos[(N-1)\omega/2] \end{bmatrix}, \qquad \boldsymbol{t}_m(\omega) = \begin{bmatrix} 1 \\ \cos\omega \\ \vdots \\ \cos[(N_i-1)\omega/2] \end{bmatrix}$$

where f_n's and m_n's are from (9.7).

Let $H_d(\omega)$ be the desired zero-phase response, the frequency-weighted minimax design of an IFIR filter amounts to finding vectors $\boldsymbol{a}_f$ and $\boldsymbol{a}_m$ that solves the problem

$$\min_{\boldsymbol{a}_f, \boldsymbol{a}_m} \max_{\omega \in \Omega} w(\omega)|H_0(\boldsymbol{a}_f, \boldsymbol{a}_m, \omega) - H_d(\omega)| \tag{9.10}$$

where $w(\omega) > 0$ is a frequency-selective weight over a frequency domain of interest, Ω. Evidently, problem (9.10) is nonconvex with respect to design variables $\boldsymbol{x}^T = [\boldsymbol{a}_f^T \ \boldsymbol{a}_m^T]$.

9.3.2 Convexification of (9.10) Using CCP

By introducing an upper bound δ for the objective function in (9.10) and treating the bound as an additional design variable, problem (9.10) becomes

$$\text{minimize } \delta \tag{9.11a}$$

$$\text{subject to } [\boldsymbol{a}_f^T \boldsymbol{t}_f(L\omega)][\boldsymbol{a}_m^T \boldsymbol{t}_m(\omega)] \le \delta_w + H_d(\omega) \tag{9.11b}$$

$$-[\boldsymbol{a}_f^T \boldsymbol{t}_f(L\omega)][\boldsymbol{a}_m^T \boldsymbol{t}_m(\omega)] \le \delta_w - H_d(\omega) \tag{9.11c}$$

for $\omega \in \Omega$, where $\delta_w = \delta/w(\omega)$. Bearing CCP in mind, we add the term $0.5s(\boldsymbol{x}, \omega)$ with

$$s(\boldsymbol{x}, \omega) = (\boldsymbol{a}_f^T \boldsymbol{t}_f(L\omega))^2 + (\boldsymbol{a}_m^T \boldsymbol{t}_m(\omega))^2 \tag{9.12}$$

to both sides of (9.11b) and (9.11c) to obtain an equivalent pair of constrains as

$$[\boldsymbol{a}_f^T \boldsymbol{t}_f(L\omega) + \boldsymbol{a}_m^T \boldsymbol{t}_m(\omega)]^2 \le s(\boldsymbol{x}, \omega) + 2\delta_w + 2H_d(\omega) \tag{9.13a}$$

$$[\boldsymbol{a}_t^T \boldsymbol{t}_f(L\omega) - \boldsymbol{a}_m^T \boldsymbol{t}_m(\omega)]^2 \le s(\boldsymbol{x}, \omega) + 2\delta_w - 2H_d(\omega) \tag{9.13b}$$

Clearly, the functions on both sides of (9.13a) and (9.13b) are convex with respect to $\boldsymbol{x}$ and δ, therefore (9.13) is suitable for application of CCP. We proceed by replacing $s(\boldsymbol{x}, \omega)$ in (9.13) by its linearization at $\boldsymbol{x}_k$, namely $s(\boldsymbol{x}_k, \omega) + \nabla s(\boldsymbol{x}_k, \omega)^T(\boldsymbol{x} - \boldsymbol{x}_k)$. This gives

$$[\boldsymbol{a}_f^T \boldsymbol{t}_f(L\omega) + (-1)^i \boldsymbol{a}_m^T \boldsymbol{t}_m(\omega)]^2 \le \eta_i(\boldsymbol{x}, \boldsymbol{x}_k, \omega), \quad i = 0, 1 \tag{9.14}$$

with

$$\eta_i(\boldsymbol{x}, \boldsymbol{x}_k, \omega) = s(\boldsymbol{x}_k, \omega) + \nabla s(\boldsymbol{x}_k, \omega)^T(\boldsymbol{x} - \boldsymbol{x}_k) + 2\delta_w + (-1)^i 2H_d(\omega)$$

where for practical purposes ω is taken from $\Omega_d = \{\omega_j | \, j = 1, 2, \cdots, K\} \subset \Omega$, a finite discrete set of frequency grids that are sufficiently dense and placed uniformly over the frequency region of interest Ω.

In summary, the convex problem to be solved in the kth iteration of CCP is given by

$$\text{minimize } \delta \tag{9.15a}$$

$$\text{subject to } [\boldsymbol{a}_f^T \boldsymbol{t}_f(L\omega_j) + (-1)^i \boldsymbol{a}_m^T \boldsymbol{t}_m(\omega_j)]^2 \le \eta_i(\boldsymbol{x}, \boldsymbol{x}_k, \omega_j) \tag{9.15b}$$

$$\text{for } i = 0, 1 \text{ and } j = 1, 2, \cdots, K$$

where $\omega_j \in \Omega_d$. It is well known that this problem of minimizing a linear function subject to convex quadratic constraints can be formulated as a semidefinite programming (SDP) or second-order cone programming (SOCP) problem [24], which can be solved efficiently [25–27]. The CCP-based algorithm may be terminated after a given number of iterates $\{\boldsymbol{x}_k\}$ have been generated or, when the difference between two consecutive error bounds, namely $|\delta_{k-1} - \delta_k|$, is less than a given convergence tolerance. In either case, the last iterate produced from (9.15) is taken to be a solution of the design problem. A step-by-step description of the proposed algorithm is given below as Algorithm 1.

Algorithm 1 for IFIR filters

Step 1: input $\boldsymbol{x}_0$, (N, N_i), Ω_d, $H_d(\omega)$ for $\omega \in \Omega_d$, $w(\omega)$, and K_i

Step 2: for $k = 0, 1, \cdots, K_i - 1$
(i) solve (9.15) for $\boldsymbol{a}_f$ and $\boldsymbol{a}_m$
(ii) construct $\boldsymbol{x}_{k+1} = [\boldsymbol{a}_f^T \; \boldsymbol{a}_m^T]^T$
end

Step 3: output $\boldsymbol{x}^* = \boldsymbol{x}_{K_i}$

9.3.3 Remarks on Convexification in (9.13)–(9.14)

Note that expressing a nonconvex function as $f(\boldsymbol{x}) = u(\boldsymbol{x}) - v(\boldsymbol{x})$ is not unique. In fact if $f(\boldsymbol{x}) = u(\boldsymbol{x}) - v(\boldsymbol{x})$ holds with $u(\boldsymbol{x})$ and $v(\boldsymbol{x})$ convex, then $f(\boldsymbol{x}) = \tilde{u}(\boldsymbol{x}) - \tilde{v}(\boldsymbol{x})$ also holds with $\tilde{u}(\boldsymbol{x}) = u(\boldsymbol{x}) + w(\boldsymbol{x})$ and $\tilde{v}(\boldsymbol{x}) = v(\boldsymbol{x}) + w(\boldsymbol{x})$ where $w(\boldsymbol{x})$ is an arbitrary convex function, hence both $\tilde{u}(\boldsymbol{x})$ and $\tilde{v}(\boldsymbol{x})$ remain convex. A natural question arising from this observation is how this non-uniqueness affects the CCP. In CCP, a nonconvex constraint $u(\boldsymbol{x}) \leq v(\boldsymbol{x})$ is replaced by

$$u(\boldsymbol{x}) \leq v(\boldsymbol{x}_k) + \nabla v(\boldsymbol{x}_k)^T(\boldsymbol{x} - \boldsymbol{x}_k) \tag{9.16}$$

Now if we treat the above nonconvex constraint as $u(\boldsymbol{x}) + w(\boldsymbol{x}) \leq v(\boldsymbol{x}) + w(\boldsymbol{x})$ with a nonlinear convex $w(\boldsymbol{x})$ (here we assume $w(\boldsymbol{x})$ is nonlinear, because adding a linear $w(\boldsymbol{x})$ does not affect the CCP at all) and apply CCP, the constraint would be replaced by

$$\begin{aligned} u(\boldsymbol{x}) + w(\boldsymbol{x}) \leq\ & v(\boldsymbol{x}_k) + \nabla v(\boldsymbol{x}_k)^T(\boldsymbol{x} - \boldsymbol{x}_k) \\ & + w(\boldsymbol{x}_k) + \nabla w(\boldsymbol{x}_k)^T(\boldsymbol{x} - \boldsymbol{x}_k) \end{aligned}$$

i.e.,

$$u(\boldsymbol{x}) + e(\boldsymbol{x}, \boldsymbol{x}_k) \leq v(\boldsymbol{x}_k) + \nabla v(\boldsymbol{x}_k)^T(\boldsymbol{x} - \boldsymbol{x}_k) \tag{9.17}$$

where, due to the convexity of $w(\boldsymbol{x})$, $e(\boldsymbol{x}, \boldsymbol{x}_k) = w(\boldsymbol{x}) - w(\boldsymbol{x}_k) - \nabla w(\boldsymbol{x}_k)^T(\boldsymbol{x} - \boldsymbol{x}_k)$ is convex and always nonnegative. On comparing (9.17) with (9.16), we note that a point $\boldsymbol{x}$ that satisfies constraint (9.17) also satisfies constraint (9.16), but the converse does not hold (unless $w(\boldsymbol{x})$ is a linear function which make $e(\boldsymbol{x}, \boldsymbol{x}_k)$ vanish). In other words, adding a redundant convex component $w(\boldsymbol{x})$ to the decomposition $f(\boldsymbol{x}) = u(\boldsymbol{x}) - v(\boldsymbol{x})$ shrinks the feasible region in a CCP-based method, hence imposing the risk of losing good solution candidates.

With above analysis in mind, we now examine the convexification steps made in Section 9.3.2. For illustration purposes, here we focus on the treatment of constraint (9.11b) since the same analysis also applies to (9.11c). By writing (9.11b) as

$$\frac{1}{2}[\boldsymbol{a}_f^T \ \boldsymbol{a}_m^T] \begin{bmatrix} \mathbf{0} & \boldsymbol{t}_f(L\omega)\boldsymbol{t}_m(\omega)^T \\ \boldsymbol{t}_m(\omega)\boldsymbol{t}_f(L\omega)^T & \mathbf{0} \end{bmatrix} \begin{bmatrix} \boldsymbol{a}_f \\ \boldsymbol{a}_m \end{bmatrix} \leq \delta_w + H_d(\omega) \tag{9.18}$$

we see that the left-hand side of (9.18) is a quadratic function with an indefinite Hessian, hence nonconvex. Therefore, constraint (9.11b) does not fit into the form in (9.4b), and an adequate convex term needs to be added to both sides of (9.18) to make CCP applicable. Intuitively, the quadratic expression in (9.18) suggests to add the term

$$\frac{1}{2}[\boldsymbol{a}_f^T \ \boldsymbol{a}_m^T] \begin{bmatrix} \boldsymbol{t}_f(L\omega)\boldsymbol{t}_f(L\omega)^T & \boldsymbol{0} \\ \boldsymbol{0} & \boldsymbol{t}_m(\omega)\boldsymbol{t}_m(\omega)^T \end{bmatrix} \begin{bmatrix} \boldsymbol{a}_f \\ \boldsymbol{a}_m \end{bmatrix} \tag{9.19}$$

which itself is convex and precisely equal to $0.5s(\boldsymbol{x}, \omega)$ (see (9.12) for the definition of $s(\boldsymbol{x}, \omega)$). In doing so, (9.18) is led to

$$\begin{aligned} &\frac{1}{2}[\boldsymbol{a}_f^T \ \boldsymbol{a}_m^T] \begin{bmatrix} \boldsymbol{t}_f(L\omega)\boldsymbol{t}_f(L\omega)^T & \boldsymbol{t}_f(L\omega)\boldsymbol{t}_m(\omega)^T \\ \boldsymbol{t}_m(\omega)\boldsymbol{t}_f(L\omega)^T & \boldsymbol{t}_m(\omega)\boldsymbol{t}_m(\omega)^T \end{bmatrix} \begin{bmatrix} \boldsymbol{a}_f \\ \boldsymbol{a}_m \end{bmatrix} \\ &\qquad \le \frac{1}{2}s(\boldsymbol{x}, \omega) + \delta_w + H_d(\omega) \end{aligned} \tag{9.20}$$

where the functions on both sides are convex, hence fitting nicely into (9.4b). It is important to realize that the function on the left-hand side of (9.20) is convex but not strictly convex because its Hessian is merely a rank-one matrix, implying that adding anything less than that of (9.19) will not lead to a formulation suitable for CCP. Clearly, (9.20) is identical to (9.13a), hence the above explains the convexification steps in Section 9.3.2.

9.4 Minimax Design of FRM Filters

9.4.1 The Design Problem

Following (9.2), let the transfer functions of the periodic filter and two masking filters of an FRM filter be given by

$$F(z) = \sum_{n=0}^{N-1} f_n z^{-n}, \quad M_a(z) = \sum_{n=0}^{N_a-1} m_n^{(a)} z^{-n}, \quad M_c(z) = \sum_{n=0}^{N_c-1} m_n^{(c)} z^{-n} \tag{9.21}$$

respectively, the zero-phase frequency response of a single-stage FRM filter can be expressed as

$$H(\boldsymbol{x}, \omega) = [\boldsymbol{a}_f^T \boldsymbol{t}_f(L\omega)][\boldsymbol{a}_a^T \boldsymbol{t}_a(\omega) - \boldsymbol{a}_c^T \boldsymbol{t}_c(\omega)] + \boldsymbol{a}_c^T \boldsymbol{t}_c(\omega) \tag{9.22}$$

where $\boldsymbol{a}_f, \boldsymbol{a}_a$, and $\boldsymbol{a}_c$ are coefficient vectors associated with filters $F(z)$, $M_a(z)$ and $M_c(z)$ in Figure 9.4 and $\boldsymbol{t}_f(\omega)$, $\boldsymbol{t}_a(\omega)$ and $\boldsymbol{t}_c(\omega)$ are respective trigonometric vectors determined by the lengths and types of the FIR filters involved. Given a desired zero-phase frequency response $H_d(\omega)$, a frequency-weighted minimax design seeks to find variable vector $\boldsymbol{x} = [\boldsymbol{a}_f^T \ \boldsymbol{a}_a^T \ \boldsymbol{a}_c^T]^T$ that solves the problem

$$\min_{\boldsymbol{x}} \max_{\omega \in \Omega} w(\omega)|H(\boldsymbol{x}, \omega) - H_d(\omega)| \tag{9.23}$$

where $w(\omega) > 0$ is a frequency-selective weight defined over a frequency region of interest, Ω. From (9.22), it is evident that (9.23) is a nonconvex problem with respect to design variable $\boldsymbol{x}$.

9.4.2 A CCP Approach to Solving (9.23)

By bounding the objective in (9.23) from above by δ and treating the bound as an additional design variable, we arrive at

$$\text{minimize } \delta \tag{9.24a}$$

$$\text{subject to } H(\boldsymbol{x}, \omega) - \delta_w - H_d(\omega) \le 0, \quad \omega \in \Omega \tag{9.24b}$$

$$-H(\boldsymbol{x}, \omega) - \delta_w + H_d(\omega) \le 0, \quad \omega \in \Omega \tag{9.24c}$$

where $\delta_w = \delta / w(\omega)$. We now take an approach similar to that of Section 9.3.2 to reformulate (9.24). By adding the convex term

$$v(\boldsymbol{x}, \omega) = [\boldsymbol{a}_f^T \boldsymbol{t}_f(L\omega)]^2 + \tfrac{1}{2}[\boldsymbol{a}_a^T \boldsymbol{t}_a(\omega)]^2 + \tfrac{1}{2}[\boldsymbol{a}_c^T \boldsymbol{t}_c(\omega)]^2 \tag{9.25}$$

to both sides of (9.24b) and (9.24c), we obtain an equivalent pair of constraints as

$$u_1(\boldsymbol{x}, \omega) \le v(\boldsymbol{x}, \omega) \tag{9.26a}$$

$$u_2(\boldsymbol{x}, \omega) \le v(\boldsymbol{x}, \omega) \tag{9.26b}$$

where

$$u_1(\boldsymbol{x}, \omega) = \frac{1}{2}(\boldsymbol{a}_{fa}^T \boldsymbol{P}_0 \boldsymbol{a}_{fa} + \boldsymbol{a}_{fc}^T \boldsymbol{Q}_1 \boldsymbol{a}_{fc}) + \boldsymbol{a}_c^T \boldsymbol{t}_c(\omega) - \delta_w - H_d(\omega) \tag{9.26c}$$

and

$$u_2(\boldsymbol{x}, \omega) = \frac{1}{2}(\boldsymbol{a}_{fa}^T \boldsymbol{P}_1 \boldsymbol{a}_{fa} + \boldsymbol{a}_{fc}^T \boldsymbol{Q}_0 \boldsymbol{a}_{fc}) - \boldsymbol{a}_c^T \boldsymbol{t}_c(\omega) - \delta_w + H_d(\omega) \tag{9.26d}$$

with

$$\boldsymbol{a}_{fa} = \begin{bmatrix} \boldsymbol{a}_f \\ \boldsymbol{a}_a \end{bmatrix}, \qquad \boldsymbol{a}_{fc} = \begin{bmatrix} \boldsymbol{a}_f \\ \boldsymbol{a}_c \end{bmatrix}$$

are convex with respect to $\boldsymbol{x}$ and δ because their Hessian matrices are characterized by positive semidefinite blocks $\boldsymbol{P}_i = \boldsymbol{p}_i\boldsymbol{p}_i^T$ and $\boldsymbol{Q}_i = \boldsymbol{q}_i\boldsymbol{q}_i^T$ with

$$\boldsymbol{p}_i = \begin{bmatrix} \boldsymbol{t}_f(L\omega) \\ (-1)^i\boldsymbol{t}_a(\omega) \end{bmatrix}, \qquad \boldsymbol{q}_i = \begin{bmatrix} \boldsymbol{t}_f(L\omega) \\ (-1)^i\boldsymbol{t}_c(\omega) \end{bmatrix} \text{ for } i = 0, 1.$$

Consequently, the constraints in (9.26a) and (9.26b) fit into the form of (9.4b). By applying CCP to (9.26), we obtain a pair of convex constraints

$$u_1(\boldsymbol{x}, \omega) \le \tilde{v}(\boldsymbol{x}, \boldsymbol{x}_k, \omega)$$

$$u_2(\boldsymbol{x}, \omega) \le \tilde{v}(\boldsymbol{x}, \boldsymbol{x}_k, \omega)$$

where

$$\tilde{v}(\boldsymbol{x}, \boldsymbol{x}_k, \omega) = v(\boldsymbol{x}_k, \omega) + \nabla^T v(\boldsymbol{x}_k, \omega)(\boldsymbol{x} - \boldsymbol{x}_k)$$

with

$$\nabla v(\boldsymbol{x}, \omega) = \begin{bmatrix} 2(\boldsymbol{a}_f^T\boldsymbol{t}_f(L\omega))\boldsymbol{t}_f(L\omega) \\ (\boldsymbol{a}_a^T\boldsymbol{t}_a(\omega))\boldsymbol{t}_a(\omega) \\ (\boldsymbol{a}_c^T\boldsymbol{t}_c(\omega))\boldsymbol{t}_c(\omega) \end{bmatrix}$$

where for practical purposes ω is taken from a finite and dense frequency grids $\Omega_d = \{\omega_j | \, j = 1, 2, \cdots, K\} \subset \Omega$. In summary, the convex problem to be solved in the kth iteration of CCP is given by

$$\text{minimize } \delta \tag{9.27a}$$

$$\text{subject to } u_1(\boldsymbol{x}, \omega_j) \le \tilde{v}(\boldsymbol{x}, \boldsymbol{x}_k, \omega_j), \;\; 1 \le j \le K \tag{9.27b}$$

$$u_2(\boldsymbol{x}, \omega_j) \le \tilde{v}(\boldsymbol{x}, \boldsymbol{x}_k, \omega_j), \;\; 1 \le j \le K \tag{9.27c}$$

Like the case of IFIR filter design, this problem of minimizing a linear function subject to convex quadratic constraints can be formulated as an SDP or SOCP problem [24], which can be solved efficiently [25–27].

Finally, we remark that the Hessian of functions $u_1(\boldsymbol{x}, \omega)$ in (9.27b) and $u_2(\boldsymbol{x}, \omega)$ in (9.27c) are of rank-two and positive semidefinite, but not positive definite as can be seen from (9.26c) and (9.26d). Therefore, an argument similar to that in Section 9.3.3 can be applied to conclude that constraints (9.27b) and (9.27c) define largest feasible region relative to other CCP-compatible options as discussed in Section 9.3.3. A step-by-step summary of the algorithm is given below.

Algorithm 2 for FRM filters

Step 1: input $\boldsymbol{x}_0$, (N, N_a, N_c), $H_d(\omega)$ for $\omega \in \Omega_d$, $w(\omega)$, and K_i
Step 2: for $k = 0, 1, \cdots, K_i - 1$
solve (9.27) for $\boldsymbol{x}_{k+1}$
end
Step 3: output $\boldsymbol{x}^* = \boldsymbol{x}_{K_i}$

9.5 FRM Filters with Reduced Complexity

FRM filters are widely considered computationally efficient [3, 4, 9]. Variants of FRM filters with further reduced complexity have also been proposed in the literature. Unlike these variants, in this section we propose a CCP algorithm which is actually a modified version of that developed in Section 9.4 for the design of FRM filters that simultaneously promotes sparsity of the impulse responses of the subfilters involved so as to reduce implementation complexity. Here an FIR filter $H_1(z)$ is said to more sparse than filter $H_2(z)$ of same length, if the impulse response of $H_1(z)$ contains more zero entries than that of $H_2(z)$. The algorithm proposed below consists of two phases:

9.5.1 Design Phase 1

The aim of phase 1 is to identify the locations (i.e. indices) of filter coefficients that may be set to zero without substantially affecting filter's performance. Motivated by the fact that for most large underdetermined systems of linear equations the minimal l_1-norm solution is also the sparsest solution [28, 29], we propose to promote the sparsity of an FRM filter by solving a modified version of (9.27) where the objective function combines upper bound with a weighted l_1-norm of design variable $\boldsymbol{x}$, namely,

$$\text{minimize } \delta + \mu \|\boldsymbol{x}\|_1 \tag{9.28a}$$

$$\text{subject to } u_1(\boldsymbol{x}, \omega_j) \leq \tilde{v}(\boldsymbol{x}, \boldsymbol{x}_k, \omega_j), \ 1 \leq j \leq K \tag{9.28b}$$

$$u_2(\boldsymbol{x}, \omega_j) \leq \tilde{v}(\boldsymbol{x}, \boldsymbol{x}_k, \omega_j), \ 1 \leq j \leq K \tag{9.28c}$$

where weight $\mu > 0$ controls the trade-off between error bound δ and sparsity of filter coefficients. Once the solution $\boldsymbol{x}_s = [\boldsymbol{a}_f^T \ \boldsymbol{a}_a^T \ \boldsymbol{a}_c^T]^T$ of problem (9.28) is obtained, a prescribed threshold $\varepsilon > 0$ is used to identify an index set $I_o = \{I_{af}, I_{aa}, I_{ac}\}$ as follows:

$$I_{af} = \{i : |\boldsymbol{a}_f(i)| \le \varepsilon\}, \quad I_{aa} = \{i : |\boldsymbol{a}_a(i)| \le \varepsilon\}, \quad I_{ac} = \{i : |\boldsymbol{a}_c(i)| \le \varepsilon\} \tag{9.29}$$

Concerning the selection of parameters μ and ε, in principle they should be small: μ should be small because otherwise the solution of (9.28) becomes less relevant to the design objective which is to minimize the error bound δ; and ε should be small because the use of a large ε would set some intrinsically nonzero coefficients to zero, hence inevitably do harm to the filter performance. We stress that the goal of phase 1 is merely to identify an index set to nullify the associated filter coefficients, and it is a collective action of μ and ε that identifies an appropriate index set: the introduction of term $\mu\|\boldsymbol{x}\|_1$ promotes the sparsity in $\boldsymbol{x}$, and the size of the index set is controlled by threshold ε.

9.5.2 Design Phase 2

The design now proceeds with a second phase in that the remaining nonzero entries of the impulse responses are optimally tuned by minimizing the same error bound as in (9.27a), but subject to additional constraints that the filter coefficients with indices in set $\{I_{af},\ I_{aa},\ I_{ac}\}$ are equal to zero. Namely, we solve

$$\text{minimize } \delta \tag{9.30a}$$

$$\text{subject to } u_1(\boldsymbol{x}, \omega_i) \le \tilde{v}(\boldsymbol{x}, \boldsymbol{x}_k, \omega_i) \text{ for } i = 1, 2, \cdots, K \tag{9.30b}$$

$$u_2(\boldsymbol{x}, \omega_i) \le \tilde{v}(\boldsymbol{x}, \boldsymbol{x}_k, \omega_i) \text{ for } i = 1, 2, \cdots, K \tag{9.30c}$$

$$\boldsymbol{a}_f(i) = 0 \text{ for } i \in I_{af} \tag{9.30d}$$

$$\boldsymbol{a}_a(i) = 0 \text{ for } i \in I_{aa} \tag{9.30e}$$

$$\boldsymbol{a}_c(i) = 0 \text{ for } i \in I_{ac} \tag{9.30f}$$

Since the additional constraints are all linear equalities, problem (9.30) remains convex and can be solved efficiently. The algorithm proposed above is outlined below.

Algorithm 3 for sparse FRM filters

Step 1: input $\boldsymbol{x}_0$, (N, N_a, N_c), Ω_d, $H_d(\omega)$ for $\omega \in \Omega_d$, $w(\omega)$, μ, ε, and K_i

Step 2: for $k = 0, 1, \cdots, K_i - 1$
solve (9.28) to obtain $\boldsymbol{x} = [\boldsymbol{a}_f^T\ \boldsymbol{a}_a^T\ \boldsymbol{a}_c^T]^T$
apply (9.29) to $\boldsymbol{x}$ to obtain I_{af}, I_{aa}, and I_{ac}
solve (9.30) to obtain $\boldsymbol{x}_{k+1}$
end

Step 3: output $\boldsymbol{x}^* = \boldsymbol{x}_{K_i}$

9.6 Design Examples

We now present several design examples to illustrate the design methods proposed above.

9.6.1 Design and Evaluation Settings

A common goal of the designs presented below is to construct a linear-phase FIR digital system with narrow transition band, whose implementation efficiency is achieved via an IFIR or FRM structure. We follow the convention to evaluate the performance of each design in terms of peak-to-peak passband ripple A_p (dB) and minimum stopband attenuation A_a (dB) [4] which are defined by

$$A_p = 20\log_{10}\left(\frac{p_{max}}{p_{min}}\right)$$

and

$$A_a = -20\log_{10}(a_{max})$$

where p_{max} and p_{min} denote the maximum and minimum $|H(e^{j\omega})|$ over passband, respectively, and a_{max} denotes the maximum $|H(e^{j\omega})|$ over stopband.

As a part of performance evaluation, each design is compared with one or more designs made by existing algorithms from the literature. In addition, these designs are compared with their counterparts obtained using conventional FIR structure, see Section 9.6.4 below.

9.6.2 Design of IFIR Filters

Example 9.1

The algorithm proposed in Section 9.3 was applied to design a lowpass IFIR filter with normalized passband edge $\omega_p = 0.15\pi$, stopband edge $\omega_a = 0.2\pi$. The sparsity factor was set to $L = 4$, and orders of $F(z)$ and $M(z)$ were set to 31 and 17, respectively. The frequency weight $w(\omega)$ was set to $w(\omega) \equiv 1$ for ω in the passband and $w(\omega) \equiv 2$ for ω in the stopband. An initial $\boldsymbol{x}_0 = [\boldsymbol{a}_{f0}^T \ \boldsymbol{a}_{m0}^T]^T$ was generated by the standard technique proposed in [3]. A total of $K = 1400$ frequency grids were uniformly placed in $[0, \ \omega_p] \bigcup [\omega_a, \ \pi]$ to form the discrete set Ω_d for problem (9.15). It took the algorithm 91 iterations to converge to an IFIR filter with $A_p = 0.0317$ dB and $A_a = 60.84$ dB. The same design problem was addressed as Example 10.29 in [4] using the method described in [30]. The method was implemented as function `ifir` in the Signal Processing Toolbox of MATLAB. With `[F,M] = ifir(4,low,`

`[0.15, 0.2], [0.002, 0.001], advanced)`, the function returns with optimized impulses of filter $F(z)$ of order 31 and $M(z)$ of order 17 (in Example 10.29 of [4], the order of $M(z)$ was said to be 16, however the order of $M(z)$ produced by the above MATLAB code was actually 17), with $A_p = 0.0340$ dB and $A_a = 60.18$ dB. The passband and stopband amplitude responses of the two designs are depicted as solid and dashed lines in Figures 9.5a and 9.5b, respectively.

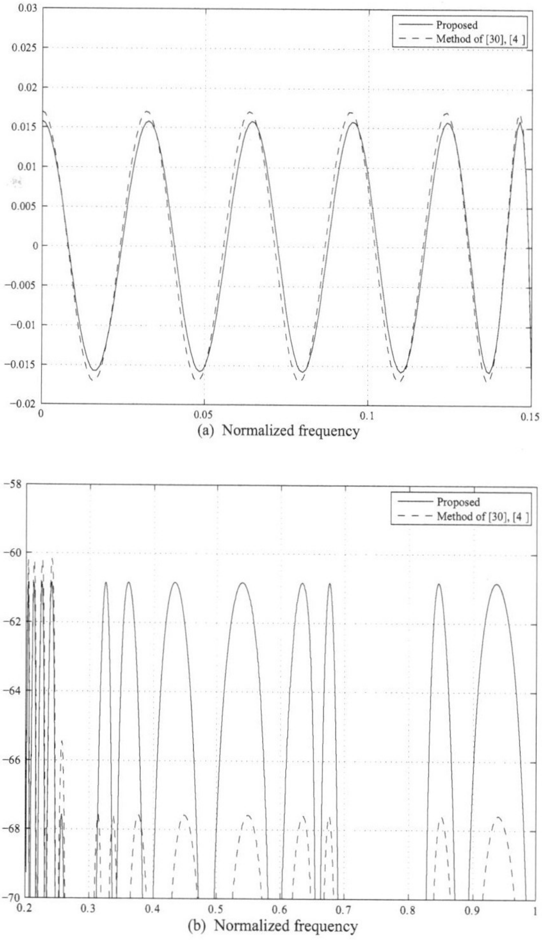

Figure 9.5 Amplitude response (in dB) of the IFIR filters for Example 9.1 by the proposed algorithm (solid line) and the method of [30] and [4] (dashed line) in (a) passband and (b) stopband.

Example 9.2
Reference [9] presents an example (Example 10.2) where a highpass IFIR filter with sampling frequency $\omega_s = 16000$ Hz, stopband edge $\omega_a = 6600$ Hz and passband edge $\omega_p = 7200$ Hz is designed. In the design, the interpolator is fixed to $M(z) = (1 - z^{-1})^4$ and sparsity factor is set to $L = 2$ while a linear-phase subfilter $F(z)$ of order 20 is optimized in minimax sense. The performance of the IFIR filter is given by $A_p = 0.9061$ dB and $A_a = 40.78$ dB.

As a follow-up of the above example, the algorithm proposed in Section 9.3 was applied to jointly optimize an $F(z)$ of order 20 and an $M(z)$ of order 4 for a highpass IFIR filter $H(z) = F(z^L)M(z)$ with the same design specifications as Example 10.2 of [9]. The frequency weight $w(\omega)$ was set to $w(\omega) \equiv 1$ for ω in the passband and $w(\omega) \equiv 4.5$ for ω in the stopband. An initial $\boldsymbol{x}_0 = [\boldsymbol{a}_{f0}^T \; \boldsymbol{a}_{m0}^T]^T$ was generated by the standard technique proposed in [3]. A total of $K = 1400$ frequency grids were uniformly placed in [0, 6600 Hz] $\bigcup$[7200 Hz, 8000 Hz] to form the discrete set Ω_d for problem (9.15). It took the algorithm 14 iterations to converge to an IFIR filter with $A_p = 0.6914$ dB and $A_a = 41.07$ dB. The amplitude responses of the two designs in the entire baseband and over its passband are depicted respectively as solid and dashed lines in Figures. 9.6a and 9.6b.

For the sake of a more efficient implementation, $M(z)$ is normalized by dividing it with the coefficient of its constant term, m_0, so that both coefficients of the constant term and the 4th-order term become unity. The transfer function $F(z^L)$ is rescaled to $m_0F(z^L)$ so that the overall transfer function $H(z)$ remains unaltered. The first eleven coefficients of $F(z)$ and first three coefficients of $M(z)$ after the normalization are shown in Table 9.1. We stress that implementing the interpolator $M(z)$ in Example 10.2 of [9] requires no multiplications, therefore the performance gain of the proposed design over that of [9] was achieved at a cost of two more multiplications per output.

9.6.3 Design of FRM Filters

Example 9.3
The algorithm proposed in Section 9.4 was applied to design a lowpass FRM filter with the same design specifications as in the first example in [3] and [10]. The normalized passband and stopband edges were $\omega_p = 0.6\pi$ and $\omega_a = 0.61\pi$. The sparsity factor was set to $L = 9$, and orders of $F(z)$, $M_a(z)$, and $M_c(z)$ were 44, 40, and 32, respectively. A trivial weight $w(\omega) \equiv 1$ was

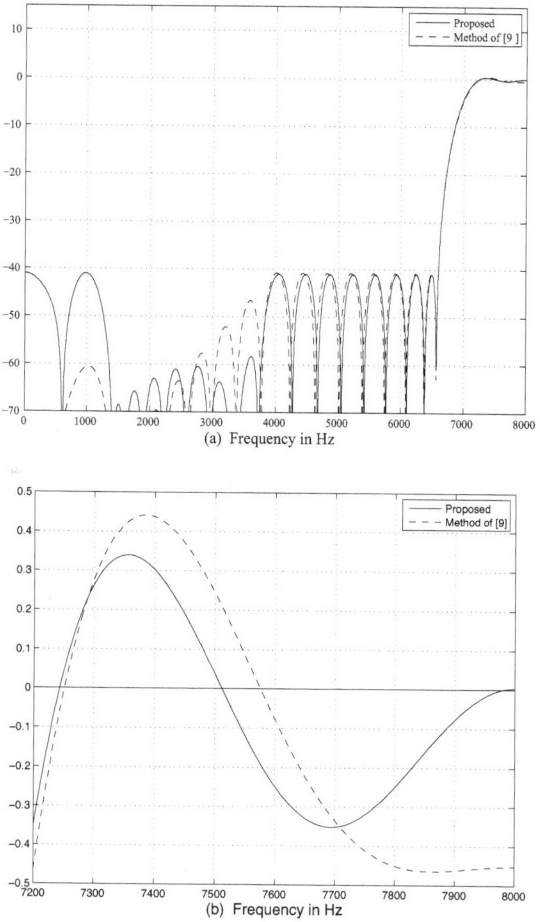

Figure 9.6 Amplitude response (in dB) of the IFIR filters in Example 9.2 by the proposed algorithm (solid line) and the method of [9] (dashed line) in (a) entire baseband and (b) passband.

utilized. With $K = 1100$, it took the algorithm 87 iterations to converge to an FRM filter with $A_p = 0.1320$ dB and $A_a = 42.49$ dB, which are favorably compared with those achieved in [3] ($A_p = 0.1792$ dB and $A_a = 40.96$ dB), which has been a benchmark for FRM filters, and those reported in [11] ($A_p = 0.1348$ dB and $A_a = 42.25$ dB). The amplitude response of the FRM filter in the entire baseband and passband is shown in Figure 9.7.

Example 9.4

The algorithm proposed in Section 9.4 was applied to design a lowpass FRM filter with the same passband, stopband, and sparsity factor as in Example 9.3,

Table 9.1 Coefficients of $F(z)$ and $M(z)$ for Example 9.2

f_i for $i = 0, 1, 2, \cdots, 10$	m_i for $i = 0, 1, 2$
0.002287418334499	1
0.001299308145379	−2.439354581245607
0.001489820681472	2.965711835392111
−0.001928211344492	
−0.004121111010416	
−0.005691224464356	
−0.002031403630543	
0.005260711732820	
0.015907596775075	
0.024282544840202	
0.028106104367025	

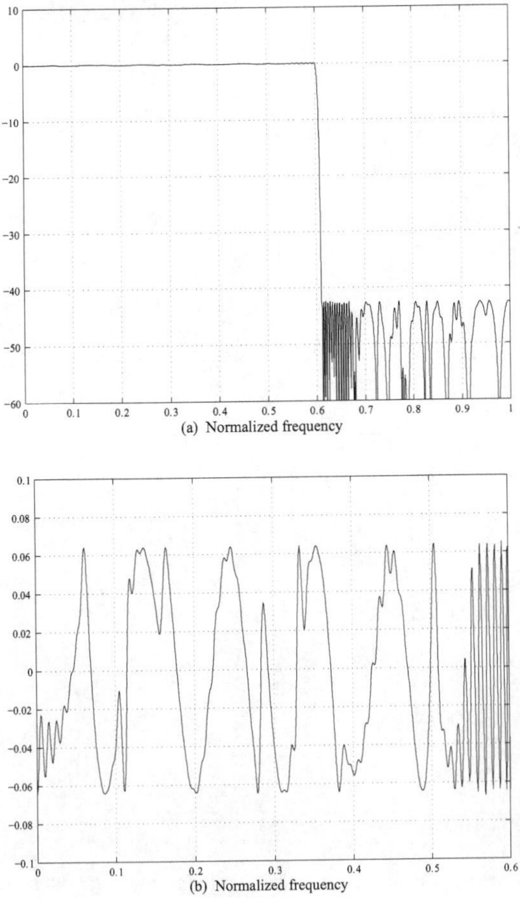

Figure 9.7 Amplitude response (in dB) of the FRM filters in Example 9.3 in (a) entire baseband and (b) passband.

i.e., $\omega_a = 0.6\pi$, $\omega_p = 0.61\pi$, and $L = 9$. However, the orders of $F(z)$, $M_a(z)$, and $M_c(z)$ were reduced to 42, 36, and 28, respectively. With $w(\omega) \equiv 2.1$ and $K = 1000$, it took the algorithm 600 iterations to converge to an FRM filter with $A_p = 0.2600$ dB and $A_a = 43.01$ dB. The amplitude response of the FRM filter in the entire baseband and passband is shown in Figure 9.8. The above design specifications coincide with a design presented in [16] based on a neural network approach. From the numerical results proved in Tables 1–3 of [16], it was found that $A_p = 0.2652$ dB (instead of 0.0672 dB as reported in [16] as there was a deep notch at passband edge 0.6π) and $A_a = 42.42$ dB.

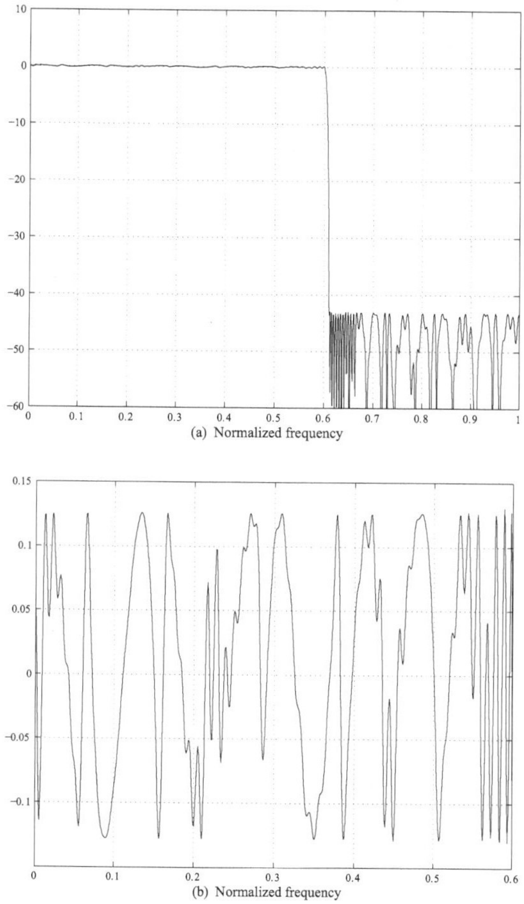

Figure 9.8 Amplitude response (in dB) of the FRM filter in Example 9.4 in (a) entire baseband and (b) passband.

Example 9.5

The algorithm proposed in Section 9.4 was also applied to design a lowpass FRM filter with orders of $F(z)$, $M_a(z)$, and $M_c(z)$ being (56, 30, 24), $L = 7$, and normalized passband and stopband edges being $\omega_p = 0.65\pi$ and $\omega_a = 0.66\pi$, respectively. With weight $w(\omega) \equiv 1$ and $K = 2400$, it took the algorithm 94 iterations to converge to an FRM filter with $A_p = 0.1510$ dB and $A_a = 41.22$ dB. The amplitude response of the FRM filter in the entire baseband and passband is shown in Figure 9.9. For comparison, in [10] lowpass FRM filters with the same passband and stopband edges and

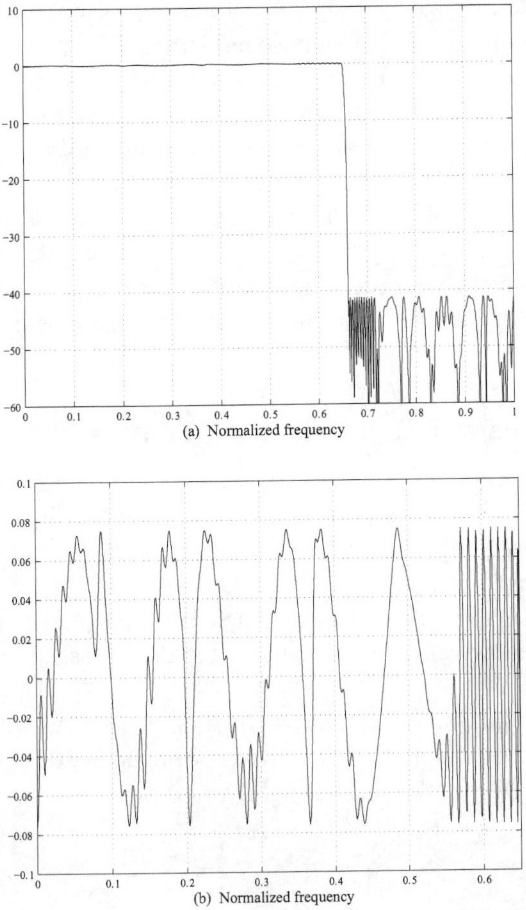

Figure 9.9 Amplitude response (in dB) of the FRM filters in Example 9.5 in (a) entire baseband and (b) passband.

interpolation factor L are designed using a weighted least-squares Chebyshev approach. With orders of $F(z)$, $M_a(z)$, and $M_c(z)$ being (56, 32, 26), the performance of the FRM filters are reported in terms of $A_p = 0.1960$ dB and $A_a = 40.11$ dB for a filter named Filter 4 and $A_p = 0.1920$ dB and $A_a = 40.44$ dB for a filter named Filter 5.

Example 9.6

The algorithm described in Section 9.5 was applied to re-design the FRM filter presented in Example 9.5 so as to reduce its complexity while maintain a comparable performance. With weight $w(\omega) \equiv 1$, $K = 1000$, and $\mu = 0.05$, phase 1 of the design was carried out using 30 iterations of (9.28) to yield a coefficient vector $\boldsymbol{x}_s = [\boldsymbol{a}_f^T \ \boldsymbol{a}_a^T \ \boldsymbol{a}_c^T]^T$. With a threshold $\varepsilon = 0.0006$, the three index sets defined in (9.29) were identified as $I_{af} = \{13, \ 18, \ 25\}$, $I_{aa} = \{11, \ 15, \ 16\}$, $I_{ac} = \{11, \ 13\}$, indicating there are a total of eight coefficients that may be set to zero values without substantially degrading the filter's performance provided that the remaining coefficients are optimized in the second phase of the design. Proceeding with design phase 2 with weight $w(\omega) \equiv 1$ and $K = 2400$, it took 10 iterations of (9.30) to converge to an FRM filter with $A_p = 0.1629$ dB and $A_a = 40.56$ dB, in that eight coefficients whose indices are identified above have been set to zero. The amplitude response of the FRM filter in the entire baseband and passband is shown in Figure 9.10.

9.6.4 Comparisons with Conventional FIR Filters

The six designs presented above are compared with their conventional counterparts that are linear-phase equiripple FIR filters designed using the Parks-McClellan (P-M) algorithm [4] with practically the same performance as those in Examples 9.1–9.6 in terms of respective A_p and A_a. The comparisons are made in terms of computational complexity measured by the number of multiplications required per output sample and the overall group delay of the filter, see Table 9.2.

It is observed from Table 9.2 that in all design instances the IFIR and FRM filters designed by the proposed algorithms offer considerably improved computational efficiency at the cost of slightly increased group delay.

9.7 Summary

We have proposed a unified approach based on CCP to the design of minimax IFIR and FRM filters. The design method is conceptually simple and produces designs that are shown to provide satisfactory performance relative to those

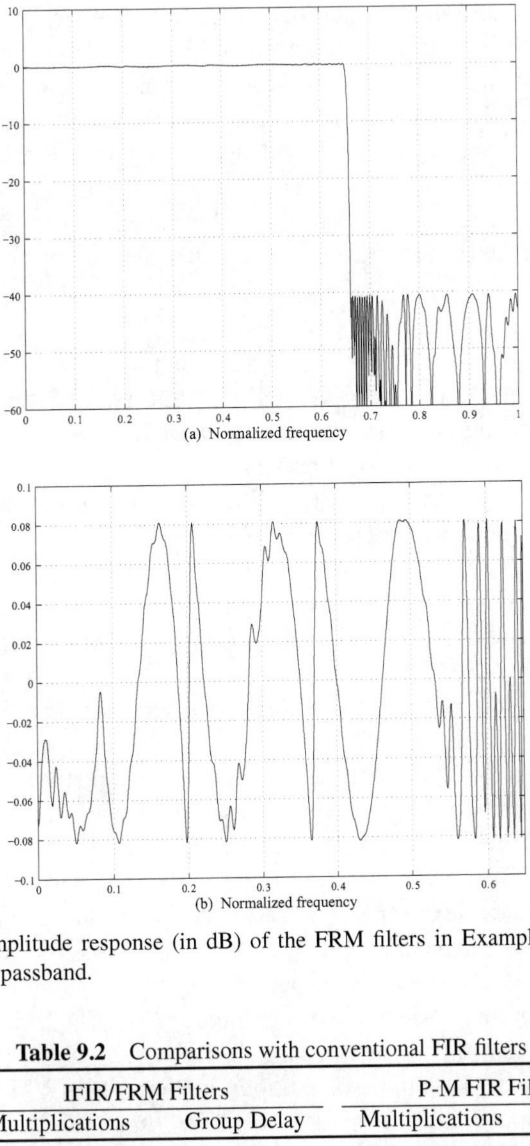

Figure 9.10 Amplitude response (in dB) of the FRM filters in Example 9.6 in (a) entire baseband and (b) passband.

Table 9.2 Comparisons with conventional FIR filters

	IFIR/FRM Filters		P-M FIR Filters [4]	
Design	Multiplications	Group Delay	Multiplications	Group Delay
1	25	70.5	65	64
2	13	22	22	21
3	60	218	210	209
4	56	207	193	192
5	58	211	202	201
6	50	211	198	197

available from the literature. In addition, the proposed design method is shown to allow extension to the design of FRM filters that simultaneously promotes sparsity of filter coefficients to reduce implementation complexity without substantially degrading filter's performance. Design examples have been presented to illustrate the proposed design algorithms that compare favorably with several known design techniques. Extensions of CCP-based designs to multistage FRM filters is possible, but developing a CCP type of convexification to maintain largest feasible region turns out to be challenging.

References

[1] Y. Neuvo, C. Y. Dong, and S. K. Mitra, "Interpolated finite impulse response filters," *IEEE Trans. Acoust., Speech, Signal Process.*, vol. ASSP-32, no. 3, pp. 563–570, June 1984.

[2] T. Saramäki, Y. Neuvo, and S. K. Mitra, "Design of computationally efficient interpolated FIR filters," *IEEE Trans. Circuits Syst.*, vol. CAS-35, no. 1, pp. 70–88, Jan. 1988.

[3] Y. C. Lim, "Frequency-response masking approach for the synthesis of sharp linear phase digital filters," *IEEE Trans. Circuits Syst.*, vol. CAS-33, no. 4, pp. 357–364, Apr. 1986.

[4] S. K. Mitra, *Digital Signal Processing – A Computer-Based Approach*, 3rd ed., McGraw Hill, 2006.

[5] Y. C. Lim and Y. Lian, "The optimum design of one- and two-dimensional FIR filters using the frequency response masking technique," *IEEE Trans. Circuits Syst. II*, vol. 40, no. 2, pp. 88–95, Feb. 1993.

[6] Y. C. Lim and Y. Lian, "Frequency-response masking approach for digital filter design: Complexity reduction via masking filter factorization," *IEEE Trans. Circuits Syst. II*, vol. 41, no. 8, pp. 518–525, Aug. 1994.

[7] T. Saramäki, Y. C. Lim, and R. Yang, "The synthesis of half-band filter using frequency-response masking technique," *IEEE Trans. Circuits Syst. II*, vol. 42, no. 1, pp. 58–60, Jan. 1995.

[8] T. Saramäki and H. Johansson, "Optimization of FIR filters using frequency-response masking approach," in *Proc. IEEE Int. Symp. Circuits Syst.*, vol. 2, pp. 177–180, Sydney, Australia, May 2001.

[9] P. S. R. Diniz, E. A. B. da Silva, and S. L. Netto, *Digital Signal Processing*, Cambridge University Press, 2002.

[10] L. C. R. de Barcellos, S. L. Netto, and P. S. R. Diniz, "Optimization of FRM filters using the WLS-Chebyshev approach," *Circuits, Syst., Signal Process.*, vol. 22, no. 2, pp. 99–113, Mar. 2003.

[11] W.-S. Lu and T. Hinamoto, "Optimal design of frequency-response-masking filters using semidefinite programming," *IEEE Trans. Circuits Syst. I*, vol. 50, no. 4, pp. 557–568, Apr. 2003.
[12] T. Saramäki, J. Yli-Kaakinen, and H. Johansson, "Optimization of frequency-response masking based FIR filters," *J. Circuits, Syst., Comput.*, vol. 12, pp. 563–589, May 2003.
[13] W.-S. Lu and T. Hinamoto, "Optimal design of frequency-response-masking filters using second-order cone programming," in *Proc. IEEE Int. Symp. Circuits Syst.*, vol. 3, pp. 878–881, Bangkok, Thailand, May 2003.
[14] Y. Liu and Z. Lin, "Optimal design of frequency-response masking filters with reduced group delays," *IEEE Trans. Circuits Syst. I*, vol. 55, no. 6, pp. 1560–1570, July 2008.
[15] Y. Wei and D. Liu, "Improved design of frequency-response masking filters using band-edge shaping filter with non-periodical frequency response," *IEEE Trans. Signal Process.*, vol. 61, no. 13, pp. 3269–3278, July 2013.
[16] X.-H. Wang and Y.-G. He, "A neural network approach to FIR filter design using frequency-response masking technique," *Signal Process.*, vol. 88, pp. 2917–2926, 2008.
[17] J. Yli-Kaakinen and T. Saramäki, "An efficient alorithm for the optimization of FIR filters synthesized using the multistage frequency-response masking approach," *Circuits, Syst., Signal Process.*, vol. 30, no. 1, pp. 157–183, 2011.
[18] Y. Wei, S. Huang, and X. Ma, "A novel approach to design low-cost two-stage frequency-response masking filters," *IEEE Trans. Circuits Syst. II*, vol. 62, no. 10, pp. 982–986, Oct. 2015.
[19] A. L. Yuille and A Rangarajan, "The concave-convex procedure," *Neural Computation*, vol. 15, no. 4, pp. 915–936, 2003.
[20] T. Lipp and S. Boyd, "Variations and extensions of the convex-concave procedure," *Research Report*, Stanford University, Aug. 2014.
[21] B. K. Sriperumbudur and G. R. Lanckriet, "On the convergence of the concave-convex procedure," in *Advances in Neural Information Processing Systems*, pp. 1759–1767, 2009.
[22] W.-S. Lu and T. Hinamoto, "A unified approach to the design of interpolated and frequency-response-masking FIR filters," *IEEE Trans. Circuits Syst. I*, vol. 63, no. 12, pp. 2257–2266, Dec. 2016.
[23] P. Hartman, "On functions representable as a difference of convex functions," *Pacific J. of Math.*, vol. 9, no. 3, pp. 707–713, 1959.

[24] A. Antoniou and W.-S. Lu, *Practical Optimization: Algorithms and Engineering Applications*, Springer, 2007.

[25] J. F. Sturm, "Using SeDuMi 1.02, a MATLAB toolbox for optimization over symmetric cones," *Optimization Methods and Software*, vol. 11–12, pp. 625–633, 1999.

[26] R. H. Tütüncü, K. C. Toh, and M. J. Todd, "Solving semidefinite-quadratic-linear programs using SDPT3," *Mathematical Programming, Series B*, vol. 9, pp. 189–217, 2003.

[27] M. Grant, S. Boyd, and Y. Ye, "Disciplined convex programming," in *Global Optimization: from Theory to Implementation, Nonconvex Optimization and Its Applications*, L. Liberti and N. Maculan, eds., Springer, 2006.

[28] D. L. Donoho, "For most largest underdetermined systems of linear equations the minimal l_1-norm solution is also the sparsest solution," *Comm. Pure Applied Math.*, vol. 59, no. 6, pp. 797–829, June 2006.

[29] E. J. Candès and M. B. Wakin, "An introduction to compressive sampling," *IEEE Signal Processing Magazine*, vol. 25, no. 2, pp. 21–30, Mar. 2008.

[30] T. Saramäki, "Finite impulse response filter design," in *Handbook for Digital Signal Processing*, S. K. Mitra and J. F. Kaiser eds., Wiley-Interscience, New York NY, 1993.

10

Design of a Class of Composite Digital Filters

10.1 Preview

A composite filter (C-filter) refers to a digital filtering system that is composed of explicit individual modules, often called subfilters, which are connected in cascade or parallel, or a mixture of both. Well known classes of C-filters include interpolated FIR filters [1, 2], frequency-response-masking filters [3], and their variants, see e.g. [4]. Over the years, analysis and design of C-filters have attracted a great deal of research interest primarily because of their ability to offer improved computational efficiency relative to their conventional counterparts when appropriate subfilter structures are chosen and their parameters are optimized in accordance with certain design criterion [5]. Recently, C-filters composed of a prototype filter and a shaping filter, connected in cascade, are shown to offer certain advantages over conventional counterparts [6, 7]. In the case of FIR C-filters [6], the shaping filter is constructed by cascading several complementary comb filters (CCFs) of the form $(1 + z^{-l})^{k_l}$ with integers $1 \leq l \leq L$ and $k_l \geq 0$. As such, the shaping filter only requires several adders and memory units to implement and is free of multiplications. Yet, as demonstrated in [6], the shaping filter is capable of effectively improving filter performance, especially for those with narrow transition bands and highly suppressed stopbands.

In this chapter, we present a new algorithm for the design of C-filters that assume the same form $H(z) = H_p(z)H_s(z)$ as in [6], however prototype filter $H_p(z)$ as well as shaping filter $H_s(z)$ are designed through different approaches in order for $H(z)$ to achieve equiripple passbands and least-squares stopbands (EPLSS). EPLSS FIR filters, especially with narrow passbands, are desirable in wireless communications, aerospace systems, and synthetic aperture radar [8]. The design algorithm proposed in this chapter uses a sequential optimization technique to deal with the design of $H_p(z)$

and $H_s(z)$ separately, yet these two design steps are coupled by alternately performing them. A strong motivation to develop such a design method is that both design steps can be formulated as convex problems hence can be solved efficiently. Numerical examples are presented to illustrate the design method and evaluate its performance.

10.2 Composite Filters and Problem Formulation

10.2.1 Composite Filters

We consider a C-filter that is composed of two subfilters, known as prototype filter $H_p(z)$ and shaping filter $H_s(z)$ respectively, that are connected in cascade as shown in Figure 10.1. Thus the transfer function of the C-filter assumes the form

$$H(z) = H_p(z) \cdot H_s(z) \tag{10.1}$$

In this chapter, we examine a class of linear-phase FIR C-filters where the prototype filter is an FIR filter of length N with transfer function

$$H_p(z) = \sum_{n=0}^{N-1} h_n z^{-n} \tag{10.2}$$

which largely determines the characteristics of $H(z)$, while the shaping filter is a CCF of the form

$$H_s(z) = \prod_{l=1}^{L} (1 + z^{-l})^{k_l} \tag{10.3}$$

where k_l are nonnegative integers, which reshapes the prototype filter for improved performance without substantial increase in implementation complexity.

For clarity of presentation, in the rest of the chapter we focus on linear-phase lowpass FIR C-filters whose frequency response can be expressed as $H(\omega) = H_p(\omega)H_s(\omega)$ with

Figure 10.1 A composite filter.

$$H_p(\omega) = \sum_{n=0}^{N-1} h_n e^{-jn\omega} = e^{-j\tau_p\omega}[\boldsymbol{x}^T \boldsymbol{c}(\omega)] \tag{10.4}$$

where $\tau_p = (N-1)/2$ is the group delay of $H_p(z)$, $\boldsymbol{x}$ is a coefficient vector determined by the impulse response of $H_p(z)$, and $\boldsymbol{c}(\omega)$ is a vector with trigonometric components, and

$$H_s(\omega) = \prod_{l=1}^{L} (1 + e^{-jl\omega})^{k_l} = e^{-j\tau_s\omega} \prod_{l=1}^{L} (2\cos\frac{l\omega}{2})^{k_l} \tag{10.5}$$

where

$$\tau_s = \frac{1}{2}\sum_{l=1}^{L} l \cdot k_l$$

Therefore, the zero-phase frequency response of the C-filter is given by

$$A(\omega) = A_p(\omega) \cdot A_s(\omega) = \boldsymbol{x}^T \boldsymbol{c}(\omega) \prod_{l=1}^{L} (2\cos\frac{l\omega}{2})^{k_l} \tag{10.6}$$

and the group delay of the C-filter is given by $\tau = \tau_p + \tau_s$.

The behavior of shaping filter $H_s(z)$ is determined by the number L of CCFs used and the power k_l for each CCF. As long as $k_l > 0$, the first notch of a single CCF $(1 + z^{-l})^{k_l}$ over the normalized baseband $[0, \ \pi]$ occurs at $\omega = \pi/l$, see for example Figure 10.2(a) for $(1 + z^{-4})$ (a scaling factor 0.5 has been used in the figure to normalize the filter gain at $\omega = 0$ to unity). By cascading several CCFs with appropriate powers k_l, a shaping filter can offer much reduced stopband energy while retaining decent passband gain. As an example, Figure 10.2(b) depicts the amplitude response of a shaping filter $H_s(z)$ with $L = 4$ and $\{k_l, \ l = 1, 2, 3, 4\} = \{4, 4, 1, 2\}$ where the filter gain at $\omega = 0$ has been normalized to unity.

10.2.2 Problem Formulation

Given a desired lowpass frequency response $H_d(\omega)$, prototype filter length N, and number of CCF's L, we seek to find a linear-phase FIR C-filter of form (10.1–10.3) such that the peak-to-peak amplitude ripple in passband is minimized subject to constraints on filter's energy as well as peak gain in stopband and total group delay τ. Note that the constraints imposed here would lead to an enhanced EPLSS C-filter in which largest peak in the stopband is under control and, in addition, the group delay is bounded so as to ensure its

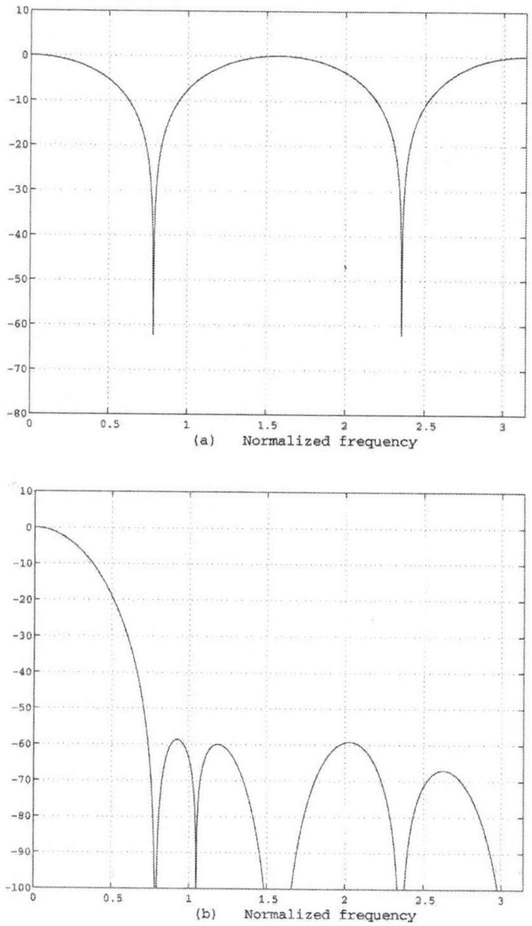

Figure 10.2 Amplitude response of (a) $1 + z^{-4}$ and (b) $(1 + z^{-1})^4(1 + z^{-2})^4\,(1 + z^{-3})^1\,(1 + z^{-4})^2$.

utility for on-line applications. The design objective is achieved by solving the constrained problem

$$\min_{\boldsymbol{x},\boldsymbol{y}} \max_{\omega \in \Omega_p} |H(\omega) - H_d(\omega)| \tag{10.7a}$$

$$\text{subject to: } \int_{\omega_a}^{\pi} |H(\omega)|^2 d\omega \le e_a \tag{10.7b}$$

$$\max_{\omega \in \Omega_a} |H(\omega)| \le \delta_a \tag{10.7c}$$

$$\frac{1}{2}\sum_{l=1}^{L} l \cdot k_l \leq D \tag{10.7d}$$

where Ω_p and Ω_a denote the passband and stopband, respectively, ω_a denotes the stopband edge, $\boldsymbol{x}$ from (10.4) and $\boldsymbol{y} = [k_1\, k_2\, \ldots\, k_L]^T$ are design variables, e_a, δ_a and D are constants representing upper bounds for stopband energy, peak filter gain in stopband, and group delay of the shaping filter, respectively. Note that the constraint in (10.7d) implies an upper bound $(N-1)/2 + D$ for the total group delay of the C-filter.

10.3 Design Method

10.3.1 Design Strategy

From (10.6), we see that design variables $\boldsymbol{x}$ and $\boldsymbol{y}$ are separate from each other. Moreover, frequency response $A(\omega)$ depends on $\boldsymbol{x}$ linearly, but on $\boldsymbol{y}$ highly nonlinearly. In addition, all components of variable $\boldsymbol{y}$ are constrained to be nonnegative integers. Under these circumstances, it is intuitively natural to optimize these design variables separately in an alternate fashion. This leads to a sequential procedure where in each step one of the variables, say $\boldsymbol{x}$, is optimized while the other variable, $\boldsymbol{y}$, is held fixed, and the solution so produced, say $\boldsymbol{x}_k$, is held fixed in the next step when variable $\boldsymbol{y}$ is updated to $\boldsymbol{y}_{k+1}$. The alternating optimization continues until a stopping criterion is met and the last pair of solutions $(\boldsymbol{y}_K, \boldsymbol{x}_K)$ is taken to be the optimal design. The rest of this section is devoted to describing the technical details involved in the design procedure.

10.3.2 Solving (10.7) with y Fixed to $y = y_k$

With a fixed that satisfies (10.7d), constraint (10.7d) can be neglected. By introducing an upper bound δ_p for the objective function as an auxiliary variable, the problem at hand becomes

$$\min \delta_p \tag{10.8a}$$

$$\text{subject to:}\quad |H(\omega) - H_d(\omega)| \leq \delta_p \;\text{ for } \omega \in \Omega_p \tag{10.8b}$$

$$\int_{\omega_a}^{\pi} |H(\omega)|^2 d\omega \leq e_a \tag{10.8c}$$

$$|H(\omega)| \leq \delta_a \quad \text{for } \omega \in \Omega_a \tag{10.8d}$$

If the desired frequency response assumes the form $H_d(\omega) = e^{-j\tau\omega}A_d(\omega)$, then constraints in (10.8b–10.8d) are reduced to

$$|\boldsymbol{x}^T\hat{\mathbf{c}}(\omega) - A_d(\omega)| \le \delta_p \quad \text{for } \omega \in \Omega_p \tag{10.9a}$$

$$\boldsymbol{x}^T\boldsymbol{Q}\boldsymbol{x} \le e_a \tag{10.9b}$$

$$|\boldsymbol{x}^T\hat{\mathbf{c}}(\omega)| \le \delta_a \quad \text{for } \omega \in \Omega_a \tag{10.9c}$$

where

$$\hat{\mathbf{c}}(\omega) = \mathbf{c}(\omega)\prod_{l=1}^{L}(2\cos\frac{l\omega}{2})^{k_l}$$

$$\boldsymbol{Q} = \int_{\omega_a}^{\pi}\hat{\mathbf{c}}(\omega)\hat{\mathbf{c}}^T(\omega)\,d\omega$$

In realistic implementation, one has to deal with a finite set of constraints and this is accomplished by replacing sets Ω_p and Ω_a with their discrete counterparts $\hat{\Omega}_p = \{\omega_{p1}, \omega_{p2}, \ldots, \omega_{pK_1}\}$ and $\hat{\Omega}_a = \{\omega_{a1}, \omega_{a2}, \ldots, \omega_{aK_2}\}$ respectively, with the frequency grids placed uniformly over the respective bands. This simplifies problem (10-8) to

$$\min \delta_p \tag{10.10a}$$

$$\text{subject to :} \quad |\boldsymbol{x}^T\hat{\mathbf{c}}(\omega) - A_d(\omega)| \le \delta_p \ \text{for } \omega \in \hat{\Omega}_p \tag{10.10b}$$

$$\boldsymbol{x}^T\boldsymbol{Q}\boldsymbol{x} \le e_a \tag{10.10c}$$

$$|\boldsymbol{x}^T\hat{\mathbf{c}}(\omega)| \le \delta_a \quad \text{for } \omega \in \hat{\Omega}_a \tag{10.10d}$$

With respect to variables $(\delta_p, \boldsymbol{x})$, the objective function and constraints (10.10b) and (10.10d) are linear, while (10.10c) is a convex quadratic constraint because $\boldsymbol{Q}$ is positive definite. As a result, (10.10) is a convex problem whose global solution can be computed using reliable and convenient solvers [9, 10]. We denote the solution of (10.10) by $(\delta_k, \boldsymbol{x}_k)$.

10.3.3 Updating y with x Fixed to $x = \boldsymbol{x}_k$

Consider the stopband energy given by

$$J(\boldsymbol{y}) = \int_{\omega_a}^{\pi}|H(\omega)|^2 d\omega = \int_{\omega_a}^{\pi}|\boldsymbol{x}^T\mathbf{c}(\omega)|^2\prod_{l=1}^{L}(4\cos^2\frac{l\omega}{2})^{k_l}d\omega \tag{10.11}$$

where $\boldsymbol{x} = \boldsymbol{x}_k$ is obtained by solving (10.10) and is fixed throughout this step. Our strategy to update $\boldsymbol{y}$ is via minimizing $J(\boldsymbol{y})$ with respect to $\boldsymbol{y}$ subject to several relevant constraints on $\boldsymbol{y}$.

A technical difficulty to utilize continuous optimization to optimize $\boldsymbol{y}$ is that the components of y, must be integers. This problem is overcome by extending the k_l's from the domain of nonnegative integers to that of nonnegative reals. When an optimizing $\boldsymbol{y}$ with non-integer components is obtained, an integer solution can be found by rounding its components to nearest integers.

Noting that for any $a > 0$, the derivative of $f = a^x$ is given by

$$\frac{\partial f}{\partial x} = a^x \log a$$

the gradient and Hessian of $J(\boldsymbol{y})$ can be evaluated by computing

$$\begin{aligned}
\frac{\partial J(\boldsymbol{y})}{\partial k_i} &= \frac{\partial}{\partial k_i} \int_{\omega_a}^{\pi} |\boldsymbol{x}^T \mathbf{c}(\omega)|^2 \prod_{l=1}^{L} \left(4\cos^2\frac{l\omega}{2}\right)^{k_l} d\omega \\
&= \int_{\omega_a}^{\pi} |\boldsymbol{x}^T \mathbf{c}(\omega)|^2 \prod_{l=1, l\neq i}^{L} \left(4\cos^2\frac{l\omega}{2}\right)^{k_l} \frac{\partial}{\partial k_i}\left(4\cos^2\frac{i\omega}{2}\right)^{k_i} d\omega \\
&= \int_{\omega_a}^{\pi} |\boldsymbol{x}^T \mathbf{c}(\omega)|^2 \prod_{l=1, l\neq i}^{L} \left(4\cos^2\frac{l\omega}{2}\right)^{k_l} \left(4\cos^2\frac{i\omega}{2}\right)^{k_i} \\
&\quad \cdot \log\left(4\cos^2\frac{i\omega}{2}\right) d\omega \\
&= \int_{\omega_a}^{\pi} |H(\omega)|^2 \log\left(4\cos^2\frac{i\omega}{2}\right) d\omega
\end{aligned} \tag{10.12a}$$

and

$$\begin{aligned}
\frac{\partial^2 J(\boldsymbol{y})}{\partial k_i \partial k_j} &= \frac{\partial^2}{\partial k_i \partial k_j} \int_{\omega_a}^{\pi} |\boldsymbol{x}^T \mathbf{c}(\omega)|^2 \prod_{l=1}^{L} \left(4\cos^2\frac{l\omega}{2}\right)^{k_l} d\omega \\
&= \int_{\omega_a}^{\pi} |\boldsymbol{x}^T \mathbf{c}(\omega)|^2 \prod_{l=1, l\neq i,j}^{L} \left(4\cos^2\frac{l\omega}{2}\right)^{k_l} \\
&\quad \cdot \frac{\partial^2}{\partial k_i \partial k_j}\left[\left(4\cos^2\frac{i\omega}{2}\right)^{k_i}\left(4\cos^2\frac{j\omega}{2}\right)^{k_j}\right] d\omega
\end{aligned} \tag{10.12b}$$

$$= \int_{\omega_a}^{\pi} |\boldsymbol{x}^T \boldsymbol{c}(\omega)|^2 \prod_{l=1, l\neq i,j}^{L} \left(4\cos^2 \frac{l\omega}{2}\right)^{k_l} \left(4\cos^2 \frac{i\omega}{2}\right)^{k_i} \left(4\cos^2 \frac{j\omega}{2}\right)^{k_j}$$

$$\cdot \log\left(4\cos^2 \frac{i\omega}{2}\right) \log\left(4\cos^2 \frac{j\omega}{2}\right) d\omega$$

$$= \int_{\omega_a}^{\pi} |H(\omega)|^2 \log\left(4\cos^2 \frac{i\omega}{2}\right) \log\left(4\cos^2 \frac{j\omega}{2}\right) d\omega$$

respectively. Note that with an arbitrary column vector $\boldsymbol{v}$ of length L

$$\begin{aligned} \boldsymbol{v}^T \nabla^2 J(\boldsymbol{y})\boldsymbol{v} &= \sum_{i=1}^{L}\sum_{j=1}^{L} \frac{\partial^2 J(\boldsymbol{y})}{\partial k_i \partial k_j} v_i v_j \\ &= \int_{\omega_a}^{\pi} |H(\omega)|^2 \sum_{i=1}^{L}\sum_{j=1}^{L} \log\left(4\cos^2 \frac{i\omega}{2}\right) \\ &\quad \cdot \log\left(4\cos^2 \frac{j\omega}{2}\right) v_i v_j d\omega \\ &= \int_{\omega_a}^{\pi} |H(\omega)|^2 \left[\sum_{i=1}^{L} \log\left(4\cos^2 \frac{i\omega}{2}\right) v_i\right] \\ &\quad \cdot \left[\sum_{j=1}^{L} \log\left(4\cos^2 \frac{j\omega}{2}\right) v_j\right] d\omega \\ &= \int_{\omega_a}^{\pi} |H(\omega)|^2 \left[\sum_{i=1}^{L} \log\left(4\cos^2 \frac{i\omega}{2}\right) v_i\right]^2 d\omega \geq 0 \end{aligned} \tag{10.13}$$

hence $J(\boldsymbol{y})$ is *convex* although it is a rather complicated function as shown in (10.11). This motivates a convex quadratic approximation of $J(\boldsymbol{y})$ as

$$\hat{J}(\boldsymbol{y}, \boldsymbol{y}_k) = J(\boldsymbol{y}_k) + \boldsymbol{d}_k^T \nabla J(\boldsymbol{y}_k) + \frac{1}{2}\boldsymbol{d}_k^T \nabla^2 J(\boldsymbol{y}_k)\boldsymbol{d}_k \tag{10.14}$$

where $\boldsymbol{d}_k = \boldsymbol{y} - \boldsymbol{y}_k$. We update $\boldsymbol{y}$ by minimizing $\hat{J}(\boldsymbol{y}, \boldsymbol{y}_k)$ subject to several constraints. These include an upper bound on group delay τ as seen in (10.7d) and nonnegativeness of the components of $\boldsymbol{y}$. An additional constraint is imposed to ensure the performance of the C-filter over the passband, especially at passband edge ω_p:

$$1 - d_p \le |H(\omega)|_{\omega=\omega_p} = |\boldsymbol{x}_k^T \boldsymbol{c}(\omega_p)| \prod_{l=1}^{L} |2\cos\frac{l\omega_p}{2}|^{k_l} \le 1 + d_p \qquad (10.15)$$

with a small $d_p > 0$. This constraint is highly nonlinear with respect to $\boldsymbol{y}$, but fortunately it is equivalent to

$$\log(1 - d_p) \le c + \sum_{l=1}^{L} k_l \log|2\cos\frac{l\omega_p}{2}| \le \log(1 + d_p) \qquad (10.16)$$

with $c = \log|\boldsymbol{x}_k^T \boldsymbol{c}(\omega_p)|$, which is *linear* with respect to $\boldsymbol{y}$. Based on above analysis, we propose to update $\boldsymbol{y}$ by solving the *convex* problem

$$\min \frac{1}{2}(\boldsymbol{y} - \boldsymbol{y}_k)^T \nabla^2 J(\boldsymbol{y}_k)(\boldsymbol{y} - \boldsymbol{y}_k) + (\boldsymbol{y} - \boldsymbol{y}_k)^T \nabla J(\boldsymbol{y}_k) \qquad (10.17a)$$

$$\text{subject to: } \text{(10.7d)}, \boldsymbol{y} \ge \boldsymbol{0}, \text{ and (10.16)} \qquad (10.17b)$$

and then rounding its solution to a nearest integer solution $\boldsymbol{y}_{k+1}$.

10.3.4 Summary of the Algorithm

The algorithm described above is outlined as Algorithm 1.

Algorithm 1 for C-Filters

Step 1 input $\boldsymbol{y}_0, (N, L, D), \omega_p, \omega_a, \delta_a, e_a, d_p$, and K.
Step 2 for $k = 0, 1, \ldots$
(i) fix $\boldsymbol{y} = \boldsymbol{y}_k$ and solve (10.10) for $\boldsymbol{x}_k$;
(ii) fix $\boldsymbol{x} = \boldsymbol{x}_k$, perform K iterations of (10.17) for $\boldsymbol{y}_{k+1}$;
(iii) if $\boldsymbol{y}_k \ne \boldsymbol{y}_{k+1}$, set $k = k + 1$ and repeat from Step (i), otherwise go to Step 3.
Step 3 output $\boldsymbol{x}^* = \boldsymbol{x}_k$, $\boldsymbol{y}^* = \boldsymbol{y}_k$.

We remark that because both (10.10) and (10.17) are convex problems, globally optimal iterates $\boldsymbol{x}_k$ and $\boldsymbol{y}_k$ can be calculated efficiently.

Since the objective function in (10.17a) represents the filter's stopband energy, minimizing it tends to increase some powers in the shaping filter until the group delay of $H_s(z)$ reaches upper bound D in constraint (10.7d). When this occurs, iterate $\boldsymbol{y}_k$ remains unchanged and Algorithm 1 terminates.

10.4 Design Example and Comparisons

We illustrate the design method proposed above by applying Algorithm 1 to design a narrow-band lowpass C-filter with linear phase response and sharp transition band specified by $\omega_p = 0.1\pi$ and $\omega_a = 0.11\pi$. The length of $H_p(z)$ was set to $N = 519$ and the peak gain of the C-filter in the stopband was set to be no greater than -60 dB, i.e., $\delta_a = 0.001$.

The performance of the filters were evaluated in terms of peak-to-peak passband ripple A_p (in dB), minimum stopband attenuation A_a (in dB), stopband energy E_a, the number of multiplications M per output sample, and group delay τ. With $L = 7$, $D = 18$, $\boldsymbol{y}_0 = [1\ 1\ 1\ 1\ 1\ 1\ 1]^T$, $d_p = 0.08$, and $e_a = 5 \times 10^{-4}$, problem (10.10) was solved and its solution $\boldsymbol{x}_0$ together with $\boldsymbol{y}_0$ defines a C-filter achieving $A_p = 0.1681$, $A_a = 60$, and $E_a = 2.15 \times 10^{-8}$. With $\boldsymbol{x}_0$ held fixed, $K = 5$ iterations of (10.17) were performed to obtain an integer solution $\boldsymbol{y}_1 = [1\ 1\ 1\ 1\ 1\ 1\ 2]^T$. Since $\boldsymbol{y}_1 \neq \boldsymbol{y}_0$, problem (10.10) was solved again with $\boldsymbol{y}$ fixed to $\boldsymbol{y}_1$, where e_a was adjusted in order obtain a solution $\boldsymbol{x}_1$ with practically the same peak gain in the stopband and stopband energy as $\boldsymbol{x}_0$ so that the two iterates $\boldsymbol{x}_0$ and $\boldsymbol{x}_1$ can be compared with each other in terms of peak passband ripple. With $e_a = 2.95 \times 10^{-4}$, the C-filter specified by $(\boldsymbol{x}_1, \boldsymbol{y}_1)$ achieved $A_p = 0.1389$, $A_a = 60.04$ and $E_a = 2.12 \times 10^{-8}$. We then run (10.17) again with $\boldsymbol{x}$ fixed to $\boldsymbol{x}_1$ and $K = 5$, this yields an integer solution $\boldsymbol{y}_2 = [1\ 1\ 1\ 1\ 1\ 1\ 2]^T$. Since $\boldsymbol{y}_2 = \boldsymbol{y}_1$, the algorithm is terminated and $(\boldsymbol{x}_1, \boldsymbol{y}_1)$ is claimed as the solution. The amplitude responses of $H_p(z)$ and $H(z)$ are shown in Figures 10.3(a) and 10.3(b), respectively, and the amplitude response of $H(z)$ in the passband is depicted in Figure 10.3(c).

The filters that are most relevant to the C-Filters addressed in this chapter are linear-phase EPLSS FIR filters with constrained peak gain in stopband and linear-phase FIR filters with equiripple passbands and stopbands obtained using the Parks-McClellan (P-M) algorithm. For comparison purposes, an EPLSS lowpass filter and a P-M lowpass filter with the same ω_p and ω_a as those in the C-filter were designed, both satisfy the same peak stopband gain of -60 dB as the C-filter. The evaluation results are summarized in Table 10.1.

It is observed that the C-filter outperforms the two conventional filters at the cost of a slight increase in group delay. Relative to the EPLSS filter, the C-filter offers reduced passband ripple and requires less number of multiplications for implementation. On comparing with the P-M filter, the C-filter also offers smaller passband ripple, reduced number of multiplications, and considerably smaller stopband energy.

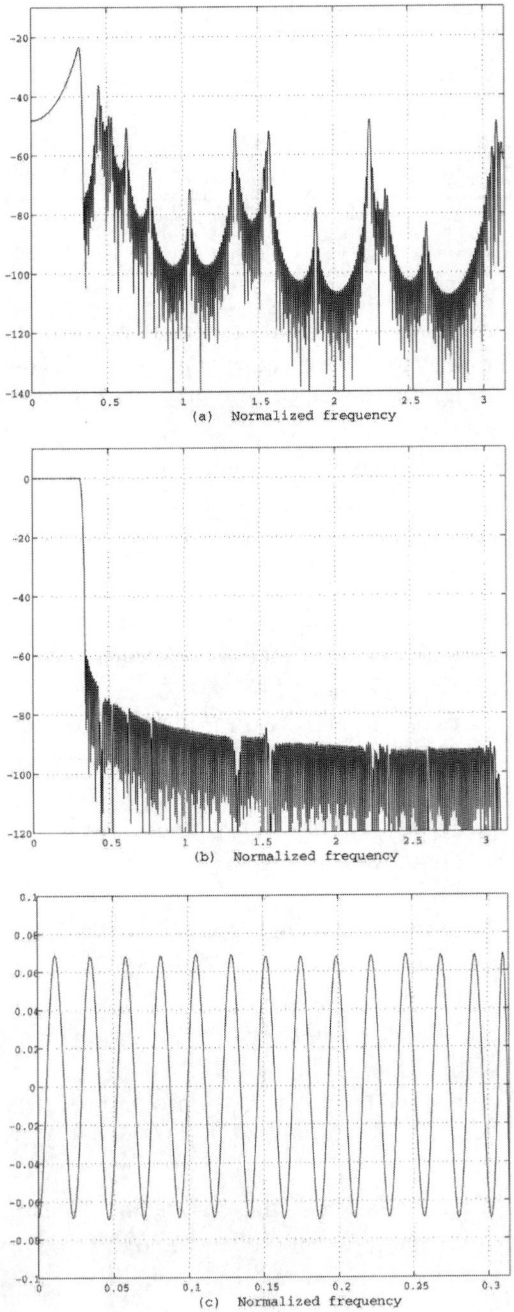

Figure 10.3 Amplitude response of (a) the prototype filter $H_p(z)$, (b) the C-filter $H(z)$ and (c) the C-filter $H(z)$ over the passband.

Table 10.1 Comparisons of C-filter with EPLSS and P-M filters

Filters	N	A_p	A_a	E_a	M	τ
C-filter	**519**	**0.1389**	**60.04**	$\mathbf{2.12 \times 10^{-8}}$	**260**	277.5
EPLSS	551	0.1465	60.03	2.13×10^{-8}	276	275
P-M	531	0.1463	60.03	1.36×10^{-6}	266	**265**

10.5 Summary

We have addressed the design of a class of composite filters by an alternating convex optimization strategy to achieve equiripple passband and least-squares stopband subject to peak-gain constraint. A design example is presented to illustrate the proposed algorithm and to demonstrate the performance of C-filter relative to the conventional EPLSS and P-M filters.

References

[1] Y. Neuvo, C. Y. Dong, and S. K. Mitra, "Interpolated finite impulse response filters," *IEEE Trans. Acoust., Speech, Signal Process.*, Vol. ASSP-32, pp. 563–570, Jun. 1984.

[2] T. Saramäki, Y. Neuvo, and S. K. Mitra, "Design of computationally efficient interpolated FIR filters," *IEEE Trans. Circuits Syst.*, vol. 35, pp. 70–88, Jan. 1988.

[3] Y. C. Lim, "Frequency-response masking approach for the synthesis of sharp linear phase digital filters," *IEEE Trans. Circuits Syst.*, vol. 33, pp. 357–364, Apr. 1986.

[4] W.-S. Lu and T. Hinamoto, "A unified approach to the design of interpolated and frequency-response-masking FIR filters," *IEEE Trans. Circuits Syst. I*, vol. 63, no. 12, pp. 2257–2266, Dec. 2016.

[5] S. K. Mitra, *Digital Signal Processing – A Computer-Based Approach*, 3rd ed., McGraw Hill, 2006.

[6] D. Shiung, Y.-Y. Yang, and C.-S. Yang, "Improving FIR filters by using cascade techniques," *IEEE Signal Processing Mag.*, vol. 33, pp. 108–114, May 2016.

[7] D. Shiung, Y.-Y. Yang, and C.-S. Yang, "Cascading tricks for designing composite filters with sharp transition bands," *IEEE Signal Processing Mag.*, Vol. 33, No. 1, pp. 151–157 and 162, Jan. 2016.

[8] J. W. Adams, "FIR digital filters with least-squares stopbands subject to peak-gain constraints," *IEEE Trans. Circuits Syst.*, vol. 39, no. 4, pp. 376–388, Apr. 1991.

[9] SeDuMi1.3: http://sedumi.ie.lehigh.edu/

[10] CVX2.1: http://cvxr.com/cvx/

11

Finite Word Length Effects

11.1 Preview

Algorithms for linear filtering are realized as programs for general-purpose digital computers or with special-purpose digital hardware in which sequence values and coefficients are stored in a binary format with finite-length registers. When a digital filter is designed with high accuracy and its coefficients are quantized, the characteristics of the resulting digital filter differ inevitably from the original design. For example, the coefficient quantization may alter a stable filter to unstable one. In addition, when a sequence to be processed is obtained by sampling a band-limited analog signal, the A/D converter produces only a finite number of possible values for each sample.

In the implementations of a digital filter, numbers are stored in finite-length registers. As a result, if sequence values and coefficients cannot be accommodated in the available registers then they must be quantized before being stored. There are three types of errors in number quantization:

(1) Coefficient-quantization errors
(2) Product-quantization errors
(3) Input-quantization errors

If *coefficient-quantization* is applied, then the frequency characteristic of the resulting digital filter might differ inevitably from the desired one. *Product-quantization errors* occur at the outputs of multipliers. For example, a b-bit data sample multiplied by a b-bit coefficient results in a product that is $2b$-bit long. If the result of arithmetic operations is not quantized in a recursive realization of a digital filter, the number of bits will increase indefinitely as data processing continues. Since a uniform register length must be used in practice throughout the filter, each multiplier must be rounded or truncated before processing continues. The errors induced by rounding or truncation propagate through the filter and rise output noise referred to output *roundoff noise*. *Input-quantization errors* occur in applications in which digital filter is utilized to process continuous-time signals.

In this chapter, we review the fixed-point and floating-point arithmetic of binary numbers and the two's complement representation of negative numbers. Limit cycles—overflow oscillations, scaling fixed-point digital filters to prevent overflow, roundoff noise, and coefficient sensitivity will be addressed. In addition, the response of a finite-word-length (FWL) state-space description and methods for obtaining limit cycle-free realization will also be examined.

11.2 Fixed-Point Arithmetic

The binary number generated by the A/D converter is assumed to be a fixed point number. Among the three forms of number representation, namely, signed magnitude, one's complement, or two's complement in fixed-point arithmetic, two's complement is most often used due to its easy implementation. A two's complement representation of a real number x is given by

$$x = \Delta\Big(-b_0 + \sum_{i=1}^{\infty} b_i 2^{-i}\Big), \qquad -\Delta \leq x \leq \Delta \tag{11.1}$$

where b_i for $i \geq 0$ is unity or zero, the first bit b_0 is the sign bit, namely, $b_0 = 1$ if $x < 0$ and $b_0 = 0$ if $x > 0$, and the value of Δ is arbitrary.

Using a finite-length register of $L + 1$ bits, the actual number stored is quantized to $Q[x]$ where

$$Q[x] = \Delta\Big(-b_0 + \sum_{i=1}^{L} b_i 2^{-i}\Big) \tag{11.2}$$

Fixed-point numbers are stored in registers, as illustrated in Figure 11.1. The quantized representation of x, denoted by $Q[x]$, must be an integral multiple of the smallest quantum q with

$$q = \Delta 2^{-L} \tag{11.3}$$

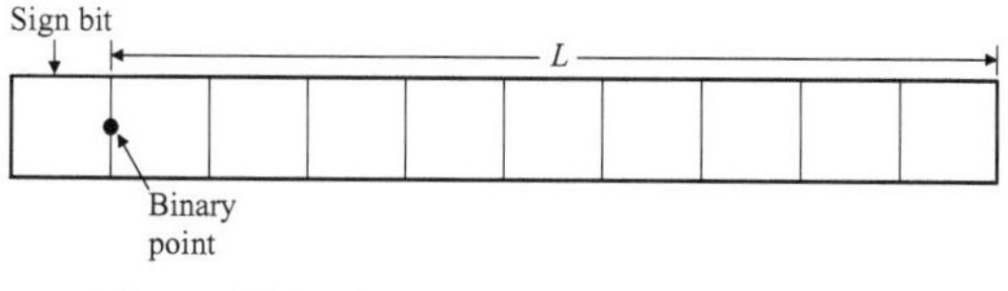

Figure 11.1 Storage of fixed-point numbers.

which is the finest separation between the 2^{L+1} numbers we can represent with $L+1$ bits. The number q can be arbitrarily defined by choosing an appropriate Δ, and is called *quantization step size*.

The error between the real number x and its finite binary representation is given by

$$e = x - Q[x] \tag{11.4}$$

It shall be assumed that in forming $Q[x]$ the number x is rounded to the nearest integer multiple of q. As a result, the quantizer that defines the relationship between x and $Q[x]$ has a characteristic as shown in Figure 11.2 where Δ is normalized to be unity, i.e., $\Delta = 1$.

From Figure 11.2, it is observed that the error e in (11.4) due to quantization lies between $-q/2$ and $q/2$, that is,

$$-\frac{q}{2} \leq e \leq \frac{q}{2} \tag{11.5}$$

Theoretical studies and numerical experiments have shown [1, 2] that the error e can be approximated as a random noise, uniformly distributed on $[-q/2,\ q/2]$ with probability density shown in Figure 11.3. Evidently, the

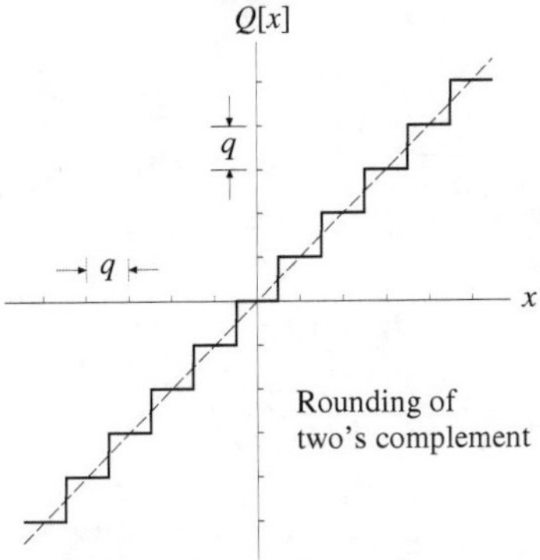

Figure 11.2 Quantizer characteristic for rounding of two's complement numbers.

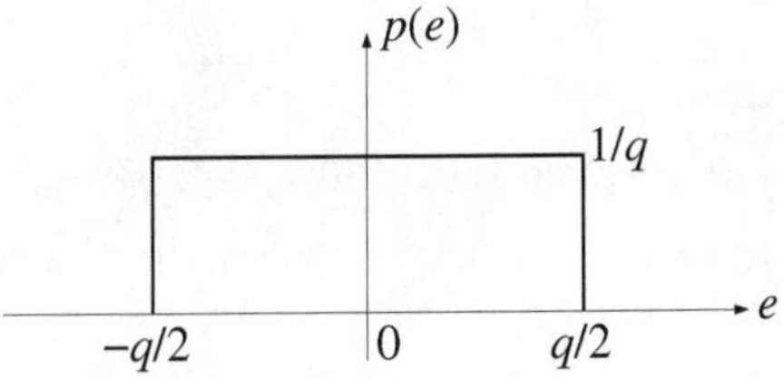

Figure 11.3 Probability density function of the quantization error.

mean value of the error e is zero, and the variance associated with this error distribution is given by

$$\sigma_e^2 = E[e^2] = \int_{-\frac{q}{2}}^{\frac{q}{2}} \frac{1}{q} e^2 de = \frac{q^2}{12} \tag{11.6}$$

If a number x is larger than the largest number representable or smaller than the smallest number representable, an overflow occurs. Overflows generally create large errors and thus must be avoided.

In two's complement representations, if nothing is done after an overflow, the overflow characteristic is periodic and a small overflow causes an error of approximately 2Δ, as shown in Figure 11.4 (a). Another way of handling overflow is to use a saturation characteristic where the overflowed register is reset to the largest or smallest number representable. However, it is not as easy to implement as the two's complement characteristic. The saturation overflow characteristic is shown in Figure 11.4 (b).

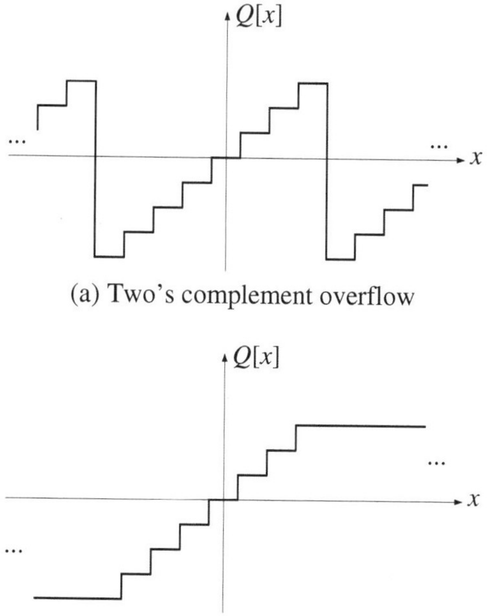

(a) Two's complement overflow

(b) Saturation overflow

Figure 11.4 Overflow characteristics. (a) Two's complement overflow characteristic. (b) Saturation overflow characteristic.

11.3 Floating-Point Arithmetic

Two basic disadvantages exist in a fixed-point arithmetic. First, the range of numbers we can handle is small. For example, the smallest number is -1 and the largest is $1-2^{-L}$ provided $\Delta = 1$ in the two's complement representation. Second, there is a tendency for the percentage error generated by truncation or rounding to increase as the magnitude of the number is decreased. These problems can be addressed to a large extent by employing a floating-point arithmetic.

In the floating-point arithmetic, a number N is expressed as

$$N = M \times 2^I \tag{11.7}$$

where $\frac{1}{2} \leq M < 1$, and I is an integer. M and I are referred to as the *mantissa* and *exponent*, respectively. Negative numbers are treated in the same manner as in fixed-point arithmetic. Floating-point numbers are stored in registers, as shown in Figure 11.5. The register is subdivided into two segments: one for the signed mantissa and another for the signed exponent.

Floating-point arithmetic leads to increased dynamic range and improved precision of processing. However, unlike fixed-point arithmetic, floating-point arithmetic also leads to increased cost of hardware and reduced speed of processing because both the mantissa and exponent must be manipulated in hardware.

11.4 Limit Cycles—Overflow Oscillations

In the sequel two's complement numbers are assumed to be used and the associated overflow characteristic is assumed to be employed. In other words, overflows are not explicitly detected and not corrected in some manner. With a two's complement overflow characteristic, a disastrous effect called the *overflow oscillations* can occur after an internal overflow. For discussion purposes, we will disregard the roundoff error caused by the quantizer and focus only on the overflow nonlinearity characteristic. Hence, two's complement and saturation overflow characteristics are depicted in Figure 11.6.

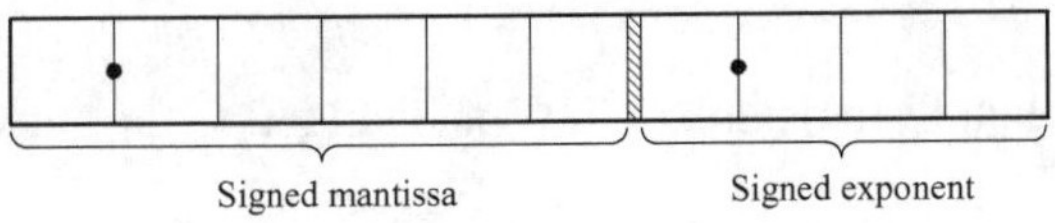

Figure 11.5 Storage of floating-point numbers.

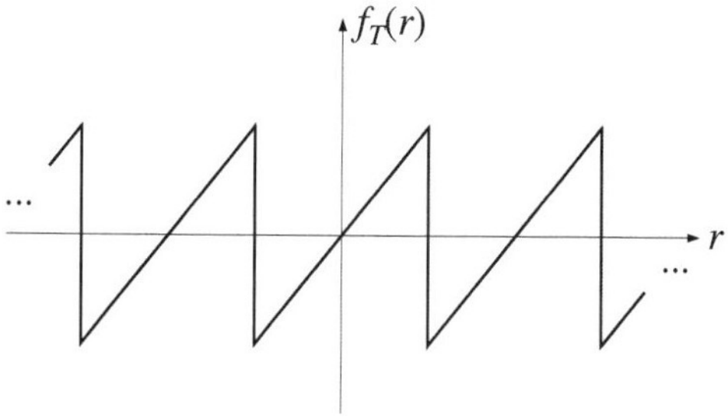

(a) Two's complement overflow

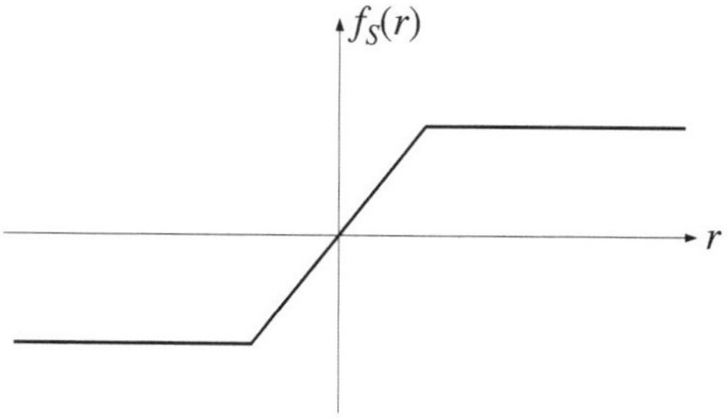

(b) Saturation overflow

Figure 11.6 Overflow characteristics. (a) Two's complement overflow characteristic. (b) Saturation overflow characteristic.

Assuming that the range of the quantizer is $(-1,\ 1)$, an overflow function f must satisfy

$$f(x) = x \text{ for } |x| < 1, \text{ and } |f(x)| \leq 1 \tag{11.8}$$

The above relations imply that the magnitude of $f(x)$ never exceeds that of x, namely,

$$|f(x)| \leq |x| \tag{11.9}$$

This is an essential property in characterizing an overflow function f.

In order to study overflow oscillations in a state-space model, we consider an idealized state equation described by

$$\boldsymbol{x}(k+1) = \boldsymbol{A}\boldsymbol{x}(k) \tag{11.10}$$

In the case of finite-length registers, (11.10) is changed to the form

$$\boldsymbol{x}(k+1) = \boldsymbol{f}\big[\boldsymbol{A}\boldsymbol{x}(k)\big] \tag{11.11}$$

with

$$f[x] = [f(x_1), f(x_2), \cdots, f(x_n)]^T$$

where f is the overflow characteristic. If the filter in (11.10) is stable, $x(k)$ approaches $\mathbf{0}$ as $k \to \infty$ in (11.10) for any initial state-variable vector $x(0)$. However, this may not occur for a system described by (11.11) which depends on A and f.

We now define the norm of a vector x as

$$||x|| = \sqrt{x^T x} = \Big(\sum_{i=1}^{n} x_i^2\Big)^{\frac{1}{2}} \tag{11.12}$$

With this definition of vector norm, the norm of a matrix A can be defined as

$$||A|| = \max_{x \neq 0} \frac{||Ax||}{||x||} = \max_{x \neq 0}\Big(\frac{x^T A^T A x}{x^T x}\Big)^{\frac{1}{2}} \tag{11.13}$$

This norm of A is the maximum increase in the length of the vector. If $||A|| < 1$, then all vectors x decrease in length under multiplication by A. In other words, if we can obtain a structure with a system matrix A such that $||A|| < 1$, then zero-input overflow oscillations will not occur.

For example, consider the following second-order transfer function:

$$\begin{aligned} H(z) &= \frac{\gamma - j\theta}{z + \alpha - j\beta} + \frac{\gamma + j\theta}{z + \alpha + j\beta} \\ &= \frac{2\gamma z + 2(\gamma\alpha + \theta\beta)}{(z+\alpha)^2 + \beta^2} \\ &= \begin{bmatrix} 1 & 1 \end{bmatrix} \left[z I_2 - \begin{bmatrix} -\alpha & \beta \\ -\beta & -\alpha \end{bmatrix} \right]^{-1} \begin{bmatrix} \gamma - \theta \\ \gamma + \theta \end{bmatrix} \end{aligned} \tag{11.14}$$

This transfer function can be realized by a state-space model

$$\begin{aligned} x(k+1) &= Ax(k) + bu(k) \\ y(k) &= cx(k) \end{aligned} \tag{11.15}$$

where $x(k)$ is a 2×1 state-variable vector, $u(k)$ is a scalar input, $y(k)$ is a scalar output, and

$$A = \begin{bmatrix} -\alpha & \beta \\ -\beta & -\alpha \end{bmatrix}, \qquad b = \begin{bmatrix} \gamma - \theta \\ \gamma + \theta \end{bmatrix}, \qquad c = \begin{bmatrix} 1 & 1 \end{bmatrix}$$

Since the relation

$$A^T A = AA^T = (\alpha^2 + \beta^2) I_2 \tag{11.16}$$

holds, it follows that

$$||A|| = \max_{x \neq 0} \left((\alpha^2 + \beta^2) \frac{x^T I_2 x}{x^T x} \right)^{\frac{1}{2}} = \sqrt{\alpha^2 + \beta^2} \tag{11.17}$$

On the other hand, from (11.14) we observe that the filter is stable if its poles are all strictly inside the unit circle, i.e., $\sqrt{\alpha^2 + \beta^2} < 1$. This in conjunction with (11.17) implies that the filter in (11.15) is stable if and only if $||A||$ is less than unity. In the literature, matrices satisfying $A^T A = AA^T$ are known as *normal matrices*, and state-space digital filters with normal system matrix A are said to be *normal* digital filters. From (11.16), we see that (11.15) is a normal digital filter. In general, it also holds that stable normal digital filters are free of zero-input overflows.

11.5 Scaling Fixed-Point Digital Filters to Prevent Overflow

Internal overflows cause large errors and therefore must be prevented. This can be achieved by appropriately scaling the realization. Scaling constraints the numerical values of internal variables in the filter to remain in a range appropriate for the hardware. The range of a filter variable is necessarily limited due to the use of finite-length registers. For fixed-point number representations, it can be expressed as a bound on internal variables $v(k)$ such that

$$|v(k)| \leq \Delta \tag{11.18}$$

where Δ is related to the quantization step size q in (11.3) and usually assumed to be unity.

A typical way of scaling is illustrated using Figures 11.7 and 11.8 where $U(z)$, $Y(z)$, $V(z)$, and $V'(z)$ are the z-transforms of input $u(k)$, output $y(k)$, internal variables $v(k)$ and $v'(k)$, respectively.

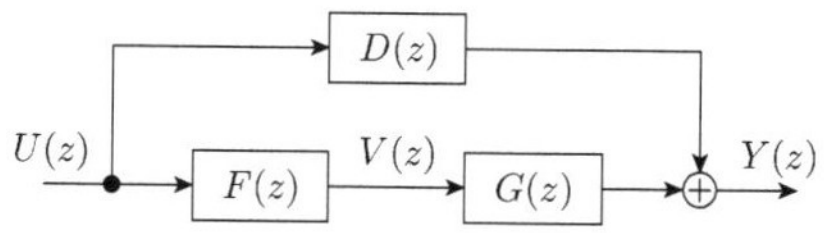

Figure 11.7 A system before scaling.

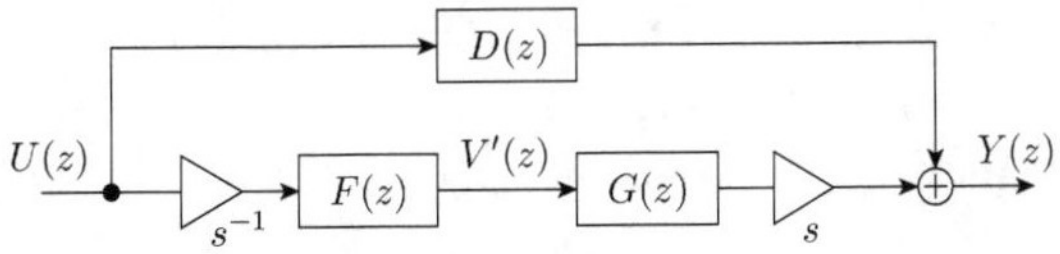

Figure 11.8 A system after scaling where s is a scaling factor.

In these figures, the transfer function of the system described by

$$H(z) = G(z)F(z) + D(z) \tag{11.19}$$

is unchanged before and after the scaling. However, the internal variable is changed from $v(k)$ to $v'(k)$. Suppose the impulse response of the transfer function $F(z)$ is denoted by $f(i)$ for $i = 0, 1, 2, \cdots$, then we can write

$$v(k) = \sum_{i=0}^{\infty} f(i)u(k-i) \tag{11.20}$$

Four typical inputs will be considered in the sequel.
(1) If the input is sinusoidal: $u(k) = \cos(k\omega)$, then

$$|v(k)| \leq \max_{\omega} |F(e^{j\omega})| \tag{11.21}$$

(2) If the input is bounded: $|u(k)| \leq 1$ for any k, then

$$|v(k)| \leq \sum_{i=0}^{\infty} |f(i)||u(k-i)| \leq \sum_{i=0}^{\infty} |f(i)| = ||\boldsymbol{f}||_1 \tag{11.22}$$

(3) If the input has finite energy: $\sum_{i=-\infty}^{k} u(i)^2 \leq 1$, then

$$|v(k)| \leq \sum_{i=0}^{\infty} |f(i)u(k-i)| \leq \left[\sum_{i=0}^{\infty} f(i)^2\right]^{\frac{1}{2}} = ||\boldsymbol{f}||_2 \tag{11.23}$$

(4) If the input is a white Gaussian with zero mean and unit variance, then

$$\sqrt{E[v^2(k)]} = \left[\sum_{i=0}^{\infty} f(i)^2\right]^{\frac{1}{2}} = ||\boldsymbol{f}||_2 \tag{11.24}$$

The first three cases stand for actual bounds on the range of variable $v(k)$ for each given input sequence. The last case is about a standard deviation of the random variable. It is known that

$$||\boldsymbol{f}||_2 \leq \max_{\omega} |F(e^{j\omega})| \leq ||\boldsymbol{f}||_1 \tag{11.25}$$

There are commonly used scaling rules to impose a certain constraint on the variable $v(k)$ in (11.18). For example, the l_1-scaling and l_2-scaling on the impulse response for internal variables are given by

$$||\boldsymbol{f}||_1 = \sum_{i=0}^{\infty} |f(i)| = 1 \tag{11.26}$$

and

$$||\boldsymbol{f}||_2 = \delta\Big[\sum_{i=0}^{\infty} f(i)^2\Big]^{\frac{1}{2}} = 1 \tag{11.27}$$

respectively, where the parameter δ is subjectively chosen. The scaling factor s in Figure 11.8 is then chosen to meet the scaling rule. For example,

$$\begin{aligned} s &= ||\boldsymbol{f}||_1 = \sum_{i=0}^{\infty} |f(i)| \quad \text{for } l_1\text{-scaling} \\ s &= \delta||\boldsymbol{f}||_2 = \delta\Big[\sum_{i=0}^{\infty} f(i)^2\Big]^{\frac{1}{2}} \quad \text{for } l_2\text{-scaling} \end{aligned} \tag{11.28}$$

11.6 Roundoff Noise

When an IIR digital filter is implemented in hardware, typically three basic operations, namely, multiplication by constants (the filter parameters), accumulation of the products, and storage into memory, are involved. Since the multiplications always increase the number of bits required to represent the products, the results of accumulations inside the filter must be quantized eventually. Suppose two B-bit numbers are multiplied together, the product is $2B$-bits long. In the case where the quantization is performed by rounding, the error caused by this quantization of internal accumulation is called *roundoff noise*.

The model used to represent roundoff noise is illustrated in Figure 11.9 where the quantizer is replaced by an additive white noise source e of variance $q^2/12$ with the quantization step size q.

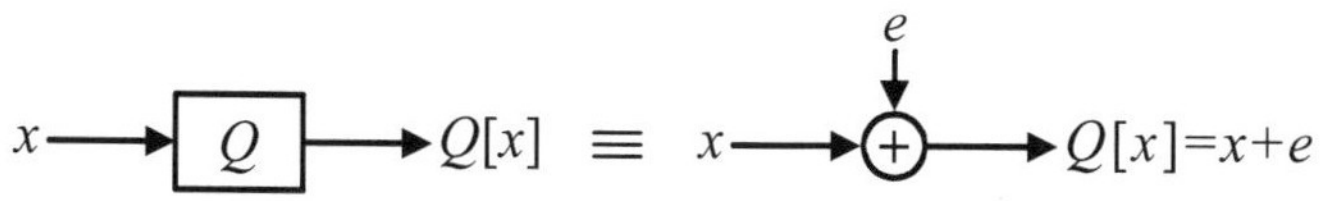

Figure 11.9 A linear equivalent model of internal quantization of product.

As shown in Figure 11.3, the roundoff noise e is assumed to be uniformly distributed on $[-q/2,\, q/2]$ with zero mean. This model also assumes that noise from different accumulators is uncorrelated, and that each noise source is uncorrelated with the input.

11.7 Coefficient Sensitivity

The quantization of filter parameters causes another effects due to the use of finite-length registers. This is proved by a deterministic change in the input-output characteristic of the filter. The effects of coefficient quantization can be evaluated using the differentiation of the transfer function with respect to the filter parameters. If serious changes in the input-output characteristic of the filter are caused by the quantization of coefficients under a certain FWL, coefficient word lengths might be lengthened accordingly.

Consider a transfer function $H(z)$ that contains N parameters $\{p_1, p_2, \cdots, p_N\}$. Let $\{\tilde{p}_i\}$ be the FWL version of $\{p_i\}$, where $\tilde{p}_i = p_i + \Delta p_i$ with Δp_i the parameter perturbation, and let $\tilde{H}(z)$ be the transfer function associated with perturbed parameters $\{\tilde{p}_i\}$. The first-order approximation of $\tilde{H}(z)$ then gives

$$\tilde{H}(z) = H(z) + \Delta H(z) \tag{11.29}$$

where $\Delta H(z)$ will be

$$\Delta H(z) = \sum_{i=1}^{N} \frac{\partial H(z)}{\partial p_i} \Delta p_i$$

Evidently, smaller $\partial H(z)/\partial p_i$ for $i = 1, 2, \cdots, N$ yields smaller transfer-function error $\Delta H(z)$. For a fixed-point implementation with B bits, the parameter perturbations are considered to be independently uniformly distributed random variables within the range $[-2^{-B-1}, 2^{-B-1}]$. Under the circumstances, a measure of the transfer function error can be statistically defined as

$$\sigma^2_{\Delta H} = \frac{1}{2\pi j} \oint_{|z|=1} E[|\Delta H(z)|^2] \frac{dz}{z} \tag{11.30}$$

where $E[\cdot]$ denotes the ensemble average operation. Since $\{\Delta p_i\}$ are independent uniformly distributed random variables, it follows that

$$E[|\Delta H(z)|^2] = \sum_{i=1}^{N} \left| \frac{\partial H(z)}{\partial p_i} \right|^2 \sigma^2 \tag{11.31}$$

where $\sigma^2 = E[(\Delta p_i)^2] = 2^{-2B}/12$. Equation (11.31) establishes an analytic relationship between variations in the transfer function induced by an FWL realization and parameter sensitivity.

11.8 State-Space Descriptions with Finite Word Length

Consider a stable, controllable and observable nth-order state-space digital filter $(\boldsymbol{A}, \boldsymbol{b}, \boldsymbol{c}, d)_n$ described by

$$\begin{aligned} \boldsymbol{x}(k+1) &= \boldsymbol{A}\boldsymbol{x}(k) + \boldsymbol{b}u(k) \\ y(k) &= \boldsymbol{c}\boldsymbol{x}(k) + du(k) \end{aligned} \tag{11.32}$$

where $\boldsymbol{x}(k)$ is an $n \times 1$ state-variable vector, $u(k)$ is a scalar input, $y(k)$ is a scalar output, and $\boldsymbol{A}, \boldsymbol{b}, \boldsymbol{c}$ and d are real constant matrices of appropriate dimensions. A block-diagram of the state-space model in (11.32) is depicted in Figure 11.10.

Taking the finite-precision nature of computer arithmetic into account, an FWL implementation of (11.32) can be obtained as

$$\begin{aligned} \tilde{\boldsymbol{x}}(k+1) &= [\boldsymbol{A} + \Delta\boldsymbol{A}]\tilde{\boldsymbol{x}}(k) + [\boldsymbol{b} + \Delta\boldsymbol{b}]u(k) + \boldsymbol{\alpha}(k) \\ \tilde{y}(k) &= [\boldsymbol{c} + \Delta\boldsymbol{c}]\tilde{\boldsymbol{x}}(k) + [d + \Delta d]u(k) + \beta(k) \end{aligned} \tag{11.33}$$

where $\tilde{\boldsymbol{x}}(k)$ is an actual state-variable vector, $\tilde{y}(k)$ is an actual output, $\Delta\boldsymbol{A}, \Delta\boldsymbol{b}, \Delta\boldsymbol{c}$ and Δd denote the quantization errors of coefficient matrices $\boldsymbol{A}, \boldsymbol{b}, \boldsymbol{c}$ and d, respectively, and $\boldsymbol{\alpha}(k)$ and $\beta(k)$ are an $n \times 1$ roundoff noise vector and a roundoff noise caused by quantization after multiplications and additions associated with $(\boldsymbol{A}, \boldsymbol{b})$ and $(\boldsymbol{c}, d)$, respectively. A block-diagram of the actual state-space model in (11.33) is depicted in Figure 11.11.

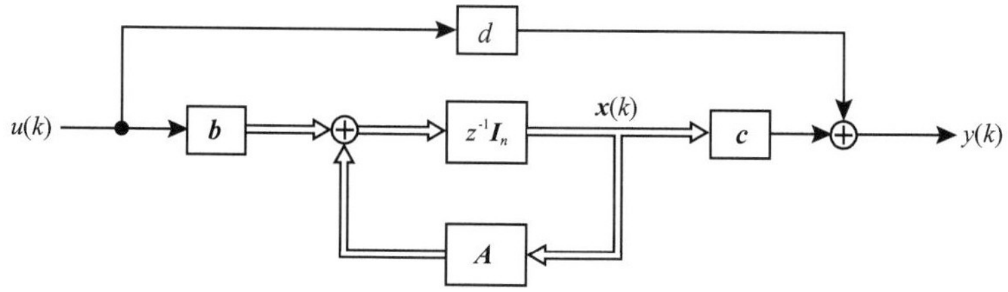

Figure 11.10 A state-space model.

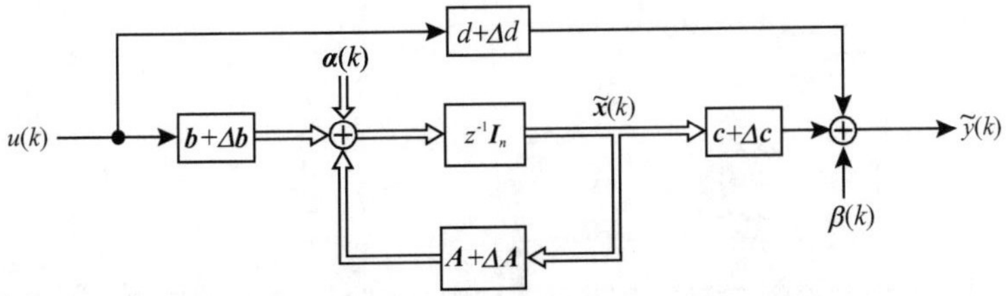

Figure 11.11 An actual state-space model.

Subtracting (11.32) from (11.33) yields

$$\begin{aligned}\Delta \boldsymbol{x}(k+1) &= \boldsymbol{A}\,\Delta \boldsymbol{x}(k) + \Delta \boldsymbol{A}\,\tilde{\boldsymbol{x}}(k) + \Delta \boldsymbol{b}\,u(k) + \boldsymbol{\alpha}(k)\\ \Delta y(k) &= \boldsymbol{c}\,\Delta \boldsymbol{x}(k) + \Delta \boldsymbol{c}\,\tilde{\boldsymbol{x}}(k) + \Delta d\,u(k) + \beta(k)\end{aligned} \tag{11.34}$$

where

$$\Delta \boldsymbol{x}(k) = \tilde{\boldsymbol{x}}(k) - \boldsymbol{x}(k), \qquad \Delta y(k) = \tilde{y}(k) - y(k)$$

Assuming that $\Delta \boldsymbol{A} \Delta \boldsymbol{x}(k) \simeq \boldsymbol{0}$ and $\Delta \boldsymbol{c} \Delta \boldsymbol{x}(k) \simeq 0$, (11.34) can be approximated as

$$\begin{aligned}\Delta \boldsymbol{x}(k+1) &= \boldsymbol{A}\,\Delta \boldsymbol{x}(k) + \Delta \boldsymbol{A}\,\boldsymbol{x}(k) + \Delta \boldsymbol{b}\,u(k) + \boldsymbol{\alpha}(k)\\ \Delta y(k) &= \boldsymbol{c}\,\Delta \boldsymbol{x}(k) + \Delta \boldsymbol{c}\,\boldsymbol{x}(k) + \Delta d\,u(k) + \beta(k)\end{aligned} \tag{11.35}$$

which leads to

$$\Delta y(k) = \Delta y_r(k) + \Delta y_c(k) \tag{11.36}$$

provided that initial state-variable vector $\Delta \boldsymbol{x}(0)$ is set to null, i.e., $\Delta \boldsymbol{x}(0) = \boldsymbol{0}$ where

$$\Delta y_r(k) = \sum_{i=0}^{k-1} \boldsymbol{c}\boldsymbol{A}^{k-i-1} \boldsymbol{\alpha}(i) + \beta(k)$$

$$\Delta y_c(k) = \sum_{i=0}^{k-1} \boldsymbol{c}\boldsymbol{A}^{k-i-1} \big[\Delta \boldsymbol{A}\,\boldsymbol{x}(i) + \Delta \boldsymbol{b}\,u(i)\big] + \Delta \boldsymbol{c}\,\boldsymbol{x}(k) + \Delta d\,u(k)$$

Equation (11.36) shows that the filter's output error $\Delta y(k)$ is represented by the sum of the roundoff error $\Delta y_r(k)$ and error $\Delta y_c(k)$ caused by coefficient quantization.

A different yet equivalent state-space description of (11.32), $(\overline{\boldsymbol{A}}, \overline{\boldsymbol{b}}, \overline{\boldsymbol{c}}, d)_n$, can be obtained via a coordinate transformation

$$\overline{\boldsymbol{x}}(k) = \boldsymbol{T}^{-1}\boldsymbol{x}(k) \tag{11.37}$$

where

$$\overline{\boldsymbol{A}} = \boldsymbol{T}^{-1}\boldsymbol{A}\boldsymbol{T}, \qquad \overline{\boldsymbol{b}} = \boldsymbol{T}^{-1}\boldsymbol{b}, \qquad \overline{\boldsymbol{c}} = \boldsymbol{c}\boldsymbol{T}$$

Accordingly, the roundoff error $\Delta y_r(k)$ and error $\Delta y_c(k)$ caused by coefficient quantization in (11.36) are transformed into

$$\begin{aligned}\Delta\overline{y}_r(k) &= \sum_{i=0}^{k-1} \boldsymbol{c}\boldsymbol{A}^{k-i-1}\boldsymbol{T}\,\overline{\boldsymbol{\alpha}}(i) + \overline{\beta}(k)\\ \Delta\overline{y}_c(k) &= \sum_{i=0}^{k-1} \boldsymbol{c}\boldsymbol{A}^{k-i-1}\boldsymbol{T}\left[\Delta\overline{\boldsymbol{A}}\,\boldsymbol{T}^{-1}\boldsymbol{x}(i) + \Delta\overline{\boldsymbol{b}}\,u(i)\right]\\ &\quad + \Delta\overline{\boldsymbol{c}}\,\boldsymbol{T}^{-1}\boldsymbol{x}(k) + \Delta d\,u(k)\end{aligned} \tag{11.38}$$

respectively, which are a function of $\boldsymbol{T}$, respectively. This reveals that the roundoff error $\Delta\overline{y}_r(k)$ and error $\Delta\overline{y}_c(k)$ caused by coefficient quantization depend on the internal structure of the state-space model.

11.9 Limit Cycle-Free Realization

This section studies conditions for a state-space realization to be free of limit cycles. We begin by setting $\Delta\boldsymbol{A} = \boldsymbol{0}$, $\Delta\boldsymbol{b} = \boldsymbol{0}$ and $u(k) = 0$ in the state equation of (11.33) which leads to

$$\tilde{\boldsymbol{x}}(k+1) = \boldsymbol{A}\tilde{\boldsymbol{x}}(k) + \boldsymbol{\alpha}(k) \tag{11.39}$$

If we perform quantization after multiplications and additions, (11.39) needs to be modified to a nonlinear equation of the form

$$\tilde{\boldsymbol{x}}(k+1) = \boldsymbol{f}[\boldsymbol{A}\tilde{\boldsymbol{x}}(k)] \tag{11.40}$$

where $\boldsymbol{f}[\cdot]$ is a nonlinear function satisfying

$$|f_i[x_i]| \le d_i|x_i| \;\text{ for }\; i = 1, 2, \cdots, n \tag{11.41}$$

with $d_i > 0$ and represents quantization after multiplications and additions, or adder's overflow. The relationship between x_i and $f_i[x_i]$ satisfying (11.41) is drawn in Figure 11.12.

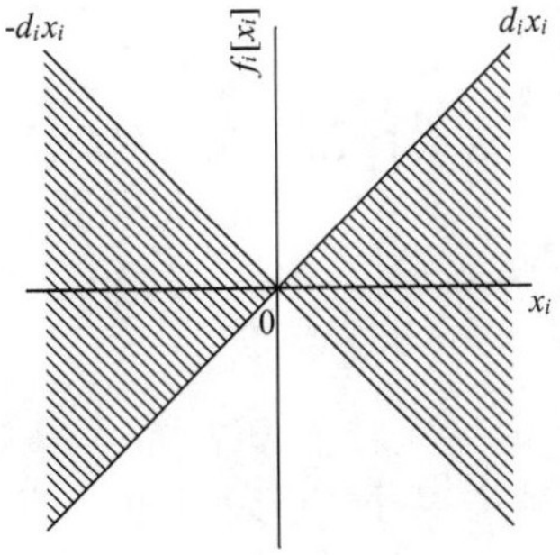

Figure 11.12 A nonlinear section satisfying (11.41).

It is noted that

$$d_i = \begin{cases} 1 & \text{for} \quad \text{quantization by truncation} \\ 1 & \text{for} \quad \text{adder's overflow} \\ 2 & \text{for} \quad \text{quantization by rounding} \end{cases} \tag{11.42}$$

If the nonlinear system in (11.40) is asymptotically stable, then the state-variable vector $\tilde{\boldsymbol{x}}(k)$ converges to **0** as k goes to infinity, hence limit cycles will not occur.

We now define a Lyapunov function as

$$V[\tilde{\boldsymbol{x}}(k)] = \tilde{\boldsymbol{x}}(k)^T \boldsymbol{P} \tilde{\boldsymbol{x}}(k) \tag{11.43}$$

where

$$\boldsymbol{P} = \text{diag}\{p_1, p_2, \cdots, p_n\} > 0$$

and compute the difference

$$\begin{aligned} \Delta V[\tilde{\boldsymbol{x}}(k)] &= V[\tilde{\boldsymbol{x}}(k+1)] - V[\tilde{\boldsymbol{x}}(k)] \\ &= \tilde{\boldsymbol{x}}(k+1)^T \boldsymbol{P} \tilde{\boldsymbol{x}}(k+1) - \tilde{\boldsymbol{x}}(k)^T \boldsymbol{P} \tilde{\boldsymbol{x}}(k) \\ &= -\tilde{\boldsymbol{x}}(k)^T [\boldsymbol{P} - (\boldsymbol{D}\boldsymbol{A})^T \boldsymbol{P}\boldsymbol{D}\boldsymbol{A}] \tilde{\boldsymbol{x}}(k) \\ &\quad + \boldsymbol{f}[\boldsymbol{A}\tilde{\boldsymbol{x}}(k)]^T \boldsymbol{P} \boldsymbol{f}[\boldsymbol{A}\tilde{\boldsymbol{x}}(k)] - [\boldsymbol{A}\tilde{\boldsymbol{x}}(k)]^T \boldsymbol{D}^T \boldsymbol{P}\boldsymbol{D}\boldsymbol{A}\tilde{\boldsymbol{x}}(k) \end{aligned} \tag{11.44}$$

where

$$\boldsymbol{D} = \text{diag}\{d_1, d_2, \cdots, d_n\}$$

From (11.41), it follows that

$$\begin{aligned} &\boldsymbol{f}[\boldsymbol{A}\tilde{\boldsymbol{x}}(k)]^T\boldsymbol{P}\boldsymbol{f}[\boldsymbol{A}\tilde{\boldsymbol{x}}(k)] - [\boldsymbol{A}\tilde{\boldsymbol{x}}(k)]^T\boldsymbol{D}^T\boldsymbol{P}\boldsymbol{D}\boldsymbol{A}\tilde{\boldsymbol{x}}(k) \\ &\quad = \sum_{i=1}^{n} p_i\{|f_i[\boldsymbol{e}_i^T\boldsymbol{A}\tilde{\boldsymbol{x}}(k)]|^2 - d_i^2|\boldsymbol{e}_i^T\boldsymbol{A}\tilde{\boldsymbol{x}}(k)|^2\} \le 0 \end{aligned} \tag{11.45}$$

where $\boldsymbol{e}_i$ denotes the ith column of the identity matrix $\boldsymbol{I}_n$ of dimension $n \times n$. Hence if there exists a positive-definite diagonal matrix $\boldsymbol{P}$ such that

$$\boldsymbol{P} - (\boldsymbol{D}\boldsymbol{A})^T\boldsymbol{P}\boldsymbol{D}\boldsymbol{A} > 0 \tag{11.46}$$

then for arbitrary state-variable vector $\tilde{\boldsymbol{x}}(k)$, we obtain

$$\Delta V[\tilde{\boldsymbol{x}}(k)] \le 0 \tag{11.47}$$

where equality holds if and only if $\tilde{\boldsymbol{x}}(k) = \boldsymbol{0}$. Consequently, the existence of a positive-definite diagonal matrix $\boldsymbol{P}$ satisfying (11.46) will ensure that the nonlinear system in (11.40) is asymptotically stable, and limit cycles will not occur for zero input $u(k) = 0$.

From (11.46), it is observed that in the case of quantization by truncation, or adder's overflow, the condition in (11.46) is changed to

$$\boldsymbol{P} - \boldsymbol{A}^T\boldsymbol{P}\boldsymbol{A} > 0 \tag{11.48}$$

and in the case of quantization by rounding, the condition in (11.46) becomes

$$\boldsymbol{P} - 4\boldsymbol{A}^T\boldsymbol{P}\boldsymbol{A} > 0 \tag{11.49}$$

For second-order filters, the condition in (11.48) reduces to a very simple condition [10] that involves the four elements of the 2×2 matrix

$$\boldsymbol{A} = \begin{bmatrix} a_{11} & a_{12} \\ a_{21} & a_{22} \end{bmatrix} \tag{11.50}$$

The condition is stated in the following theorem.

Theorem 11.1

Suppose that the magnitudes of the eigenvalues of the 2×2 matrix $\boldsymbol{A}$ in (11.50) are strictly less than unity. There exists a positive-definite diagonal matrix $\boldsymbol{P}$ for which $\boldsymbol{P} - \boldsymbol{A}^T\boldsymbol{P}\boldsymbol{A}$ is positive definite if and only if one of the following two sets of conditions holds:

(a) $a_{12}a_{21} \geq 0$ (11.51)

(b) $a_{12}a_{21} < 0$ and $|a_{11} - a_{22}| + \det[\boldsymbol{A}] < 1$ (11.52)

Proof

The characteristic polynomial of matrix $\boldsymbol{A}$ is given by

$$\det(z\boldsymbol{I}_2 - \boldsymbol{A}) = z^2 + a_1 z + a_2 \tag{11.53}$$

where

$$a_1 = -\mathrm{tr}[\boldsymbol{A}] = -(a_{11} + a_{22})$$

$$a_2 = \det[\boldsymbol{A}] = a_{11}a_{22} - a_{12}a_{21}$$

Recall the well-known conditions called *stability triangle* as described in Section 3.2.6 that the roots of the characteristic equation $z^2 + a_1 z + a_2 = 0$ satisfy $|\lambda| < 1$ if and only if the following conditions hold:

$$1 - a_2 > 0 \tag{11.54a}$$

$$1 + a_1 + a_2 > 0 \tag{11.54b}$$

$$1 - a_1 + a_2 > 0 \tag{11.54c}$$

It can readily be verified that there exists a positive-definite diagonal matrix $\boldsymbol{P}$ for which $\boldsymbol{P} - \boldsymbol{A}^T\boldsymbol{P}\boldsymbol{A}$ is positive definite if and only if there exists a nonsingular diagonal matrix $\boldsymbol{T}$ for which $\boldsymbol{I}_2 - \boldsymbol{M}$ is positive definite, where

$$\boldsymbol{M} = (\boldsymbol{T}^{-1}\boldsymbol{A}\boldsymbol{T})^T\boldsymbol{T}^{-1}\boldsymbol{A}\boldsymbol{T} \tag{11.55}$$

with $\boldsymbol{P} = \boldsymbol{T}^{-T}\boldsymbol{T}^{-1}$. The matrix $\boldsymbol{I}_2 - \boldsymbol{M}$ is positive definite if and only if both its eigenvalues are positive, and this is true if and only if

$$\mathrm{tr}[\boldsymbol{I}_2 - \boldsymbol{M}] > 0 \ \text{ and } \ \det[\boldsymbol{I}_2 - \boldsymbol{M}] > 0 \tag{11.56}$$

because these are the sum and product of the eigenvalues. Hence we consider

$$\begin{aligned}
\det[\boldsymbol{I}_2 - \boldsymbol{M}] &= \det[z\boldsymbol{I}_2 - \boldsymbol{M}]|_{z=1} \\
&= 1 - \mathrm{tr}[\boldsymbol{M}] + \det[\boldsymbol{M}] \\
&= 1 - \mathrm{tr}[\boldsymbol{M}] + (\det[\boldsymbol{A}])^2 \\
\mathrm{tr}[\boldsymbol{I}_2 - \boldsymbol{M}] &= 2 - \mathrm{tr}[\boldsymbol{M}] \\
&> 1 - \mathrm{tr}[\boldsymbol{M}] + (\det[\boldsymbol{A}])^2 \\
&= \det[\boldsymbol{I}_2 - \boldsymbol{M}]
\end{aligned} \tag{11.57}$$

which can be obtained from the fact that $(\det[\boldsymbol{A}])^2 < 1$ for stable filters. From (11.57), we need consider only the inequality $\det[\boldsymbol{I}_2 - \boldsymbol{M}] > 0$, since it will then follow that $\mathrm{tr}[\boldsymbol{I}_2 - \boldsymbol{M}] > 0$. In terms of the elements of $\boldsymbol{T}$, we can derive from (11.57) that

$$\det[\boldsymbol{I}_2 - \boldsymbol{M}] = 1 - (a_{11}^2 + a_{12}^2\tau^2 + a_{21}^2\tau^{-2} + a_{22}^2) + (\det[\boldsymbol{A}])^2 \quad (11.58)$$

where $\boldsymbol{T} = \mathrm{diag}\{t_1,\ t_2\}$ and $\tau = t_2/t_1$. Applying the arithmetic-geometric mean inequality

$$\frac{a_{12}^2\tau^2 + a_{21}^2\tau^{-2}}{2} \geq |a_{12}a_{21}| \quad (11.59)$$

to (11.58) results in

$$\begin{aligned}\det[\boldsymbol{I}_2 - \boldsymbol{M}] &\leq 1 - (a_{11}^2 + 2\,|a_{12}a_{21}| + a_{22}^2) + (\det[\boldsymbol{A}])^2 \\ &= (1 + \det[\boldsymbol{A}])^2 - (\mathrm{tr}[\boldsymbol{A}])^2 - 2(a_{12}a_{21} - |a_{12}a_{21}|)\end{aligned} \quad (11.60)$$

where equality holds if and only if $\tau^2 = |a_{21}/a_{12}|$. If $a_{12}a_{21} \geq 0$, then the right side of (11.60) is the product of the left sides of (11.54b) and (11.54c), and is therefore positive. If $a_{12}a_{21} < 0$, then the right side of (11.60) becomes

$$1-(a_{11}^2-2a_{12}a_{21}+a_{22}^2)+(\det[\boldsymbol{A}])^2 = (1-\det[\boldsymbol{A}])^2-(a_{11}-a_{22})^2 \quad (11.61)$$

which is positive if and only if (11.52) holds. This completes the proof of Theorem 11.1. ■

11.10 Summary

In this chapter, we have reviewed the fixed-point and floating-point arithmetic of binary numbers and the two's complement representation of negative numbers. Limit cycles—overflow oscillations, scaling fixed-point digital filters to prevent overflow, roundoff noise, and coefficient sensitivity have also been addressed. Finally, the response of a FWL state-space description and methods for obtaining limit cycle-free realization have been explored. The material studied in this chapter provides a basis for the techniques to be presented in the remaining chapters.

References

[1] W. R. Bennet, "Spectra of quantized signals," *Bell Syst. Tech. J.*, vol. 27, pp. 446–472, July 1948.

[2] B. Widrow, "A study of rough amplitude quantization by means of Nyquist sampling theory," *IRE Trans. Circuit Theory*, vol. CT-3, pp. 266–276, Dec. 1956.

[3] A. V. Oppenheim and R. W. Schafer, *Digital Signal Processing*, NJ: Prentice-Hall, 1975.

[4] T. Higuchi, *Fundamentals of Digital Signal Processing*, Tokyo, Japan, Shokodo Co., 1986.

[5] R. A. Roberts and C. T. Mullis, *Digital Signal Processing*, Addison-Wesley, 1987.

[6] A. Antoniou, *Digital Filters*, 2nd ed., NJ: McGraw-Hill, 1993.

[7] T. Hinamoto, S. Yokoyama, T. Inoue, W. Zeng and W.-S. Lu, "Analysis and minimization of l_2-sensitivity for linear systems and two-dimensional state-space filters using General controllability and observability Grammians," *IEEE Trans. Circuits Syst. I*, vol. 49, no. 9, pp. 1279–1289, Sept. 2002.

[8] S. K. Mitra, *Digital Signal Processing*, 3rd ed., NJ: McGraw-Hill, 2006.

[9] C. W. Barnes and A. T. Fam, "Minimum norm recursive digital filters that are free of overflow limit cycles" *IEEE Trans. Circuits Syst.*, vol. CAS-24, no. 10, pp. 569–574, Oct. 1977.

[10] W. L. Mills, C. T. Mullis and R. A. Roberts, "Digital filter realizations without overflow oscillations" *IEEE Trans. Acoust., Speech, Signal Process.*, vol. ASSP-26, no. 4, pp. 334–338, Aug. 1978.

12

l_2-Sensitivity Analysis and Minimization

12.1 Preview

It is of practical significance in many applications to construct a filter structure so that the coefficient sensitivity of the digital filter is minimum or nearly minimum in a certain sense. Due to finite-word-length (FWL) effects caused by coefficient truncation or rounding, poor sensitivity may lead to degradation of the transfer characteristics in an FWL implementation of the digital filter. For instance, the characteristics of an originally stable filter might be so altered that the filter becomes unstable. This motivates the study of the coefficient sensitivity minimization problem for digital filters. Techniques for synthesizing the state-space filter structures that minimize the deviation of a transfer function caused by coefficient quantization can be divided into two main classed, namely l_1/l_2-mixed sensitivity minimization [1–5] and l_2-sensitivity minimization [6–9]. In [6–9], it has been argued that the sensitivity measure based on the l_2-norm only is more natural and reasonable relative to the l_1/l_2-mixed sensitivity minimization. More recently, the problem of minimizing l_2-sensitivity subject to l_2-scaling constraints has been examined for state-space digital filters [10, 11]. It is known that the use of scaling constraints can be beneficial for suppressing overflow oscillations [12, 13].

In this chapter, the l_2-sensitivities with respect to the coefficient matrix for a state-space digital filter is analyzed, and an l_2-sensitivity measure is derived. Next, a simple method for minimizing the l_2-sensitivity measure is presented by using a recursive matrix equation. In addition, two techniques for minimizing the l_2-sensitivity measure subject to l_2-scaling constraints are described: one employs a quasi-Newton algorithm and the other relies on a Lagrange function.

Numerical experiments are presented to illustrate the validity and effectiveness of these algorithms and demonstrate their performance.

12.2 l_2-Sensitivity Analysis

Consider a stable, controllable and observable state-space digital filter $(\boldsymbol{A}, \boldsymbol{b}, \boldsymbol{c}, d)_n$ described by

$$\begin{aligned} \boldsymbol{x}(k+1) &= \boldsymbol{A}\boldsymbol{x}(k) + \boldsymbol{b}u(k) \\ y(k) &= \boldsymbol{c}\boldsymbol{x}(k) + du(k) \end{aligned} \tag{12.1}$$

where $\boldsymbol{x}(k)$ is an $n \times 1$ state-variable vector, $u(k)$ is a scalar input, $y(k)$ is a scalar output, and $\boldsymbol{A}$, $\boldsymbol{b}$, $\boldsymbol{c}$ and d are $n \times n$, $n \times 1$, $1 \times n$ and 1×1 real constant matrices, respectively, and these matrices are given by

$$\boldsymbol{A} = \begin{bmatrix} a_{11} & a_{12} & \cdots & a_{1n} \\ a_{21} & a_{22} & \cdots & a_{2n} \\ \vdots & \vdots & \ddots & \vdots \\ a_{n1} & a_{n2} & \cdots & a_{nn} \end{bmatrix}, \qquad \boldsymbol{b} = \begin{bmatrix} b_1 \\ b_2 \\ \vdots \\ b_n \end{bmatrix}$$

$$\boldsymbol{c} = \begin{bmatrix} c_1 & c_2 & \cdots & c_n \end{bmatrix}$$

The transfer function of the filter in (12.1) can be expressed as

$$H(z) = \boldsymbol{c}(z\boldsymbol{I}_n - \boldsymbol{A})^{-1}\boldsymbol{b} + d \tag{12.2}$$

The l_2-sensitivities of the transfer function with respect to coefficient matrices $\boldsymbol{A}$, $\boldsymbol{b}$, $\boldsymbol{c}$ and d are computed as follows.

Definition 12.1

Let $\boldsymbol{X}$ and $f(\boldsymbol{X})$ be an $m \times n$ real matrix and a scalar complex function of $\boldsymbol{X}$ differentiable with respect to all entries of $\boldsymbol{X}$, respectively. The sensitivity function of $f(\boldsymbol{X})$ with respect to $\boldsymbol{X}$ is then defined as [5]

$$\boldsymbol{S}_{\boldsymbol{X}} = \frac{\partial f(\boldsymbol{X})}{\partial \boldsymbol{X}}, \qquad (\boldsymbol{S}_{\boldsymbol{X}})_{ij} = \frac{\partial f(\boldsymbol{X})}{\partial x_{ij}} \tag{12.3}$$

where both x_{ij} and $(\boldsymbol{X})_{ij}$ denote the (i, j)th entry of matrix $\boldsymbol{X}$, respectively. By virtue of (12.2), Definition 12.1 and the formula

$$\frac{\partial \boldsymbol{A}^{-1}}{\partial a_{ij}} = -\boldsymbol{A}^{-1}\frac{\partial \boldsymbol{A}}{\partial a_{ij}}\boldsymbol{A}^{-1} \tag{12.4}$$

the sensitivities of $H(z)$ with respect to elements a_{ij}, b_i, c_j and d are evaluated by

$$\frac{\partial H(z)}{\partial a_{ij}} = g_i(z) f_j(z), \qquad \frac{\partial H(z)}{\partial b_i} = g_i(z)$$
$$\frac{\partial H(z)}{\partial c_j} = f_j(z), \qquad \frac{\partial H(z)}{\partial d} = 1 \tag{12.5}$$

respectively, where

$$\boldsymbol{f}(z) = (z\boldsymbol{I}_n - \boldsymbol{A})^{-1}\boldsymbol{b} = \begin{bmatrix} f_1(z) \\ f_2(z) \\ \vdots \\ f_n(z) \end{bmatrix}$$

$$\boldsymbol{g}(z) = \boldsymbol{c}(z\boldsymbol{I}_n - \boldsymbol{A})^{-1} = [g_1(z), g_2(z), \cdots, g_n(z)]$$

Equation (12.5) is equivalent to

$$\frac{\partial H(z)}{\partial \boldsymbol{A}} = [\boldsymbol{f}(z)\boldsymbol{g}(z)]^T, \qquad \frac{\partial H(z)}{\partial \boldsymbol{b}} = \boldsymbol{g}^T(z)$$
$$\frac{\partial H(z)}{\partial \boldsymbol{c}^T} = \boldsymbol{f}(z), \qquad \frac{\partial H(z)}{\partial d} = 1 \tag{12.6}$$

Definition 12.2
Let $\boldsymbol{X}(z)$ be an $m \times n$ complex-valued matrix function of complex variable z, and $x_{pq}(z)$ be the (p,q)th entry of $\boldsymbol{X}(z)$. The l_2-norm of $\boldsymbol{X}(z)$ is defined as

$$\begin{aligned} \|\boldsymbol{X}(z)\|_2 &= \left[\frac{1}{2\pi}\int_0^{2\pi} \sum_{p=1}^{m}\sum_{q=1}^{n} \left|x_{pq}(e^{j\omega})\right|^2 d\omega\right]^{\frac{1}{2}} \\ &= \left(\mathrm{tr}\left[\frac{1}{2\pi j}\oint_{|z|=1} \boldsymbol{X}(z)\boldsymbol{X}^H(z)\frac{dz}{z}\right]\right)^{\frac{1}{2}} \end{aligned} \tag{12.7}$$

Using (12.6) and (12.7), the overall l_2-sensitivity measure for the transfer function in (12.2) is defined by

$$\begin{aligned} S_o &= \left\|\frac{\partial H(z)}{\partial \boldsymbol{A}}\right\|_2^2 + \left\|\frac{\partial H(z)}{\partial \boldsymbol{b}}\right\|_2^2 + \left\|\frac{\partial H(z)}{\partial \boldsymbol{c}^T}\right\|_2^2 \\ &= \left\|[\boldsymbol{f}(z)\boldsymbol{g}(z)]^T\right\|_2^2 + \left\|\boldsymbol{g}^T(z)\right\|_2^2 + \|\boldsymbol{f}(z)\|_2^2 \end{aligned} \tag{12.8}$$

The term d in (12.2) and the sensitivity with respect to it are coordinate independent, and therefore they are neglected here.

It is easy to show that the l_2-sensitivity measure in (12.8) can be expressed as

$$S_o = \mathrm{tr}[\boldsymbol{N}(\boldsymbol{I}_n)] + \mathrm{tr}[\boldsymbol{W}_o] + \mathrm{tr}[\boldsymbol{K}_c] \tag{12.9}$$

where

$$\boldsymbol{K}_c = \frac{1}{2\pi j}\oint_{|z|=1} \boldsymbol{f}(z)\boldsymbol{f}^T(z^{-1})\frac{dz}{z}$$

$$\boldsymbol{W}_o = \frac{1}{2\pi j}\oint_{|z|=1} \boldsymbol{g}^T(z)\boldsymbol{g}(z^{-1})\frac{dz}{z}$$

$$\boldsymbol{N}(\boldsymbol{P}) = \frac{1}{2\pi j}\oint_{|z|=1} [\boldsymbol{f}(z)\boldsymbol{g}(z)]^T\boldsymbol{P}^{-1}\boldsymbol{f}(z^{-1})\boldsymbol{g}(z^{-1})\frac{dz}{z}$$

Here $\boldsymbol{P}$ is an $n \times n$ nonsingular matrix that will later be related to a coordinate transformation for the state-variable vector of the digital filter (see Section 12.3 below). Noting that

$$\begin{aligned}(z\boldsymbol{I}_n - \boldsymbol{A})^{-1} &= z^{-1}\boldsymbol{I}_n + \boldsymbol{A}z^{-2} + \boldsymbol{A}^2z^{-3} + \cdots \\ &= \sum_{i=0}^{\infty}\boldsymbol{A}^i z^{-(i+1)}\end{aligned} \tag{12.10}$$

and utilizing *Cauchy's integral theorem*

$$\frac{1}{2\pi j}\oint_C z^k\frac{dz}{z} = \begin{cases} 1, & k = 0 \\ 0, & k \neq 0 \end{cases} \tag{12.11}$$

where C is a counterclockwise contour that encircles the origin, matrices $\boldsymbol{K}_c$ and $\boldsymbol{W}_o$ in (12.9) can be written as

$$\begin{aligned}\boldsymbol{K}_c &= \sum_{k=0}^{\infty}\boldsymbol{A}^k\boldsymbol{b}\boldsymbol{b}^T(\boldsymbol{A}^k)^T \\ \boldsymbol{W}_o &= \sum_{k=0}^{\infty}(\boldsymbol{A}^k)^T\boldsymbol{c}^T\boldsymbol{c}\boldsymbol{A}^k\end{aligned} \tag{12.12}$$

respectively, and they can be obtained by solving the Lyapunov equations

$$\begin{aligned}\boldsymbol{K}_c &= \boldsymbol{A}\boldsymbol{K}_c\boldsymbol{A}^T + \boldsymbol{b}\boldsymbol{b}^T \\ \boldsymbol{W}_o &= \boldsymbol{A}^T\boldsymbol{W}_o\boldsymbol{A} + \boldsymbol{c}^T\boldsymbol{c}\end{aligned} \tag{12.13}$$

respectively. The matrices $\boldsymbol{K}_c$ and $\boldsymbol{W}_o$ in (12.12) are called the *controllability Grammian* and *observability Grammian*, respectively. Similarly, matrix $\boldsymbol{N}(\boldsymbol{P})$ in (12.9) can be derived from

$$\boldsymbol{N}(\boldsymbol{P}) = \sum_{i=0}^{\infty} \boldsymbol{H}^T(i)\boldsymbol{P}^{-1}\boldsymbol{H}(i) \tag{12.14}$$

where

$$\boldsymbol{H}(i) = \sum_{l=0}^{i} \boldsymbol{A}^l \boldsymbol{b}\boldsymbol{c}\boldsymbol{A}^{i-l}$$

It is noted that $\boldsymbol{N}(\boldsymbol{P})$ can also be obtained by solving the Lyapunov equation in closed form, as shown in (12.21) and (12.22) later.

12.3 Realization with Minimal l_2-Sensitivity

If a coordinate transformation defined by

$$x(k) = \boldsymbol{T}^{-1}\boldsymbol{x}(k) \tag{12.15}$$

is applied to the filter in (12.1), we obtain a new realization $(\overline{\boldsymbol{A}}, \overline{\boldsymbol{b}}, \overline{\boldsymbol{c}}, d)_n$ characterized by

$$\begin{aligned} \overline{\boldsymbol{x}}(k+1) &= \overline{\boldsymbol{A}}\overline{\boldsymbol{x}}(k) + \overline{\boldsymbol{b}}u(k) \\ y(k) &= \overline{\boldsymbol{c}}\,\overline{\boldsymbol{x}}(k) + du(k) \end{aligned} \tag{12.16}$$

where

$$\overline{\boldsymbol{A}} = \boldsymbol{T}^{-1}\boldsymbol{A}\boldsymbol{T}, \qquad \overline{\boldsymbol{b}} = \boldsymbol{T}^{-1}\boldsymbol{b}, \qquad \overline{\boldsymbol{c}} = \boldsymbol{c}\boldsymbol{T}$$

Accordingly, the controllability and observability Grammians relating to $(\overline{\boldsymbol{A}}, \overline{\boldsymbol{b}}, \overline{\boldsymbol{c}}, d)_n$ can be expressed as

$$\overline{\boldsymbol{K}}_c = \boldsymbol{T}^{-1}\boldsymbol{K}_c\boldsymbol{T}^{-T}, \qquad \overline{\boldsymbol{W}}_o = \boldsymbol{T}^T\boldsymbol{W}_o\boldsymbol{T} \tag{12.17}$$

respectively. From (12.2) and (12.16), it is obvious that the transfer function $H(z)$ is invariant under the coordinate transformation (12.15). In addition, under the coordinate transformation (12.15), matrix $\boldsymbol{N}(\boldsymbol{I}_n)$ becomes $\boldsymbol{T}^T\boldsymbol{N}(\boldsymbol{P})\boldsymbol{T}$ and the l_2-sensitivity measure in (12.9) is changed to

$$S_o(\boldsymbol{T}) = \mathrm{tr}[\boldsymbol{T}^T\boldsymbol{N}(\boldsymbol{T}\boldsymbol{T}^T)\boldsymbol{T}] + \mathrm{tr}[\boldsymbol{T}^T\boldsymbol{W}_o\boldsymbol{T}] + \mathrm{tr}[\boldsymbol{T}^{-1}\boldsymbol{K}_c\boldsymbol{T}^{-T}] \tag{12.18}$$

which is equivalent to

$$S(\boldsymbol{P}) = \mathrm{tr}[\boldsymbol{N}(\boldsymbol{P})\boldsymbol{P}] + \mathrm{tr}[\boldsymbol{W}_o\boldsymbol{P}] + \mathrm{tr}[\boldsymbol{K}_c\boldsymbol{P}^{-1}] \tag{12.19}$$

where $\boldsymbol{P} = \boldsymbol{T}\boldsymbol{T}^T$. By noting that

$$\begin{aligned}\overline{\boldsymbol{f}}(z)\overline{\boldsymbol{g}}(z) &= \boldsymbol{T}^{-1}\boldsymbol{f}(z)\boldsymbol{g}(z)\boldsymbol{T} \\ &= \begin{bmatrix} \boldsymbol{T}^{-1} & \boldsymbol{0} \end{bmatrix} \begin{bmatrix} z\boldsymbol{I}_n - \boldsymbol{A} & -\boldsymbol{b}\boldsymbol{c} \\ \boldsymbol{0} & z\boldsymbol{I}_n - \boldsymbol{A} \end{bmatrix}^{-1} \begin{bmatrix} \boldsymbol{0} \\ \boldsymbol{T} \end{bmatrix}\end{aligned} \tag{12.20}$$

where

$$\overline{\boldsymbol{f}}(z) = (z\boldsymbol{I}_n - \overline{\boldsymbol{A}})^{-1}\overline{\boldsymbol{b}}, \qquad \overline{\boldsymbol{g}}(z) = \overline{\boldsymbol{c}}(z\boldsymbol{I}_n - \overline{\boldsymbol{A}})^{-1}$$

and denoting the observability Grammian of a composite system $\overline{\boldsymbol{f}}(z)\overline{\boldsymbol{g}}(z)$ in (12.20) by $\boldsymbol{Y}$, it is easy to show that for an arbitrary $\boldsymbol{P} = \boldsymbol{T}\boldsymbol{T}^T$, matrix $\boldsymbol{N}(\boldsymbol{P})$ can be obtained by solving the Lyapunov equation

$$\boldsymbol{Y} = \begin{bmatrix} \boldsymbol{A} & \boldsymbol{b}\boldsymbol{c} \\ \boldsymbol{0} & \boldsymbol{A} \end{bmatrix}^T \boldsymbol{Y} \begin{bmatrix} \boldsymbol{A} & \boldsymbol{b}\boldsymbol{c} \\ \boldsymbol{0} & \boldsymbol{A} \end{bmatrix} + \begin{bmatrix} \boldsymbol{P}^{-1} & \boldsymbol{0} \\ \boldsymbol{0} & \boldsymbol{0} \end{bmatrix} \tag{12.21}$$

and then taking the lower-right $n \times n$ block of $\boldsymbol{Y}$ as $\boldsymbol{N}(\boldsymbol{P})$, namely,

$$\boldsymbol{N}(\boldsymbol{P}) = \begin{bmatrix} \boldsymbol{0} & \boldsymbol{I}_n \end{bmatrix} \boldsymbol{Y} \begin{bmatrix} \boldsymbol{0} \\ \boldsymbol{I}_n \end{bmatrix} \tag{12.22}$$

It is well known that the solution of minimizing $S(\boldsymbol{P})$ in (12.19) with respect to $\boldsymbol{P}$ must satisfy the Karush-Kuhn-Tucker (KKT) condition $\partial S(\boldsymbol{P})/\partial \boldsymbol{P} = \boldsymbol{0}$. Using the formula for evaluating the matrix gradient [5, 14]

$$\begin{aligned}\frac{\partial\, \mathrm{tr}[\boldsymbol{M}\boldsymbol{X}]}{\partial \boldsymbol{X}} &= \boldsymbol{M}^T \\ \frac{\partial\, \mathrm{tr}[\boldsymbol{M}\boldsymbol{X}^{-1}]}{\partial \boldsymbol{X}} &= -\left[\boldsymbol{X}^{-1}\boldsymbol{M}\boldsymbol{X}^{-1}\right]^T\end{aligned} \tag{12.23}$$

the gradient of $S(\boldsymbol{P})$ in (12.19) with respect to $\boldsymbol{P}$ is found to be

$$\frac{\partial S(\boldsymbol{P})}{\partial \boldsymbol{P}} = \boldsymbol{N}(\boldsymbol{P}) - \boldsymbol{P}^{-1}\boldsymbol{M}(\boldsymbol{P})\boldsymbol{P}^{-1} + \boldsymbol{W}_o - \boldsymbol{P}^{-1}\boldsymbol{K}_c\boldsymbol{P}^{-1} \tag{12.24}$$

where

$$\boldsymbol{M}(\boldsymbol{P}) = \frac{1}{2\pi j}\oint_{|z|=1} \boldsymbol{F}(z^{-1})\boldsymbol{G}(z^{-1})\boldsymbol{P}\left[\boldsymbol{F}(z)\boldsymbol{G}(z)\right]^T \frac{dz}{z}$$

We remark that $\mathrm{tr}[\boldsymbol{N}(\boldsymbol{P})\boldsymbol{P}] = \mathrm{tr}[\boldsymbol{M}(\boldsymbol{P})\boldsymbol{P}^{-1}]$. If the controllability Grammian of a composite system $\overline{\boldsymbol{f}}(z)\overline{\boldsymbol{g}}(z)$ in (12.20) is denoted by $\boldsymbol{X}$, it is easily shown that for an arbitrary $\boldsymbol{P} = \boldsymbol{T}\boldsymbol{T}^T$, matrix $\boldsymbol{M}(\boldsymbol{P})$ can be obtained by solving the Lyapunov equation

$$\boldsymbol{X} = \begin{bmatrix} \boldsymbol{A} & \boldsymbol{b}\boldsymbol{c} \\ \boldsymbol{0} & \boldsymbol{A} \end{bmatrix} \boldsymbol{X} \begin{bmatrix} \boldsymbol{A} & \boldsymbol{b}\boldsymbol{c} \\ \boldsymbol{0} & \boldsymbol{A} \end{bmatrix}^T + \begin{bmatrix} \boldsymbol{0} & \boldsymbol{0} \\ \boldsymbol{0} & \boldsymbol{P} \end{bmatrix} \tag{12.25}$$

and then taking the upper-left $n \times n$ block of $\boldsymbol{X}$ as $\boldsymbol{M}(\boldsymbol{P})$, namely,

$$\boldsymbol{M}(\boldsymbol{P}) = \begin{bmatrix} \boldsymbol{I}_n & \boldsymbol{0} \end{bmatrix} \boldsymbol{X} \begin{bmatrix} \boldsymbol{I}_n \\ \boldsymbol{0} \end{bmatrix} \tag{12.26}$$

Therefore, the KKT condition becomes

$$\boldsymbol{P}\boldsymbol{F}(\boldsymbol{P})\boldsymbol{P} = \boldsymbol{G}(\boldsymbol{P}) \tag{12.27}$$

where

$$\boldsymbol{F}(\boldsymbol{P}) = \boldsymbol{N}(\boldsymbol{P}) + \boldsymbol{W}_o, \qquad \boldsymbol{G}(\boldsymbol{P}) = \boldsymbol{M}(\boldsymbol{P}) + \boldsymbol{K}_c$$

Equation (12.27) is highly nonlinear with respect to $\boldsymbol{P}$. An effective approach for solving (12.27) is to relax it into the recursive second-order matrix equation

$$\boldsymbol{P}_{k+1}\boldsymbol{F}(\boldsymbol{P}_k)\boldsymbol{P}_{k+1} = \boldsymbol{G}(\boldsymbol{P}_k) \tag{12.28}$$

where $\boldsymbol{P}_k$ is assumed to be known from the previous recursion. Note that if the matrix sequence $\{\boldsymbol{P}_k\}$ converges to its limit matrix, say $\boldsymbol{P}$, then (12.28) converges to (12.27) as k goes infinity.

Noting that for a positive definite $\boldsymbol{W}$ and a semi-positive definite $\boldsymbol{M}$, matrix equation $\boldsymbol{P}\boldsymbol{W}\boldsymbol{P} = \boldsymbol{M}$ has the unique solution [5]

$$\boldsymbol{P} = \boldsymbol{W}^{-\frac{1}{2}}[\boldsymbol{W}^{\frac{1}{2}}\boldsymbol{M}\boldsymbol{W}^{\frac{1}{2}}]^{\frac{1}{2}}\boldsymbol{W}^{-\frac{1}{2}} \tag{12.29}$$

the solution $\boldsymbol{P}_{k+1}$ of (12.28) is given by

$$\boldsymbol{P}_{k+1} = \boldsymbol{F}(\boldsymbol{P}_k)^{-\frac{1}{2}}[\boldsymbol{F}(\boldsymbol{P}_k)^{\frac{1}{2}}\boldsymbol{G}(\boldsymbol{P}_k)\boldsymbol{F}(\boldsymbol{P}_k)^{\frac{1}{2}}]^{\frac{1}{2}}\boldsymbol{F}(\boldsymbol{P}_k)^{-\frac{1}{2}} \tag{12.30}$$

where $\boldsymbol{P}_k$ is the solution of the previous iteration, and the initial estimate $\boldsymbol{P}_0$ is often chosen as $\boldsymbol{P}_0 = \boldsymbol{I}_n$. This iteration process continues until

$$|S(\boldsymbol{P}_{k+1}) - S(\boldsymbol{P}_k)| < \varepsilon \tag{12.31}$$

is satisfied where $\varepsilon > 0$ is a prescribed tolerance. If the iteration is terminated at step k, $\boldsymbol{P}_k$ is claimed to be a solution point.

We now obtain the optimal coordinate transformation matrix $\boldsymbol{T}$ that solves the problem of minimizing $S(\boldsymbol{P})$ in (12.19). As analyzed earlier, the optimal $\boldsymbol{T}$ assumes the form

$$\boldsymbol{T} = \boldsymbol{P}^{\frac{1}{2}}\boldsymbol{U} \tag{12.32}$$

where $\boldsymbol{P}^{1/2}$ is square root of the matrix $\boldsymbol{P}$ obtained above, and $\boldsymbol{U}$ is an arbitrary $n \times n$ orthogonal matrix. The optimal realization with minimal l_2-sensitivity can readily be constructed by substituting (12.32) into (12.16).

12.4 l_2-Sensitivity Minimization Subject to l_2-Scaling Constraints Using Quasi-Newton Algorithm

12.4.1 l_2-Scaling and Problem Formulation

Let $\boldsymbol{X}(z)$ and $U(z)$ be the z-transforms of the state-variable vector $\boldsymbol{x}(k)$ and the input $u(k)$ in (12.1), respectively. Then the relation of the state variables to the input in the frequency domain can be expressed as

$$\boldsymbol{X}(z) = \boldsymbol{f}(z)U(z) \tag{12.33}$$

where $\boldsymbol{f}(z)$ is a transfer function from the input $u(k)$ to the state-variable vector $\boldsymbol{x}(k)$, and described by

$$\boldsymbol{f}(z) = (z\boldsymbol{I}_n - \boldsymbol{A})^{-1}\boldsymbol{b} = \sum_{k=1}^{\infty} \boldsymbol{A}^{k-1}\boldsymbol{b}z^{-k}$$

whose impulse response is seen as the sequence $\{\boldsymbol{b}, \boldsymbol{A}\boldsymbol{b}, \cdots, \boldsymbol{A}^{k-1}\boldsymbol{b}, \cdots\}$. If the input has finite energy, that is,

$$\sum_{l=-\infty}^{k-1} u^2(l) \le 1 \tag{12.34}$$

then we have

$$\begin{aligned} |x_i(k)| &\le \sum_{l=1}^{\infty} |\boldsymbol{e}_i^T \boldsymbol{A}^{l-1}\boldsymbol{b}u(k-l)| \\ &\le \left[\sum_{l=1}^{\infty} \boldsymbol{e}_i^T \boldsymbol{A}^{l-1}\boldsymbol{b}\boldsymbol{b}^T(\boldsymbol{A}^T)^{l-1}\boldsymbol{e}_i\right]^{\frac{1}{2}} \\ &= \boldsymbol{e}_i^T \boldsymbol{K}_c \boldsymbol{e}_i \end{aligned} \tag{12.35}$$

where $x_i(k)$ denotes the ith element of the state-variable vector $\boldsymbol{x}(k)$, $\boldsymbol{e}_i$ is the ith column of an identity matrix $\boldsymbol{I}_n$ of dimension $n \times n$, and matrix

$$\boldsymbol{K}_c = \sum_{k=0}^{\infty} \boldsymbol{A}^k \boldsymbol{b}\boldsymbol{b}^T (\boldsymbol{A}^k)^T$$

is the *controllability Grammian* of the filter in (12.1) that can be obtained by solving the Lyapunov equation in (12.13). In the above, the l_2-scaling rule is given by

$$\delta^2 \boldsymbol{e}_i^T \boldsymbol{K}_c \boldsymbol{e}_i = 1 \ \text{ for } \ i = 1, 2, \cdots, n \tag{12.36}$$

where there is no loss of generality in assuming that scalar δ can be chosen as $\delta = 1$.

We are now in a position to apply the l_2-scaling rule to the new realization in (12.16), i.e.,

$$\boldsymbol{e}_i^T \overline{\boldsymbol{K}}_c \boldsymbol{e}_i = \boldsymbol{e}_i^T \boldsymbol{T}^{-1} \boldsymbol{K}_c \boldsymbol{T}^{-T} \boldsymbol{e}_i = 1 \ \text{ for } \ i = 1, 2, \cdots, n \tag{12.37}$$

As a result, the problem of l_2-sensitivity minimization subject to l_2-scaling constraints is now formulated as follows: *Given the matrices* **A**, **b**, *and* **c**, *obtain an* $n \times n$ *nonsingular matrix* $\boldsymbol{T}$ *which minimizes the* l_2*-sensitivity measure* $S_o(\boldsymbol{T})$ *in (12.18) subject to the* l_2*-scaling constraints in (12.37).*

12.4.2 Minimization of (12.18) Subject to l_2-Scaling Constraints — Using Quasi-Newton Algorithm

When the state-space model in (12.1) is assumed to be stable and controllable, the controllability Grammian $\boldsymbol{K}_c$ in (12.12) is symmetric and positive-definite. This implies that $\boldsymbol{K}_c^{\frac{1}{2}}$ satisfying $\boldsymbol{K}_c = \boldsymbol{K}_c^{\frac{1}{2}} \boldsymbol{K}_c^{\frac{1}{2}}$ is also symmetric and positive-definite.

By defining

$$\hat{\boldsymbol{T}} = \boldsymbol{T}^T \boldsymbol{K}_c^{-\frac{1}{2}} \tag{12.38}$$

the l_2-scaling constraints in (12.37) can be written as

$$\boldsymbol{e}_i^T \hat{\boldsymbol{T}}^{-T} \hat{\boldsymbol{T}}^{-1} \boldsymbol{e}_i = 1 \ \text{ for } \ i = 1, 2, \cdots, n \tag{12.39}$$

The constraints in (12.39) simply state that each column in matrix $\hat{\boldsymbol{T}}^{-1}$ must be a unit vector. If matrix $\hat{\boldsymbol{T}}^{-1}$ is assumed to have the form

$$\hat{\boldsymbol{T}}^{-1} = \left[\frac{\boldsymbol{t}_1}{||\boldsymbol{t}_1||}, \frac{\boldsymbol{t}_2}{||\boldsymbol{t}_2||}, \cdots, \frac{\boldsymbol{t}_n}{||\boldsymbol{t}_n||} \right] \tag{12.40}$$

so that (12.39) is always satisfied. From (12.18) and (12.38), it follows that

$$\begin{aligned} J(\hat{\boldsymbol{T}}) &= \mathrm{tr}[\hat{\boldsymbol{T}}\hat{\boldsymbol{N}}(\hat{\boldsymbol{T}})\hat{\boldsymbol{T}}^T] + \mathrm{tr}[\hat{\boldsymbol{T}}\hat{\boldsymbol{W}}_o\hat{\boldsymbol{T}}^T] + n \\ &= \mathrm{tr}[\hat{\boldsymbol{T}}^{-T}\hat{\boldsymbol{M}}(\hat{\boldsymbol{T}})\hat{\boldsymbol{T}}^{-1}] + \mathrm{tr}[\hat{\boldsymbol{T}}\hat{\boldsymbol{W}}_o\hat{\boldsymbol{T}}^T] + n \end{aligned} \tag{12.41}$$

where

$$\hat{\boldsymbol{N}}(\hat{\boldsymbol{T}}) = \boldsymbol{K}_c^{\frac{1}{2}}\boldsymbol{N}(\boldsymbol{K}_c^{\frac{1}{2}}\hat{\boldsymbol{T}}^T\hat{\boldsymbol{T}}\boldsymbol{K}_c^{\frac{1}{2}})\boldsymbol{K}_c^{\frac{1}{2}}, \qquad \hat{\boldsymbol{W}}_o = \boldsymbol{K}_c^{\frac{1}{2}}\boldsymbol{W}_o\boldsymbol{K}_c^{\frac{1}{2}}$$

$$\hat{\boldsymbol{M}}(\hat{\boldsymbol{T}}) = \boldsymbol{K}_c^{-\frac{1}{2}}\boldsymbol{M}(\boldsymbol{K}_c^{\frac{1}{2}}\hat{\boldsymbol{T}}^T\hat{\boldsymbol{T}}\boldsymbol{K}_c^{\frac{1}{2}})\boldsymbol{K}_c^{-\frac{1}{2}}$$

From the foregoing arguments, the problem of obtaining an $n \times n$ nonsingular matrix $\boldsymbol{T}$ which minimizes $S_o(\boldsymbol{T})$ in (12.18) subject to the l_2-scaling constraints in (12.37) can be converted into an unconstrained optimization problem of obtaining an $n \times n$ nonsingular matrix $\hat{\boldsymbol{T}}$ which minimizes $J(\hat{\boldsymbol{T}})$ in (12.41).

We now apply a quasi-Newton algorithm [15] to minimize (12.41) with respect to matrix $\hat{\boldsymbol{T}}$ in (12.40). Let $\boldsymbol{x}$ be the column vector that collects the independent variables in matrix $\hat{\boldsymbol{T}}$, i.e.,

$$\boldsymbol{x} = (\boldsymbol{t}_1^T, \boldsymbol{t}_2^T, \cdots, \boldsymbol{t}_n^T)^T \tag{12.42}$$

Then, $J(\hat{\boldsymbol{T}})$ is a function of $\boldsymbol{x}$ and is denoted by $J(\boldsymbol{x})$. The algorithm starts with a trivial initial point $\boldsymbol{x}_0$ obtained from an initial assignment $\hat{\boldsymbol{T}} = \boldsymbol{I}_n$. Then, in the kth iteration, a quasi-Newton algorithm updates the most recent point $\boldsymbol{x}_k$ to point $\boldsymbol{x}_{k+1}$ as

$$\boldsymbol{x}_{k+1} = \boldsymbol{x}_k + \alpha_k \boldsymbol{d}_k \tag{12.43}$$

where

$$\boldsymbol{d}_k = -\boldsymbol{S}_k \nabla J(\boldsymbol{x}_k), \qquad \alpha_k = arg\big[\min_{\alpha} J(\boldsymbol{x}_k + \alpha \boldsymbol{d}_k)\big]$$

$$\boldsymbol{S}_{k+1} = \boldsymbol{S}_k + \left(1 + \frac{\boldsymbol{\gamma}_k^T \boldsymbol{S}_k \boldsymbol{\gamma}_k}{\boldsymbol{\gamma}_k^T \boldsymbol{\delta}_k}\right)\frac{\boldsymbol{\delta}_k \boldsymbol{\delta}_k^T}{\boldsymbol{\gamma}_k^T \boldsymbol{\delta}_k} - \frac{\boldsymbol{\delta}_k \boldsymbol{\gamma}_k^T \boldsymbol{S}_k + \boldsymbol{S}_k \boldsymbol{\gamma}_k \boldsymbol{\delta}_k^T}{\boldsymbol{\gamma}_k^T \boldsymbol{\delta}_k}$$

$$\boldsymbol{S}_0 = \boldsymbol{I}, \qquad \boldsymbol{\delta}_k = \boldsymbol{x}_{k+1} - \boldsymbol{x}_k, \qquad \boldsymbol{\gamma}_k = \nabla J(\boldsymbol{x}_{k+1}) - \nabla J(\boldsymbol{x}_k)$$

In the above, $\nabla J(\boldsymbol{x})$ is the gradient of $J(\boldsymbol{x})$ with respect to $\boldsymbol{x}$, and $\boldsymbol{S}_k$ is a positive-definite approximation of the inverse Hessian matrix of $J(\boldsymbol{x})$.

This iteration process continues until

$$|J(\boldsymbol{x}_{k+1}) - J(\boldsymbol{x}_k)| < \varepsilon \tag{12.44}$$

is satisfied where $\varepsilon > 0$ is a prescribed tolerance. If the iteration is terminated at step k, the $\boldsymbol{x}_k$ is viewed as a solution point.

12.4.3 Gradient of $J(\boldsymbol{x})$

The implementation of (12.43) requires the gradient of $J(\boldsymbol{x})$, which can be efficiently evaluated using the closed-form expressions derived below.

Each term of the objective function in (12.41) has the form $J(\boldsymbol{x}) = \mathrm{tr}[\hat{\boldsymbol{T}}\boldsymbol{N}\hat{\boldsymbol{T}}^T]$ (or $J(\boldsymbol{x}) = \mathrm{tr}[\hat{\boldsymbol{T}}^{-T}\boldsymbol{M}\hat{\boldsymbol{T}}^{-1}]$) which, in the light of (12.40), can be expressed as

$$J(\boldsymbol{x}) = \mathrm{tr}\left[\left[\frac{\boldsymbol{t}_1}{||\boldsymbol{t}_1||}, \frac{\boldsymbol{t}_2}{||\boldsymbol{t}_2||}, \cdots, \frac{\boldsymbol{t}_n}{||\boldsymbol{t}_n||}\right]^{-1} \boldsymbol{N} \left[\frac{\boldsymbol{t}_1}{||\boldsymbol{t}_1||}, \frac{\boldsymbol{t}_2}{||\boldsymbol{t}_2||}, \cdots, \frac{\boldsymbol{t}_n}{||\boldsymbol{t}_n||}\right]^{-T}\right] \tag{12.45}$$

In order to compute $\partial J(\boldsymbol{x})/\partial t_{ij}$, we perturb the ith component of vector $\boldsymbol{t}_j$ by a small amount, say Δ, and keep the rest of $\hat{\boldsymbol{T}}$ unchanged. If we denote the perturbed j the column of $\hat{\boldsymbol{T}}^{-1}$ by $\tilde{\boldsymbol{t}}_j/||\tilde{\boldsymbol{t}}_j||$, then a linear approximation of $\tilde{\boldsymbol{t}}_j/||\tilde{\boldsymbol{t}}_j||$ can be obtained as

$$\frac{\tilde{\boldsymbol{t}}_j}{||\tilde{\boldsymbol{t}}_j||} \simeq \frac{\boldsymbol{t}_j}{||\boldsymbol{t}_j||} + \Delta\, \partial\left\{\frac{\boldsymbol{t}_j}{||\boldsymbol{t}_j||}\right\}/\partial t_{ij} = \frac{\boldsymbol{t}_j}{||\boldsymbol{t}_j||} - \Delta\, \boldsymbol{g}_{ij} \tag{12.46}$$

where

$$\boldsymbol{g}_{ij} = -\partial\left\{\frac{\boldsymbol{t}_j}{||\boldsymbol{t}_j||}\right\}/\partial t_{ij} = \frac{1}{||\boldsymbol{t}_j||^3}(t_{ij}\boldsymbol{t}_j - ||\boldsymbol{t}_j||^2\boldsymbol{e}_i)$$

Now let $\hat{\boldsymbol{T}}_{ij}$ be the matrix obtained from $\hat{\boldsymbol{T}}$ with a perturbed (i,j)th component, then we obtain

$$\hat{\boldsymbol{T}}_{ij}^{-1} = \hat{\boldsymbol{T}}^{-1} - \Delta\boldsymbol{g}_{ij}\boldsymbol{e}_j^T \tag{12.47}$$

and up to the first-order, the matrix inversion formula [16, p. 655] gives

$$\hat{\boldsymbol{T}}_{ij} = \hat{\boldsymbol{T}} + \frac{\Delta\hat{\boldsymbol{T}}\boldsymbol{g}_{ij}\boldsymbol{e}_j^T\hat{\boldsymbol{T}}}{1 - \Delta\boldsymbol{e}_j^T\hat{\boldsymbol{T}}\boldsymbol{g}_{ij}} \simeq \hat{\boldsymbol{T}} + \Delta\hat{\boldsymbol{T}}\boldsymbol{g}_{ij}\boldsymbol{e}_j^T\hat{\boldsymbol{T}} \tag{12.48}$$

For convenience, we define $\hat{\boldsymbol{T}}_{ij} = \hat{\boldsymbol{T}} + \Delta \boldsymbol{S}$ and write

$$
\begin{aligned}
\hat{\boldsymbol{T}}_{ij}\boldsymbol{N}\hat{\boldsymbol{T}}_{ij}^{T} &= (\hat{\boldsymbol{T}} + \Delta\boldsymbol{S})\boldsymbol{N}(\hat{\boldsymbol{T}} + \Delta\boldsymbol{S})^{T} \\
&= \hat{\boldsymbol{T}}\boldsymbol{N}\hat{\boldsymbol{T}}^{T} + \Delta\boldsymbol{S}\boldsymbol{N}\hat{\boldsymbol{T}}^{T} + \Delta\hat{\boldsymbol{T}}\boldsymbol{N}\boldsymbol{S}^{T} + \Delta^{2}\boldsymbol{S}\boldsymbol{N}\boldsymbol{S}^{T}
\end{aligned}
\tag{12.49}
$$

which implies that

$$
\mathrm{tr}\big[\hat{\boldsymbol{T}}_{ij}\boldsymbol{N}\hat{\boldsymbol{T}}_{ij}^{T}\big] - \mathrm{tr}\big[\hat{\boldsymbol{T}}\boldsymbol{N}\hat{\boldsymbol{T}}^{T}\big] \simeq \Delta\,\mathrm{tr}\big[\boldsymbol{S}(\boldsymbol{N} + \boldsymbol{N}^{T})\hat{\boldsymbol{T}}^{T}\big] \tag{12.50}
$$

provided that Δ is sufficiently small. Hence

$$
\begin{aligned}
\frac{\partial\,\mathrm{tr}\big[\hat{\boldsymbol{T}}\boldsymbol{N}\hat{\boldsymbol{T}}^{T}\big]}{\partial t_{ij}} &= \lim_{\Delta\to 0}\frac{\mathrm{tr}\big[\hat{\boldsymbol{T}}_{ij}\boldsymbol{N}\hat{\boldsymbol{T}}_{ij}^{T}\big] - \mathrm{tr}\big[\hat{\boldsymbol{T}}\boldsymbol{N}\hat{\boldsymbol{T}}^{T}\big]}{\Delta} \\
&= \mathrm{tr}\big[\boldsymbol{S}(\boldsymbol{N} + \boldsymbol{N}^{T})\hat{\boldsymbol{T}}^{T}\big]
\end{aligned}
\tag{12.51}
$$

Similarly, if we define $\hat{\boldsymbol{T}}_{ij}^{-1} = \hat{\boldsymbol{T}}^{-1} - \Delta\boldsymbol{S}$, then we arrive at

$$
\begin{aligned}
\frac{\partial\mathrm{tr}\big[\hat{\boldsymbol{T}}^{-T}\boldsymbol{M}\hat{\boldsymbol{T}}^{-1}\big]}{\partial t_{ij}} &= \lim_{\Delta\to 0}\frac{\mathrm{tr}\big[\hat{\boldsymbol{T}}_{ij}^{-T}\boldsymbol{M}\hat{\boldsymbol{T}}_{ij}^{-1}\big] - \mathrm{tr}\big[\hat{\boldsymbol{T}}^{-T}\boldsymbol{M}\hat{\boldsymbol{T}}^{-1}\big]}{\Delta} \\
&= -\mathrm{tr}\big[\boldsymbol{S}^{T}(\boldsymbol{M} + \boldsymbol{M}^{T})\hat{\boldsymbol{T}}^{-1}\big]
\end{aligned}
\tag{12.52}
$$

Referring to (12.51), it follows from (12.41) and (12.48) that

$$
\begin{aligned}
\frac{\partial\mathrm{tr}\big[\hat{\boldsymbol{T}}\hat{\boldsymbol{N}}(\hat{\boldsymbol{T}})\hat{\boldsymbol{T}}^{T}\big]}{\partial t_{ij}} &= 2\,\mathrm{tr}\big[\hat{\boldsymbol{T}}\boldsymbol{g}_{ij}\boldsymbol{e}_{j}^{T}\hat{\boldsymbol{T}}\hat{\boldsymbol{N}}(\hat{\boldsymbol{T}})\hat{\boldsymbol{T}}^{T}\big] \\
&= 2\,\boldsymbol{e}_{j}^{T}\hat{\boldsymbol{T}}\hat{\boldsymbol{N}}(\hat{\boldsymbol{T}})\hat{\boldsymbol{T}}^{T}\hat{\boldsymbol{T}}\boldsymbol{g}_{ij}
\end{aligned}
\tag{12.53}
$$

and

$$
\begin{aligned}
\frac{\partial\mathrm{tr}\big[\hat{\boldsymbol{T}}\hat{\boldsymbol{W}}_{o}\hat{\boldsymbol{T}}^{T}\big]}{\partial t_{ij}} &= 2\,\mathrm{tr}\big[\hat{\boldsymbol{T}}\boldsymbol{g}_{ij}\boldsymbol{e}_{j}^{T}\hat{\boldsymbol{T}}\hat{\boldsymbol{W}}_{o}\hat{\boldsymbol{T}}^{T}\big] \\
&= 2\,\boldsymbol{e}_{j}^{T}\hat{\boldsymbol{T}}\hat{\boldsymbol{W}}_{o}\hat{\boldsymbol{T}}^{T}\hat{\boldsymbol{T}}\boldsymbol{g}_{ij}
\end{aligned}
\tag{12.54}
$$

Referring to (12.52), it follows from (12.41) and (12.47) that

$$\begin{aligned}\frac{\partial \mathrm{tr}\big[\hat{\boldsymbol{T}}^{-T}\hat{\boldsymbol{M}}(\hat{\boldsymbol{T}})\hat{\boldsymbol{T}}^{-1}\big]}{\partial t_{ij}} &= -2\,\mathrm{tr}\big[\boldsymbol{e}_j\boldsymbol{g}_{ij}^T\hat{\boldsymbol{M}}(\hat{\boldsymbol{T}})\hat{\boldsymbol{T}}^{-1}\big] \\ &= -2\,\boldsymbol{g}_{ij}^T\hat{\boldsymbol{M}}(\hat{\boldsymbol{T}})\hat{\boldsymbol{T}}^{-1}\boldsymbol{e}_j \\ &= -2\,\boldsymbol{e}_j^T\hat{\boldsymbol{T}}^{-T}\hat{\boldsymbol{M}}(\hat{\boldsymbol{T}})\boldsymbol{g}_{ij}\end{aligned} \tag{12.55}$$

As a result, the gradient of $J(\boldsymbol{x})$ now can be evaluated in closed-form as

$$\frac{\partial J(\hat{\boldsymbol{T}})}{\partial t_{ij}} = \lim_{\Delta\to 0}\frac{J(\hat{\boldsymbol{T}}_{ij}) - J(\hat{\boldsymbol{T}})}{\Delta} = 2\big(\beta_1 - \beta_2 + \beta_3\big) \tag{12.56}$$

where

$$\beta_1 = \boldsymbol{e}_j^T\hat{\boldsymbol{T}}\hat{\boldsymbol{N}}(\hat{\boldsymbol{T}})\hat{\boldsymbol{T}}^T\hat{\boldsymbol{T}}\boldsymbol{g}_{ij}, \qquad \beta_2 = \boldsymbol{e}_j^T\hat{\boldsymbol{T}}^{-T}\hat{\boldsymbol{M}}(\hat{\boldsymbol{T}})\boldsymbol{g}_{ij}$$
$$\beta_3 = \boldsymbol{e}_j^T\hat{\boldsymbol{T}}\hat{\boldsymbol{W}}_o\hat{\boldsymbol{T}}^T\hat{\boldsymbol{T}}\boldsymbol{g}_{ij}$$

12.5 l_2-Sensitivity Minimization Subject to l_2-Scaling Constraints Using Lagrange Function

12.5.1 Minimization of (12.19) Subject to l_2-Scaling Constraints — Using Lagrange Function

The problem of minimizing $S(\boldsymbol{P})$ in (12.19) subject to l_2-scaling constraints in (12.37) is a constrained nonlinear optimization problem where matrix $\boldsymbol{P}$ is the variable. If we sum the n constraints in (12.37) up, then we have

$$\mathrm{tr}[\boldsymbol{T}^{-1}\boldsymbol{K}_c\boldsymbol{T}^{-T}] = \mathrm{tr}[\boldsymbol{K}_c\boldsymbol{P}^{-1}] = n \tag{12.57}$$

Consequently, the problem of minimizing (12.19) subject to the constraints in (12.37) can be *relaxed* into the following problem:

$$\begin{aligned}&\text{minimize} \quad S(\boldsymbol{P}) \text{ in (12.19) with respect to } \boldsymbol{P} \\ &\text{subject to} \quad \mathrm{tr}[\boldsymbol{K}_c\boldsymbol{P}^{-1}] = n\end{aligned} \tag{12.58}$$

Although clearly a solution of problem (12.58) is not necessarily a solution of the problem of minimizing (12.19) subject to l_2-scaling constraints in (12.37), it is important to stress that the ultimate solution we seek for is not matrix $\boldsymbol{P}$ but a nonsingular matrix $\boldsymbol{T}$ that is related to the solution of the problem of minimizing (12.19) subject to l_2-scaling constraints in (12.37) as $\boldsymbol{P} = \boldsymbol{T}\boldsymbol{T}^T$. If matrix $\boldsymbol{P}$ is a solution of problem (12.58) and $\boldsymbol{P}^{1/2}$ denotes a matrix square root of $\boldsymbol{P}$, i.e., $\boldsymbol{P} = \boldsymbol{P}^{1/2}\boldsymbol{P}^{1/2}$, then it is easy to see that any matrix $\boldsymbol{T}$ of the form $\boldsymbol{T} = \boldsymbol{P}^{1/2}\boldsymbol{U}$ where $\boldsymbol{U}$ is an arbitrary orthogonal matrix still holds the relation $\boldsymbol{P} = \boldsymbol{T}\boldsymbol{T}^T$. As will be shown shortly, under the constraint $\mathrm{tr}[\boldsymbol{K}_c\boldsymbol{P}^{-1}] = n$ in (12.58) there exists an orthogonal matrix $\boldsymbol{U}$ such that matrix $\boldsymbol{T} = \boldsymbol{P}^{1/2}\boldsymbol{U}$ satisfies l_2-scaling constraints in (12.37), where $\boldsymbol{P}^{1/2}$ is a square root of the solution matrix $\boldsymbol{P}$ for problem (12.58).

For these reasons, we now address problem (12.58) as the first step of our solution procedure. To solve (12.58), we define the Lagrange function of the problem as

$$\begin{aligned} J(\boldsymbol{P},\lambda) &= \mathrm{tr}[\boldsymbol{N}(\boldsymbol{P})\boldsymbol{P}] + \mathrm{tr}[\boldsymbol{W}_o\boldsymbol{P}] + \mathrm{tr}[\boldsymbol{K}_c\boldsymbol{P}^{-1}] \\ &\quad + \lambda(\mathrm{tr}[\boldsymbol{K}_c\boldsymbol{P}^{-1}] - n) \end{aligned} \tag{12.59}$$

where λ is a Lagrange multiplier. It is well known that the solution of problem (12.58) must satisfy the Karush-Kuhn-Tucker (KKT) conditions $\partial J(\boldsymbol{P},\lambda)/\partial \boldsymbol{P} = \boldsymbol{0}$ and $\partial J(\boldsymbol{P},\lambda)/\partial\lambda = 0$. From (12.23), the gradients are found to be

$$\begin{aligned} \frac{\partial J(\boldsymbol{P},\lambda)}{\partial \boldsymbol{P}} &= \boldsymbol{N}(\boldsymbol{P}) - \boldsymbol{P}^{-1}\boldsymbol{M}(\boldsymbol{P})\boldsymbol{P}^{-1} \\ &\quad + \boldsymbol{W}_o - (\lambda+1)\boldsymbol{P}^{-1}\boldsymbol{K}_c\boldsymbol{P}^{-1} \\ \frac{\partial J(\boldsymbol{P},\lambda)}{\partial \lambda} &= \mathrm{tr}[\boldsymbol{K}_c\boldsymbol{P}^{-1}] - n \end{aligned} \tag{12.60}$$

where $\boldsymbol{M}(\boldsymbol{P})$ can be obtained by solving (12.25) and (12.26). Hence the KKT conditions become

$$\boldsymbol{P}\boldsymbol{F}(\boldsymbol{P})\boldsymbol{P} = \boldsymbol{G}(\boldsymbol{P},\lambda), \qquad \mathrm{tr}[\boldsymbol{K}_c\boldsymbol{P}^{-1}] = n \tag{12.61}$$

where

$$\begin{aligned} \boldsymbol{F}(\boldsymbol{P}) &= \boldsymbol{N}(\boldsymbol{P}) + \boldsymbol{W}_o \\ \boldsymbol{G}(\boldsymbol{P},\lambda) &= \boldsymbol{M}(\boldsymbol{P}) + (\lambda+1)\boldsymbol{K}_c \end{aligned}$$

The first equation in (12.61) is highly nonlinear with respect to $\boldsymbol{P}$. An effective approach for solving the first equation in (12.61) is to *relax* it into the recursive second-order matrix equation

$$\boldsymbol{P}_{k+1}\boldsymbol{F}(\boldsymbol{P}_k)\boldsymbol{P}_{k+1} = \boldsymbol{G}(\boldsymbol{P}_k,\lambda_{k+1}) \tag{12.62}$$

where $\boldsymbol{P}_k$ is assumed to be known from the previous recursion. Recalling the solution given by (12.29), the solution $\boldsymbol{P}_{k+1}$ of (12.62) is given by

$$\boldsymbol{P}_{k+1}=\boldsymbol{F}(\boldsymbol{P}_k)^{-\frac{1}{2}}[\boldsymbol{F}(\boldsymbol{P}_k)^{\frac{1}{2}}\boldsymbol{G}(\boldsymbol{P}_k,\lambda_{k+1})\boldsymbol{F}(\boldsymbol{P}_k)^{\frac{1}{2}}]^{\frac{1}{2}}\boldsymbol{F}(\boldsymbol{P}_k)^{-\frac{1}{2}} \quad (12.63)$$

To derive a recursive formula for the Lagrange multiplier λ, we employ (12.61) to write

$$\mathrm{tr}[\boldsymbol{P}\boldsymbol{F}(\boldsymbol{P})] = \mathrm{tr}[\boldsymbol{M}(\boldsymbol{P})\boldsymbol{P}^{-1}] + n(\lambda+1) \quad (12.64)$$

which naturally suggests the recursion for λ

$$\lambda_{k+1} = \frac{\mathrm{tr}[\boldsymbol{P}_k\boldsymbol{F}(\boldsymbol{P}_k)] - \mathrm{tr}[\boldsymbol{M}(\boldsymbol{P}_k)\boldsymbol{P}_k^{-1}]}{n} - 1 \quad (12.65)$$

The iteration process starts with an initial estimate $\boldsymbol{P}_0$, and continues until

$$\left|S(\boldsymbol{P}_{k+1}) - S(\boldsymbol{P}_k)\right| + \left|n - \mathrm{tr}[\boldsymbol{K}_c\boldsymbol{P}_{k+1}^{-1}]\right| < \varepsilon \quad (12.66)$$

is satisfied for a prescribed tolerance $\varepsilon > 0$. If the iteration is terminated at step k, $\boldsymbol{P}_k$ is claimed to be a solution point.

12.5.2 Derivation of Nonsingular T from P to Satisfy l_2-Scaling Constraints

As the second step of the solution procedure, having obtained an optimal $\boldsymbol{P}$, we now turn our attention to the construction of the optimal coordinate transformation matrix $\boldsymbol{T}$ that solves the problem of minimizing (12.19) subject to l_2-scaling constraints in (12.37). As is given in (12.32), the optimal $\boldsymbol{T}$ assumes the form

$$\boldsymbol{T} = \boldsymbol{P}^{\frac{1}{2}}\boldsymbol{U} \quad (12.67)$$

where $\boldsymbol{U}$ is an $n \times n$ orthogonal matrix that can be determined to satisfy the l_2-scaling constraints as follows. From (12.17) and (12.67), it follows that

$$\overline{\boldsymbol{K}}_c = \boldsymbol{U}^T\boldsymbol{P}^{-\frac{1}{2}}\boldsymbol{K}_c\boldsymbol{P}^{-\frac{1}{2}}\boldsymbol{U} \quad (12.68)$$

To find an $n \times n$ orthogonal matrix $\boldsymbol{U}$ such that the matrix $\overline{\boldsymbol{K}}_c$ satisfies the l_2-scaling constraints in (12.37), we perform the eigenvalue-eigenvector decomposition for the symmetric positive-definite matrix $\boldsymbol{P}^{-\frac{1}{2}}\boldsymbol{K}_c\boldsymbol{P}^{-\frac{1}{2}}$ as

$$\boldsymbol{P}^{-\frac{1}{2}}\boldsymbol{K}_c\boldsymbol{P}^{-\frac{1}{2}} = \boldsymbol{R}\boldsymbol{\Theta}\boldsymbol{R}^T \quad (12.69)$$

where $\boldsymbol{\Theta} = \mathrm{diag}\{\theta_1, \theta_2, \cdots, \theta_n\}$ with $\theta_i > 0$ for all i and $\boldsymbol{R}$ is an $n \times n$ orthogonal matrix. Next, an $n \times n$ orthogonal matrix $\boldsymbol{S}$ such that

$$\boldsymbol{e}_i^T \boldsymbol{S}\boldsymbol{\Theta}\boldsymbol{S}^T \boldsymbol{e}_i = 1 \ \text{ for } i = 1, 2, \cdots, n \tag{12.70}$$

can be obtained by a numerical procedure [13, p. 278]. Using (12.68)–(12.70), it can be readily verified that the orthogonal matrix $\boldsymbol{U} = \boldsymbol{R}\boldsymbol{S}^T$ leads to a $\overline{\boldsymbol{K}}_c$ in (12.68) whose diagonal elements are equal to unity, hence the l_2-scaling constraints in (12.37) are satisfied. This matrix $\boldsymbol{U}$ together with (12.67) yields a solution for the problem of minimizing (12.19) subject to the l_2-scaling constraints in (12.37) as

$$\boldsymbol{T} = \boldsymbol{P}^{\frac{1}{2}} \boldsymbol{R}\boldsymbol{S}^T \tag{12.71}$$

12.6 Numerical Experiments

12.6.1 Filter Description and Initial l_2-Sensitivity

Consider a third-order lowpass IIR digital filter described by

$$H(z) = 10^{-1} \frac{0.15940z^3 + 0.47821z^2 + 0.47825z + 0.15937}{z^3 - 1.97486z^2 + 1.55616z - 0.45377}$$

Magnitude response of this filter is depicted in Figure 12.1.

This filter can be realized by a state-space model $(\boldsymbol{A}, \boldsymbol{b}, \boldsymbol{c}, d)_3$ in (12.1) as

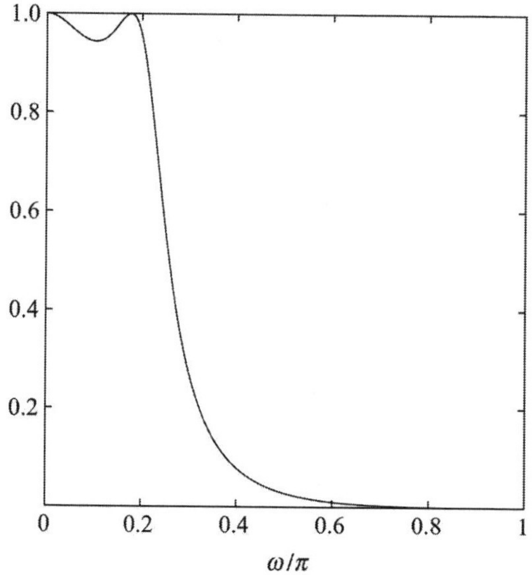

Figure 12.1 The magnitude response of a lowpass digital filter.

$$A = \begin{bmatrix} 0 & 1 & 0 \\ 0 & 0 & 1 \\ 0.45377 & -1.55616 & 1.97486 \end{bmatrix}, \qquad b = \begin{bmatrix} 0 \\ 0 \\ 1 \end{bmatrix}$$

$$c = 10^{-1} \begin{bmatrix} 0.2317 & 0.2302 & 0.7930 \end{bmatrix}, \qquad d = 0.01594$$

Carrying out the computation of (12.13), (12.21), (12.22), (12.25) and (12.26), the controllability and observability Grammians K_c and W_o, and matrices $N(I_3)$ and $M(I_3)$ were computed as

$$K_c = \begin{bmatrix} 17.061835 & 14.886464 & 9.602768 \\ 14.886464 & 17.061835 & 14.886464 \\ 9.602768 & 14.886464 & 17.061835 \end{bmatrix}$$

$$W_o = \begin{bmatrix} 0.048104 & -0.119291 & 0.095427 \\ -0.119291 & 0.311061 & -0.249968 \\ 0.095427 & -0.249968 & 0.231012 \end{bmatrix}$$

$$N(I_3) = \begin{bmatrix} 8.921384 & -22.046468 & 17.916293 \\ -22.046468 & 55.671739 & -46.052035 \\ 17.916293 & -46.052035 & 42.522104 \end{bmatrix}$$

$$M(I_3) = \begin{bmatrix} 35.705076 & 32.086502 & 22.476116 \\ 32.086502 & 35.705076 & 32.086502 \\ 22.476116 & 32.086502 & 35.705076 \end{bmatrix}$$

and the l_2-sensitivity measure in (12.9) was found to be

$$S_o = 158.890911$$

Performing the l_2-scaling to the above state-space model $(A, b, c, d)_3$ with a diagonal coordinate-transformation matrix given by

$$T_o = \text{diag}\{4.130597 \quad 4.130597 \quad 4.130597\}$$

led the controllability Grammian to

$$\overline{K}_c = T_o^{-1} K_c T_o^{-T} = \begin{bmatrix} 1.000000 & 0.872501 & 0.562821 \\ 0.872501 & 1.000000 & 0.872501 \\ 0.562821 & 0.872501 & 1.000000 \end{bmatrix}$$

and the l_2-sensitivity subject to l_2-scaling was found to be

$$S'_o = 120.184738$$

12.6.2 l_2-Sensitivity Minimization

Choosing $\boldsymbol{P}_0 = \boldsymbol{I}_3$ in (12.30) as the initial estimate, and setting tolerance to $\varepsilon = 10^{-8}$ in (12.31), it took the algorithm addressed in Section 12.3 forty-one iterations to converge to

$$\boldsymbol{P} = \begin{bmatrix} 81.462531 & 48.380250 & 17.978611 \\ 48.380250 & 39.029141 & 24.141251 \\ 17.978611 & 24.141251 & 23.949438 \end{bmatrix}$$

which yields

$$\boldsymbol{T} = \boldsymbol{P}^{\frac{1}{2}} = \begin{bmatrix} 8.189112 & 3.732952 & 0.682678 \\ 3.732952 & 4.301777 & 2.566890 \\ 0.682678 & 2.566890 & 4.110287 \end{bmatrix}$$

where $\boldsymbol{U}$ in (12.32) was chosen as $\boldsymbol{U} = \boldsymbol{I}_3$. The profile of the l_2-sensitivity measure in (12.19) during the first 41 iterations of the algorithm is shown in Figure 12.2.

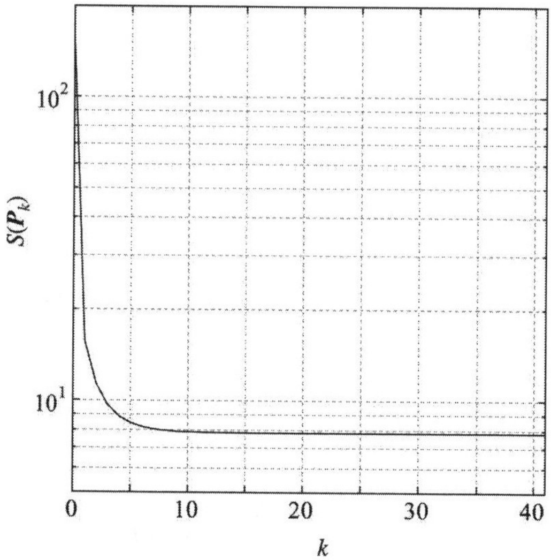

Figure 12.2 Profile of $S(\boldsymbol{P}_k)$ during the first 41 iterations.

An equivalent realization was then obtained from (12.16) as

$$\overline{A} = \begin{bmatrix} 0.608861 & 0.297687 & 0.046275 \\ -0.320667 & 0.570460 & 0.440629 \\ -0.082097 & -0.388935 & 0.795539 \end{bmatrix}, \quad \overline{b} = \begin{bmatrix} 0.148865 \\ -0.413801 \\ 0.476987 \end{bmatrix}$$

$$\overline{c} = \begin{bmatrix} 0.329811 & 0.389074 & 0.400853 \end{bmatrix}$$

Moreover, the controllability and observability Grammians $\overline{K}_c$ and $\overline{W}_o$, and matrices $N(P)$ and $M(P)$ were computed from (12.17), (12.21), (12.22), (12.25) and (12.26) as

$$\overline{K}_c = \begin{bmatrix} 0.121737 & 0.026720 & 0.058892 \\ 0.026720 & 0.455129 & -0.024594 \\ 0.058892 & -0.024594 & 0.834658 \end{bmatrix}$$

$$\overline{W}_o = \begin{bmatrix} 0.167774 & 0.125597 & 0.069289 \\ 0.125597 & 0.425887 & 0.065911 \\ 0.069289 & 0.065911 & 0.817617 \end{bmatrix}$$

$$N(P) = \begin{bmatrix} 0.189796 & -0.479778 & 0.376453 \\ -0.479778 & 1.249598 & -1.002489 \\ 0.376453 & -1.002489 & 0.921438 \end{bmatrix}$$

$$M(P) = \begin{bmatrix} 68.481534 & 59.708525 & 36.979189 \\ 59.708525 & 68.481534 & 59.708525 \\ 36.979189 & 59.708525 & 68.481534 \end{bmatrix}$$

With realization $(\overline{A}, \overline{b}, \overline{c}, d)_3$, the l_2-sensitivity measure in (12.19) was minimized to

$$S(P) = 7.832683$$

12.6.3 l_2-Sensitivity Minimization Subject to l_2-Scaling Constraints Using Quasi-Newton Algorithm

The quasi-Newton algorithm was applied to minimize (12.41) by choosing $\hat{T} = I_3$ as an initial assignment, and setting tolerance to $\varepsilon = 10^{-8}$ in (12.44). It took the algorithm addressed in Section 12.4 twelve iterations to converge to the solution

$$\hat{T} = \begin{bmatrix} 1.399512 & -0.399054 & 0.519219 \\ -0.909296 & 0.955306 & 0.384011 \\ -1.034718 & -0.498614 & 0.380517 \end{bmatrix}$$

which is equivalent to

$$T = \begin{bmatrix} 4.425883 & -0.877680 & -4.360270 \\ 2.866741 & 1.652464 & -2.805767 \\ 2.021688 & 2.654510 & -0.495080 \end{bmatrix}$$

The profile of $J(\boldsymbol{x})$ during the first 12 iterations of the algorithm is depicted in Figure 12.3.

As a result, an equivalent realization was derived from (12.16) as

$$\overline{\boldsymbol{A}} = \begin{bmatrix} 0.664789 & 0.072468 & 0.067395 \\ 0.074199 & 0.717025 & 0.590431 \\ 0.002388 & -0.449754 & 0.593046 \end{bmatrix}, \quad \overline{\boldsymbol{b}} = \begin{bmatrix} 0.668510 \\ -0.005654 \\ 0.679708 \end{bmatrix}$$

$$\overline{\boldsymbol{c}} = \begin{bmatrix} 0.328860 & 0.228207 & -0.204876 \end{bmatrix}$$

In addition, the controllability and observability Grammians $\overline{\boldsymbol{K}}_c$ and $\overline{\boldsymbol{W}}_o$, and matrices $\boldsymbol{N}(\boldsymbol{P})$ and $\boldsymbol{M}(\boldsymbol{P})$ were computed from (12.17), (12.21), (12.22), (12.25) and (12.26) as

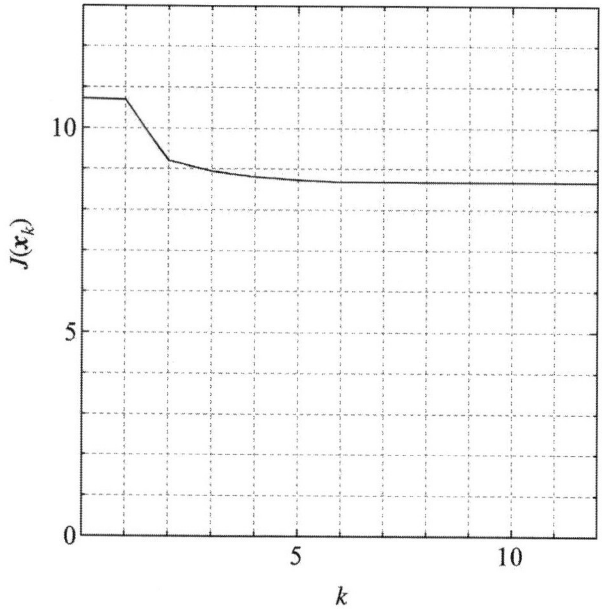

Figure 12.3 Profile of $J(\boldsymbol{x}_k)$ during the first 12 iterations.

$$\overline{\boldsymbol{K}}_c = \begin{bmatrix} 1.000000 & 0.622354 & 0.458665 \\ 0.622354 & 1.000000 & 0.029904 \\ 0.458665 & 0.029904 & 1.000000 \end{bmatrix}$$

$$\overline{\boldsymbol{W}}_o = \begin{bmatrix} 0.226005 & 0.168659 & 0.033336 \\ 0.168659 & 0.222675 & 0.007233 \\ 0.033336 & 0.007233 & 0.218695 \end{bmatrix}$$

$$\boldsymbol{N}(\boldsymbol{P}) = \begin{bmatrix} 0.401254 & -1.014637 & 0.795713 \\ -1.014637 & 2.643913 & -2.120091 \\ 0.795713 & -2.120091 & 1.948564 \end{bmatrix}$$

$$\boldsymbol{M}(\boldsymbol{P}) = \begin{bmatrix} 32.235281 & 28.094505 & 17.370799 \\ 28.094505 & 32.235281 & 28.094505 \\ 17.370799 & 28.094505 & 32.235281 \end{bmatrix}$$

and the minimized l_2-sensitivity measure in (12.41) was found to be

$$J(\hat{\boldsymbol{T}}) = 8.672132$$

12.6.4 l_2-Sensitivity Minimization Subject to l_2-Scaling Constraints Using Lagrange Function

The recursive matrix equation in (12.63) together with (12.65) was applied to minimize (12.59) by choosing $\boldsymbol{P}_0 = \boldsymbol{I}_3$ in (12.63) and (12.65) as an initial assignment, and setting tolerance to $\varepsilon = 10^{-6}$ in (12.66). It took the algorithm addressed in Section 12.5 two-hundred sixty-seven iterations to converge to the solution

$$\boldsymbol{P} = \begin{bmatrix} 39.370678 & 23.471408 & 8.776617 \\ 23.471408 & 18.821159 & 11.571209 \\ 8.776617 & 11.571209 & 11.378742 \end{bmatrix}$$

which in conjunction with (12.71) led to

$$\boldsymbol{T} = \begin{bmatrix} -3.372610 & 5.212122 & 0.911024 \\ -0.385158 & 4.309519 & -0.317577 \\ 1.806782 & 2.848438 & 0.026109 \end{bmatrix}$$

By using the coordinate transformation matrix $\boldsymbol{T}$ obtained above, an equivalent realization was derived from (12.16) as

$$\overline{\boldsymbol{A}} = \begin{bmatrix} 0.737652 & -0.291843 & 0.388074 \\ 0.460932 & 0.635801 & 0.085061 \\ -0.329056 & 0.012485 & 0.601407 \end{bmatrix}, \quad \overline{\boldsymbol{b}} = \begin{bmatrix} 0.385940 \\ 0.098326 \\ 0.866211 \end{bmatrix}$$

$$\overline{\boldsymbol{c}} = \begin{bmatrix} 0.056268 & 0.445851 & 0.015868 \end{bmatrix}$$

In addition, the controllability and observability Grammians $\overline{\boldsymbol{K}}_c$ and $\overline{\boldsymbol{W}}_o$, and matrices $\boldsymbol{N}(\boldsymbol{P})$ and $\boldsymbol{M}(\boldsymbol{P})$ were computed from (12.17), (12.21), (12.22), (12.25) and (12.26) as

$$\overline{\boldsymbol{K}}_c = \begin{bmatrix} 1.000000 & 0.546707 & 0.546707 \\ 0.546707 & 1.000000 & 0.028559 \\ 0.546707 & 0.028559 & 1.000000 \end{bmatrix}$$

$$\overline{\boldsymbol{W}}_o = \begin{bmatrix} 0.222437 & 0.110113 & 0.109850 \\ 0.110113 & 0.295772 & 0.007599 \\ 0.109850 & 0.007599 & 0.149166 \end{bmatrix}$$

$$\boldsymbol{N}(\boldsymbol{P}) = \begin{bmatrix} 0.401254 & -1.014637 & 0.795713 \\ -1.014637 & 2.643914 & -2.120092 \\ 0.795713 & -2.120092 & 1.948565 \end{bmatrix}$$

$$\boldsymbol{M}(\boldsymbol{P}) = \begin{bmatrix} 32.235271 & 28.094497 & 17.370795 \\ 28.094497 & 32.235271 & 28.094497 \\ 17.370795 & 28.094497 & 32.235271 \end{bmatrix}$$

and the minimized l_2-sensitivity measure in (12.19) was found to be

$$S(\boldsymbol{P}) = 8.672133$$

where $J(\boldsymbol{P}, \lambda) = 8.672132$ and $\lambda = -0.777542$.

The profiles of $S(\boldsymbol{P})$ in (12.19) and $\mathrm{tr}[\boldsymbol{K}_c \boldsymbol{P}^{-1}]$ during the first 267 iterations of the algorithm are depicted in Figures 12.4.

12.7 Summary

The minimization problem of l_2-sensitivity for state-space digital filters has been considered with or without l_2-scaling constraints. The problem free from l_2-scaling constraints has been solved by employing a recursive matrix equation. The constrained optimization problem has been solved by two iterative methods: one converts the constrained optimization problem at hand into an unconstrained problem and solves it using a quasi-Newton algorithm,

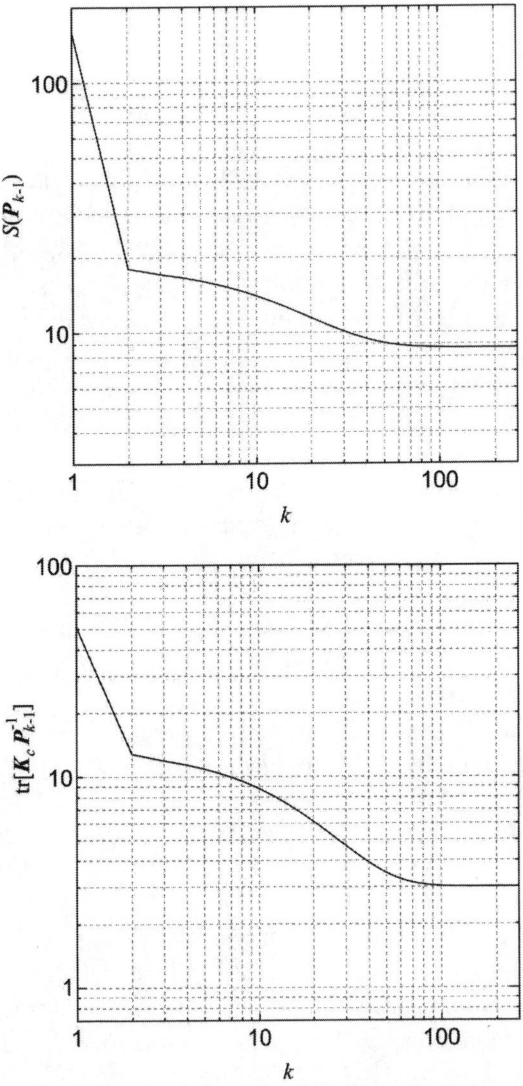

Figure 12.4 Profiles of $S(\boldsymbol{P}_k)$ and $\mathrm{tr}[\boldsymbol{K}_c\boldsymbol{P}_k^{-1}]$ during the first 267 iterations.

while the other relaxes the constraints into a single constraint on matrix trace and solves the relaxed problem with an efficient matrix iteration scheme based on the Lagrange function. Simulation results in numerical experiments have demonstrated the validity and effectiveness of these techniques.

References

[1] L. Thiele, "Design of sensitivity and round-off noise optimal state-space discrete systems," *Int. J. Circuit Theory Appl.*, vol. 12, pp. 39–46, Jan. 1984.

[2] V. Tavsanoglu and L. Thiele, "Optimal design of state-space digital filter by simultaneous minimization of sensitivity and roundoff noise," *IEEE Trans. Circuits Syst.*, vol. CAS-31, no. 10, pp. 884–888, Oct. 1984.

[3] L. Thiele, "On the sensitivity of linear state-space systems," *IEEE Trans. Circuits Syst.*, vol. CAS-33, no. 5, pp. 502–510, May 1986.

[4] M. Iwatsuki, M. Kawamata and T. Higuchi, "Statistical sensitivity and minimum sensitivity structures with fewer coefficients in discrete time linear systems," *IEEE Trans. Circuits Syst.*, vol. CAS-37, no. 1, pp. 72–80, Jan. 1989.

[5] G. Li, B. D. O. Anderson, M. Gevers and J. E. Perkins, "Optimal FWL design of state-space digital systems with weighted sensitivity minimization and sparseness consideration," *IEEE Trans. Circuits Syst. I*, vol. 39, no. 5, pp. 365–377, May 1992.

[6] W.-Y. Yan and J. B. Moore, "On L_2-sensitivity minimization of linear state-space systems," *IEEE Trans. Circuits Syst. I*, vol. 39, no. 8, pp. 641–648, Aug. 1992.

[7] G. Li and M. Gevers, "Optimal synthetic FWL design of state-space digital filters," in *Proc. ICASSP 1992*, vol. 4, pp. 429–432.

[8] M. Gevers and G. Li, *Parameterizations in Control, Estimation and Filtering Problems: Accuracy Aspects*. New York: Springer-Verlag, 1993.

[9] T. Hinamoto, S. Yokoyama, T. Inoue, W. Zeng and W.-S. Lu, "Analysis and minimization of L_2-sensitivity for linear systems and two-dimensional state-space filters using general controllability and observability Grammians," *IEEE Trans. Circuits Syst. I*, vol. 49, no. 9, pp. 1279–1289, Sept. 2002.

[10] T. Hinamoto, H. Ohnishi and W.-S. Lu, "Minimization of L_2-sensitivity for state-space digital filters subject to L_2-dynamic-range scaling constraints," *IEEE Trans. Circuits Syst.-II*, vol. 52, no. 10, pp. 641–645, Oct. 2005.

[11] T. Hinamoto, K. Iwata and W.-S. Lu, "L_2-sensitivity Minimization of one- and two-dimensional state-space digital filters subject to L_2-scaling

constraints," *IEEE Trans. Signal Process.*, vol. 54, no. 5, pp. 1804–1812, May 2006.

[12] C. T. Mullis and R. A. Roberts, "Synthesis of minimum roundoff noise fixed-point digital filters," *IEEE Trans. Circuits Syst.*, vol. CAS-23, no. 9, pp. 551–562, Sept. 1976.

[13] S. Y. Hwang, "Minimum uncorrelated unit noise in state-space digital filtering," *IEEE Trans. Acoust., Speech, Signal Process.*, vol. ASSP-25, no. 4, pp. 273–281, Aug. 1977.

[14] L. L. Scharf, *Statistical Signal Processing*, Reading, MA: Addison-Wesley, 1991.

[15] R. Fletcher, *Practical Methods of Optimization*, 2nd ed., Wiley, New York, 1987.

[16] T. Kailath, *Linear System*, Englewood Cliffs, N.J.: Prentice-Hall, 1980.

13

Pole and Zero Sensitivity Analysis and Minimization

13.1 Preview

When a transfer function with infinite accuracy coefficients is designed so as to meet the filter specification requirements and the transfer function is implemented by a state-space model with a finite binary representation, the state-space parameters must be truncated or rounded to fit the finite-word-length (FWL) constraints. This coefficient quantization inevitably changes the characteristics of the digital filter. For instance, it may alter a stable filter to an unstable one. This motivates the study of minimizing coefficients sensitivity. As is well known, there are several ways to define sensitivity of a filter with respect to its coefficients. Two of them are based on a mixed l_1/l_2 norm and a pure l_2 norm, respectively. One of these sensitivity definitions measures changes of a certain transfer function, while the other is defined in terms of the poles and zeros of a filter. Several techniques concerning minimization of the l_1/l_2- and l_2-sensitivity measures have been proposed [1–7]. Alternatively, pole and zero sensitivity of a filter with respect to state-space parameters has been analyzed and its reduction and minimization have been addressed [7–12]. A method for minimizing a zero sensitivity measure subject to minimal pole sensitivity for state-space digital filters has also been explored without taking l_2-scaling into account [11]. Recently, techniques for minimizing a weighted pole and zero sensitivity measure subject to l_2-scaling constraints have been developed [12].

In this chapter, a weighted measure for pole and zero sensitivity for state-space digital filters is introduced, and the problem of minimizing this measure is studied. To this end, an iterative technique for minimizing this measure is presented by employing a recursive matrix equation. A simple method for minimizing a zero sensitivity measure subject to minimal pole sensitivity is also given by pursuing an optimal coordinate transformation. Furthermore, the minimization of the above sensitivity measure subject to l_2-scaling constraints is performed by extending the aforementioned solution method. This method relaxes the constraints into a single constraint on matrix trace and solves the relaxed problem with an efficient matrix iteration scheme where the Lagrange multiplier is determined via a bisection method. The problem of minimizing the above sensitivity measure subject to l_2-scaling constraints is also explored by applying a quasi-Newton algorithm where we convert the constrained optimization problem at hand into an unconstrained problem and then solve it iteratively by using a quasi-Newton algorithm.

Numerical experiments are included to demonstrate the validity and effectiveness of the above techniques.

13.2 Pole and Zero Sensitivity Analysis

Consider a stable, controllable, and observable state-space digital filter $(\boldsymbol{A}, \boldsymbol{b}, \boldsymbol{c}, d)_n$ of order n described by

$$\begin{aligned} \boldsymbol{x}(k+1) &= \boldsymbol{A}\boldsymbol{x}(k) + \boldsymbol{b}u(k) \\ y(k) &= \boldsymbol{c}\boldsymbol{x}(k) + du(k) \end{aligned} \tag{13.1}$$

where $\boldsymbol{x}(k)$ is an $n \times 1$ state-variable vector, $u(k)$ is a scalar input, $y(k)$ is a scalar output, and $\boldsymbol{A}, \boldsymbol{b}, \boldsymbol{c}$ and d are real constant matrices of appropriate dimensions. The transfer function of the digital filter in (13.1) can be expressed as

$$H(z) = \boldsymbol{c}(z\boldsymbol{I}_n - \boldsymbol{A})^{-1}\boldsymbol{b} + d \tag{13.2}$$

Assuming that a direct path from the input to the output exists in (13.1), i.e., $d \neq 0$, the poles and zeros of $H(z)$ are given by $\{\lambda_l\} = \lambda(\boldsymbol{A})$ and $\{v_l\} = \lambda(\boldsymbol{Z})$, respectively, where an $n \times n$ matrix $\boldsymbol{Z}$ is defined by

$$\boldsymbol{Z} = \boldsymbol{A} - d^{-1}\boldsymbol{b}\boldsymbol{c} \tag{13.3}$$

Notice that if $d \neq 0$ then we obtain [13]

$$\begin{aligned} H(z)^{-1} &= d^{-1} - d^{-1}\boldsymbol{c}\big[z\boldsymbol{I}_n - \big(\boldsymbol{A} - d^{-1}\boldsymbol{b}\boldsymbol{c}\big)\big]^{-1}\boldsymbol{b}\,d^{-1} \\ &= \frac{1}{\boldsymbol{c}(z\boldsymbol{I}_n - \boldsymbol{A})^{-1}\boldsymbol{b} + d} \end{aligned} \tag{13.4}$$

This reveals that the zeros of the filter in (13.1) coincides with the eigenvalues of matrix $\boldsymbol{Z} = \boldsymbol{A} - d^{-1}\boldsymbol{b}\boldsymbol{c}$.

The pole sensitivity matrix for the lth eigenvalue λ_l of $\boldsymbol{A}$ is defined by

$$\frac{\partial \lambda_l}{\partial \boldsymbol{A}} = \begin{bmatrix} \frac{\partial \lambda_l}{\partial a_{11}} & \cdots & \frac{\partial \lambda_l}{\partial a_{1n}} \\ \vdots & \ddots & \vdots \\ \frac{\partial \lambda_l}{\partial a_{n1}} & \cdots & \frac{\partial \lambda_l}{\partial a_{nn}} \end{bmatrix} \tag{13.5}$$

and the zero sensitivity matrix for the lth eigenvalue v_l of $\boldsymbol{Z}$ is defined by

$$\frac{\partial v_l}{\partial \boldsymbol{Z}} = \begin{bmatrix} \frac{\partial v_l}{\partial z_{11}} & \cdots & \frac{\partial v_l}{\partial z_{1n}} \\ \vdots & \ddots & \vdots \\ \frac{\partial v_l}{\partial z_{n1}} & \cdots & \frac{\partial v_l}{\partial z_{nn}} \end{bmatrix} \tag{13.6}$$

Lemma 13.1
Let $\boldsymbol{x}_p(l)$ be a right eigenvector of $\boldsymbol{A}$ corresponding to λ_l and $\boldsymbol{y}_p(l)$ be the reciprocal left eigenvector of $\boldsymbol{A}$ that corresponds to $\boldsymbol{x}_p(l)$. If $\boldsymbol{A}$ has a full set of n linearly independent eigenvectors, then the following holds:

$$\left(\frac{\partial \lambda_l}{\partial \boldsymbol{A}}\right)^T = \boldsymbol{x}_p(l)\boldsymbol{y}_p^H(l) \ \text{ for } \ l = 1, 2, \cdots, n \tag{13.7}$$

Proof
Since $\boldsymbol{A}$ has n linearly independent eigenvectors $\{\boldsymbol{x}_p(l), l = 1, 2, \cdots, n\}$, we have

$$\boldsymbol{A}\boldsymbol{x}_p(l) = \lambda_l \boldsymbol{x}_p(l) \ \text{ for } \ l = 1, 2, \cdots, n \tag{13.8}$$

The reciprocal basis vectors $\boldsymbol{y}_p(l)$ for $l = 1, 2, \cdots, n$ are defined by

$$\begin{aligned} \boldsymbol{X}_p &= \left[\boldsymbol{x}_p(1), \boldsymbol{x}_p(2), \cdots, \boldsymbol{x}_p(n)\right] \\ \boldsymbol{Y}_p &= \left[\boldsymbol{y}_p(1), \boldsymbol{y}_p(2), \cdots, \boldsymbol{y}_p(n)\right] = \boldsymbol{X}_p^{-H} \end{aligned} \tag{13.9}$$

Hence

$$\boldsymbol{Y}_p^H \boldsymbol{X}_p = \boldsymbol{I}_n \iff \boldsymbol{y}_p^H(k)\boldsymbol{x}_p(l) = \delta_{kl} \tag{13.10}$$

Multiplying (13.8) from the left by $\boldsymbol{y}_p(l)^H$ and using (13.10) yields

$$\lambda_l = \boldsymbol{y}_p^H(l)\boldsymbol{A}\boldsymbol{x}_p(l) \tag{13.11}$$

By virtue of the two identities from linear algebra

$$\mathrm{tr}[\boldsymbol{AB}] = \mathrm{tr}[\boldsymbol{BA}]$$
$$\frac{\partial}{\partial \boldsymbol{A}}\mathrm{tr}[\boldsymbol{AB}] = \frac{\partial}{\partial \boldsymbol{A}}\mathrm{tr}[\boldsymbol{BA}] = \boldsymbol{B}^T \tag{13.12}$$

the differentiation of (13.11) with respect to $\boldsymbol{A}$ becomes

$$\frac{\partial \lambda_l}{\partial \boldsymbol{A}} = \frac{\partial}{\partial \boldsymbol{A}}\boldsymbol{y}_p(l)^H \boldsymbol{A}\boldsymbol{x}_p(l) = \frac{\partial}{\partial \boldsymbol{A}}\mathrm{tr}[\boldsymbol{A}\boldsymbol{x}_p(l)\boldsymbol{y}_p(l)^H] = (\boldsymbol{x}_p(l)\boldsymbol{y}_p(l)^H)^T \tag{13.13}$$

This completes the proof of Lemma 13.1. ■

Lemma 13.2

Let $\boldsymbol{x}_z(l)$ be a right eigenvector of $\boldsymbol{Z}$ corresponding to v_l and $\boldsymbol{y}_z(l)$ be the reciprocal left eigenvector that corresponds to $\boldsymbol{x}_z(l)$. If $\boldsymbol{Z}$ has n linearly independent eigenvectors, then we have

$$\left(\frac{\partial v_l}{\partial \boldsymbol{Z}}\right)^T = \boldsymbol{x}_z(l)\boldsymbol{y}_z^H(l) \ \text{ for } l = 1, 2, \cdots, n \tag{13.14}$$

Proof

The proof of this lemma is identical to that of Lemma 13.1. ■

Lemma 13.3

If (13.8) holds, it can be verified that

$$\begin{aligned}
\frac{\partial v_l}{\partial \boldsymbol{A}} &= \frac{\partial v_l}{\partial \boldsymbol{Z}}, & \frac{\partial v_l}{\partial \boldsymbol{b}} &= -d^{-1}\frac{\partial v_l}{\partial \boldsymbol{Z}}\boldsymbol{c}^T \\
\frac{\partial v_l}{\partial \boldsymbol{c}} &= -d^{-1}\boldsymbol{b}^T\frac{\partial v_l}{\partial \boldsymbol{Z}}, & \frac{\partial v_l}{\partial d} &= d^{-2}\boldsymbol{b}^T\frac{\partial v_l}{\partial \boldsymbol{Z}}\boldsymbol{c}^T
\end{aligned} \tag{13.15}$$

Proof

Let a_{ij} and z_{ij} be the (i,j)th entry of $\boldsymbol{A}$ and $\boldsymbol{Z}$, respectively. By virtue of $v_l = \boldsymbol{y}_z^H(l)\boldsymbol{Z}\boldsymbol{x}_z(l)$, $\boldsymbol{Z}\boldsymbol{x}_z(l) = v_l\boldsymbol{x}_z(l)$, $\boldsymbol{y}_z^H(l)\boldsymbol{Z} = v_l\boldsymbol{y}_z^H(l)$ and $\boldsymbol{y}_z^H(l)\boldsymbol{x}_z(l) = 1$, we can write

$$\begin{aligned}
\frac{\partial v_l}{\partial a_{ij}} &= \frac{\partial \boldsymbol{y}_z^H(l)}{\partial a_{ij}}\boldsymbol{Z}\boldsymbol{x}_z(l) + \boldsymbol{y}_z^H(l)\frac{\partial \boldsymbol{Z}}{\partial a_{ij}}\boldsymbol{x}_z(l) + \boldsymbol{y}_z^H(l)\boldsymbol{Z}\frac{\partial \boldsymbol{x}_z(l)}{\partial a_{ij}} \\
&= \boldsymbol{y}_z^H(l)\frac{\partial \boldsymbol{Z}}{\partial a_{ij}}\boldsymbol{x}_z(l) + v_l\left\{\frac{\partial \boldsymbol{y}_z^H(l)}{\partial a_{ij}}\boldsymbol{x}_z(l) + \boldsymbol{y}_z^H(l)\frac{\partial \boldsymbol{x}_z(l)}{\partial a_{ij}}\right\} \\
&= \boldsymbol{y}_z^H(l)\frac{\partial \boldsymbol{Z}}{\partial a_{ij}}\boldsymbol{x}_z(l) = \boldsymbol{y}_z^H(l)\frac{\partial \boldsymbol{Z}}{\partial z_{ij}}\boldsymbol{x}_z(l) = \frac{\partial v_l}{\partial z_{ij}}
\end{aligned} \tag{13.16}$$

which is equivalent to $\partial v_l/\partial \boldsymbol{A} = \partial v_l/\partial \boldsymbol{Z}$. Similarly, letting b_i and c_j be the ith entry of $\boldsymbol{b}$ and the jth entry of $\boldsymbol{c}$, respectively, we can write

$$\begin{aligned}\frac{\partial v_l}{\partial b_i} &= \boldsymbol{y}_z^H(l)\frac{\partial \boldsymbol{Z}}{\partial b_i}\boldsymbol{x}_z(l) = -d^{-1}\boldsymbol{y}_z^H(l)\boldsymbol{e}_i\boldsymbol{c}\boldsymbol{x}_z(l) \\ &= -d^{-1}\boldsymbol{e}_i^T\boldsymbol{y}_z^H(l)^T\boldsymbol{x}_z(l)^T\boldsymbol{c}^T = -d^{-1}\boldsymbol{e}_i^T\frac{\partial v_l}{\partial \boldsymbol{Z}}\boldsymbol{c}^T\end{aligned} \tag{13.17}$$

$$\begin{aligned}\frac{\partial v_l}{\partial c_j} &= \boldsymbol{y}_z^H(l)\frac{\partial \boldsymbol{Z}}{\partial c_j}\boldsymbol{x}_z(l) = -d^{-1}\boldsymbol{y}_z^H(l)\boldsymbol{b}\boldsymbol{e}_j^T\boldsymbol{x}_z(l) \\ &= -d^{-1}\boldsymbol{b}^T\boldsymbol{y}_z^H(l)^T\boldsymbol{x}_z(l)^T\boldsymbol{e}_j = -d^{-1}\boldsymbol{b}^T\frac{\partial v_l}{\partial \boldsymbol{Z}}\boldsymbol{e}_j\end{aligned} \tag{13.18}$$

$$\begin{aligned}\frac{\partial v_l}{\partial d} &= \boldsymbol{y}_z^H(l)\frac{\partial \boldsymbol{Z}}{\partial d}\boldsymbol{x}_z(l) = d^{-2}\boldsymbol{y}_z^H(l)\boldsymbol{b}\boldsymbol{c}\boldsymbol{x}_z(l) \\ &= d^{-2}\boldsymbol{b}^T\boldsymbol{y}_z^H(l)^T\boldsymbol{x}_z(l)^T\boldsymbol{c}^T = d^{-2}\boldsymbol{b}^T\frac{\partial v_l}{\partial \boldsymbol{Z}}\boldsymbol{c}^T\end{aligned} \tag{13.19}$$

because

$$\begin{aligned}\frac{\partial v_l}{\partial z_{ij}} &= \boldsymbol{y}_z^H(l)\frac{\partial \boldsymbol{Z}}{\partial z_{ij}}\boldsymbol{x}_z(l) = \boldsymbol{y}_z^H(l)\boldsymbol{e}_i\boldsymbol{e}_j^T\boldsymbol{x}_z(l) \\ &= \boldsymbol{e}_i^T\boldsymbol{y}_z^H(l)^T\boldsymbol{x}_z(l)^T\boldsymbol{e}_j = (\boldsymbol{y}_z^H(l)^T\boldsymbol{x}_z(l)^T)_{ij}\end{aligned}$$

which is equivalent to

$$\frac{\partial v_l}{\partial \boldsymbol{Z}} = \boldsymbol{y}_z^H(l)^T\boldsymbol{x}_z(l)^T$$

This completes the proof of Lemma 13.3. ■

We now define the pole sensitivity measure J_p and the zero sensitivity measure J_z for the digital filter in (13.1) as

$$J_p = \sum_{l=1}^{n}\left\|\frac{\partial \lambda_l}{\partial \boldsymbol{A}}\right\|_F^2 = \sum_{l=1}^{n}\left\|\left(\frac{\partial \lambda_l}{\partial \boldsymbol{A}}\right)^T\right\|_F^2 \tag{13.20a}$$

$$J_z = \sum_{l=1}^{n}\left\{\left\|\frac{\partial v_l}{\partial \boldsymbol{A}}\right\|_F^2 + \left\|\frac{\partial v_l}{\partial \boldsymbol{b}}\right\|_F^2 + \left\|\frac{\partial v_l}{\partial \boldsymbol{c}}\right\|_F^2 + \left\|\frac{\partial v_l}{\partial d}\right\|_F^2\right\} \tag{13.20b}$$

respectively, where $||\boldsymbol{M}||_F$ denotes the Frobenius norm of an $m \times n$ complex matrix $\boldsymbol{M}$, which is defined by

$$||\boldsymbol{M}||_F = \Big(\sum_{i=1}^{m}\sum_{j=1}^{n}|(\boldsymbol{M})_{ij}|^2\Big)^{\frac{1}{2}}$$

Here, $(\boldsymbol{M})_{ij}$ stands for the (i, j)th element of $\boldsymbol{M}$. By noting that

$$||\boldsymbol{M}||_F^2 = \mathrm{tr}[\boldsymbol{M}\boldsymbol{M}^H] = \mathrm{tr}[\boldsymbol{M}^H\boldsymbol{M}] = ||\boldsymbol{M}^H||_F^2 \qquad (13.21)$$

and using (13.7), (13.14) and (13.15), we can write (13.20a) and (13.20b) as

$$J_p = \sum_{l=1}^{n} \left(\boldsymbol{x}_p^H(l)\boldsymbol{x}_p(l)\right)\left(\boldsymbol{y}_p^H(l)\boldsymbol{y}_p(l)\right) \qquad (13.22a)$$

$$J_z = \sum_{l=1}^{n} \left(\boldsymbol{x}_z^H(l)\boldsymbol{x}_z(l) + \alpha_l^2\right)\left(\boldsymbol{y}_z^H(l)\boldsymbol{y}_z(l) + \beta_l^2\right) \qquad (13.22b)$$

respectively, where

$$\alpha_l = |d^{-1}\boldsymbol{c}\boldsymbol{x}_z(l)|, \qquad \beta_l = |d^{-1}\boldsymbol{y}_z^H(l)\boldsymbol{b}| \qquad (13.22c)$$

Lemma 13.4

The pole sensitivity measure J_p in (13.22a) is lower bounded by

$$J_p \geq n \qquad (13.23)$$

where the equality holds if and only if matrix $\boldsymbol{A}$ is normal, i.e., $\boldsymbol{A}\boldsymbol{A}^T = \boldsymbol{A}^T\boldsymbol{A}$.

Proof

By the Schwarz inequality and (13.10), it follows from (13.22a) that

$$J_p = \sum_{l=1}^{n} ||\boldsymbol{x}_p(l)||_2^2\, ||\boldsymbol{y}_p(l)||_2^2 \geq \sum_{l=1}^{n} |\boldsymbol{y}_p(l)^H\boldsymbol{x}_p(l)|^2 = \sum_{l=1}^{n} \delta_{ll} = n \qquad (13.24)$$

If $\boldsymbol{y}_p(l) = \mu_l\boldsymbol{x}_p(l)$ with $\mu_l \neq 0$ holds for $l = 1, 2, \cdots, n$ then $J_p = n$. In this case, it follows from (13.10) that

$$\boldsymbol{D}_\mu\boldsymbol{X}_p^H\boldsymbol{X}_p = \boldsymbol{I}_n \qquad (13.25)$$

where $\boldsymbol{D}_\mu = \mathrm{diag}\{\mu_1, \mu_2, \cdots, \mu_n\}$. This means that $\boldsymbol{X}_p^{-1} = \boldsymbol{D}_\mu\boldsymbol{X}_p^H$ holds. On the other hand, $\boldsymbol{A}\boldsymbol{x}_p(l) = \lambda_l\boldsymbol{x}_p(l)$ for $l = 1, 2, \cdots, n$ can be written as

$$\boldsymbol{A} = \boldsymbol{X}_p\boldsymbol{D}_\lambda\boldsymbol{X}_p^{-1} \qquad (13.26)$$

where $\boldsymbol{X}_p = [\boldsymbol{x}_p(1), \boldsymbol{x}_p(2), \cdots, \boldsymbol{x}_p(n)]$ and $\boldsymbol{D}_\lambda = \mathrm{diag}\{\lambda_1, \lambda_2, \cdots, \lambda_n\}$. By substituting $\boldsymbol{X}_p^{-1} = \boldsymbol{D}_\mu\boldsymbol{X}_p^H$ into (13.26), it can be verified that

$\boldsymbol{A}\boldsymbol{A}^H = \boldsymbol{A}^H\boldsymbol{A}$. Conversely, if $\boldsymbol{A}$ is normal, there exists an $n \times n$ unitary matrix $\boldsymbol{Q}^H$ such that $\boldsymbol{Q}^H\boldsymbol{A}\boldsymbol{Q} = \boldsymbol{D}_\lambda$ which is equivalent to $\boldsymbol{Q} = \boldsymbol{X}_p$. Alternatively, it follows from (13.10) that

$$\boldsymbol{Y}_p^H\boldsymbol{X}_p = \boldsymbol{I}_n \tag{13.27}$$

Since $\boldsymbol{Q}^H\boldsymbol{Q} = \boldsymbol{X}_p^H\boldsymbol{X}_p = \boldsymbol{I}_n$, we arrive at $\boldsymbol{Y}_p = \boldsymbol{X}_p$ which means that $\boldsymbol{y}_p(l) = \boldsymbol{x}_p(l)$ for $l = 1, 2, \cdots, n$. As a result, we obtain $J_p = n$. This completes the proof of Lemma 13.4. ■

Lemma 13.5

The zero sensitivity measure J_p in (13.22b) is lower bounded by

$$J_z \geq \sum_{l=1}^{n} (1 + |\alpha_l\beta_l|)^2 \tag{13.28a}$$

where the equality holds if and only if matrix $\boldsymbol{Z} = \boldsymbol{A} - d^{-1}\boldsymbol{b}\boldsymbol{c}$ is normal, i.e., $\boldsymbol{Z}\boldsymbol{Z}^T = \boldsymbol{Z}^T\boldsymbol{Z}$ subject to its right eigenvector matrix $\boldsymbol{X}_z$ satisfying

$$\boldsymbol{X}_z^H\boldsymbol{X}_z = \mathrm{diag}\left\{\frac{|\alpha_1|}{|\beta_1|}, \frac{|\alpha_2|}{|\beta_2|}, \cdots, \frac{|\alpha_n|}{|\beta_n|}\right\} \tag{13.28b}$$

with $\boldsymbol{X}_z = \left[\boldsymbol{x}_z(1), \boldsymbol{x}_z(2), \cdots, \boldsymbol{x}_z(n)\right]$.

Proof

From (13.22b) and the arithmetic-geometric mean inequality, we obtain

$$\begin{aligned} J_z &= \sum_{l=1}^{n} \Big(||\boldsymbol{x}_z(l)||_2^2 \, ||\boldsymbol{y}_z(l)||_2^2 + \alpha_l^2 \, ||\boldsymbol{y}_z(l)||_2^2 + \beta_l^2 \, ||\boldsymbol{x}_z(l)||_2^2 + \alpha_l^2\beta_l^2 \Big) \\ &\geq \sum_{l=1}^{n} \Big(||\boldsymbol{x}_z(l)||_2^2 \, ||\boldsymbol{y}_z(l)||_2^2 + 2|\alpha_l\beta_l| \, ||\boldsymbol{y}_z(l)||_2 \, ||\boldsymbol{x}_z(l)||_2 + \alpha_l^2\beta_l^2 \Big) \\ &\triangleq J_z' \end{aligned} \tag{13.29a}$$

where the equality holds if and only if

$$|\alpha_l| \, ||\boldsymbol{y}_z(l)||_2 = |\beta_l| \, ||\boldsymbol{x}_z(l)||_2 \ \text{ for } l = 1, 2, \cdots, n \tag{13.29b}$$

Moreover, from (13.10) and (13.29a) it can be derived that

$$\begin{aligned}
J_z' &= \sum_{l=1}^{n}\Big(||\boldsymbol{x}_z(l)||_2^2\,||\boldsymbol{y}_z(l)||_2^2 + 2|\alpha_l\beta_l|\,||\boldsymbol{y}_z(l)||_2\,||\boldsymbol{x}_z(l)||_2 + \alpha_l^2\beta_l^2\Big) \\
&\geq \sum_{l=1}^{n}\Big(|\boldsymbol{y}_z(l)^H\boldsymbol{x}_z(l)|^2 + 2|\alpha_l\beta_l|\,|\boldsymbol{y}_z(l)^H\boldsymbol{x}_z(l)| + \alpha_l^2\beta_l^2\Big) \qquad (13.30\text{a}) \\
&= \sum_{l=1}^{n}(1+|\alpha_l\beta_l|)^2
\end{aligned}$$

where the equality holds if and only if

$$\boldsymbol{y}_z(l) = \kappa_l\boldsymbol{x}_z(l),\ \kappa_l \neq 0 \ \text{ for } l = 1, 2, \cdots, n \qquad (13.30\text{b})$$

In order for the equality in both (13.29a) and (13.30a) to hold simultaneously, both (13.29b) and (13.30b) must be satisfied. The conditions in (13.29b) and (13.30b) are satisfied provided that

$$\boldsymbol{Y}_z = \boldsymbol{X}_z\text{diag}\Big\{\frac{|\alpha_1|}{|\beta_1|}, \frac{|\alpha_2|}{|\beta_2|}, \cdots, \frac{|\alpha_n|}{|\beta_n|}\Big\}^{-1} \qquad (13.31)$$

with $\boldsymbol{Y}_z = [\boldsymbol{y}_z(1), \boldsymbol{y}_z(2), \cdots, \boldsymbol{y}_z(n)] = \boldsymbol{X}_z^{-H}$. Evidently, (13.31) is equivalent to (13.28b) and this completes the proof of Lemma 13.5. ■

13.3 Realization with Minimal Pole and Zero Sensitivity

13.3.1 Weighted Pole and Zero Sensitivity Minimization Without Imposing l_2-Scaling Constraints

Now if a coordinate transformation defined by

$$\overline{\boldsymbol{x}}(k) = \boldsymbol{T}^{-1}\boldsymbol{x}(k) \qquad (13.32)$$

is applied to the digital filter in (13.1), the new realization $(\overline{\boldsymbol{A}}, \overline{\boldsymbol{b}}, \overline{\boldsymbol{c}}, d)_n$ can be characterized by

$$\overline{\boldsymbol{A}} = \boldsymbol{T}^{-1}\boldsymbol{A}\boldsymbol{T}, \qquad \overline{\boldsymbol{b}} = \boldsymbol{T}^{-1}\boldsymbol{b}, \qquad \overline{\boldsymbol{c}} = \boldsymbol{c}\boldsymbol{T} \qquad (13.33)$$

From (13.2) and (13.33), it is observed that the transfer function $H(z)$ is invariant under the coordinate transformation in (13.32). Note that the right eigenvectors of the original and transformed system matrices $\boldsymbol{A}$ and $\overline{\boldsymbol{A}}$

corresponding to λ_l, namely $\boldsymbol{x}_p(l)$ and $\overline{\boldsymbol{x}}_p(l)$, and their counterpart reciprocal left eigenvectors, namely $\boldsymbol{y}_p(l)$ and $\overline{\boldsymbol{y}}_p(l)$, are related as

$$\overline{\boldsymbol{x}}_p(l) = \boldsymbol{T}^{-1}\boldsymbol{x}_p(l), \qquad \overline{\boldsymbol{y}}_p(l) = \boldsymbol{T}^T\boldsymbol{y}_p(l) \tag{13.34}$$

for $l = 1, 2, \cdots, n$, respectively. Similarly, the right eigenvectors of the original and transformed system matrices $\boldsymbol{Z}$ and $\overline{\boldsymbol{Z}}$, and their counterpart reciprocal left eigenvectors are related as

$$\overline{\boldsymbol{x}}_z(l) = \boldsymbol{T}^{-1}\boldsymbol{x}_z(l), \qquad \overline{\boldsymbol{y}}_z(l) = \boldsymbol{T}^T\boldsymbol{y}_z(l) \tag{13.35}$$

for $l = 1, 2, \cdots, n$, respectively. Therefore, for the new realization $(\overline{\boldsymbol{A}}, \overline{\boldsymbol{b}}, \overline{\boldsymbol{c}}, d)_n$ specified in (13.33), the pole and zero sensitivity measures in (13.22a) and (13.22b) can be expressed as

$$J_p(\boldsymbol{T}) = \sum_{l=1}^{n} \left(\boldsymbol{x}_p^H(l)\boldsymbol{T}^{-T}\boldsymbol{T}^{-1}\boldsymbol{x}_p(l)\right)\left(\boldsymbol{y}_p^H(l)\boldsymbol{T}\boldsymbol{T}^T\boldsymbol{y}_p(l)\right) \tag{13.36a}$$

$$J_z(\boldsymbol{T}) = \sum_{l=1}^{n} \left(\boldsymbol{x}_z^H(l)\boldsymbol{T}^{-T}\boldsymbol{T}^{-1}\boldsymbol{x}_z(l) + \alpha_l^2\right)\left(\boldsymbol{y}_z^H(l)\boldsymbol{T}\boldsymbol{T}^T\boldsymbol{y}_z(l) + \beta_l^2\right) \tag{13.36b}$$

respectively.

Under these circumstances, we examine a weighted pole and zero sensitivity measure $J_\gamma(\boldsymbol{T})$ defined by

$$J_\gamma(\boldsymbol{T}) = \gamma J_p(\boldsymbol{T}) + (1-\gamma) J_z(\boldsymbol{T}) \tag{13.37}$$

where $0 \le \gamma \le 1$ is a weighting factor to control the trade-off between the two component sensitivities in the sense that a $J_\gamma(\boldsymbol{T})$ with a greater γ represents a measure that places more emphasis on pole sensitivity, while a $J_\gamma(\boldsymbol{T})$ with a smaller γ serves as a measure that weights more heavily on zero sensitivity. In addition, by setting γ to unity or zero $J_\gamma(\boldsymbol{T})$ becomes $J_p(\boldsymbol{T})$ or $J_z(\boldsymbol{T})$, respectively.

The problem of minimizing the weighted pole and zero sensitivity measure is now formulated as follows: *For given* $\boldsymbol{A}$, $\boldsymbol{b}$, $\boldsymbol{c}$ *and* d, *obtain an* $n \times n$ *transformation matrix* $\boldsymbol{T}$ *which minimizes* $J_\gamma(\boldsymbol{T})$ *in (13.37) with* γ *specified by the designer.*

The pole and zero sensitivity measures in (13.36a) and (13.36b) can also be expressed in terms of matrix $\boldsymbol{P} = \boldsymbol{T}\boldsymbol{T}^T$ as

$$J_p(\boldsymbol{T}) = \sum_{l=1}^{n} \left(\operatorname{tr}\left[\boldsymbol{x}_p(l)\boldsymbol{x}_p^H(l)\boldsymbol{P}^{-1}\right]\right)\left(\operatorname{tr}\left[\boldsymbol{y}_p(l)\boldsymbol{y}_p^H(l)\boldsymbol{P}\right]\right) \tag{13.38a}$$

$$J_z(\boldsymbol{T}) = \sum_{l=1}^{n} \left(\text{tr}\left[\boldsymbol{x}_z(l)\boldsymbol{x}_z^H(l)\boldsymbol{P}^{-1}\right] + \alpha_l^2 \right) \left(\text{tr}\left[\boldsymbol{y}_z(l)\boldsymbol{y}_z^H(l)\boldsymbol{P}\right] + \beta_l^2 \right) \tag{13.38b}$$

respectively. To minimize $J_\gamma(\boldsymbol{T})$ in (13.37) with respect to an $n \times n$ positive-definite symmetric matrix $\boldsymbol{P}$, we compute

$$\frac{\partial J_\gamma(\boldsymbol{T})}{\partial \boldsymbol{P}} = \boldsymbol{M}_\gamma(\boldsymbol{P}) - \boldsymbol{P}^{-1}\boldsymbol{N}_\gamma(\boldsymbol{P})\boldsymbol{P}^{-1} \tag{13.39}$$

where

$$\begin{aligned}\boldsymbol{M}_\gamma(\boldsymbol{P}) = {} & \gamma \sum_{l=1}^{n} \text{tr}\left[\boldsymbol{x}_p(l)\boldsymbol{x}_p^H(l)\boldsymbol{P}^{-1}\right]\boldsymbol{y}_p(l)\boldsymbol{y}_p^H(l) \\ & + (1-\gamma)\sum_{l=1}^{n}\left(\text{tr}\left[\boldsymbol{x}_z(l)\boldsymbol{x}_z^H(l)\boldsymbol{P}^{-1}\right] + \alpha_l^2\right)\boldsymbol{y}_z(l)\boldsymbol{y}_z^H(l) \\ \boldsymbol{N}_\gamma(\boldsymbol{P}) = {} & \gamma \sum_{l=1}^{n} \text{tr}\left[\boldsymbol{y}_p(l)\boldsymbol{y}_p^H(l)\boldsymbol{P}\right]\boldsymbol{x}_p(l)\boldsymbol{x}_p^H(l) \\ & + (1-\gamma)\sum_{l=1}^{n}\left(\text{tr}\left[\boldsymbol{y}_z(l)\boldsymbol{y}_z^H(l)\boldsymbol{P}\right] + \beta_l^2\right)\boldsymbol{x}_z(l)\boldsymbol{x}_z^H(l)\end{aligned}$$

By setting $\partial J_\gamma(\boldsymbol{T})/\partial \boldsymbol{P} = \boldsymbol{0}$, we obtain

$$\boldsymbol{P}\boldsymbol{M}_\gamma(\boldsymbol{P})\boldsymbol{P} = \boldsymbol{N}_\gamma(\boldsymbol{P}) \tag{13.40}$$

The equation in (13.40) is highly nonlinear with respect to $\boldsymbol{P}$. An effective approach for solving (13.40) is to relax it into the recursive second-order matrix equation

$$\boldsymbol{P}_{k+1}\boldsymbol{M}_\gamma(\boldsymbol{P}_k)\boldsymbol{P}_{k+1} = \boldsymbol{N}_\gamma(\boldsymbol{P}_k) \tag{13.41}$$

The recursion starts with an initial estimate $\boldsymbol{P}_0$. Thus $\boldsymbol{P}_k$ in (13.41) is assumed to be known from the previous recursion, and the next iterate $\boldsymbol{P}_{k+1}$ is given by

$$\boldsymbol{P}_{k+1} = \boldsymbol{M}_\gamma(\boldsymbol{P}_k)^{-\frac{1}{2}}[\boldsymbol{M}_\gamma(\boldsymbol{P}_k)^{\frac{1}{2}}\boldsymbol{N}_\gamma(\boldsymbol{P}_k)\boldsymbol{M}_\gamma(\boldsymbol{P}_k)^{\frac{1}{2}}]^{\frac{1}{2}}\boldsymbol{M}_\gamma(\boldsymbol{P}_k)^{-\frac{1}{2}} \tag{13.42}$$

This iteration process continues until

$$\left| J_\gamma(\boldsymbol{P}_{k+1}^{\frac{1}{2}}) - J_\gamma(\boldsymbol{P}_k^{\frac{1}{2}}) \right| < \varepsilon \tag{13.43}$$

is satisfied for a prescribed tolerance $\varepsilon > 0$ where $\boldsymbol{P}^{\frac{1}{2}}$ stands for the square root of matrix $\boldsymbol{P}$. If the iteration is terminated at step k, $\boldsymbol{P}_k$ is claimed to be a solution point.

We now turn our attention to the construction of a coordinate transformation matrix $\boldsymbol{T}$ that solves the problem of minimizing the weighted pole and zero sensitivity measure in (13.37). Since $\boldsymbol{P} = \boldsymbol{T}\boldsymbol{T}^T$, an optimizing $\boldsymbol{T}$ assumes the form

$$\boldsymbol{T} = \boldsymbol{P}^{\frac{1}{2}}\boldsymbol{U} \tag{13.44}$$

where $\boldsymbol{U}$ is an arbitrary $n \times n$ orthogonal matrix.

13.3.2 Zero Sensitivity Minimization Subject to Minimal Pole Sensitivity

Suppose the coordinate transformation matrix $\boldsymbol{T}_o$ yields $J_p(\boldsymbol{T}_o) = n$, it can be verified that $J_p(\pm\sqrt{\zeta}\,\boldsymbol{T}_o) = n$ holds for any scalar $\pm\sqrt{\zeta}$. Therefore, one can reduce the zero sensitivity as much as possible while keep the pole sensitivity unaltered by adequately tuning the value of ζ. To this end, we compute

$$\begin{aligned}\frac{\partial J_z(\pm\sqrt{\zeta}\,\boldsymbol{T}_o)}{\partial \zeta} = \zeta^{-2}\sum_{l=1}^{n}\Big\{&\zeta^2\alpha_l^2\boldsymbol{y}_z^H(l)\boldsymbol{T}_o\boldsymbol{T}_o^T\boldsymbol{y}_z(l)\\ &-\beta_l^2\boldsymbol{x}_z^H(l)\boldsymbol{T}_o^{-T}\boldsymbol{T}_o^{-1}\boldsymbol{x}_z(l)\Big\}\end{aligned} \tag{13.45}$$

By solving $\partial J_z(\pm\sqrt{\zeta}\,\boldsymbol{T}_o)/\partial\zeta = 0$, the optimal value of ζ is found to be

$$\zeta = \sqrt{\frac{\displaystyle\sum_{l=1}^{n}\beta_l^2\boldsymbol{x}_z^H(l)\boldsymbol{T}_o^{-T}\boldsymbol{T}_o^{-1}\boldsymbol{x}_z(l)}{\displaystyle\sum_{l=1}^{n}\alpha_l^2\boldsymbol{y}_z^H(l)\boldsymbol{T}_o\boldsymbol{T}_o^T\boldsymbol{y}_z(l)}} \tag{13.46}$$

This implies that the optimal coordinate transformation matrix $\boldsymbol{T}^{opt}$ that minimizes $J_z(\pm\sqrt{\zeta}\,\boldsymbol{T}_o)$ with respect to ζ subject to $J_p(\boldsymbol{T}_o) = n$ can be obtained as

$$\boldsymbol{T}^{opt} = \pm\sqrt{\zeta}\,\boldsymbol{T}_o \tag{13.47}$$

where ζ is given by (13.46). Notice that by applying the arithmetic-geometric mean inequality, the following holds:

$$
\begin{aligned}
J_z(\boldsymbol{T}_o) - J_z(\boldsymbol{T}^{opt}) \\
&= \sum_{l=1}^{n} \alpha_l^2 \boldsymbol{y}_z^H(l)\boldsymbol{T}_o\boldsymbol{T}_o^T\boldsymbol{y}_z(l) + \sum_{l=1}^{n} \beta_l^2 \boldsymbol{x}_z^H(l)\boldsymbol{T}_o^{-T}\boldsymbol{T}_o^{-1}\boldsymbol{x}_z(l) \\
&\quad -2\sqrt{\sum_{l=1}^{n} \alpha_l^2 \boldsymbol{y}_z^H(l)\boldsymbol{T}_o\boldsymbol{T}_o^T\boldsymbol{y}_z(l) \sum_{l=1}^{n} \beta_l^2 \boldsymbol{x}_z^H(l)\boldsymbol{T}_o^{-T}\boldsymbol{T}_o^{-1}\boldsymbol{x}_z(l)} \\
&\geq 0
\end{aligned}
\tag{13.48}
$$

Therefore, as expected, by using the optimal $\boldsymbol{T}^{opt}$ instead of matrix $\boldsymbol{T}_o$, the zero sensitivity always gets reduced.

13.4 Pole Zero Sensitivity Minimization Subject to l_2-Scaling Constraints Using Lagrange Function

13.4.1 l_2-Scaling Constraints and Problem Formulation

The controllability Grammian $\boldsymbol{K}_c$ of the digital filter in (13.1) plays an important role in the dynamic-range scaling of the state-variable vector $\overline{\boldsymbol{x}}(k)$, and $\boldsymbol{K}_c$ can be obtained by solving the Lyapunov equation

$$\boldsymbol{K}_c = \boldsymbol{A}\boldsymbol{K}_c\boldsymbol{A}^T + \boldsymbol{b}\boldsymbol{b}^T \tag{13.49}$$

With an equivalent state-space realization as specified in (13.33), the controllability Grammian for the transformed system assumes the form

$$\overline{\boldsymbol{K}}_c = \boldsymbol{T}^{-1}\boldsymbol{K}_c\boldsymbol{T}^{-T} \tag{13.50}$$

If l_2-scaling constraints are imposed on the new state-variable vector $\overline{\boldsymbol{x}}(k)$ in (13.32), it is required that

$$\boldsymbol{e}_i^T\overline{\boldsymbol{K}}_c\boldsymbol{e}_i = \boldsymbol{e}_i^T\boldsymbol{T}^{-1}\boldsymbol{K}_c\boldsymbol{T}^{-T}\boldsymbol{e}_i = 1 \;\text{ for }\; i = 1, 2, \cdots, n \tag{13.51}$$

The problem being considered here is to obtain an $n \times n$ transformation matrix $\boldsymbol{T}$ that minimizes (13.37) subject to the l_2-scaling constraints in (13.51).

13.4.2 Minimization of (13.37) Subject to l_2-Scaling Constraints — Using Lagrange Function

To minimize (13.37), where $J_p(\boldsymbol{T})$ and $J_z(\boldsymbol{T})$ are given by (13.38a) and (13.38b), respectively, with respect to an $n \times n$ symmetric positive-definite matrix $\boldsymbol{P}$ subject to l_2-scaling constraints in (13.51), we define the Lagrange function

$$I_\gamma(\boldsymbol{P},\xi) = J_\gamma(\boldsymbol{T}) + \xi\big(\mathrm{tr}[\boldsymbol{K}_c\boldsymbol{P}^{-1}] - n\big) \tag{13.52}$$

where ξ is a Lagrange multiplier, and compute the gradients

$$\begin{aligned}\frac{\partial I_\gamma(\boldsymbol{P},\xi)}{\partial \boldsymbol{P}} &= \boldsymbol{M}_\gamma(\boldsymbol{P}) - \boldsymbol{P}^{-1}\boldsymbol{N}_\gamma(\boldsymbol{P})\boldsymbol{P}^{-1} - \xi\boldsymbol{P}^{-1}\boldsymbol{K}_c\boldsymbol{P}^{-1}\\ \frac{\partial I_\gamma(\boldsymbol{P},\xi)}{\partial \xi} &= \mathrm{tr}[\boldsymbol{K}_c\boldsymbol{P}^{-1}] - n\end{aligned} \tag{13.53}$$

where $\boldsymbol{M}_\gamma(\boldsymbol{P})$ and $\boldsymbol{N}_\gamma(\boldsymbol{P})$ are defined in (13.39). Set $\partial I_\gamma(\boldsymbol{P},\xi)/\partial\boldsymbol{P} = \boldsymbol{0}$ and $\partial I_\gamma(\boldsymbol{P},\xi)/\partial\xi = 0$ to yield

$$\boldsymbol{P}\boldsymbol{M}_\gamma(\boldsymbol{P})\boldsymbol{P} = \boldsymbol{G}_\gamma(\boldsymbol{P},\xi), \qquad \mathrm{tr}[\boldsymbol{K}_c\boldsymbol{P}^{-1}] = n \tag{13.54}$$

where

$$\boldsymbol{G}_\gamma(\boldsymbol{P},\xi) = \boldsymbol{N}_\gamma(\boldsymbol{P}) + \xi\boldsymbol{K}_c$$

To solve the highly nonlinear equations in (13.54), we propose an iterative technique which starts with an initial estimate $\boldsymbol{P}_0$ and relaxes the first equation in (13.54) into a recursive second-order matrix equation

$$\boldsymbol{P}_{k+1}\boldsymbol{M}_\gamma(\boldsymbol{P}_k)\boldsymbol{P}_{k+1} = \boldsymbol{G}_\gamma(\boldsymbol{P}_k,\xi_{k+1}) \tag{13.55}$$

where $\boldsymbol{P}_k$ is known from the previous recursion. Solving (13.55) for a symmetric and positive-definite $\boldsymbol{P}_{k+1}$, we obtain

$$\boldsymbol{P}_{k+1} = \boldsymbol{M}_\gamma(\boldsymbol{P}_k)^{-\frac{1}{2}}[\boldsymbol{M}_\gamma(\boldsymbol{P}_k)^{\frac{1}{2}}\boldsymbol{G}_\gamma(\boldsymbol{P}_k,\xi_{k+1})\boldsymbol{M}_\gamma(\boldsymbol{P}_k)^{\frac{1}{2}}]^{\frac{1}{2}}\boldsymbol{M}_\gamma(\boldsymbol{P}_k)^{-\frac{1}{2}} \tag{13.56}$$

where the Lagrange multiplier ξ_{k+1} can be efficiently obtained using a bisection method so that

$$f(\gamma,\xi_{k+1}) = \Big|n - \mathrm{tr}\big[\tilde{\boldsymbol{K}}_k\tilde{\boldsymbol{G}}_k(\gamma,\xi_{k+1})\big]\Big| < \varepsilon \tag{13.57}$$

with

$$\tilde{\boldsymbol{G}}_k(\gamma,\xi_{k+1}) = [\boldsymbol{M}_\gamma(\boldsymbol{P}_k)^{\frac{1}{2}}\boldsymbol{G}_\gamma(\boldsymbol{P}_k,\xi_{k+1})\boldsymbol{M}_\gamma(\boldsymbol{P}_k)^{\frac{1}{2}}]^{-\frac{1}{2}}$$

$$\tilde{\boldsymbol{K}}_k = \boldsymbol{M}_\gamma(\boldsymbol{P}_k)^{\frac{1}{2}}\boldsymbol{K}_c\boldsymbol{M}_\gamma(\boldsymbol{P}_k)^{\frac{1}{2}}$$

This iteration process continues until

$$\Big|I_\gamma\big(\boldsymbol{P}_{k+1},\xi_{k+1}\big) - I_\gamma\big(\boldsymbol{P}_k,\xi_k\big)\Big| < \varepsilon \tag{13.58}$$

is satisfied for a prescribed tolerance $\varepsilon > 0$. If the iteration is terminated at step k, we set $\boldsymbol{P} = \boldsymbol{P}_k$ and claim it to be a solution.

13.4.3 Derivation of Nonsingular T from P to Satisfy l_2-Scaling Constraints

Having obtained an optimal $\boldsymbol{P}$, we now turn our attention to the construction of the optimal coordinate transformation matrix $\boldsymbol{T}$ that solves the problem of minimizing (13.37) subject to l_2-scaling constraints in (13.51). As is analyzed earlier, the optimal $\boldsymbol{T}$ assumes the form

$$\boldsymbol{T} = \boldsymbol{P}^{\frac{1}{2}}\boldsymbol{U} \tag{13.59}$$

where $\boldsymbol{U}$ is an $n \times n$ orthogonal matrix that can be determined to satisfy the l_2-scaling constraints as follows. From (13.50) and (13.59), it follows that

$$\overline{\boldsymbol{K}}_c = \boldsymbol{U}^T\boldsymbol{P}^{-\frac{1}{2}}\boldsymbol{K}_c\boldsymbol{P}^{-\frac{1}{2}}\boldsymbol{U} \tag{13.60}$$

To find an $n \times n$ orthogonal matrix $\boldsymbol{U}$ such that the matrix $\overline{\boldsymbol{K}}_c$ satisfies l_2-scaling constraints in (13.51), we perform the eigen-decomposition for the symmetric positive-definite matrix $\boldsymbol{P}^{-\frac{1}{2}}\boldsymbol{K}_c\boldsymbol{P}^{-\frac{1}{2}}$ as

$$\boldsymbol{P}^{-\frac{1}{2}}\boldsymbol{K}_c\boldsymbol{P}^{-\frac{1}{2}} = \boldsymbol{R}\boldsymbol{\Theta}\boldsymbol{R}^T \tag{13.61}$$

where $\boldsymbol{\Theta} = \mathrm{diag}\{\theta_1, \theta_2, \cdots, \theta_n\}$ with $\theta_i > 0$ for all i and $\boldsymbol{R}$ is an $n \times n$ orthogonal matrix. Next, an $n \times n$ orthogonal matrix $\boldsymbol{S}$ such that

$$\boldsymbol{e}_i^T\boldsymbol{S}\boldsymbol{\Theta}\boldsymbol{S}^T\boldsymbol{e}_i = 1 \quad \text{for } i = 1, 2, \cdots, n \tag{13.62}$$

can be obtained by a numerical procedure [18, p.278]. Using (13.60)-(13.62), it can be readily verified that the orthogonal matrix $\boldsymbol{U} = \boldsymbol{R}\boldsymbol{S}^T$ leads to a $\overline{\boldsymbol{K}}_c$ in (13.60) whose diagonal elements are equal to unity, hence the l_2-scaling constraints in (13.51) are satisfied. This matrix $\boldsymbol{U}$ together with (13.59) yields the solution for the problem of minimizing (13.37) subject to the l_2-scaling constraints in (13.51) as

$$\boldsymbol{T} = \boldsymbol{P}^{\frac{1}{2}}\boldsymbol{R}\boldsymbol{S}^T \tag{13.63}$$

13.5 Pole and Zero Sensitivity Minimization Subject to l_2-Scaling Constraints Using Quasi-Newton Algorithm

13.5.1 l_2-Scaling and Problem Formulation

This section explores a technique for pole and zero sensitivity minimization subject to l_2-scaling constraints, where the coordinate transformation matrix is optimized using an unconstrained optimization approach known as quasi-Newton algorithm.

We start by defining

$$\hat{\boldsymbol{T}} = \boldsymbol{T}^T \boldsymbol{K}_c^{-\frac{1}{2}} \tag{13.64}$$

which leads (13.51) to

$$\boldsymbol{e}_i^T \hat{\boldsymbol{T}}^{-T} \hat{\boldsymbol{T}}^{-1} \boldsymbol{e}_i = 1 \;\text{ for } i = 1, 2, \cdots, n \tag{13.65}$$

Evidently, these constraints are always satisfied if matrix $\hat{\boldsymbol{T}}^{-1}$ assumes the form

$$\hat{\boldsymbol{T}}^{-1} = \left[\frac{\boldsymbol{t}_1}{||\boldsymbol{t}_1||}, \frac{\boldsymbol{t}_2}{||\boldsymbol{t}_2||}, \cdots, \frac{\boldsymbol{t}_n}{||\boldsymbol{t}_n||} \right] \tag{13.66}$$

Defining an $n^2 \times 1$ vector $\boldsymbol{x} = (\boldsymbol{t}_1^T, \boldsymbol{t}_2^T, \cdots, \boldsymbol{t}_n^T)^T$ that consists of independent variables in $\hat{\boldsymbol{T}}^{-1}$, the pole and zero sensitivity measures $J_p(\boldsymbol{T})$ and $J_z(\boldsymbol{T})$ in (13.36a) and (13.36b) can be written as

$$\begin{aligned} J_p(\boldsymbol{x}) &= \sum_{l=1}^{n} \left(\hat{\boldsymbol{x}}_p^H(l) \hat{\boldsymbol{T}}^{-1} \hat{\boldsymbol{T}}^{-T} \hat{\boldsymbol{x}}_p(l) \right) \left(\hat{\boldsymbol{y}}_p^H(l) \hat{\boldsymbol{T}}^T \hat{\boldsymbol{T}} \hat{\boldsymbol{y}}_p(l) \right) \\ J_z(\boldsymbol{x}) &= \sum_{l=1}^{n} \left(\hat{\boldsymbol{x}}_z^H(l) \hat{\boldsymbol{T}}^{-1} \hat{\boldsymbol{T}}^{-T} \hat{\boldsymbol{x}}_z(l) + \alpha_l^2 \right) \left(\hat{\boldsymbol{y}}_z^H(l) \hat{\boldsymbol{T}}^T \hat{\boldsymbol{T}} \hat{\boldsymbol{y}}_z(l) + \beta_l^2 \right) \end{aligned} \tag{13.67}$$

respectively, where

$$\hat{\boldsymbol{x}}_p(l) = \boldsymbol{K}_c^{-\frac{1}{2}} \boldsymbol{x}_p(l), \qquad \hat{\boldsymbol{y}}_p(l) = \boldsymbol{K}_c^{\frac{1}{2}} \boldsymbol{y}_p(l)$$

$$\hat{\boldsymbol{x}}_z(l) = \boldsymbol{K}_c^{-\frac{1}{2}} \boldsymbol{x}_z(l), \qquad \hat{\boldsymbol{y}}_z(l) = \boldsymbol{K}_c^{\frac{1}{2}} \boldsymbol{y}_z(l)$$

The original constrained optimization problem formulated in Section 13.4.1 can now be converted into an unconstrained optimization problem of obtaining an $n^2 \times 1$ vector $\boldsymbol{x}$ which minimizes

$$J_\gamma(\boldsymbol{x}) = \gamma J_p(\boldsymbol{x}) + (1 - \gamma) J_z(\boldsymbol{x}), \quad 0 \le \gamma \le 1 \tag{13.68}$$

13.5.2 Minimization of (13.68) Subject to l_2-Scaling Constraints — Using Quasi-Newton Algorithm

We now solve the unconstrained problem by a quasi-Newton algorithm that starts with an initial point $\boldsymbol{x}_0$ obtained from the assignment $\hat{\boldsymbol{T}}^{-1} = \boldsymbol{I}_n$. In the

kth iteration, the quasi-Newton algorithm, known as the Broyden-Fletcher-Goldfarb-Shanno (BFGS) algorithm [14], updates the current point $\boldsymbol{x}_k$ to point $\boldsymbol{x}_{k+1}$ as [15]

$$\boldsymbol{x}_{k+1} = \boldsymbol{x}_k + \alpha_k \boldsymbol{d}_k \tag{13.69}$$

where

$$\boldsymbol{d}_k = -\boldsymbol{S}_k \nabla J_\gamma(\boldsymbol{x}_k), \qquad \alpha_k = arg\left[\min_{\alpha} \; J_\gamma(\boldsymbol{x}_k + \alpha \boldsymbol{d}_k)\right]$$

$$\boldsymbol{S}_{k+1} = \boldsymbol{S}_k + \left(1 + \frac{\boldsymbol{\varphi}_k^T \boldsymbol{S}_k \boldsymbol{\varphi}_k}{\boldsymbol{\varphi}_k^T \boldsymbol{\delta}_k}\right)\frac{\boldsymbol{\delta}_k \boldsymbol{\delta}_k^T}{\boldsymbol{\varphi}_k^T \boldsymbol{\delta}_k} - \frac{\boldsymbol{\delta}_k \boldsymbol{\varphi}_k^T \boldsymbol{S}_k + \boldsymbol{S}_k \boldsymbol{\varphi}_k \boldsymbol{\delta}_k^T}{\boldsymbol{\varphi}_k^T \boldsymbol{\delta}_k}, \quad \boldsymbol{S}_0 = \boldsymbol{I}$$

$$\boldsymbol{\delta}_k = \boldsymbol{x}_{k+1} - \boldsymbol{x}_k, \qquad \boldsymbol{\varphi}_k = \nabla J_\gamma(\boldsymbol{x}_{k+1}) - \nabla J_\gamma(\boldsymbol{x}_k)$$

Here, $\nabla J_\gamma(\boldsymbol{x})$ denotes the gradient of $J_\gamma(\boldsymbol{x})$ with respect to $\boldsymbol{x}$, and $\boldsymbol{S}_k$ is a positive-definite approximation of the inverse Hessian matrix of $J_\gamma(\boldsymbol{x}_k)$.

This iteration process continues until

$$|J_\gamma(\boldsymbol{x}_{k+1}) - J_\gamma(\boldsymbol{x}_k)| < \varepsilon \tag{13.70}$$

is satisfied where $\varepsilon > 0$ is a prescribed tolerance.

13.5.3 Gradient of $J(\boldsymbol{x})$

The gradient of $J_\gamma(\boldsymbol{x})$ can be evaluated using closed-form expressions as

$$\begin{aligned} \frac{\partial J_\gamma(\boldsymbol{x})}{\partial t_{ij}} &= \lim_{\Delta \to 0} \frac{J_\gamma(\hat{\boldsymbol{T}}_{ij}) - J_\gamma(\hat{\boldsymbol{T}})}{\Delta} \\ &= \gamma \frac{\partial J_p(\boldsymbol{x})}{\partial t_{ij}} + (1-\gamma)\frac{\partial J_z(\boldsymbol{x})}{\partial t_{ij}} \end{aligned} \tag{13.71}$$

where $\hat{\boldsymbol{T}}_{ij}$ is the matrix obtained from $\hat{\boldsymbol{T}}$ with a perturbed (i, j)th component by Δ. It follows that [16, p. 655]

$$\begin{aligned} \hat{\boldsymbol{T}}_{ij} &= \hat{\boldsymbol{T}} + \frac{\Delta \hat{\boldsymbol{T}} \boldsymbol{g}_{ij} \boldsymbol{e}_j^T \hat{\boldsymbol{T}}}{1 - \Delta \boldsymbol{e}_j^T \hat{\boldsymbol{T}} \boldsymbol{g}_{ij}} \simeq \hat{\boldsymbol{T}} + \Delta \hat{\boldsymbol{T}} \boldsymbol{g}_{ij} \boldsymbol{e}_j^T \hat{\boldsymbol{T}} \\ \hat{\boldsymbol{T}}_{ij}^{-1} &= \hat{\boldsymbol{T}}^{-1} - \Delta \boldsymbol{g}_{ij} \boldsymbol{e}_j^T \\ \boldsymbol{g}_{ij} &= -\partial \left\{ \frac{\boldsymbol{t}_j}{||\boldsymbol{t}_j||} \right\} / \partial t_{ij} = \frac{1}{||\boldsymbol{t}_j||^3}(t_{ij}\boldsymbol{t}_j - ||\boldsymbol{t}_j||^2 e_i) \end{aligned} \tag{13.72}$$

and

$$\begin{aligned}\frac{\partial J_p(\boldsymbol{x})}{\partial t_{ij}} &= \sum_{l=1}^{n}\Big\{\Big(\hat{\boldsymbol{x}}_p^H(l)\hat{\boldsymbol{M}}(\hat{\boldsymbol{T}})\hat{\boldsymbol{x}}_p(l)\Big)\Big(\hat{\boldsymbol{y}}_p^H(l)\hat{\boldsymbol{T}}^T\hat{\boldsymbol{T}}\hat{\boldsymbol{y}}_p(l)\Big)\\ &\quad+\Big(\hat{\boldsymbol{x}}_p^H(l)\hat{\boldsymbol{T}}^{-1}\hat{\boldsymbol{T}}^{-T}\hat{\boldsymbol{x}}_p(l)\Big)\Big(\hat{\boldsymbol{y}}_p^H(l)\hat{\boldsymbol{N}}(\hat{\boldsymbol{T}})\hat{\boldsymbol{y}}_p(l)\Big)\Big\}\\ \frac{\partial J_z(\boldsymbol{x})}{\partial t_{ij}} &= \sum_{l=1}^{n}\Big\{\hat{\boldsymbol{x}}_z^H(l)\hat{\boldsymbol{M}}(\hat{\boldsymbol{T}})\hat{\boldsymbol{x}}_z(l)\Big(\hat{\boldsymbol{y}}_z^H(l)\hat{\boldsymbol{T}}^T\hat{\boldsymbol{T}}\hat{\boldsymbol{y}}_z(l)+\beta_l^2\Big)\\ &\quad+\Big(\hat{\boldsymbol{x}}_z^H(l)\hat{\boldsymbol{T}}^{-1}\hat{\boldsymbol{T}}^{-T}\hat{\boldsymbol{x}}_z(l)+\alpha_l^2\Big)\hat{\boldsymbol{y}}_z^H(l)\hat{\boldsymbol{N}}(\hat{\boldsymbol{T}})\hat{\boldsymbol{y}}_z(l)\Big\}\end{aligned} \tag{13.73}$$

where

$$\hat{\boldsymbol{M}}(\hat{\boldsymbol{T}}) = -\,[\boldsymbol{g}_{ij}\boldsymbol{e}_j^T\hat{\boldsymbol{T}}^{-T} + \hat{\boldsymbol{T}}^{-1}\boldsymbol{e}_j\boldsymbol{g}_{ij}^T]$$

$$\hat{\boldsymbol{N}}(\hat{\boldsymbol{T}}) = \hat{\boldsymbol{T}}^T[\boldsymbol{e}_j\boldsymbol{g}_{ij}^T\hat{\boldsymbol{T}}^T + \hat{\boldsymbol{T}}\boldsymbol{g}_{ij}\boldsymbol{e}_j^T]\hat{\boldsymbol{T}}$$

13.6 Numerical Experiments

13.6.1 Filter Description and Initial Pole and Zero Sensitivity

Consider a state-space digital filter $(\boldsymbol{A}, \boldsymbol{b}, \boldsymbol{c}, d)_4$ in (13.1) described by

$$\boldsymbol{A} = \begin{bmatrix} 3.7183 & 1 & 0 & 0\\ -5.2153 & 0 & 1 & 0\\ 3.2689 & 0 & 0 & 1\\ -0.7724 & 0 & 0 & 0\end{bmatrix}, \qquad \boldsymbol{b} = 10^{-3}\begin{bmatrix}0.1755\\ 0.0178\\ 0.1652\\ 0.0052\end{bmatrix}$$

$$\boldsymbol{c} = \begin{bmatrix} 1 & 0 & 0 & 0\end{bmatrix}, \qquad d = 0.0227$$

for which the eigenvalues of matrices $\boldsymbol{A}$ and $\boldsymbol{Z}$ were found to be

$$\lambda = 0.963556 \pm j0.156437, \quad 0.895594 \pm j0.092081$$

$$v = 1.147681 \pm j0.329541, \quad 0.707604 \pm j0.202980$$

respectively. By using (13.22c), we obtained

$$\begin{bmatrix}\alpha_1\\ \alpha_2\\ \alpha_3\\ \alpha_4\end{bmatrix} = \begin{bmatrix}12.215615\\ 12.215615\\ 9.687213\\ 9.687213\end{bmatrix}, \qquad \begin{bmatrix}\beta_1\\ \beta_2\\ \beta_3\\ \beta_4\end{bmatrix} = \begin{bmatrix}0.386463\\ 0.386463\\ 0.310390\\ 0.310390\end{bmatrix}$$

The original pole and zero sensitivity measures in (13.22a) and (13.22b) were computed as

$$J_p = 7.209829 \times 10^6, \qquad J_z = 1.135788 \times 10^6$$

and the controllability Grammian was computed using (13.49) as

$$\boldsymbol{K}_c = \begin{bmatrix} 0.071608 & -0.195274 & 0.178646 & -0.054830 \\ -0.195274 & 0.533746 & -0.489317 & 0.150472 \\ 0.178646 & -0.489317 & 0.449437 & -0.138452 \\ -0.054830 & 0.150472 & -0.138452 & 0.042721 \end{bmatrix}$$

13.6.2 Weighted Pole and Zero Sensitivity Minimization Without Imposing l_2-Scaling Constraints

By choosing $\gamma = 1$ in (13.37), $\varepsilon = 10^{-8}$ in (13.43), and $\boldsymbol{P}_0 = \boldsymbol{I}_4$ in (13.42) as an initial estimate, it took the algorithm addressed in Section 13.3 two iterations to converge to

$$\boldsymbol{P} = 10^3 \begin{bmatrix} 0.318569 & -0.890465 & 0.835038 & -0.262540 \\ -0.890465 & 2.494146 & -2.343591 & 0.738320 \\ 0.835038 & -2.343591 & 2.206461 & -0.696498 \\ -0.262540 & 0.738320 & -0.696498 & 0.220301 \end{bmatrix}$$

which yields

$$\boldsymbol{T} = \boldsymbol{P}^{\frac{1}{2}} = 10 \begin{bmatrix} 0.504660 & -1.279387 & 1.093553 & -0.313564 \\ -1.279387 & 3.491572 & -3.188277 & 0.973880 \\ 1.093553 & -3.188277 & 3.110342 & -1.014605 \\ -0.313564 & 0.973880 & -1.014605 & 0.356118 \end{bmatrix}$$

From (13.36a) and (13.36b), it follows that

$$J_p(\boldsymbol{T}) = 4.000000, \qquad J_z(\boldsymbol{T}) = 1.819305 \times 10^5$$

and from (13.33) we obtained

$$\overline{\boldsymbol{A}} = \begin{bmatrix} 0.938212 & 0.125295 & -0.023034 & 0.057956 \\ -0.112124 & 0.961707 & 0.106409 & 0.016275 \\ -0.039345 & -0.091772 & 0.920683 & 0.074085 \\ -0.073949 & -0.006215 & -0.060044 & 0.897698 \end{bmatrix}$$

$$\overline{\boldsymbol{b}} = 10^{-3}\begin{bmatrix} 6.180513 \\ 6.309750 \\ 6.330612 \\ 6.224428 \end{bmatrix}$$

$$\overline{\boldsymbol{c}} = \begin{bmatrix} 5.046598 & -12.793870 & 10.935529 & -3.135635 \end{bmatrix}$$

It was observed that $\overline{\boldsymbol{A}}\,\overline{\boldsymbol{A}}^T = \overline{\boldsymbol{A}}^T\overline{\boldsymbol{A}}$ holds, hence matrix $\overline{\boldsymbol{A}}$ is normal and the lower bound $n = 4$ of $J_p(\boldsymbol{T})$ is achieved. The weighted pole and zero sensitivity performance $J_\gamma(\boldsymbol{P}_k^{\frac{1}{2}})$ for $k = 0, 1, 2$ were obtained as

$$\begin{aligned} &\left[J_\gamma(\boldsymbol{P}_0^{\frac{1}{2}}) \quad J_\gamma(\boldsymbol{P}_1^{\frac{1}{2}}) \quad J_\gamma(\boldsymbol{P}_2^{\frac{1}{2}}) \right] \\ &= \begin{bmatrix} 7.209829 \times 10^6 & 4.000000 & 4.000000 \end{bmatrix} \end{aligned}$$

from which it is seen that the iterative algorithm converges to the optimal solution with two iterations.

Since matrix $\boldsymbol{T}$ shown above yields $J_p(\boldsymbol{T}) = 4$, we set $\boldsymbol{T} = \boldsymbol{T}_o$. Then, by applying (13.46) and (13.47) it was found that

$$\boldsymbol{T}^{opt} = \begin{bmatrix} 0.133686 & -0.338914 & 0.289686 & -0.083064 \\ -0.338914 & 0.924930 & -0.844586 & 0.257984 \\ 0.289686 & -0.844586 & 0.823940 & -0.268772 \\ -0.083064 & 0.257984 & -0.268772 & 0.094337 \end{bmatrix}$$

and

$$J_z(\boldsymbol{T}^{opt}) = 593.186873, \qquad \zeta = 7.017385 \times 10^{-4}$$

It is noted that $\boldsymbol{T}^{opt}$ obtained above corresponds to the optimal coordinate transformation matrix that minimizes a zero sensitivity measure $J_z(\pm\sqrt{\zeta}\,\boldsymbol{T}_o)$ with respect to ζ subject to minimal pole sensitivity satisfying $J_p(\boldsymbol{T}_o) = 4$.

The optimal realization specified by $\boldsymbol{A}^{opt} = (\boldsymbol{T}^{opt})^{-1}\boldsymbol{A}\boldsymbol{T}^{opt}$, $\boldsymbol{b}^{opt} = (\boldsymbol{T}^{opt})^{-1}\boldsymbol{b}$, $\boldsymbol{c}^{opt} = \boldsymbol{c}\boldsymbol{T}^{opt}$ became

$$\boldsymbol{A}^{opt} = \begin{bmatrix} 0.938212 & 0.125295 & -0.023034 & 0.057956 \\ -0.112124 & 0.961707 & 0.106409 & 0.016275 \\ -0.039345 & -0.091772 & 0.920683 & 0.074085 \\ -0.073949 & -0.006215 & -0.060044 & 0.897698 \end{bmatrix}$$

$$b^{opt} = \begin{bmatrix} 0.233312 \\ 0.238191 \\ 0.238978 \\ 0.234970 \end{bmatrix}$$

$$c^{opt} = \begin{bmatrix} 0.133686 & -0.338914 & 0.289686 & -0.083064 \end{bmatrix}$$

The optimized pole and zero sensitivity measures corresponding to various weight values γ are summarized in Table 13.1 where

$$J_z(\boldsymbol{T}) \geq \sum_{l=1}^{4}(1 + |\alpha_l\beta_l|)^2 = 97.566165$$

always holds. From Table 13.1, it is observed that the lower bound of $J_z(\boldsymbol{T})$ was achieved with $\gamma = 0$. It is also observed that the sum of pole sensitivity and zero sensitivity reaches minimum when the weight value $\gamma = 0.5$ is chosen. Moreover, with γ as an adjustable design parameter, the proposed weighted optimization provides a variety of options to suit the need of a particular application where the designer may prefer smaller pole sensitivity or smaller zero sensitivity.

13.6.3 Weighted Pole and Zero Sensitivity Minimization Subject to l_2-Scaling Constraints Using Lagrange Function

By setting $\gamma = 1$ in (13.37), $\varepsilon = 10^{-8}$ in (13.58), and choosing $\boldsymbol{P}_0 = \boldsymbol{I}_4$ in (13.56) as an initial estimate, it took the algorithm addressed in Section 13.4 eight iterations to converge to

Table 13.1 Performance comparison in Section 13.6.2

γ	$J_\gamma(\boldsymbol{T})$	$J_p(\boldsymbol{T})$	$J_z(\boldsymbol{T})$	$J_p(\boldsymbol{T})+J_z(\boldsymbol{T})$
1.0	4.000000	4.000000	1.819305×10^5	1.819345×10^5
0.9	26.138641	9.756561	173.577365	183.333926
0.8	40.878190	13.926822	148.683663	162.610485
0.7	53.474454	18.020813	136.199616	154.220429
0.6	64.638191	22.453675	127.914966	150.368641
0.5	74.608206	27.598918	121.617494	149.216412
0.4	83.435795	34.001550	116.391959	150.393509
0.3	91.030285	42.698223	111.744026	154.442249
0.2	97.096840	56.211294	107.318227	163.529521
0.1	100.813973	83.685603	102.717125	186.402729
0.0	97.566165	283.698424	97.566165	381.264589

$$\boldsymbol{P} = 10 \begin{bmatrix} 0.336146 & -0.948542 & 0.899849 & -0.286556 \\ -0.948542 & 2.679553 & -2.544864 & 0.811398 \\ 0.899849 & -2.544864 & 2.419739 & -0.772465 \\ -0.286556 & 0.811398 & -0.772465 & 0.246930 \end{bmatrix}$$

which in conjunction with (13.63) led to

$$\boldsymbol{T} = \begin{bmatrix} 1.163353 & -0.501460 & 1.162615 & -0.636344 \\ -3.325136 & 1.323961 & -3.211571 & 1.916231 \\ 3.196696 & -1.163301 & 2.981285 & -1.933181 \\ -1.032291 & 0.337201 & -0.927826 & 0.655060 \end{bmatrix}$$

where

$$J_p(\boldsymbol{T}) = 4.000000, \qquad J_z(\boldsymbol{T}) = 5452.531195.$$

By using (13.33), we obtained

$$\overline{\boldsymbol{A}} = \begin{bmatrix} 0.927393 & -0.117544 & -0.055973 & 0.012748 \\ 0.127480 & 0.931340 & -0.021222 & 0.036623 \\ 0.003557 & 0.063032 & 0.935066 & 0.122093 \\ 0.029127 & 0.015908 & -0.123729 & 0.924501 \end{bmatrix}$$

$$\overline{\boldsymbol{b}} = \begin{bmatrix} 0.320110 \\ 0.041111 \\ -0.168643 \\ 0.244432 \end{bmatrix}$$

$$\overline{\boldsymbol{c}} = \begin{bmatrix} 1.163353 & -0.501460 & 1.162615 & -0.636344 \end{bmatrix}$$

It was observed that $\overline{\boldsymbol{A}}\,\overline{\boldsymbol{A}}^T = \overline{\boldsymbol{A}}^T\overline{\boldsymbol{A}}$ holds. This means that matrix $\overline{\boldsymbol{A}}$ is normal and the lower bound $n = 4$ of $J_p(\boldsymbol{T})$ is achieved. For the new realization, the controllability Grammian was found from (13.50) to be

$$\overline{\boldsymbol{K}}_c = \begin{bmatrix} 1.000000 & 0.296865 & -0.585417 & 0.804018 \\ 0.296865 & 1.000000 & 0.585417 & 0.804018 \\ -0.585417 & 0.585417 & 1.000000 & 0.000000 \\ 0.804018 & 0.804018 & 0.000000 & 1.000000 \end{bmatrix}$$

where the l_2-scaling constraints are satisfied. $I_\gamma(\boldsymbol{P}, \xi)$ performance of 8 iterations is shown in Figure 13.1, from which it is seen that the iterative algorithm converges with 8 iterations where

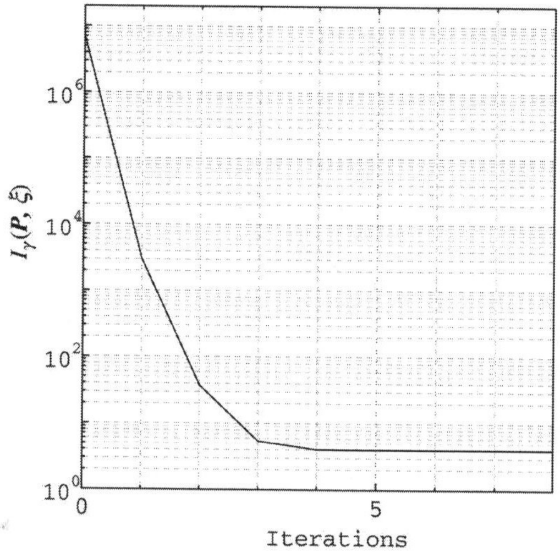

Figure 13.1 Profile of $I_\gamma(\boldsymbol{P}, \xi)$ during the first 8 iterations with $\gamma = 1$.

$$\begin{bmatrix} \xi_1 & \xi_2 \\ \xi_3 & \xi_4 \\ \xi_5 & \xi_6 \\ \xi_7 & \xi_8 \end{bmatrix} = \begin{bmatrix} -1.069263 \times 10^6 & -4.213525 \times 10^2 \\ -7.117870 \times 10^0 & -4.496284 \times 10^{-1} \\ -1.817527 \times 10^{-2} & -5.577404 \times 10^{-4} \\ -1.676641 \times 10^{-5} & -5.025417 \times 10^{-7} \end{bmatrix}$$

The optimized pole and zero sensitivity measures subject to l_2-scaling constraints corresponding to various values of γ are summarized in Table 13.2.

Table 13.2 Lagrange function method subject to scaling constraints

γ	$J_\gamma(\boldsymbol{T})$	$J_p(\boldsymbol{T})$	$J_z(\boldsymbol{T})$	$J_p(\boldsymbol{T})+J_z(\boldsymbol{T})$
1.0	4.000000	4.000000	5452.531195	5456.531195
0.9	26.798631	10.315471	175.147077	185.462547
0.8	41.489644	14.827471	148.138333	162.965804
0.7	53.902744	19.064263	135.192533	154.256796
0.6	64.863774	23.442505	126.995679	150.438183
0.5	74.672573	28.277118	121.068028	149.345146
0.4	83.435914	33.963990	116.417196	150.381186
0.3	91.139040	41.177416	112.551165	153.728580
0.2	97.629962	51.388289	109.190380	160.578669
0.1	102.474584	69.082077	106.184863	175.266940
0.0	104.016409	125.930718	104.016409	229.947127

13.6.4 Weighted Pole and Zero Sensitivity Minimization Subject to l_2-Scaling Constraints Using Quasi-Newton Algorithm

By choosing $\gamma = 1$ in (13.68), $\varepsilon = 10^{-8}$ in (13.70), and starting with an initial point $\boldsymbol{x}_0$ obtained from the assignment $\hat{\boldsymbol{T}}^{-1} = \boldsymbol{I}_4$ in (13.66), it took the algorithm addressed in Section 13.5 nineteen iterations to converge to

$$\hat{\boldsymbol{T}}^{-1} = \begin{bmatrix} 0.864243 & 0.900469 & 0.786701 & 0.542671 \\ -0.490448 & 0.391427 & 0.345446 & 0.722078 \\ -0.094086 & -0.168073 & 0.507431 & -0.076744 \\ 0.060769 & -0.087704 & 0.065444 & 0.422163 \end{bmatrix}$$

or equivalently,

$$\boldsymbol{T} = \begin{bmatrix} 0.149803 & -0.161958 & 0.261061 & -0.203040 \\ -0.366793 & 0.433947 & -0.742778 & 0.584686 \\ 0.299031 & -0.385952 & 0.706504 & -0.563004 \\ -0.080834 & 0.113909 & -0.224536 & 0.181708 \end{bmatrix}$$

where

$$J_p(\boldsymbol{T}) = 4.000000, \qquad J_z(\boldsymbol{T}) = 936.412522.$$

By using (13.33), we obtained

$$\overline{\boldsymbol{A}} = \begin{bmatrix} 0.955963 & 0.084559 & -0.070022 & 0.105312 \\ -0.107373 & 0.934715 & 0.072300 & 0.045359 \\ 0.033913 & -0.107775 & 0.913094 & 0.026735 \\ -0.102298 & 0.006930 & -0.057878 & 0.914528 \end{bmatrix}$$

$$\overline{\boldsymbol{b}} = \begin{bmatrix} 0.068313 \\ 0.257916 \\ 0.497924 \\ 0.484019 \end{bmatrix}$$

$$\overline{\boldsymbol{c}} = \begin{bmatrix} 0.149803 & -0.161958 & 0.261061 & -0.203040 \end{bmatrix}$$

It was observed that $\overline{\boldsymbol{A}}\,\overline{\boldsymbol{A}}^T = \overline{\boldsymbol{A}}^T\overline{\boldsymbol{A}}$ holds. This implies that matrix $\overline{\boldsymbol{A}}$ is normal and the lower bound $n = 4$ of $J_p(\boldsymbol{T})$ is achieved. For the new realization, the controllability Grammian was found from (13.50) to be

$$\overline{\boldsymbol{K}}_c = \begin{bmatrix} 1.000000 & 0.596733 & 0.466712 & 0.147733 \\ 0.596733 & 1.000000 & 0.752591 & 0.747172 \\ 0.466712 & 0.752591 & 1.000000 & 0.665045 \\ 0.147733 & 0.747172 & 0.665045 & 1.000000 \end{bmatrix}$$

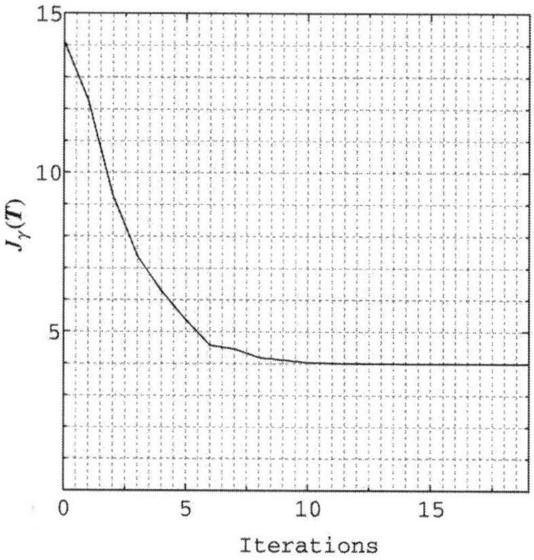

Figure 13.2 Profile of $J_\gamma(\boldsymbol{T})$ during the first 19 iterations with $\gamma = 1$.

where the l_2-scaling constraints are satisfied. The pole sensitivity performance of 19 iterations is shown in Figure 13.2 where $\gamma = 1$, i.e., $J_\gamma(\boldsymbol{T}) = J_p(\boldsymbol{T})$, from which it is seen that the iterative algorithm converges with 19 iterations.

The optimized pole and zero sensitivity measures subject to the l_2-scaling constraints corresponding to various values of γ are summarized in Table 13.3.

Table 13.3 Quasi-Newton method subject to scaling constraints

γ	$J_\gamma(\boldsymbol{T})$	$J_p(\boldsymbol{T})$	$J_z(\boldsymbol{T})$	$J_p(\boldsymbol{T})+J_z(\boldsymbol{T})$
1.0	4.000000	4.000000	936.412522	940.412522
0.9	26.798631	10.315452	175.147242	185.462694
0.8	41.489644	14.827462	148.138371	162.965833
0.7	53.902744	19.064221	135.192631	154.256852
0.6	64.863774	23.442437	126.995780	150.438217
0.5	74.672573	28.277078	121.068068	149.345146
0.4	83.435914	33.963942	116.417229	150.381170
0.3	91.139040	41.177323	112.551204	153.728527
0.2	97.629962	51.388153	109.190414	160.578567
0.1	102.474584	69.081932	106.184879	175.266811
0.0	104.016409	125.930606	104.016409	229.947015

Table 13.4 Performance comparison among four methods

Method	l_2-Sensitivity	J_p	J_z
Quasi-Newton ($\gamma = 1$)	143.884664	4	936.412522
Quasi-Newton ($\gamma = 0$)	995.595955	125.930606	104.016409
Lagrange Function ($\gamma = 1$)	657.722872	4	5452.531195
Lagrange Function ($\gamma = 0$)	995.596747	125.930718	104.016409
Hinamoto et al. [5]	92.906421	4.628256	398.788778
Hinamoto et al. [6]	92.906418	4.628253	398.789252

We now conclude this section with a remark on the numerical results summarized in Tables 13.2 and 13.3. Concerning the case of $\gamma = 1$, it follows from (13.37) that the use of $\gamma = 1$ simply excludes $J_z(\boldsymbol{T})$ in the optimization procedure and, as a result, the zero sensitivity went wildly large as shown in Tables 13.2 and 13.3. For practical system implementations, therefore, the use of γ in the range $0 \leq \gamma < 1$ is recommended.

The problem of minimizing the l_2-sensitivity of a transfer function subject to l_2-scaling constraints was solved by different techniques in Chapter 12 [5, 6]. In Table 13.4 the performances of the methods presented in this chapter are compared with those achieved by the methods in Chapter 12 for four implementation settings. From this table it is observed that in case $\gamma = 1$, the methods presented in this chapter provide reduced pole sensitivity than those in Chapter 12, and in case $\gamma = 0$, the techniques in this chapter produce reduced zero sensitivity than those in Chapter 12.

13.7 Summary

Three iterative techniques for minimizing a weighted pole and zero sensitivity measure have been presented. The problem free from l_2-scaling constraints has been solved by employing a recursive matrix equation. A simple method has also been given to obtain the optimal coordinate transformation matrix which minimizes a zero sensitivity measure subject to minimal pole sensitivity. Moreover, two iterative methods for minimizing the weighted pole and zero sensitivity measure subject to l_2-scaling constraints have been introduced. One relaxes the constraints into a single constraint on matrix trace and solves the relaxed problem with an efficient matrix iteration scheme based on the Lagrange function and a bisection method, while the other converts the constrained optimization problem at hand into an unconstrained problem and solves it using a quasi-Newton algorithm.

Simulation results in numerical experiments have demonstrated the validity and effectiveness of the above techniques.

References

[1] L. Thiele, "Design of sensitivity and round-off noise optimal state-space discrete systems," *Int. J. Circuit Theory Appl.*, vol. 12, pp. 39–46, Jan. 1984.

[2] L. Thiele, "On the sensitivity of linear state-space systems," *IEEE Trans. Circuits Syst.*, vol. CAS-33, no. 5, pp. 502–510, May 1986.

[3] G. Li, B. D. O. Anderson, M. Gevers and J. E. Perkins, "Optimal FWL design of state-space digital systems with weighted sensitivity minimization and sparseness consideration," *IEEE Trans. Circuits Syst. I*, vol. 39, no. 5, pp. 365–377, May 1992.

[4] W.-Y. Yan and J. B. Moore, "On L^2-sensitivity minimization of linear state-space systems," *IEEE Trans. Circuits Syst. I*, vol. 39, no. 8, pp. 641–648, Aug. 1992.

[5] T. Hinamoto, H. Ohnishi and W.-S. Lu, "Minimization of L_2-sensitivity for state-space digital filters subject to L_2-dynamic-range scaling constraints," *IEEE Trans. Circuits Syst. II*, vol. 52, no. 10, pp. 641–645, Oct. 2005.

[6] T. Hinamoto, K. Iwata and W.-S. Lu, "L_2-sensitivity minimization of one- and two-dimensional state-space digital filters subject to L_2-scaling constraints," *IEEE Trans. Signal Process.*, vol. 54, no. 5, pp. 1804–1812, May 2006.

[7] M. Gevers and G. Li, *Parameterizations in Control, Estimation and Filtering Problems: Accuracy Aspects*, New York: Springer-Verlag, 1993.

[8] P. E. Mantey, "Eigenvalue sensitivity and state-variable selection," *IEEE Trans. Automatic Contr.*, vol. AC-13, no. 3, pp. 263–269, Jun. 1968.

[9] R. E. Skelton and D. A. Wagie, "Minimal root sensitivity in linear systems," *J. Guidance Contr.*, vol. 7, no. 5, pp. 570–574, Sep.–Oct. 1984.

[10] D. Williamson, "Roundoff noise minimization and pole-zero sensitivity in fixed-point digital filters using residue feedback," *IEEE Trans. Acoust., Speech, Signal Process.*, vol. ASSP-34, no. 5, pp. 1210–1220, Oct. 1986.

[11] G. Li, "On pole and zero sensitivity of linear systems," *IEEE Trans. Circuits Syst. I*, vol. 44, no. 7, pp. 583–590, Jul. 1997.

[12] T. Hinamoto, A. Doi and W.-S. Lu "Minimization of weighted pole and zero sensitivity for state-space digital filters," *IEEE Trans. Circuits Syst. I*, vol. 63, no. 1, pp. 103–113, Jan. 2016.
[13] R. A. Roberts and C. T. Mullis, *Digital Signal Processing*, Addison-Wesley Publishing Company, Inc., 1987.
[14] J. E. Dennis and J. J. More, "Quasi-Newton methods, motivation and theory," *SIAM Rev.*, vol. 19, no. 1, pp. 46–89, 1977.
[15] R. Fletcher, *Practical Methods of optimization*, 2nd ed. New York: Wiley, 1987.
[16] T. Kailath, *Linear System*. Englewood Cliffs, NJ: Prentice-Hall, 1980.
[17] H. Togawa, *Handbook of Numerical Methods*, Tokyo, Japan, Saience-sha, 1992.
[18] S. Y. Hwang, "Minimum uncorrelated unit noise in state-space digital filtering," *IEEE Trans. Acoust., Speech, Signal Process.*, vol. ASSP-25, no. 4, pp. 273–281, Aug. 1977.

14

Error Spectrum Shaping

14.1 Preview

Error feedback is also called error spectrum shaping, and it is known as an effective method for the reduction of quantization error generated in finite-word-length (FWL) implementations of IIR digital filters, and it is especially so when dealing with fixed-point implementations of narrow-band lowpass filters. Error feedback can be achieved by extracting the quantization error after multiplication and addition, and then feeding the error signal back through simple filters. When error feedback is applied to an IIR digital filter with either external or internal description, it only affects the transfer function of the quantization error signal, but not the input-output characteristic of the filter. As a result, error feedback neither alters coefficient sensitivities nor enhances overflow properties of the filter. It is well known that the level of the filter's output quantization noise of an IIR filter tends to become high when the poles lie close to the unit circle. This problem can also be addressed effectively by error feedback whose parameters are chosen appropriately so that the zeros of the transfer function from the quantization error to the filter's output are tuned via error spectrum shaping so as to reduce the effects of the quantization noise at the filter's output.

In this chapter, we study the problem of minimizing the roundoff noise at the filter's output by applying high-order error feedback for both external and internal descriptions of IIR digital filters. The optimal solution of high-order error feedback is obtained for IIR digital filters as well as state-space digital filters. As alternatives to the optimal solution, suboptimal solutions for the high-order error feedback with symmetric or antisymmetric coefficients (matrices) are then derived. Finally, we present numerical experiments to demonstrate the validity and effectiveness of the techniques addressed in this chapter.

14.2 IIR Digital Filters with High-Order Error Feedback

14.2.1 Nth-Order Optimal Error Feedback

The error feedback is implemented by modifying the quantizer in the filter structure. In a fixed-point implementation, quantization is usually performed by discarding the lower bits of the double-precision accumulator (two's complement truncation). Hence, the quantization error is equal to the residue left in the lower part. Figure 14.1 depicts a quantizer with Nth-order error feedback where the error is fed back through a simple FIR filter. Referring to Figure 14.1, we obtain

$$\begin{aligned} x(k) &= u(k) + \beta_1 e(k-1) + \beta_2 e(k-2) + \cdots + \beta_N e(k-N) \\ e(k) &= \tilde{x}(k) - x(k) \end{aligned} \tag{14.1}$$

Substituting the first equation in (14.1) into the second one yields

$$\tilde{x}(k) = u(k) + e(k) + \beta_1 e(k-1) + \beta_2 e(k-2) + \cdots + \beta_N e(k-N) \tag{14.2}$$

By taking the z-transform on the both sides of (14.2), we have

$$\tilde{X}(z) = U(z) + \left(1 + \beta_1 z^{-1} + \beta_2 z^{-2} + \cdots + \beta_N z^{-N}\right) E(z) \tag{14.3}$$

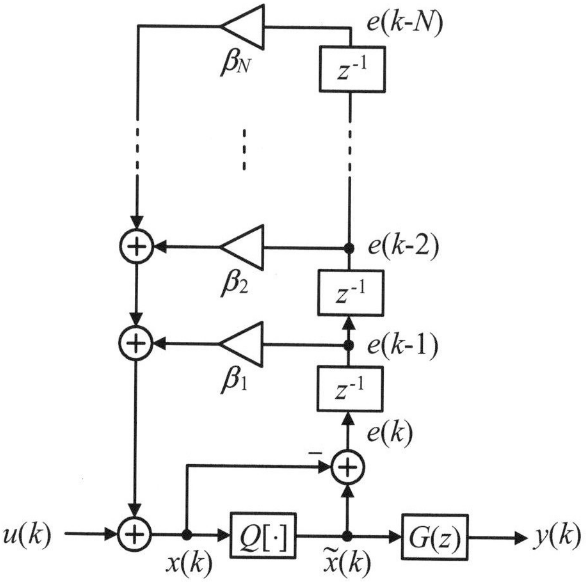

Figure 14.1 A quantizer with Nth-order error feedback.

where $\tilde{X}(z)$, $U(z)$ and $E(z)$ are the z-transforms of signals $\tilde{x}(k)$, $u(k)$ and $e(k)$, respectively. Let the transfer function from the quantization point to the filter's output be denoted by $G(z)$. In general, $G(z)$ is a rational transfer function of a linear, time-invariant, causal, and stable system of order usually higher than N. Under the circumstance, we obtain

$$Y(z) = G(z)\tilde{X}(z) = G(z)U(z) + G(z)B(z)E(z) \qquad (14.4)$$

where

$$B(z) = 1 + \beta_1 z^{-1} + \beta_2 z^{-2} + \cdots + \beta_N z^{-N}$$

and $Y(z)$ indicates the z-transform of the output signal $y(k)$. It is standard to assume that each quantizer is modeled as an independent additive white noise source with variance $\sigma^2 = 2^{-2b}/12$ where $(1+b)$ is the wordlength (1 bit for sign). The normalized noise gain (noise variance) from the noise source to the filter's output can be written as

$$I(\boldsymbol{\beta}) = \frac{\sigma_{out}^2}{\sigma^2} = \frac{1}{2\pi j}\oint_{|z|=1} G(z)B(z)G(z^{-1})B(z^{-1})\frac{dz}{z} \qquad (14.5)$$

which is equivalent to

$$I(\boldsymbol{\beta}) = \frac{1}{\pi}\int_0^{\pi} |B(e^{j\omega})|^2 Q(\omega)d\omega \qquad (14.6)$$

where

$$\boldsymbol{\beta} = \begin{bmatrix}\beta_1 & \beta_2 & \cdots & \beta_N\end{bmatrix}^T, \qquad Q(\omega) = |G(e^{j\omega})|^2$$

Note that

$$\begin{aligned} B(z^{-1})B(z) &= 1 + \sum_{i=1}^{N}\beta_i(z^i + z^{-i}) + \sum_{i=1}^{N}\sum_{l=1}^{N}\beta_i\beta_l z^{i-l} \\ &= 1 + \sum_{i=1}^{N}\beta_i(z^i + z^{-i}) + \sum_{i=1}^{N}\beta_i^2 + \sum_{l=1}^{N-1}\sum_{i=1}^{N-l}\beta_i\beta_{i+l}(z^l + z^{-l}) \end{aligned} \qquad (14.7)$$

which can be written as

$$|B(e^{j\omega})|^2 = 1 + 2\sum_{i=1}^{N}\beta_i\cos(i\omega) + \sum_{i=1}^{N}\sum_{l=1}^{N}\beta_i\beta_l\cos(i-l)\omega \qquad (14.8)$$

because

$$e^{j\omega l} + e^{-j\omega l} = 2\cos(l\omega) = \cos(l\omega) + \cos(-l\omega)$$

By referring to (14.8) and defining

$$q_i = \frac{1}{\pi} \int_0^{\pi} Q(\omega) \cos(i\omega) d\omega \tag{14.9}$$

the normalized noise gain in (14.6) can be expressed as

$$\begin{aligned} I(\boldsymbol{\beta}) &= \sum_{i=1}^{N} \sum_{l=1}^{N} \beta_i \beta_l q_{|i-l|} + 2 \sum_{i=1}^{N} \beta_i q_i + q_0 \\ &= \boldsymbol{\beta}^T \boldsymbol{R} \boldsymbol{\beta} + 2\boldsymbol{\beta}^T \boldsymbol{p} + q_0 \end{aligned} \tag{14.10}$$

where

$$\boldsymbol{R} = \begin{bmatrix} q_0 & q_1 & \cdots & q_{N-1} \\ q_1 & q_0 & \cdots & q_{N-2} \\ \vdots & \vdots & \ddots & \vdots \\ q_{N-1} & q_{N-2} & \cdots & q_0 \end{bmatrix}, \qquad \boldsymbol{p} = \begin{bmatrix} q_1 \\ q_2 \\ \vdots \\ q_N \end{bmatrix}$$

The matrix $\boldsymbol{R}$ is recognized as the $N \times N$ autocorrelation matrix of the output error, which is a symmetric, positive-definite Toeplitz matrix. The vector $\boldsymbol{p}$ is the crosscorrelation vector between the input and output error. To find the optimal $\boldsymbol{\beta}$ that minimizes the normalized noise gain $I(\boldsymbol{\beta})$, we use (14.10) to compute the gradient of $I(\boldsymbol{\beta})$ with respect to $\boldsymbol{\beta}$ and set it to null, which leads to

$$\frac{\partial I(\boldsymbol{\beta})}{\partial \boldsymbol{\beta}} = 2\boldsymbol{R}\boldsymbol{\beta} + 2\boldsymbol{p} = \boldsymbol{0} \tag{14.11}$$

Therefore, the optimal $\boldsymbol{\beta}$ is found to be

$$\boldsymbol{\beta}^{opt} = -\boldsymbol{R}^{-1}\boldsymbol{p} \tag{14.12}$$

14.2.2 Computation of Autocorrelation Coefficients

For the present problem, the autocorrelation coefficients q_i's depend only on the given rational transfer function $G(z)$ and can thus be determined exactly. The z-domain version of (14.9) gives the autocorrelation coefficients via the inverse z transform as

$$q_i = \frac{1}{2\pi j} \oint_{|z|=1} z^i G(z) G(z^{-1}) \frac{dz}{z} \tag{14.13}$$

Since the autocorrelation sequence is symmetric, i.e., $q_i = q_{-i}$ for any integer i, (14.13) is as well given in the form

$$q_i = \frac{1}{2\pi j} \oint_{|z|=1} \frac{z^i + z^{-i}}{2} G(z)G(z^{-1}) \frac{dz}{z} \tag{14.14}$$

By denoting the impulse response of $G(z)$ by $\{g_k | \, k = 0, 1, 2, \cdots\}$ and utilizing *Cauchy's integral theorem*

$$\frac{1}{2\pi j} \oint_C z^k \frac{dz}{z} = \begin{cases} 1, & k = 0 \\ 0, & k \neq 0 \end{cases} \tag{14.15}$$

where C is a counterclockwise contour that encircles the origin, (14.14) can be written as

$$\begin{aligned} q_i &= \frac{1}{2\pi j} \oint_{|z|=1} \frac{z^i + z^{-i}}{2} \sum_{k=0}^{\infty} g_k z^{-k} \sum_{l=0}^{\infty} g_l z^l \frac{dz}{z} \\ &= \frac{1}{2} \left[\frac{1}{2\pi j} \oint_{|z|=1} \sum_{k=0}^{\infty} \sum_{l=0}^{\infty} g_k g_l z^{l+i-k} \frac{dz}{z} \right. \\ &\quad \left. + \frac{1}{2\pi j} \oint_{|z|=1} \sum_{k=0}^{\infty} \sum_{l=0}^{\infty} g_k g_l z^{l-i-k} \frac{dz}{z} \right] \\ &= \frac{1}{2} \left[\sum_{k=0}^{\infty} g_k g_{k-i} + \sum_{k=0}^{\infty} g_k g_{k+i} \right] = \sum_{k=0}^{\infty} g_k g_{k+i} \end{aligned} \tag{14.16}$$

The above equation leads to

$$\begin{aligned} q_i &= g_0 g_i + \sum_{k=1}^{\infty} \boldsymbol{c} \boldsymbol{A}^{k-1} \boldsymbol{b} \boldsymbol{c} \boldsymbol{A}^{k+i-1} \boldsymbol{b} \\ &= g_0 g_i + \boldsymbol{c} \sum_{k=1}^{\infty} \boldsymbol{A}^{k-1} \boldsymbol{b} \boldsymbol{b}^T (\boldsymbol{A}^{k-1})^T (\boldsymbol{A}^i)^T \boldsymbol{c}^T \\ &= g_0 g_i + \boldsymbol{c} \boldsymbol{K}_c (\boldsymbol{A}^i)^T \boldsymbol{c}^T \end{aligned} \tag{14.17}$$

where

$$G(z) = d + \boldsymbol{c}(z\boldsymbol{I}_n - \boldsymbol{A})^{-1}\boldsymbol{b} = \sum_{k=0}^{\infty} g_k z^{-k}$$

$$g_k = \boldsymbol{c}\boldsymbol{A}^{k-1}\boldsymbol{b} \ \text{ for } \ k \geq 1, \qquad g_0 = d$$

and $\boldsymbol{K}_c$ is the controllability Grammian of the state-space realization $(\boldsymbol{A}, \boldsymbol{b}, \boldsymbol{c}, d)_n$ of $G(z)$ that can be obtained by solving the Lyapunov equation

$$\boldsymbol{K}_c = \boldsymbol{A}\boldsymbol{K}_c\boldsymbol{A}^T + \boldsymbol{b}\boldsymbol{b}^T$$

Similarly, the autocorrelation coefficients can also be derived from

$$q_i = g_0 g_i + \boldsymbol{b}^T \boldsymbol{W}_o \boldsymbol{A}^i \boldsymbol{b} \tag{14.18}$$

instead of (14.17) where $\boldsymbol{W}_o$ is the observability Grammian of the state-space realization $(\boldsymbol{A}, \boldsymbol{b}, \boldsymbol{c}, d)_n$ of $G(z)$ which can be obtained by solving the Lyapunov equation

$$\boldsymbol{W}_o = \boldsymbol{A}^T \boldsymbol{W}_o \boldsymbol{A} + \boldsymbol{c}^T \boldsymbol{c}$$

14.2.3 Error Feedback with Symmetric or Antisymmetric Coefficients

In practice, the implementation of Nth-order optimal error feedback is often too costly because of the N explicit multiplications required. One way for reducing the number of multiplications is to constrain $B(z)$ to be symmetric or antisymmetric. This halves the number of required multiplications. The symmetry constrains the zeroes of the filter to be exactly on the unit circle in most cases.

A. Odd-Order Error Feedback with Symmetric Coefficients

Suppose the order of an error feedback filter is odd, say $N = 2M + 1$, and the coefficients are symmetric, then the error transfer function $B(z)$ in (14.4) can be written as

$$B(z) = 1 + z^{-(2M+1)} + \sum_{i=1}^{M} \beta_i \left(z^{-i} + z^{-(2M+1)+i}\right) \tag{14.19}$$

The polynomial in (14.19) leads to

$$\begin{aligned} B(z^{-1})B(z) = {} & 2 + z^{2M+1} + z^{-(2M+1)} \\ & + 2\sum_{i=1}^{M} \beta_i \left(z^i + z^{-i} + z^{(2M+1)-i} + z^{-(2M+1)+i}\right) \\ & + \sum_{i=1}^{M}\sum_{l=1}^{M} \beta_i \beta_l \left(z^{i-l} + z^{-(i-l)} + z^{2M+1-(i+l)} \right. \\ & \left. + z^{-(2M+1)+i+l}\right) \end{aligned} \tag{14.20}$$

which is equivalent to

$$|B(e^{j\omega})|^2 = 2\Big[1 + \cos(2M+1)\omega + 2\sum_{i=1}^{M} \beta_i\{\cos i\omega + \cos(2M+1-i)\omega\}$$
$$+ \sum_{i=1}^{M}\sum_{l=1}^{M} \beta_i\beta_l\{\cos(i-l)\omega + \cos(2M+1-i-l)\omega\}\Big] \tag{14.21}$$

Substituting (14.21) into (14.6) yields

$$\begin{aligned} I(\boldsymbol{\beta}_o) &= 2\Big[q_0 + q_{2M+1} + 2\sum_{i=1}^{M}\beta_i\big(q_i + q_{2M+1-i}\big) \\ &\quad + \sum_{i=1}^{M}\sum_{l=1}^{M}\beta_i\beta_l\big(q_{i-l} + q_{2M+1-i-l}\big)\Big] \\ &= 2\Big[q_0 + q_{2M+1} + 2\boldsymbol{\beta}_o^T(\boldsymbol{p}_o + \tilde{\boldsymbol{p}}_o) + \boldsymbol{\beta}_o^T(\boldsymbol{R}_o + \tilde{\boldsymbol{R}}_o)\boldsymbol{\beta}_o\Big] \end{aligned} \tag{14.22}$$

where

$$\boldsymbol{\beta}_o = \begin{bmatrix} \beta_1 \\ \beta_2 \\ \vdots \\ \beta_M \end{bmatrix}, \quad \boldsymbol{p}_o = \begin{bmatrix} q_1 \\ q_2 \\ \vdots \\ q_M \end{bmatrix}, \quad \boldsymbol{R}_o = \begin{bmatrix} q_0 & q_1 & \cdots & q_{M-1} \\ q_1 & q_0 & \cdots & q_{M-2} \\ \vdots & \vdots & \ddots & \vdots \\ q_{M-1} & q_{M-2} & \cdots & q_0 \end{bmatrix}$$

$$\tilde{\boldsymbol{p}}_o = \begin{bmatrix} q_{2M} \\ q_{2M-1} \\ \vdots \\ q_{M+1} \end{bmatrix}, \quad \tilde{\boldsymbol{R}}_o = \begin{bmatrix} q_{2M-1} & q_{2M-2} & \cdots & q_M \\ q_{2M-2} & q_{2M-3} & \cdots & q_{M-1} \\ \vdots & \vdots & \ddots & \vdots \\ q_M & q_{M-1} & \cdots & q_1 \end{bmatrix}$$

The optimal $\boldsymbol{\beta}_o$ is found by setting $\partial I(\boldsymbol{\beta}_o)/\partial\boldsymbol{\beta}_o = \mathbf{0}$ as

$$\boldsymbol{\beta}_o^{opt} = -\big(\boldsymbol{R}_o + \tilde{\boldsymbol{R}}_o\big)^{-1}\big(\boldsymbol{p}_o + \tilde{\boldsymbol{p}}_o\big) \tag{14.23}$$

B. Odd-Order Error Feedback with Antisymmetric Coefficients

Suppose the order of an error feedback filter is odd, say $N = 2M+1$, and the coefficients are antisymmetric, then the error transfer function $B(z)$ in (14.4) can be written as

$$B(z) = 1 - z^{-(2M+1)} + \sum_{i=1}^{M} \beta_i \big(z^{-i} - z^{-(2M+1)+i}\big) \tag{14.24}$$

The polynomial in (14.24) leads to

$$\begin{aligned} B(z^{-1})B(z) = 2 - \big(&z^{2M+1} + z^{-(2M+1)}\big) \\ &+ 2\sum_{i=1}^{M} \beta_i \big(z^{i} + z^{-i} - z^{(2M+1)-i} - z^{-(2M+1)+i}\big) \\ &+ \sum_{i=1}^{M}\sum_{l=1}^{M} \beta_i\beta_l \big(z^{i-l} + z^{-(i-l)} - z^{2M+1-(i+l)} \\ &- z^{-(2M+1)+i+l}\big) \end{aligned} \tag{14.25}$$

which is equivalent to

$$\begin{aligned} |B(e^{j\omega})|^2 = 2\Big[1 - \cos(2M+1)\omega + 2\sum_{i=1}^{M} \beta_i\{\cos i\omega - \cos(2M+1-i)\omega\} \\ + \sum_{i=1}^{M}\sum_{l=1}^{M} \beta_i\beta_l\{\cos(i-l)\omega - \cos(2M+1-i-l)\omega\}\Big] \end{aligned} \tag{14.26}$$

Substituting (14.26) into (14.6) yields

$$\begin{aligned} I(\boldsymbol{\beta}_o) &= 2\Big[q_0 - q_{2M+1} + 2\sum_{i=1}^{M} \beta_i \big(q_i - q_{2M+1-i}\big) \\ &\quad + \sum_{i=1}^{M}\sum_{l=1}^{M} \beta_i\beta_l \big(q_{i-l} - q_{2M+1-i-l}\big)\Big] \\ &= 2\Big[q_0 - q_{2M+1} + 2\boldsymbol{\beta}_o^T(\boldsymbol{p}_o - \tilde{\boldsymbol{p}}_o) + \boldsymbol{\beta}_o^T(\boldsymbol{R}_o - \tilde{\boldsymbol{R}}_o)\boldsymbol{\beta}_o\Big] \end{aligned} \tag{14.27}$$

where $\boldsymbol{\beta}_o$, $\boldsymbol{p}_o$, $\tilde{\boldsymbol{p}}_o$, $\boldsymbol{R}_o$ and $\tilde{\boldsymbol{R}}_o$ are defined as in (14.22). The optimal $\boldsymbol{\beta}_o$ is found by setting $\partial I(\boldsymbol{\beta}_o)/\partial \boldsymbol{\beta}_o = \boldsymbol{0}$ as

$$\boldsymbol{\beta}_o^{opt} = -\big(\boldsymbol{R}_o - \tilde{\boldsymbol{R}}_o\big)^{-1}\big(\boldsymbol{p}_o - \tilde{\boldsymbol{p}}_o\big) \tag{14.28}$$

C. Even-Order Error Feedback with Symmetric Coefficients

Suppose the order of an error feedback filter is even, say $N = 2L$, and the coefficients are symmetric, then the error transfer function $B(z)$ in (14.4) can be written as

$$B(z) = 1 + \sum_{i=1}^{L-1} \beta_i \big(z^{-i} + z^{-2L+i}\big) + \beta_L z^{-L} + z^{-2L} \tag{14.29}$$

The above polynomial leads to

$$\begin{aligned} B(z^{-1})B(z) = 2 + z^{2L} + z^{-2L} + 2\beta_L\big(z^L + z^{-L}\big) \\ + 2\sum_{i=1}^{L-1} \beta_i\big(z^i + z^{-i} + z^{2L-i} + z^{-2L+i}\big) \\ + \sum_{i=1}^{L-1}\sum_{l=1}^{L-1} \beta_i\beta_l\big(z^{i-l} + z^{-(i-l)} + z^{2L-i-l} + z^{-2L+i+l}\big) \\ + 2\sum_{i=1}^{L-1} \beta_L\beta_i\big(z^{L-i} + z^{-L+i}\big) + \beta_L^2 \end{aligned} \tag{14.30}$$

which is equivalent to

$$\begin{aligned} |B(e^{j\omega})|^2 = 2 + 2\cos 2L\omega + 4\beta_L \cos L\omega \\ + 4\sum_{i=1}^{L-1} \beta_i\{\cos i\omega + \cos(2L-i)\omega\} \\ + 2\sum_{i=1}^{L-1}\sum_{l=1}^{L-1} \beta_i\beta_l\{\cos(i-l)\omega + \cos(2L-i-l)\omega\} \\ + 4\sum_{i=1}^{L-1} \beta_L\beta_i \cos(L-i)\omega + \beta_L^2 \end{aligned} \tag{14.31}$$

Substituting (14.31) into (14.6) yields

$$I(\boldsymbol{\beta}_{se}) = 2\big(q_0 + q_{2L}\big) + 4\beta_L q_L + 4\sum_{i=1}^{L-1} \beta_i\big(q_i + q_{2L-i}\big)$$

$$+\,2\sum_{i=1}^{L-1}\sum_{l=1}^{L-1}\beta_i\beta_l\big(q_{i-l}+q_{2L-i-l}\big)+4\sum_{i=1}^{L-1}\beta_L\beta_i q_{L-i}+\beta_L^2 q_o$$

$$=\;2\big(q_0+q_{2L}\big)+2\boldsymbol{\beta}_{se}^T\begin{bmatrix}2\big(\boldsymbol{p}_e+\tilde{\boldsymbol{p}}_e\big)\\ 2q_L\end{bmatrix}+\boldsymbol{\beta}_{se}^T\begin{bmatrix}2(\boldsymbol{R}_e+\tilde{\boldsymbol{R}}_e) & 2\boldsymbol{r}\\ 2\boldsymbol{r}^T & q_0\end{bmatrix}\boldsymbol{\beta}_{se} \tag{14.32}$$

where

$$\boldsymbol{\beta}_{se}=\begin{bmatrix}\beta_1\\ \beta_2\\ \vdots\\ \beta_L\end{bmatrix},\quad \boldsymbol{p}_e=\begin{bmatrix}q_1\\ q_2\\ \vdots\\ q_{L-1}\end{bmatrix},\quad \tilde{\boldsymbol{p}}_e=\begin{bmatrix}q_{2L-1}\\ q_{2L-2}\\ \vdots\\ q_{L+1}\end{bmatrix},\quad \boldsymbol{r}=\begin{bmatrix}q_{L-1}\\ q_{L-2}\\ \vdots\\ q_1\end{bmatrix}$$

$$\boldsymbol{R}_e=\begin{bmatrix}q_0 & q_1 & \cdots & q_{L-2}\\ q_1 & q_0 & \cdots & q_{L-3}\\ \vdots & \vdots & \ddots & \vdots\\ q_{L-2} & q_{L-3} & \cdots & q_0\end{bmatrix},\quad \tilde{\boldsymbol{R}}_e=\begin{bmatrix}q_{2L-2} & q_{2L-3} & \cdots & q_L\\ q_{2L-3} & q_{2L-4} & \cdots & q_{L-1}\\ \vdots & \vdots & \ddots & \vdots\\ q_L & q_{L-1} & \cdots & q_2\end{bmatrix}$$

The optimal $\boldsymbol{\beta}_{se}$ is found by setting $\partial I(\boldsymbol{\beta}_{se})/\partial\boldsymbol{\beta}_{se}=\mathbf{0}$ as

$$\boldsymbol{\beta}_{se}^{opt}=-\begin{bmatrix}2(\boldsymbol{R}_e+\tilde{\boldsymbol{R}}_e) & 2\boldsymbol{r}\\ 2\boldsymbol{r}^T & q_0\end{bmatrix}^{-1}\begin{bmatrix}2\big(\boldsymbol{p}_e+\tilde{\boldsymbol{p}}_e\big)\\ 2q_L\end{bmatrix} \tag{14.33}$$

D. Even-Order Error Feedback with Antisymmetric Coefficients

Suppose the order of an error feedback filter is even, say $N=2L$, and the coefficients are antisymmetric, then the error transfer function $B(z)$ in (14.4) can be written as

$$B(z)=1+\sum_{i=1}^{L-1}\beta_i\big(z^{-i}-z^{-2L+i}\big)-z^{-2L},\qquad \beta_L=0 \tag{14.34}$$

The above polynomial leads to

$$\begin{aligned}B(z^{-1})B(z)&=2-\big(z^{2L}+z^{-2L}\big)\\ &\quad+2\sum_{i=1}^{L-1}\beta_i\big[z^i+z^{-i}-\big(z^{2L-i}+z^{-2L+i}\big)\big]\\ &\quad+\sum_{i=1}^{L-1}\sum_{l=1}^{L-1}\beta_i\beta_l\big[z^{i-l}+z^{-(i-l)}-\big(z^{2L-i-l}+z^{-2L+i+l}\big)\big]\end{aligned} \tag{14.35}$$

which is equivalent to

$$|B(e^{j\omega})|^2 = 2 - 2\cos 2L\omega + 4\sum_{i=1}^{L-1}\beta_i\big[\cos i\omega - \cos(2L-i)\omega\big] \\ + 2\sum_{i=1}^{L-1}\sum_{l=1}^{L-1}\beta_i\beta_l\big[\cos(i-l)\omega - \cos(2L-i-l)\omega\big] \tag{14.36}$$

Substituting (14.36) into (14.6) yields

$$\begin{aligned} I(\boldsymbol{\beta}_{ae}) &= 2\big(q_0 - q_{2L}\big) + 4\sum_{i=1}^{L-1}\beta_i\big(q_i - q_{2L-i}\big) \\ &\quad + 2\sum_{i=1}^{L-1}\sum_{l=1}^{L-1}\beta_i\beta_l\big(q_{i-l} - q_{2L-i-l}\big) \\ &= 2\big(q_0 - q_{2L}\big) + 4\boldsymbol{\beta}_{ae}^T\big(\boldsymbol{p}_e - \tilde{\boldsymbol{p}}_e\big) + 2\boldsymbol{\beta}_{ae}^T\big(\boldsymbol{R}_e - \tilde{\boldsymbol{R}}_e\big)\boldsymbol{\beta}_{ae} \end{aligned} \tag{14.37}$$

where $\boldsymbol{p}_e$, $\tilde{\boldsymbol{p}}_e$, $\boldsymbol{R}_e$ and $\tilde{\boldsymbol{R}}_e$ are defined as in (14.32), and

$$\boldsymbol{\beta}_{ae} = \begin{bmatrix}\beta_1 & \beta_2 & \cdots & \beta_{L-1}\end{bmatrix}^T$$

The optimal $\boldsymbol{\beta}_{ae}$ is found by setting $\partial I(\boldsymbol{\beta}_{ae})/\partial\boldsymbol{\beta}_{ae} = \mathbf{0}$ as

$$\boldsymbol{\beta}_{ae}^{opt} = -\big(\boldsymbol{R}_e - \tilde{\boldsymbol{R}}_e\big)^{-1}\big(\boldsymbol{p}_e - \tilde{\boldsymbol{p}}_e\big) \tag{14.38}$$

Table 14.1 Suboptimal symmetric and antisymmetric error feedback coefficients

Order	Symmetric $B(z)$	Antisymmetric $B(z)$
$N=1$	$1+z^{-1}$	$1-z^{-1}$
$N=2$	$1+\beta_1 z^{-1}+z^{-2}$ $\beta_1 = \dfrac{-2q_1}{q_0}$	$1-z^{-2}$
$N=3$	$1+\beta_1 z^{-1}+\beta_1 z^{-2}+z^{-3}$ $\beta_1 = \dfrac{-(q_1+q_2)}{q_0+q_1}$	$1+\beta_1 z^{-1}-\beta_1 z^{-2}-z^{-3}$ $\beta_1 = \dfrac{-(q_1-q_2)}{q_0-q_1}$
$N=4$	$1+\beta_1 z^{-1}+\beta_2 z^{-2}+\beta_1 z^{-3}+z^{-4}$ $\beta_1 = \dfrac{2q_1q_2 - q_0(q_1+q_3)}{q_0(q_0+q_2)-2q_1^2}$ $\beta_2 = \dfrac{2q_1(q_1+q_3)-2q_2(q_0+q_2)}{q_0(q_0+q_2)-2q_1^2}$	$1+\beta_1 z^{-1}-\beta_1 z^{-3}-z^{-4}$ $\beta_1 = \dfrac{-(q_1-q_3)}{q_0-q_2}$

Symmetric and antisymmetric solutions of the order 1 to 4 are summarized in Table 14.1. From the table, it is observed that the first-order solutions have no free parameters but they possess a fixed real zero at $z = \pm 1$, thus being suitable for narrow-band, lowpass, or highpass filters when only moderate noise reduction is required. The second- to fourth-order solutions contain at most 2 free parameters which control the locations of the complex-conjugate zeros, thus more capable of efficient noise reduction.

14.3 State-Space Filter with High-Order Error Feedback

14.3.1 Nth-Order Optimal Error Feedback

Consider a stable, controllable and observable nth-order state-space digital filter $(\boldsymbol{A}, \boldsymbol{b}, \boldsymbol{c}, d)_n$ described by

$$\begin{aligned} \boldsymbol{x}(k+1) &= \boldsymbol{A}\boldsymbol{x}(k) + \boldsymbol{b}u(k) \\ y(k) &= \boldsymbol{c}\boldsymbol{x}(k) + du(k) \end{aligned} \tag{14.39}$$

where $\boldsymbol{x}(k)$ is an $n \times 1$ state-variable vector, $u(k)$ is a scalar input, $y(k)$ is a scalar output, and $\boldsymbol{A}, \boldsymbol{b}, \boldsymbol{c}$ and d are real constant matrices of appropriate dimensions. By taking quantization performed before matrix-vector multiplication into account, a finite-word-length implementation of (14.39) with *high-order* error feedback can be obtained as

$$\begin{aligned} \tilde{\boldsymbol{x}}(k+1) &= \boldsymbol{A}\boldsymbol{Q}[\tilde{\boldsymbol{x}}(k)] + \boldsymbol{b}u(k) + \sum_{i=1}^{N} \boldsymbol{F}_i \boldsymbol{e}(k-i+1) \\ \tilde{y}(k) &= \boldsymbol{c}\boldsymbol{Q}[\tilde{\boldsymbol{x}}(k)] + du(k) \end{aligned} \tag{14.40}$$

where $\boldsymbol{F}_1, \boldsymbol{F}_2, \cdots, \boldsymbol{F}_N$ are referred to as $n \times n$ *high-order* error feedback matrices and

$$\boldsymbol{e}(k) = \boldsymbol{Q}[\tilde{\boldsymbol{x}}(k)] - \tilde{\boldsymbol{x}}(k)$$

A block diagram illustrating a state-space model with high-order error feedback is shown in Figure 14.2.

The coefficient matrices $\boldsymbol{A}, \boldsymbol{b}, \boldsymbol{c}$, and d in (14.40) are assumed to have exact fractional B_c-bit representations. The FWL state-variable vector $\tilde{\boldsymbol{x}}(k)$ and each output $\tilde{y}(k)$ has B-bit fractional representations, while the input $u(k)$ is a $(B - B_c)$-bit fraction. The quantizer $\boldsymbol{Q}[\cdot]$ in (14.40) rounds the B-bit fraction $\tilde{\boldsymbol{x}}(k)$ to $(B - B_c)$-bit after the multiplications and additions, where

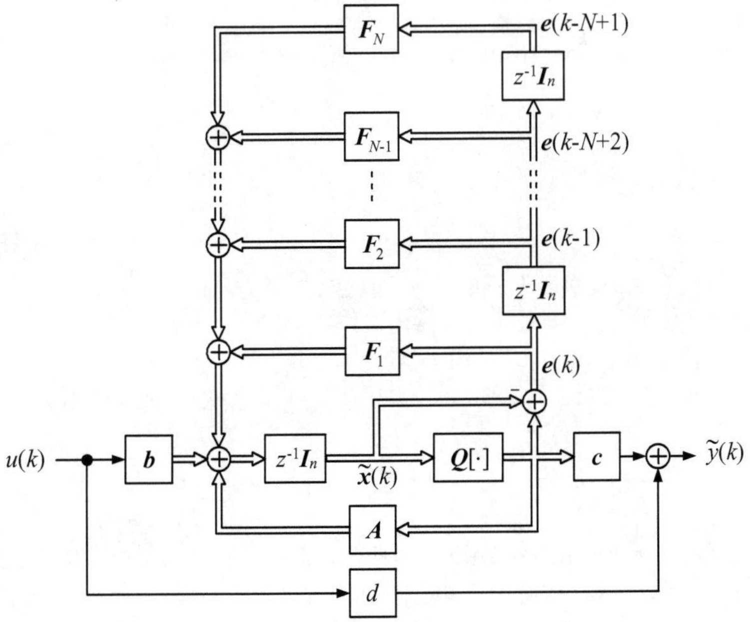

Figure 14.2 A state-space model with high-order error feedback.

the sign bit is not counted. It is assumed that the roundoff error $\boldsymbol{e}(k)$ can be modeled as a Gaussian random process with zero mean and covariance $\sigma^2 \boldsymbol{I}_n$.

By subtracting (14.39) from (14.40), we obtain

$$\begin{aligned}\Delta\boldsymbol{x}(k+1) &= \boldsymbol{A}\Delta\boldsymbol{x}(k) + \boldsymbol{A}\boldsymbol{e}(k) + \sum_{i=1}^{N} \boldsymbol{F}_i \boldsymbol{e}(k-i+1) \\ \Delta y(k) &= \boldsymbol{c}\Delta\boldsymbol{x}(k) + \boldsymbol{c}\boldsymbol{e}(k)\end{aligned} \tag{14.41}$$

where

$$\Delta\boldsymbol{x}(k) = \tilde{\boldsymbol{x}}(k) - \boldsymbol{x}(k), \qquad \Delta y(k) = \tilde{y}(k) - y(k)$$

By taking the z-transform on both sides of (14.41), we have

$$\begin{aligned}z\left[\Delta\boldsymbol{X}(z) - \Delta\boldsymbol{x}(0)\right] &= \boldsymbol{A}\Delta\boldsymbol{X}(z) + \boldsymbol{A}\boldsymbol{E}(z) + \sum_{i=1}^{N} \boldsymbol{F}_i\, z^{-i+1} \boldsymbol{E}(z) \\ \Delta Y(z) &= \boldsymbol{c}\Delta\boldsymbol{X}(z) + \boldsymbol{c}\boldsymbol{E}(z)\end{aligned} \tag{14.42}$$

where $\Delta \boldsymbol{X}(z)$, $\Delta Y(z)$ and $\boldsymbol{E}(z)$ represent the z-transforms of $\Delta \boldsymbol{x}(k)$, $\Delta y(k)$ and $\boldsymbol{e}(k)$, respectively. By setting $\Delta \boldsymbol{x}(0) = \boldsymbol{0}$, it follows from (14.42) that

$$\begin{aligned}
\Delta Y(z) &= \boldsymbol{H}_e(z)\boldsymbol{E}(z) \\
\boldsymbol{H}_e(z) &= \boldsymbol{c}(z\boldsymbol{I}_n - \boldsymbol{A})^{-1}\left(\boldsymbol{A} + \sum_{i=1}^{N} \boldsymbol{F}_i z^{-i+1}\right) + \boldsymbol{c} \\
&= \boldsymbol{c}(z\boldsymbol{I}_n - \boldsymbol{A})^{-1}\left(\boldsymbol{A} + \sum_{i=1}^{N} \boldsymbol{F}_i z^{-i+1} + z\boldsymbol{I}_n - \boldsymbol{A}\right) \\
&= z\,\boldsymbol{G}(z)\boldsymbol{B}(z)
\end{aligned} \tag{14.43}$$

where

$$\boldsymbol{G}(z) = \boldsymbol{c}(z\boldsymbol{I}_n - \boldsymbol{A})^{-1}, \qquad \boldsymbol{B}(z) = \boldsymbol{I}_n + \sum_{i=1}^{N} \boldsymbol{F}_i z^{-i}$$

We now define the normalized noise gain $J(\boldsymbol{F}) = \sigma_{out}^2/\sigma^2$ with $\boldsymbol{F} = [\boldsymbol{F}_1^T, \boldsymbol{F}_2^T, \cdots, \boldsymbol{F}_N^T]^T$ in terms of the transfer function $\boldsymbol{H}_e(z) = z\boldsymbol{G}(z)\boldsymbol{B}(z)$ as

$$J(\boldsymbol{F}) = \mathrm{tr}\left[\frac{1}{2\pi j}\oint_{|z|=1} \boldsymbol{B}^H(z)\boldsymbol{G}^H(z)\boldsymbol{G}(z)\boldsymbol{B}(z)\frac{dz}{z}\right] \tag{14.44}$$

where $\boldsymbol{B}^H(z)$ and $\boldsymbol{G}^H(z)$ denote the conjugate transpose of $\boldsymbol{B}(z)$ and $\boldsymbol{G}(z)$, respectively. The problem being considered here is to obtain the error feedback matrices $\boldsymbol{F}_1$, $\boldsymbol{F}_2$, $\cdots$, $\boldsymbol{F}_N$ which minimize the normalized noise gain $J(\boldsymbol{F})$ in (14.44).

By substituting $\boldsymbol{G}(z)$ and $\boldsymbol{B}(z)$ in (14.43) into (14.44), we obtain

$$\begin{aligned}
J(\boldsymbol{F}) &= \mathrm{tr}\left[\frac{1}{2\pi j}\oint_{|z|=1} \Big\{\boldsymbol{I}_n + \sum_{i=1}^{N} \boldsymbol{F}_i^T z^i\Big\}\boldsymbol{Q}(z)\Big\{\boldsymbol{I}_n + \sum_{l=1}^{N} \boldsymbol{F}_l z^{-l}\Big\}\frac{dz}{z}\right] \\
&= \mathrm{tr}\left[\frac{1}{2\pi j}\oint_{|z|=1} \boldsymbol{Q}(z)\frac{dz}{z} + \sum_{i=1}^{N}\Big\{\frac{1}{2\pi j}\oint_{|z|=1}(z^i + z^{-i})\boldsymbol{Q}(z)\frac{dz}{z}\Big\}\boldsymbol{F}_i\right. \\
&\quad \left. + \sum_{i=1}^{N}\sum_{l=1}^{N} \boldsymbol{F}_i^T\Big\{\frac{1}{2\pi j}\oint_{|z|=1} z^{i-l}\boldsymbol{Q}(z)\frac{dz}{z}\Big\}\boldsymbol{F}_l\right]
\end{aligned} \tag{14.45}$$

where

$$\boldsymbol{Q}(z) = \boldsymbol{G}^H(z)\boldsymbol{G}(z)$$

By defining

$$\boldsymbol{Q}_i = \frac{1}{2\pi j} \oint_{|z|=1} \frac{z^i + z^{-i}}{2} \boldsymbol{Q}(z) \frac{dz}{z} \quad \text{for } i = 0, 1, \cdots, N \tag{14.46}$$

the normalized noise gain in (14.45) can be expressed as

$$\begin{aligned} J(\boldsymbol{F}) &= \text{tr}\left[\boldsymbol{Q}_0 + 2\sum_{i=1}^{N} \boldsymbol{Q}_i \boldsymbol{F}_i + \sum_{i=1}^{N}\sum_{l=1}^{N} \boldsymbol{F}_i^T \boldsymbol{Q}_{|i-l|} \boldsymbol{F}_l\right] \\ &= \text{tr}\left[\boldsymbol{Q}_0 + 2\boldsymbol{S}^T \boldsymbol{F} + \boldsymbol{F}^T \boldsymbol{R} \boldsymbol{F}\right] \end{aligned} \tag{14.47}$$

where

$$\boldsymbol{S} = \begin{bmatrix} \boldsymbol{Q}_1 \\ \boldsymbol{Q}_2 \\ \vdots \\ \boldsymbol{Q}_N \end{bmatrix}, \qquad \boldsymbol{R} = \begin{bmatrix} \boldsymbol{Q}_0 & \boldsymbol{Q}_1 & \cdots & \boldsymbol{Q}_{N-1} \\ \boldsymbol{Q}_1 & \boldsymbol{Q}_0 & \cdots & \boldsymbol{Q}_{N-2} \\ \vdots & \vdots & \ddots & \vdots \\ \boldsymbol{Q}_{N-1} & \boldsymbol{Q}_{N-2} & \cdots & \boldsymbol{Q}_0 \end{bmatrix}$$

The optimal solution is found by setting $\partial J(\boldsymbol{F})/\partial \boldsymbol{F} = \boldsymbol{0}$ and solving the equation for $\boldsymbol{F} = [\boldsymbol{F}_1^T, \boldsymbol{F}_2^T, \cdots, \boldsymbol{F}_N^T]^T$, which gives

$$\boldsymbol{F}^{opt} = -\boldsymbol{R}^+ \boldsymbol{S} \tag{14.48}$$

The optimal solution obtained above minimizes (14.47) as

$$J_{min}(\boldsymbol{F}^{opt}) = \text{tr}\left[\boldsymbol{Q}_0 - \boldsymbol{S}^T \boldsymbol{R}^+ \boldsymbol{S}\right] \tag{14.49}$$

where $\boldsymbol{R}^+$ is the pseudoinverse matrix of $\boldsymbol{R}$.

14.3.2 Computation of $\boldsymbol{Q}_i$ for $i = 0, 1, \cdots, N-1$

By substituting

$$\boldsymbol{G}(z) = \boldsymbol{c}(z\boldsymbol{I}_n - \boldsymbol{A})^{-1} = \sum_{k=0}^{\infty} \boldsymbol{c}\boldsymbol{A}^k z^{-(k+1)} \tag{14.50}$$

into (14.46), we obtain

$$\begin{aligned} \boldsymbol{Q}_i &= \frac{1}{2\pi j} \oint_{|z|=1} \frac{z^i + z^{-i}}{2} \sum_{k=0}^{\infty}\sum_{l=0}^{\infty} (\boldsymbol{A}^k)^T \boldsymbol{c}^T \boldsymbol{c} \boldsymbol{A}^l z^{k-l} \frac{dz}{z} \\ &= \frac{1}{2\pi j} \oint_{|z|=1} \sum_{k=0}^{\infty}\sum_{l=0}^{\infty} (\boldsymbol{A}^k)^T \boldsymbol{c}^T \boldsymbol{c} \boldsymbol{A}^l \left(\frac{z^{k+i-l} + z^{k-i-l}}{2}\right) \frac{dz}{z} \end{aligned} \tag{14.51}$$

Applying the *Cauchy integral theorem* in (14.15) to (14.51) yields

$$\begin{aligned} \boldsymbol{Q}_i &= \frac{1}{2}\sum_{k=0}^{\infty}\Big[(\boldsymbol{A}^k)^T\boldsymbol{c}^T\boldsymbol{c}\boldsymbol{A}^{k+i} + (\boldsymbol{A}^{k+i})^T\boldsymbol{c}^T\boldsymbol{c}\boldsymbol{A}^i\Big] \\ &= \frac{1}{2}\left[\boldsymbol{W}_o\boldsymbol{A}^i + (\boldsymbol{A}^i)^T\boldsymbol{W}_o\right] \end{aligned} \tag{14.52}$$

where

$$\boldsymbol{W}_o = \sum_{k=0}^{\infty}(\boldsymbol{A}^k)^T\boldsymbol{c}^T\boldsymbol{c}\boldsymbol{A}^k$$

is the observability Grammian of the filter in (14.39), that can be obtained by solving the Lyapunov equation

$$\boldsymbol{W}_o = \boldsymbol{A}^T\boldsymbol{W}_o\boldsymbol{A} + \boldsymbol{c}^T\boldsymbol{c}$$

14.3.3 Error Feedback with Symmetric or Antisymmetric Matrices

In order to reduce the number of multiplications, the coefficient matrices of the error feedback filter $\boldsymbol{B}(z)$ in (14.43) is constrained to be symmetric or antisymmetric. In this way, the number of multiplications required will be reduced by a half.

A. Odd-Order Error Feedback with Symmetric Matrices

Suppose the error feedback filter $\boldsymbol{B}(z)$ has an odd order, say $N = 2M + 1$, then a condition for the coefficient matrices of the error feedback filter $\boldsymbol{B}(z)$ to be symmetric is given by

$$\boldsymbol{F}_i = \boldsymbol{F}_{2M+1-i} \ \text{ for } \ i = 0, 1, 2, \cdots, M \tag{14.53}$$

where $\boldsymbol{F}_0 = \boldsymbol{I}_n$. The error feedback filter $\boldsymbol{B}(z)$ in (14.43) can then be written as

$$\boldsymbol{B}(z) = \boldsymbol{I}_n + \sum_{i=1}^{M}\boldsymbol{F}_i\Big(z^{-i} + z^{-(2M+1-i)}\Big) + z^{-(2M+1)}\boldsymbol{I}_n \tag{14.54}$$

which yields

$$\boldsymbol{B}^H(z)\boldsymbol{Q}(z)\boldsymbol{B}(z) = 2\boldsymbol{Q}(z) + \Big(z^{2M+1} + z^{-(2M+1)}\Big)\boldsymbol{Q}(z)$$

$$
\begin{aligned}
&+\sum_{l=1}^{M}\Big(z^{l}+z^{-l}+z^{2M+1-l}+z^{-(2M+1-l)}\Big)\boldsymbol{Q}(z)\boldsymbol{F}_l \\
&+\sum_{i=1}^{M}\Big(z^{i}+z^{-i}+z^{2M+1-i}+z^{-(2M+1-i)}\Big)\boldsymbol{F}_i^T\boldsymbol{Q}(z) \\
&+\sum_{i=1}^{M}\sum_{l=1}^{M}\Big(z^{i-l}+z^{-(i-l)}+z^{2M+1-i-l} \\
&+z^{-(2M+1-i-l)}\Big)\boldsymbol{F}_i^T\boldsymbol{Q}(z)\boldsymbol{F}_l
\end{aligned}
\tag{14.55}
$$

By substituting (14.55) into (14.44), we obtain

$$
\begin{aligned}
J(\boldsymbol{F}_o) &= 2\,\mathrm{tr}\Bigg[\boldsymbol{Q}_0+\boldsymbol{Q}_{2M+1}+2\sum_{i=1}^{M}\Big(\boldsymbol{Q}_i+\boldsymbol{Q}_{2M+1-i}\Big)\boldsymbol{F}_i \\
&\qquad +\sum_{i=1}^{M}\sum_{l=1}^{M}\boldsymbol{F}_i^T\Big(\boldsymbol{Q}_{|i-l|}+\boldsymbol{Q}_{2M+1-i-l}\Big)\boldsymbol{F}_l\Bigg] \\
&= 2\,\mathrm{tr}\Big[\boldsymbol{Q}_0+\boldsymbol{Q}_{2M+1}+2(\boldsymbol{S}_o+\tilde{\boldsymbol{S}}_o)^T\boldsymbol{F}_o+\boldsymbol{F}_o^T(\boldsymbol{R}_o+\tilde{\boldsymbol{R}}_o)\boldsymbol{F}_o\Big]
\end{aligned}
\tag{14.56}
$$

where

$$
\boldsymbol{F}_o=\begin{bmatrix}\boldsymbol{F}_1\\ \boldsymbol{F}_2\\ \vdots\\ \boldsymbol{F}_M\end{bmatrix},\quad
\boldsymbol{S}_o=\begin{bmatrix}\boldsymbol{Q}_1\\ \boldsymbol{Q}_2\\ \vdots\\ \boldsymbol{Q}_M\end{bmatrix},\quad
\boldsymbol{R}_o=\begin{bmatrix}\boldsymbol{Q}_0 & \boldsymbol{Q}_1 & \cdots & \boldsymbol{Q}_{M-1}\\ \boldsymbol{Q}_1 & \boldsymbol{Q}_0 & \cdots & \boldsymbol{Q}_{M-2}\\ \vdots & \vdots & \ddots & \vdots\\ \boldsymbol{Q}_{M-1} & \boldsymbol{Q}_{M-2} & \cdots & \boldsymbol{Q}_0\end{bmatrix}
$$

$$
\tilde{\boldsymbol{S}}_o=\begin{bmatrix}\boldsymbol{Q}_{2M}\\ \boldsymbol{Q}_{2M-1}\\ \vdots\\ \boldsymbol{Q}_{M+1}\end{bmatrix},\qquad
\tilde{\boldsymbol{R}}_o=\begin{bmatrix}\boldsymbol{Q}_{2M-1} & \boldsymbol{Q}_{2M-2} & \cdots & \boldsymbol{Q}_{M}\\ \boldsymbol{Q}_{2M-2} & \boldsymbol{Q}_{2M-3} & \cdots & \boldsymbol{Q}_{M-1}\\ \vdots & \vdots & \ddots & \vdots\\ \boldsymbol{Q}_{M} & \boldsymbol{Q}_{M-1} & \cdots & \boldsymbol{Q}_1\end{bmatrix}
$$

The optimal solution is found by setting $\partial J(\boldsymbol{F}_o)/\partial \boldsymbol{F}_o = \boldsymbol{0}$ and solving the equation for $\boldsymbol{F}_o$, which gives

$$
\boldsymbol{F}_o^{opt} = -(\boldsymbol{R}_o+\tilde{\boldsymbol{R}}_o)^{+}(\boldsymbol{S}_o+\tilde{\boldsymbol{S}}_o) \tag{14.57}
$$

The optimal solution obtained above minimizes (14.56) as

$$J_{min}(\boldsymbol{F}_o^{opt}) = 2\,\mathrm{tr}\left[\boldsymbol{Q}_0 + \boldsymbol{Q}_{2M+1} - (\boldsymbol{S}_o + \tilde{\boldsymbol{S}}_o)^T(\boldsymbol{R}_o + \tilde{\boldsymbol{R}}_o)^+(\boldsymbol{S}_o + \tilde{\boldsymbol{S}}_o)\right] \tag{14.58}$$

B. Odd-Order Error Feedback with Antisymmetric Matrices

Suppose the error feedback filter $\boldsymbol{B}(z)$ has an odd order, say $N = 2M + 1$, then a condition for the coefficient matrices of the error feedback filter $\boldsymbol{B}(z)$ to be antisymmetric is given by

$$\boldsymbol{F}_i = -\boldsymbol{F}_{2M+1-i} \quad \text{for} \quad i = 0, 1, 2, \cdots, M \tag{14.59}$$

where $\boldsymbol{F}_0 = \boldsymbol{I}_n$. The error feedback filter $\boldsymbol{B}(z)$ in (14.43) can then be written as

$$\boldsymbol{B}(z) = \boldsymbol{I}_n + \sum_{i=1}^{M} \boldsymbol{F}_i\left(z^{-i} - z^{-(2M+1-i)}\right) - z^{-(2M+1)}\boldsymbol{I}_n \tag{14.60}$$

which yields

$$\begin{aligned}
\boldsymbol{B}^H(z)\boldsymbol{Q}(z)\boldsymbol{B}(z) = 2\boldsymbol{Q}(z) &- \left(z^{2M+1} + z^{-(2M+1)}\right)\boldsymbol{Q}(z)\\
&+ \sum_{l=1}^{M}\left(z^{l} + z^{-l} - z^{2M+1-l} - z^{-(2M+1-l)}\right)\boldsymbol{Q}(z)\boldsymbol{F}_l\\
&+ \sum_{i=1}^{M}\left(z^{i} + z^{-i} - z^{2M+1-i} - z^{-(2M+1-i)}\right)\boldsymbol{F}_i^T\boldsymbol{Q}(z)\\
&+ \sum_{i=1}^{M}\sum_{l=1}^{M}\left(z^{i-l} + z^{-(i-l)} - z^{2M+1-i-l}\right.\\
&\left. - z^{-(2M+1-i-l)}\right)\boldsymbol{F}_i^T\boldsymbol{Q}(z)\boldsymbol{F}_l
\end{aligned} \tag{14.61}$$

By substituting (14.61) into (14.44), we obtain

$$J(\boldsymbol{F}_o) = 2\,\mathrm{tr}\left[\boldsymbol{Q}_0 - \boldsymbol{Q}_{2M+1} + 2\sum_{i=1}^{M}\left(\boldsymbol{Q}_i - \boldsymbol{Q}_{2M+1-i}\right)\boldsymbol{F}_i\right.$$

$$+\sum_{i=1}^{M}\sum_{l=1}^{M}\boldsymbol{F}_i^T\Big(\boldsymbol{Q}_{|i-l|}-\boldsymbol{Q}_{2M+1-i-l}\Big)\boldsymbol{F}_l\Bigg] \tag{14.62}$$

$$=2\,\mathrm{tr}\Big[\boldsymbol{Q}_0-\boldsymbol{Q}_{2M+1}+2(\boldsymbol{S}_o-\tilde{\boldsymbol{S}}_o)^T\boldsymbol{F}_o+\boldsymbol{F}_o^T(\boldsymbol{R}_o-\tilde{\boldsymbol{R}}_o)\boldsymbol{F}_o\Big]$$

where $\boldsymbol{F}_o$, $\boldsymbol{S}_o$, $\tilde{\boldsymbol{S}}_o$, $\boldsymbol{R}_o$ and $\tilde{\boldsymbol{R}}_o$ are defined as in (14.56).

The optimal solution is found by setting $\partial J(\boldsymbol{F}_o)/\partial\boldsymbol{F}_o=\boldsymbol{0}$ and solving the equation for $\boldsymbol{F}_o$, which gives

$$\boldsymbol{F}_o^{opt}=-(\boldsymbol{R}_o-\tilde{\boldsymbol{R}}_o)^+(\boldsymbol{S}_o-\tilde{\boldsymbol{S}}_o) \tag{14.63}$$

The optimal solution obtained above minimizes (14.62) as

$$J_{min}(\boldsymbol{F}_o^{opt})=2\,\mathrm{tr}\Big[\boldsymbol{Q}_0-\boldsymbol{Q}_{2M+1}-(\boldsymbol{S}_o-\tilde{\boldsymbol{S}}_o)^T(\boldsymbol{R}_o-\tilde{\boldsymbol{R}}_o)^+(\boldsymbol{S}_o-\tilde{\boldsymbol{S}}_o)\Big] \tag{14.64}$$

C. Even-Order Error Feedback with Symmetric Matrices

Suppose the error feedback filter $\boldsymbol{B}(z)$ has an even order, say $N=2L$, then a condition for the coefficient matrices of the error feedback filter $\boldsymbol{B}(z)$ to be symmetric is given by

$$\boldsymbol{F}_i=\boldsymbol{F}_{2L-i}\ \text{ for }\ i=0,1,2,\cdots,L-1 \tag{14.65}$$

where $\boldsymbol{F}_0=\boldsymbol{I}_n$. The error feedback filter $\boldsymbol{B}(z)$ in (14.43) can then be written as

$$\boldsymbol{B}(z)=\boldsymbol{I}_n+\sum_{i=1}^{L-1}\boldsymbol{F}_i\Big(z^{-i}+z^{-(2L-i)}\Big)+\boldsymbol{F}_Lz^{-L}+z^{-2L}\boldsymbol{I}_n \tag{14.66}$$

which yields

$$\begin{aligned}\boldsymbol{B}^H(z)\boldsymbol{Q}(z)\boldsymbol{B}(z)&=2\,\boldsymbol{Q}(z)+\big(z^{2L}+z^{-2L}\big)\boldsymbol{Q}(z)\\&\quad+\sum_{l=1}^{L-1}\big(z^l+z^{-l}+z^{2L-l}+z^{-(2L-l)}\big)\boldsymbol{Q}(z)\boldsymbol{F}_l\\&\quad+\sum_{i=1}^{L-1}\boldsymbol{F}_i^T\big(z^i+z^{-i}+z^{2L-i}+z^{-(2L-i)}\big)\boldsymbol{Q}(z)\end{aligned}$$

$$
\begin{aligned}
&+ \big(z^L + z^{-L}\big)\boldsymbol{Q}(z)\boldsymbol{F}_L + \boldsymbol{F}_L^T\big(z^L + z^{-L}\big)\boldsymbol{Q}(z) \\
&+ \sum_{l=1}^{L-1} \boldsymbol{F}_L^T\big(z^{L-l} + z^{-(L-l)}\big)\boldsymbol{Q}(z)\boldsymbol{F}_l \\
&+ \sum_{i=1}^{L-1} \boldsymbol{F}_i^T\big(z^{L-i} + z^{-(L-i)}\big)\boldsymbol{Q}(z)\boldsymbol{F}_L + \boldsymbol{F}_L^T\boldsymbol{Q}(z)\boldsymbol{F}_L \qquad (14.67) \\
&+ \sum_{i=1}^{L-1}\sum_{l=1}^{L-1} \boldsymbol{F}_i^T\big(z^{i-l} + z^{-(i-l)} + z^{2L-l-i} \\
&+ z^{-(2L-l-i)}\big)\boldsymbol{Q}(z)\boldsymbol{F}_l
\end{aligned}
$$

By substituting (14.67) into (14.44), we have

$$
\begin{aligned}
J(\boldsymbol{F}_{se}) = \mathrm{tr}\Big[&2\big(\boldsymbol{Q}_0 + \boldsymbol{Q}_{2L}\big) + 4\,\boldsymbol{Q}_L\boldsymbol{F}_L + 4\sum_{i=1}^{L-1}\big(\boldsymbol{Q}_i + \boldsymbol{Q}_{2L-i}\big)\boldsymbol{F}_i \\
&+ 2\sum_{i=1}^{L-1}\sum_{l=1}^{L-1}\boldsymbol{F}_i^T\big(\boldsymbol{Q}_{|i-l|} + \boldsymbol{Q}_{2L-i-l}\big)\boldsymbol{F}_l\Big] + 4\sum_{i=1}^{L-1}\boldsymbol{F}_i^T\boldsymbol{Q}_{L-i}\boldsymbol{F}_L \\
&+ \boldsymbol{F}_L^T\boldsymbol{Q}_0\boldsymbol{F}_L
\end{aligned}
\qquad (14.68)
$$

where

$$
\boldsymbol{F}_{se} = \begin{bmatrix}\boldsymbol{F}_1^T & \boldsymbol{F}_2^T & \cdots & \boldsymbol{F}_L^T\end{bmatrix}^T
$$

The normalized noise gain in (14.68) can be written as

$$
\begin{aligned}
J(\boldsymbol{F}_{se}) = \mathrm{tr}\Bigg[&2\big(\boldsymbol{Q}_0 + \boldsymbol{Q}_{2L}\big) + 4\boldsymbol{F}_{se}^T\begin{bmatrix}\boldsymbol{P}_e + \tilde{\boldsymbol{P}}_e \\ \boldsymbol{Q}_L\end{bmatrix} \\
&+ \boldsymbol{F}_{se}^T\begin{bmatrix}2(\boldsymbol{R}_e + \tilde{\boldsymbol{R}}_e) & 2\boldsymbol{\Gamma} \\ 2\boldsymbol{\Gamma}^T & \boldsymbol{Q}_0\end{bmatrix}\boldsymbol{F}_{se}\Bigg]
\end{aligned}
\qquad (14.69)
$$

where

$$
\boldsymbol{P}_e = \begin{bmatrix}\boldsymbol{Q}_1 \\ \boldsymbol{Q}_2 \\ \vdots \\ \boldsymbol{Q}_{L-1}\end{bmatrix}, \quad \tilde{\boldsymbol{P}}_e = \begin{bmatrix}\boldsymbol{Q}_{2L-1} \\ \boldsymbol{Q}_{2L-2} \\ \vdots \\ \boldsymbol{Q}_{L+1}\end{bmatrix}, \quad \boldsymbol{R}_e = \begin{bmatrix}\boldsymbol{Q}_0 & \boldsymbol{Q}_1 & \cdots & \boldsymbol{Q}_{L-2} \\ \boldsymbol{Q}_1 & \boldsymbol{Q}_0 & \cdots & \boldsymbol{Q}_{L-3} \\ \vdots & \vdots & \ddots & \vdots \\ \boldsymbol{Q}_{L-2} & \boldsymbol{Q}_{L-3} & \cdots & \boldsymbol{Q}_0\end{bmatrix}
$$

$$\tilde{R}_e = \begin{bmatrix} Q_{2L-2} & Q_{2L-3} & \cdots & Q_L \\ Q_{2L-3} & Q_{2L-4} & \cdots & Q_{L-1} \\ \vdots & \vdots & \ddots & \vdots \\ Q_L & Q_{L-1} & \cdots & Q_2 \end{bmatrix}, \qquad \Gamma = \begin{bmatrix} Q_{L-1} \\ Q_{L-2} \\ \vdots \\ Q_1 \end{bmatrix}$$

The optimal solution is found by setting $\partial J(F_{se})/\partial F_{se} = 0$ and solving the equation for F_{se}, which gives

$$F_{se}^{opt} = -2 \begin{bmatrix} 2(R_e + \tilde{R}_e) & 2\Gamma \\ 2\Gamma^T & Q_0 \end{bmatrix}^+ \begin{bmatrix} P_e + \tilde{P}_e \\ Q_L \end{bmatrix} \tag{14.70}$$

The optimal solution obtained above minimizes (14.69) as

$$J_{min}(F_{se}^{opt}) = \mathrm{tr}\left[2(Q_0 + Q_{2L}) - 4 \begin{bmatrix} P_e + \tilde{P}_e \\ Q_L \end{bmatrix}^T \begin{bmatrix} 2(R_e + \tilde{R}_e) & 2\Gamma \\ 2\Gamma^T & Q_0 \end{bmatrix}^+ \begin{bmatrix} P_e + \tilde{P}_e \\ Q_L \end{bmatrix} \right] \tag{14.71}$$

D. Even-Order Error Feedback with Antisymmetric Matrices

Suppose the order of an error feedback filter is even, say $N = 2L$, and the coefficient matrices are antisymmetric, then a condition for the coefficient matrices of the error feedback filter $B(z)$ to be antisymmetric is given by

$$F_i = -F_{2L-i} \ \text{ for } i = 0, 1, 2, \cdots, L-1 \tag{14.72}$$

where $F_0 = I_n$ and $F_L = 0$. The error feedback filter $B(z)$ in (14.43) can then be written as

$$B(z) = I_n + \sum_{i=1}^{L-1} F_i \left(z^{-i} - z^{-(2L-i)} \right) - z^{-2L} I_n \tag{14.73}$$

which yields

$$B^H(z)Q(z)B(z) = 2\,Q(z) - (z^{2L} + z^{-2L})Q(z) + \sum_{l=1}^{L-1} (z^l + z^{-l} - z^{2L-l} - z^{-(2L-l)})Q(z)F_l$$

$$\begin{aligned}&+\sum_{i=1}^{L-1} \boldsymbol{F}_i^T\big(z^i + z^{-i} - z^{2L-i} - z^{-(2L-i)}\big)\boldsymbol{Q}(z)\\&+\sum_{i=1}^{L-1}\sum_{l=1}^{L-1} \boldsymbol{F}_i^T\big(z^{i-l} + z^{-(i-l)} - z^{2L-l-i}\\&- z^{-(2L-l-i)}\big)\boldsymbol{Q}(z)\boldsymbol{F}_l\end{aligned} \tag{14.74}$$

By substituting (14.74) into (14.44), we have

$$\begin{aligned}J(\boldsymbol{F}_{ae}) = 2\,\mathrm{tr}\Bigg[&\boldsymbol{Q}_0 - \boldsymbol{Q}_{2L} + 2\sum_{i=1}^{L-1}\big(\boldsymbol{Q}_i - \boldsymbol{Q}_{2L-i}\big)\boldsymbol{F}_i\\&+\sum_{i=1}^{L-1}\sum_{l=1}^{L-1} \boldsymbol{F}_i^T\big(\boldsymbol{Q}_{|i-l|} - \boldsymbol{Q}_{2L-i-l}\big)\boldsymbol{F}_l\Bigg]\end{aligned} \tag{14.75}$$

where

$$\boldsymbol{F}_{ae} = \begin{bmatrix}\boldsymbol{F}_1^T & \boldsymbol{F}_2^T & \cdots & \boldsymbol{F}_{L-1}^T\end{bmatrix}^T$$

The normalized noise gain in (14.75) can be written as

$$J(\boldsymbol{F}_{ae}) = 2\,\mathrm{tr}\Big[\boldsymbol{Q}_0 - \boldsymbol{Q}_{2L} + 2\boldsymbol{F}_{ae}^T\big(\boldsymbol{P}_e - \tilde{\boldsymbol{P}}_e\big) + \boldsymbol{F}_{ae}^T\big(\boldsymbol{R}_e - \tilde{\boldsymbol{R}}_e\big)\boldsymbol{F}_{ae}\Big] \tag{14.76}$$

Table 14.2 Suboptimal symmetric and antisymmetric error feedback matrices

Order	Symmetric ***B(z)***	Antisymmetric ***B(z)***
$N=1$	$\boldsymbol{I}_n + z^{-1}\boldsymbol{I}_n$	$\boldsymbol{I}_n - z^{-1}\boldsymbol{I}_n$
$N=2$	$\boldsymbol{I}_n + \boldsymbol{F}_1 z^{-1} + z^{-2}\boldsymbol{I}_n$ $\boldsymbol{F}_1 = -2\boldsymbol{Q}_0^+\boldsymbol{Q}_1$	$\boldsymbol{I}_n - z^{-2}\boldsymbol{I}_n$
$N=3$	$\boldsymbol{I}_n + \boldsymbol{F}_1 z^{-1} + \boldsymbol{F}_1 z^{-2} + z^{-3}\boldsymbol{I}_n$ $\boldsymbol{F}_1 = -(\boldsymbol{Q}_0+\boldsymbol{Q}_1)^+(\boldsymbol{Q}_1+\boldsymbol{Q}_2)$	$\boldsymbol{I}_n + \boldsymbol{F}_1 z^{-1} - \boldsymbol{F}_1 z^{-2} - z^{-3}\boldsymbol{I}_n$ $\boldsymbol{F}_1 = -(\boldsymbol{Q}_0-\boldsymbol{Q}_1)^+(\boldsymbol{Q}_1-\boldsymbol{Q}_2)$
$N=4$	$\boldsymbol{I}_n + \boldsymbol{F}_1 z^{-1} + \boldsymbol{F}_2 z^{-2} + \boldsymbol{F}_1 z^{-3} + z^{-4}\boldsymbol{I}_n$ $\begin{bmatrix}\boldsymbol{F}_1\\\boldsymbol{F}_2\end{bmatrix} = -2\begin{bmatrix}2(\boldsymbol{Q}_0+\boldsymbol{Q}_2) & 2\boldsymbol{Q}_1\\2\boldsymbol{Q}_1 & \boldsymbol{Q}_0\end{bmatrix}^+\begin{bmatrix}\boldsymbol{Q}_1+\boldsymbol{Q}_3\\\boldsymbol{Q}_2\end{bmatrix}$	$\boldsymbol{I}_n + \boldsymbol{F}_1 z^{-1} - \boldsymbol{F}_1 z^{-3} - z^{-4}\boldsymbol{I}_n$ $\boldsymbol{F}_1 = -(\boldsymbol{Q}_0-\boldsymbol{Q}_2)^+(\boldsymbol{Q}_1-\boldsymbol{Q}_3)$

where $\boldsymbol{P}_e$, $\tilde{\boldsymbol{P}}_e$, $\boldsymbol{R}_e$, and $\tilde{\boldsymbol{R}}_e$ are defined in (14.69). The optimal solution is found by setting $\partial J(\boldsymbol{F}_{ae})/\partial \boldsymbol{F}_{ae} = \boldsymbol{0}$ and solving the equation for $\boldsymbol{F}_{ae}$, which gives

$$\boldsymbol{F}_{ae}^{opt} = -\left(\boldsymbol{R}_e - \tilde{\boldsymbol{R}}_e\right)^{+}\left(\boldsymbol{P}_e - \tilde{\boldsymbol{P}}_e\right) \tag{14.77}$$

The optimal solution obtained above minimizes (14.76) as

$$J_{min}(\boldsymbol{F}_{ae}^{opt}) = 2\,\mathrm{tr}\left[\boldsymbol{Q}_0 - \boldsymbol{Q}_{2L} - \left(\boldsymbol{P}_e - \tilde{\boldsymbol{P}}_e\right)^T\left(\boldsymbol{R}_e - \tilde{\boldsymbol{R}}_e\right)^{+}\left(\boldsymbol{P}_e - \tilde{\boldsymbol{P}}_e\right)\right] \tag{14.78}$$

Symmetric and antisymmetric solutions for order $N = 1$ to $N = 4$ are summarized in Table 14.2.

14.4 Numerical Experiments

14.4.1 Example 1 : An IIR Digital Filter

As an example, consider a 4th-order elliptic lowpass filter whose transfer function is given by

$$G(z) = \frac{1}{(1 - 1.773152z^{-1} + 0.801564z^{-2})(1 - 1.833400z^{-1} + 0.927062z^{-2})}$$

$$= \frac{1}{1 - 3.606552z^{-1} + 4.979522z^{-2} - 3.113409z^{-3} + 0.743099z^{-4}}$$

Assuming that signal quantization is performed after the accumulation of products, the normalized noise gain in (14.10) of this filter in direct form implementation without error feedback was found to be 43.5068 dB. Table 14.3 summarizes the normalized noise gains of the elliptic IIR filter when optimal, symmetric, and antisymmetric error feedbacks of order N for $N = 1, 2, 3,$ and 4 were applied. The parameters characterizing the error feedback loops are also included in the table.

It is observed that increasing the order of $B(z)$ reduces the optimal noise gain and, as expected, with $N = 4$ the solution $B(z) = 1/G(z)$ achieves complete noise cancellation, i.e., a zero normalized noise gain. We remark that the same solution can be approached by solving a higher order error feedback from (14.12). This solution can also be interpreted as a double-precision implementation of the filter [5, 6]. In the case where the noise transfer function $G(z)$ is not purely recursive, complete cancellation is no longer possible. However, the solution asymptotically approaches the 0 dB level when the order of the error feedback increases.

Table 14.3 Error feedback for a 4th-order elliptic lowpass filter

N		Optimal	Symmetric	Antisymmetric
1	Noise (dB)	30.2480	49.4751	30.3002
	β_1	–0.976105	1	–1
2	Noise (dB)	15.4705	15.5071	36.2320
	β_1	–1.935827	–1.952210	0
	β_2	0.983216	1	–1
3	Noise (dB)	3.4891	21.4206	36.1761
	β_1	–2.887382	–0.952611	–2.919043
	β_2	2.856706	–0.952611	2.919043
	β_3	–0.967798	1	–1
4	Noise (dB)	0.0000	0.5971	8.9126
	β_1	–3.606552	–3.855180	–1.919584
	β_2	4.979522	5.713413	0
	β_3	–3.113409	–3.855180	1.919584
	β_4	0.743099	1	–1

The implementation of error feedback is often the most efficient if explicit multiplications are not needed at all. For example, if the error-feedback coefficients are quantized to powers of two, only additions or subtractions with shift are needed for implementation. The results of rounding the optimal error feedback coefficients to integers or a power-of-two representation with 3 bits after the binary point are summarized in Table 14.4. We remark that improved results may be achieved by discrete optimization in conjunction with dynamic programming.

14.4.2 Example 2 : A State-Space Digital Filter

Consider a state-space digital filter $(\boldsymbol{A}_o, \boldsymbol{b}_o, \boldsymbol{c}_o, d)_3$ described by

$$\boldsymbol{A}_o = \begin{bmatrix} 0 & 0 & 0.4537681 \\ 1 & 0 & -1.5561612 \\ 0 & 1 & 1.9748611 \end{bmatrix}, \qquad \boldsymbol{b}_o = \begin{bmatrix} 1 \\ 0 \\ 0 \end{bmatrix}$$

$$\boldsymbol{c}_o = 10^{-1} \begin{bmatrix} 0.7930672 & 1.7963671 & 2.5451875 \end{bmatrix}$$

$$d = 1.5941494 \times 10^{-2}$$

By applying the coordinate transformation matrix given by

$$\boldsymbol{T}_o = \text{diag}\{2.1244192,\ 4.9806829,\ 4.1306156\}$$

Table 14.4 Powers-of-two error feedback for a 4th-order elliptic lowpass filter

N		Infinite Precision	Integer Quantization	3-Bit Quantization
2	Noise (dB)	15.4705	19.3827	22.2841
	β_1	–1.935827	–2	–1.875
	β_2	0.983216	1	1.000
3	Noise (dB)	3.4891	9.6814	6.3832
	β_1	–2.887382	–3	–2.875
	β_2	2.856706	3	2.875
	β_3	–0.967798	–1	–1.000
4	Noise (dB)	0.0000	30.2714	1.9722
	β_1	–3.606552	–4	–3.625
	β_2	4.979522	5	5.000
	β_3	–3.113409	–3	–3.125
	β_4	0.743099	1	0.750

to the original filter $(\boldsymbol{A}_o, \boldsymbol{b}_o, \boldsymbol{c}_o, d)_3$, a new realization specified by $\boldsymbol{A} = \boldsymbol{T}_o^{-1}\boldsymbol{A}_o\boldsymbol{T}_o$, $\boldsymbol{b} = \boldsymbol{T}_o^{-1}\boldsymbol{b}_o$, $\boldsymbol{c} = \boldsymbol{c}_o\boldsymbol{T}_o$ and d was constructed as

$$\boldsymbol{A} = \begin{bmatrix} 0 & 0 & 0.8822843 \\ 0.4265317 & 0 & -1.2905667 \\ 0 & 1.2057968 & 1.9748611 \end{bmatrix}, \qquad \boldsymbol{b} = \begin{bmatrix} 0.4707169 \\ 0 \\ 0 \end{bmatrix}$$

$$\boldsymbol{c} = \begin{bmatrix} 0.1684807 & 0.8947135 & 1.0513191 \end{bmatrix}$$

$$d = 1.5941494 \times 10^{-2}$$

The controllability Grammian $\boldsymbol{K}_c$ and the observability Grammian $\boldsymbol{W}_o$ were then computed by solving the Lyapunov equations $\boldsymbol{K}_c = \boldsymbol{A}\boldsymbol{K}_c\boldsymbol{A}^T + \boldsymbol{b}\boldsymbol{b}^T$ and $\boldsymbol{W}_o = \boldsymbol{A}^T\boldsymbol{W}_o\boldsymbol{A} + \boldsymbol{c}^T\boldsymbol{c}$ as

$$\boldsymbol{K}_c = \begin{bmatrix} 1.000000 & -0.848957 & 0.769793 \\ -0.848957 & 1.000000 & -0.914218 \\ 0.769793 & -0.914218 & 1.000000 \end{bmatrix}$$

$$\boldsymbol{W}_o = \begin{bmatrix} 1.042736 & 2.182632 & 1.257200 \\ 2.182632 & 5.575521 & 3.950707 \\ 1.257200 & 3.950707 & 3.284172 \end{bmatrix}$$

respectively, and the normalized noise gain of the filter $(\boldsymbol{A}, \boldsymbol{b}, \boldsymbol{c}, d)_3$ without error feedback was found from (14.47) to be

$$J(\boldsymbol{0}) = \text{tr}[\boldsymbol{Q}_0] = \text{tr}[\boldsymbol{W}_o] = 9.902430 \ \ (9.957418 \text{ dB})$$

A. Optimal Error Feedback

As an example, we consider the problem of applying high-order error feedback to the filter $(\boldsymbol{A}, \boldsymbol{b}, \boldsymbol{c}, d)_3$, and seek to find the optimal solutions which utilize error feedback with $N = 1, 2$, and 3 (partially 4).

(1) Case $N = 1$: The optimal error feedback matrix was obtained using (14.48) as

$$\boldsymbol{F}_1 = \begin{bmatrix} -1.480125 & -0.318225 & 1.163136 \\ 0.499455 & -0.784328 & -1.242740 \\ -0.379979 & 0.076911 & 0.289592 \end{bmatrix}$$

and the normalized noise gain of the filter was computed from (14.47) as

$$J(\boldsymbol{F}) = 2.968791 \ \ (4.725796 \text{ dB})$$

(2) Case $N = 2$: The optimal error feedback matrices were derived from (14.48) as

$$\boldsymbol{F}_1 = \begin{bmatrix} -2.581439 & 1.096233 & 4.790081 \\ 0.758043 & -2.656162 & -4.057192 \\ -0.561834 & 1.074577 & 1.862577 \end{bmatrix}$$

$$\boldsymbol{F}_2 = \begin{bmatrix} 0.217390 & -3.681772 & -5.053548 \\ 0.607375 & 3.863745 & 3.919715 \\ -0.504042 & -2.412000 & -2.240139 \end{bmatrix}$$

and the normalized noise gain of the filter was calculated from (14.47) as

$$J(\boldsymbol{F}) = 1.259031 \ \ (1.000363 \text{ dB})$$

(3) Case $N = 3$: The optimal error feedback matrices were found from (14.48) to be

$$\boldsymbol{F}_1 = \begin{bmatrix} -2.352663 & 0.319526 & 3.478115 \\ 0.555381 & -3.777495 & -5.153576 \\ -0.440912 & 2.064015 & 2.916668 \end{bmatrix}$$

$$\boldsymbol{F}_2 = \begin{bmatrix} -0.0893259 & 1.394268 & 1.741796 \\ 0.8827037 & 2.564305 & 1.885559 \\ -0.6555847 & -1.520423 & -0.900386 \end{bmatrix}$$

$$\boldsymbol{F}_3 = \begin{bmatrix} 0.281060 & -0.463009 & -0.998642 \\ -0.241377 & -1.353617 & -1.328390 \\ 0.142676 & 1.119958 & 1.184455 \end{bmatrix},$$

and the normalized noise gain of the filter was found from (14.47) to be

$$J(\boldsymbol{F}) = 1.212530 \ \ (0.836924 \text{ dB})$$

B. Suboptimal Symmetric or Antisymmetric Error Feedback

Now we present suboptimal symmetric or antisymmetric solutions which employ error feedback with $N = 2, 3$, and 4.

(1) Case $N = 2$: The suboptimal symmetric error feedback matrix was obtained using (14.70) (or Table 14.2) as

$$\boldsymbol{F}_1 = \begin{bmatrix} -2.960249 & -0.636449 & 2.326272 \\ 0.998910 & -1.568656 & -2.485479 \\ -0.759957 & 0.153823 & 0.579183 \end{bmatrix}$$

and the normalized noise gain of the filter was computed from (14.69) as

$$J(\boldsymbol{F}_1) = 2.082088 \ \ (3.184990 \text{ dB})$$

(2) Case $N = 3$: The suboptimal symmetric and antisymmetric error feedback matrices were derived from (14.57) and (14.63) (or Table 14.2) as

$$\boldsymbol{F}_1 = \begin{bmatrix} -2.364050 & -2.585538 & -0.263467 \\ 1.365418 & 1.207583 & -0.137477 \\ -1.065876 & -1.337423 & -0.377562 \end{bmatrix} \ : \text{symmetric}$$

$$\boldsymbol{F}_1 = \begin{bmatrix} -2.798829 & 4.778005 & 9.843629 \\ 0.150669 & -6.519907 & -7.976907 \\ -0.057792 & 3.486577 & 4.102716 \end{bmatrix} \ : \text{antisymmetric}$$

and the normalized noise gains of the filter were calculated from (14.56) and (14.62) as

$$J(\boldsymbol{F}_1) = 2.815986 \ \ (4.496304 \text{ dB}) \ : \text{symmetric}$$

$$J(\boldsymbol{F}_1) = 2.220137 \ \ (3.463798 \text{ dB}) \ : \text{antisymmetric}$$

(3) Case $N = 4$: The suboptimal symmetric and antisymmetric error feedback matrices were found from (14.70) and (14.77) (or Table 14.2) to be

$$F_1 = \begin{bmatrix} -2.218375 & -0.165585 & 2.641467 \\ 0.320466 & -5.130139 & -6.489098 \\ -0.280215 & 3.186687 & 4.081233 \end{bmatrix} \quad : \text{ symmetric}$$

$$F_2 = \begin{bmatrix} 0.293638 & 2.859654 & 2.962318 \\ 1.563867 & 5.098261 & 3.993561 \\ -1.188198 & -3.022329 & -1.936498 \end{bmatrix} \quad : \text{ symmetric}$$

$$F_1 = \begin{bmatrix} -2.633723 & 0.782534 & 4.476757 \\ 0.796758 & -2.423879 & -3.825186 \\ -0.583588 & 0.944057 & 1.732213 \end{bmatrix} \quad : \text{ antisymmetric}$$

respectively, and from (14.69) and (14.76), the normalized noise gains of the filter were found to be

$$J(F) = 2.082088 \;\; (3.184990 \text{ dB}) \;\; : \text{ symmetric}$$

$$J(F_1) = 2.768031 \;\; (4.421709 \text{ dB}) \;\; : \text{ antisymmetric}$$

respectively, where $F = [F_1^T, F_2^T]^T$.

When $N = 1, 2, 3$, and 4, optimal, symmetric, and antisymmetric error feedbacks were applied to the state-space digital filter in Example 2, the results obtained are summarized in Table 14.5.

From the table, it is observed that an increase of N in $B(z)$ tends to reduce the normalized noise gain in optimal solutions, as expected, while it does not always reduce the normalized noise gain in symmetric or antisymmetric solutions.

The results of rounding the optimal and suboptimal error feedback coefficients to integers or a power-of-two representation with 3 bits after the binary point are summarized in Table 14.6.

Table 14.5 Error feedback noise gain (dB) for a 4th-order state-space lowpass filter

N	Optimal	Symmetric	Antisymmetric
1	4.7258 (dB)	15.5855 (dB)	5.3438 (dB)
2	1.0004 (dB)	3.1850 (dB)	9.9092 (dB)
3	0.8369 (dB)	4.4963 (dB)	3.4638 (dB)
4	0.5852 (dB)	3.1850 (dB)	4.4217 (dB)

Table 14.6 Powers-of-two error feedback noise gain (dB) obtained by rounding optimal and suboptimal solutions

	N	Infinite Precision	Integer Quantization	3-Bit Quantization
Optimal	1	4.7258 (dB)	5.0011 (dB)	4.7690 (dB)
	2	1.0004 (dB)	5.6358 (dB)	1.3982 (dB)
	3	0.8369 (dB)	8.3475 (dB)	1.5807 (dB)
Symmetric	2	3.1850 (dB)	8.7398 (dB)	3.2645 (dB)
	3	4.4963 (dB)	9.9118 (dB)	4.5565 (dB)
	4	3.1850 (dB)	9.5234 (dB)	3.4388 (dB)
Antisymmetric	3	3.4638 (dB)	5.5044 (dB)	3.4718 (dB)
	4	4.4217 (dB)	6.8862 (dB)	3.4388 (dB)

14.5 Summary

The optimal solution of general high-order error feedback has been presented for both external and state-space descriptions of IIR digital filters. As alternatives for efficient implementations, suboptimal schemes with symmetric or antisymmetric coefficients has been examined. In addition, numerical experiments have been presented to demonstrate the validity and effectiveness of the present techniques, where the error feedback quantizer with power-of-two coefficients has been considered.

References

[1] T. I. Laakso and I. O. Hartimo, "Noise reduction in recursive digital filters using high-order error feedback," *IEEE Trans. Signal Process.*, vol. 40, no. 5, pp. 1096–1107, May 1992.

[2] T. Hinamoto and S. Karino "Noise reduction in state-space digital filters using high-order error feedback," *IEICE Trans. Part A*, vol. J77-A, no. 9, pp. 1214–1222, Sep. 1994.

[3] T. Hinamoto and S. Karino "High-order error feedback for noise reduction in state-space digital filters," in *Proc. Int. Conf. Acoust., Speech, Signal Process. (ICASSP'94)*, May, 1994, vol. 3, pp. 1387–1390.

[4] P. P. Vaidyanathan, "On error-spectrum shaping in state-space digital filters," *IEEE Trans. Circuits Syst.*, vol. CAS-32, no. 1, pp. 88–92, Jan. 1985.

[5] W. E. Higgins and D. C. Munson, "Noise reduction strategies for digital filters: Error spectrum shaping versus the optimal linear state-space formulation," *IEEE Trans. Acoust., Speech, Signal Process.*, vol. ASSP-30, no. 6, pp. 963–973, Dec. 1982.
[6] C. T. Mullis and R. A. Roberts, "An interpretation of error spectrum shaping in digital filters," *IEEE Trans. Acoust., Speech, Signal Process.*, vol. ASSP-30, no. 6, pp. 1013–1015, Dec. 1982.

15

Roundoff Noise Analysis and Minimization

15.1 Preview

In the implementation of IIR digital filters with fixed-point arithmetic, it is of critical significance to reduce the effects of roundoff noise at the filter's output. An approach is to synthesize the optimal state-space filter structure for the roundoff noise gain to be minimized by applying a linear transformation to state-space coordinates subject to l_2-scaling constraints [1–4]. As another approach, error feedback is found effective for reducing finite-word-length (FWL) effects in IIR digital filters, and many error feedback methods have been proposed in the past [5–14]. Alternatively, the roundoff noise can also be reduced by introducing delta operators to IIR digital filters [15, 16], or by adopting a new structure based on the concept of polynomial operators for digital filter implementation [17]. As a natural extension of the aforementioned methods, novel methods which combine state-space realization and error feedback have been developed for achieving better performance [18–20]. Separately and jointly optimized scalar or general error feedback matrix for state-space filters have been explored [18]. A jointly-optimized iterative algorithm with a general, diagonal, or scalar error feedback matrix using a quasi-Newton method has been developed for state-space digital filters [19]. In addition, the use of *high-order* error feedback and its effect on noise-reduction performance for state-space digital filters have been investigated [20].

In the first half of this chapter, following the work by Mullis-Roberts and Hwang [1–3], a method for synthesizing the optimal internal structure that minimizes the roundoff noise subject to l_2-scaling constraints is presented. Unlike the method by Mullis-Roberts and Hwang, however, the present method relaxes the l_2-scaling constraints into a single constraint on matrix trace and solves the relaxed problem with an effective closed-form matrix solution.

In the second half of this chapter, a joint optimization technique of high-order error feedback and state-space realization for minimizing the roundoff noise subject to l_2-scaling constraints is introduced [20]. The objective function is minimized by employing a quasi-Newton algorithm.

Numerical experiments are included to illustrate the validity and effectiveness of these algorithms and demonstrate their performance.

15.2 Filters Quantized after Multiplications

15.2.1 Roundoff Noise Analysis and Problem Formulation

Consider a stable, controllable and observable state-space digital filter $(\boldsymbol{A}, \boldsymbol{b}, \boldsymbol{c}, d)_n$ described by

$$\begin{aligned} \boldsymbol{x}(k+1) &= \boldsymbol{A}\boldsymbol{x}(k) + \boldsymbol{b}u(k) \\ y(k) &= \boldsymbol{c}\boldsymbol{x}(k) + du(k) \end{aligned} \tag{15.1}$$

where $\boldsymbol{x}(k)$ is an $n \times 1$ state-variable vector, $u(k)$ is a scalar input, $y(k)$ is a scalar output, and $\boldsymbol{A}$, $\boldsymbol{b}$, $\boldsymbol{c}$, and d are $n \times n$, $n \times 1$, $1 \times n$, and 1×1 real constant matrices, respectively. A block diagram of a state-space digital filter in (15.1) is shown in Figure 15.1.

Due to product quantization, the actual filter implemented by a FWL machine is

$$\begin{aligned} \tilde{\boldsymbol{x}}(k+1) &= \boldsymbol{A}\tilde{\boldsymbol{x}}(k) + \boldsymbol{b}u(k) + \boldsymbol{\alpha}(k) + \boldsymbol{\beta}(k) \\ \tilde{y}(k) &= \boldsymbol{c}\tilde{\boldsymbol{x}}(k) + du(k) + \gamma(k) + \delta(k) \end{aligned} \tag{15.2}$$

where $\tilde{\boldsymbol{x}}(k)$ is the actual state-variable vector, $\tilde{y}(k)$ is the actual output, and $\boldsymbol{\alpha}(k)$, $\boldsymbol{\beta}(k)$, $\gamma(k)$, and $\delta(k)$ are $n \times 1$, $n \times 1$, 1×1, and 1×1 error vectors

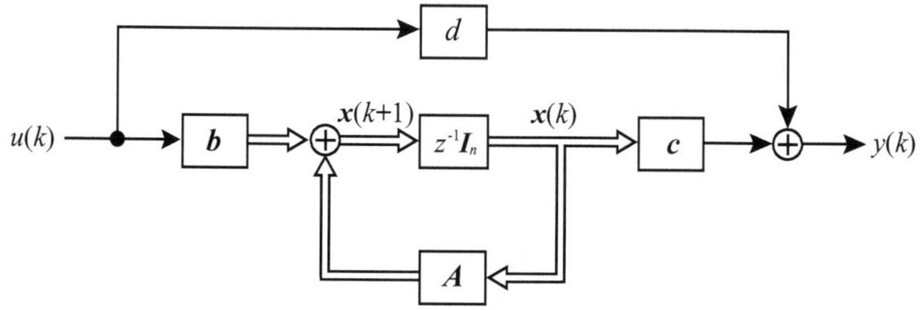

Figure 15.1 Block diagram of a state-space digital filter.

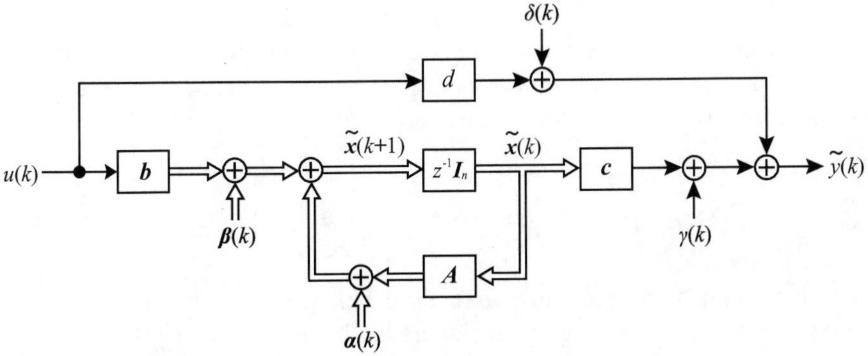

Figure 15.2 Block diagram of an actual state-space digital filter with several noise sources.

generated due to product quantization associated with the $\boldsymbol{A}$, $\boldsymbol{b}$, $\boldsymbol{c}$, and d matrices, respectively. A block diagram of an actual state-space digital filter with several noise sources in (15.2) is illustrated in Figure 15.2.

Subtracting (15.1) from (15.2), we obtain

$$\begin{aligned} \Delta\boldsymbol{x}(k+1) &= \boldsymbol{A}\Delta\boldsymbol{x}(k) + \boldsymbol{\alpha}(k) + \boldsymbol{\beta}(k) \\ \Delta y(k) &= \boldsymbol{c}\Delta\boldsymbol{x}(k) + \gamma(k) + \delta(k) \end{aligned} \tag{15.3}$$

where $\Delta\boldsymbol{x}(k) = \tilde{\boldsymbol{x}}(k) - \boldsymbol{x}(k)$ is the state error vector and $\Delta y(k) = \tilde{y}(k) - y(k)$ is the output noise. A block diagram of a state-space model for noise propagation in (15.3) is depicted in Figure 15.3.

Assuming that $\Delta\boldsymbol{x}(0) = \boldsymbol{0}$ in (15.3), we have

$$\Delta y(k) = \boldsymbol{c}\sum_{l=0}^{k-1} \boldsymbol{A}^{k-l-1}\big[\boldsymbol{\alpha}(l) + \boldsymbol{\beta}(l)\big] + \gamma(k) + \delta(k) \tag{15.4}$$

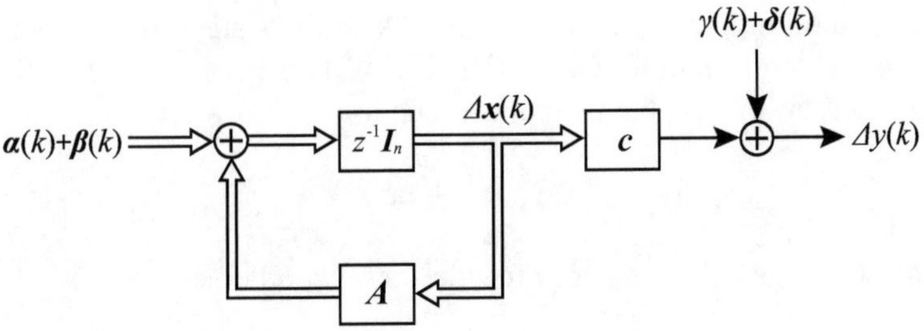

Figure 15.3 Block diagram of a state-space model for noise propagation.

One way for measuring the noise component in the above state-space model is to estimate its average power or variance. Under the usual assumption that the product quantization errors are white noises being statistically independent from source to source, and from time to time, the expected square error is

$$E[\Delta y(k)^2] = \left[\sum_{l=0}^{k-1} \boldsymbol{c}\boldsymbol{A}^l \boldsymbol{Q} (\boldsymbol{c}\boldsymbol{A}^l)^T + \mu + \nu \right] \frac{E_0^2}{12} \tag{15.5}$$

where $\boldsymbol{Q}$ is a diagonal matrix whose ith diagonal element q_i is the number of coefficients in the ith rows of $\boldsymbol{A}$ and $\boldsymbol{b}$ that are neither 0 nor ± 1, $E_0^2/12$ is the variance of each noise source, and μ and ν are the number of neither 0 nor ± 1 constants in $\boldsymbol{c}$ and d, respectively.

For stable digital filters with distinct natural frequencies, the sequence in (15.5) is shown to converge, and the variance of the output noise $\Delta y(k)$ is given by

$$\begin{aligned} E[\Delta y^2] &= \lim_{M\to\infty} E\left[\frac{1}{M+1} \sum_{k=0}^{M} \Delta y(k)^2 \right] \\ &= \left[\sum_{l=0}^{\infty} \boldsymbol{c}\boldsymbol{A}^l \boldsymbol{Q} (\boldsymbol{c}\boldsymbol{A}^l)^T + \mu + \nu \right] \frac{E_0^2}{12} \\ &= \left[\mathrm{tr}[\boldsymbol{Q}\boldsymbol{W}_o] + \mu + \nu \right] \frac{E_0^2}{12} \end{aligned} \tag{15.6}$$

where $\boldsymbol{W}_o$ is the observability Grammian of the filter in (15.1), that can be obtained by solving the Lyapunov equation

$$\boldsymbol{W}_o = \boldsymbol{A}^T \boldsymbol{W}_o \boldsymbol{A} + \boldsymbol{c}^T \boldsymbol{c} \tag{15.7}$$

It should be noted that the l_2-scaling constraints on the state-variable vector $\boldsymbol{x}(k)$ involve the controllability Grammian $\boldsymbol{K}_c$ of the filter in (15.1), which can be computed by solving the Lyapunov equation

$$\boldsymbol{K}_c = \boldsymbol{A}\boldsymbol{K}_c\boldsymbol{A}^T + \boldsymbol{b}\boldsymbol{b}^T \tag{15.8}$$

Applying a coordinate transformation for the state-variable vector

$$\overline{\boldsymbol{x}}(k) = \boldsymbol{T}^{-1}\boldsymbol{x}(k) \tag{15.9}$$

to the filter in (15.1), we obtain a new realization $(\overline{\boldsymbol{A}}, \overline{\boldsymbol{b}}, \overline{\boldsymbol{c}}, d)_n$ described by

$$\begin{aligned} \overline{\boldsymbol{x}}(k+1) &= \overline{\boldsymbol{A}}\overline{\boldsymbol{x}}(k) + \overline{\boldsymbol{b}}u(k) \\ y(k) &= \overline{\boldsymbol{c}}\,\overline{\boldsymbol{x}}(k) + du(k) \end{aligned} \tag{15.10}$$

where

$$\overline{\boldsymbol{A}} = \boldsymbol{T}^{-1}\boldsymbol{A}\boldsymbol{T}, \qquad \overline{\boldsymbol{b}} = \boldsymbol{T}^{-1}\boldsymbol{b}, \qquad \overline{\boldsymbol{c}} = \boldsymbol{c}\boldsymbol{T}$$

A block diagram illustrating an equivalent state-space digital filter is shown in Figure 15.4.

Accordingly, the controllability and observability Grammians relating to $(\overline{\boldsymbol{A}}, \overline{\boldsymbol{b}}, \overline{\boldsymbol{c}}, d)_n$ can be expressed as

$$\overline{\boldsymbol{K}}_c = \boldsymbol{T}^{-1}\boldsymbol{K}_c\boldsymbol{T}^{-T}, \qquad \overline{\boldsymbol{W}}_o = \boldsymbol{T}^T\boldsymbol{W}_o\boldsymbol{T} \tag{15.11}$$

respectively. In this case, (15.6) can be written as

$$E[\Delta\overline{y}^2] = \left[\text{tr}[\overline{\boldsymbol{Q}}\,\overline{\boldsymbol{W}}_o] + \overline{\mu} + \nu \right] \frac{E_0^2}{12} \tag{15.12}$$

where $\overline{\boldsymbol{Q}}$ is a diagonal matrix whose ith diagonal element $\overline{q}_i$ is the number of coefficients in the ith rows of $\overline{\boldsymbol{A}}$ and $\overline{\boldsymbol{b}}$ that are neither 0 nor ± 1, $\overline{\mu}$ is the number of nonzero-or-unity elements in $\overline{\boldsymbol{c}}$. Also, the l_2-scaling constraints are imposed on the state-variable vector $\overline{\boldsymbol{x}}(k)$ so that

$$\boldsymbol{e}_i^T\overline{\boldsymbol{K}}_c\boldsymbol{e}_i = 1 \text{ for } i = 1, 2, \cdots, n \tag{15.13}$$

The problem of roundoff noise minimization is now formulated as follows. Given an arbitrary initial realization $(\boldsymbol{A}, \boldsymbol{b}, \boldsymbol{c}, d)_n$ with associated $\boldsymbol{K}_c$ and $\boldsymbol{W}_o$ matrices, find an $n \times n$ nonsingular transformation matrix $\boldsymbol{T}$, and a new

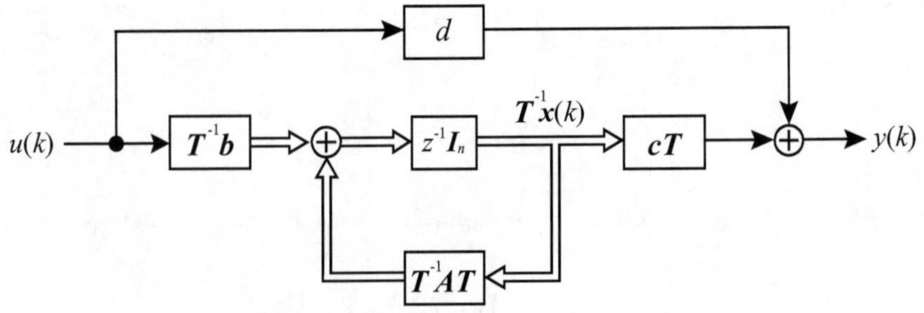

Figure 15.4 Block diagram of an equivalent state-space digital filter.

realization $(\overline{\boldsymbol{A}}, \overline{\boldsymbol{b}}, \overline{\boldsymbol{c}}, d)_n$ such that $\mathrm{tr}[\overline{\boldsymbol{Q}}\,\boldsymbol{T}^T\boldsymbol{W}_o\boldsymbol{T}]$ is minimized subject to the l_2-scaling constraints in (15.13).

However, it turns out that the diagonal matrix $\overline{\boldsymbol{Q}}$ is extremely difficult to be explicitly expressed as a function of $\boldsymbol{A}$, $\boldsymbol{b}$ and an arbitrary $\boldsymbol{T}$.

Hence, under the most pessimistic assumption that $\overline{\boldsymbol{Q}} = (n+1)\boldsymbol{I}_n$ and $\overline{\mu} = n$, i.e., the coefficients of $\boldsymbol{A}$, $\boldsymbol{b}$, and $\boldsymbol{c}$ are neither 0 nor ± 1, below we shall investigate the problem of minimizing $\mathrm{tr}[\boldsymbol{T}^T\boldsymbol{W}_o\boldsymbol{T}]$ subject to the l_2-scaling constraints in (15.13).

15.2.2 Roundoff Noise Minimization Subject to l_2-Scaling Constraints

First, we develop an analytical method for minimizing $\mathrm{tr}[\boldsymbol{T}^T\boldsymbol{W}_o\boldsymbol{T}]$ with respect to a nonsingular $\boldsymbol{T}$ matrix subject to the l_2-scaling constraints in (15.13). To this end, we define the Lagrange function

$$J(\boldsymbol{P}, \lambda) = \mathrm{tr}[\boldsymbol{W}_o\boldsymbol{P}] + \lambda\left(\mathrm{tr}[\boldsymbol{K}_c\boldsymbol{P}^{-1}] - n\right) \tag{15.14}$$

where $\boldsymbol{P} = \boldsymbol{T}\boldsymbol{T}^T$ and λ is a Lagrange multiplier. The optimal coordinate transformation matrix $\boldsymbol{T}$ can be determined by solving the equations

$$\begin{aligned} \frac{\partial J(\boldsymbol{P}, \lambda)}{\partial \boldsymbol{P}} &= \boldsymbol{W}_o - \lambda\boldsymbol{P}^{-1}\boldsymbol{K}_c\boldsymbol{P}^{-1} = \boldsymbol{0} \\ \frac{\partial J(\boldsymbol{P}, \lambda)}{\partial \lambda} &= \mathrm{tr}[\boldsymbol{K}_c\boldsymbol{P}^{-1}] - n = 0 \end{aligned} \tag{15.15}$$

which lead to

$$\boldsymbol{P}\boldsymbol{W}_o\boldsymbol{P} = \lambda\boldsymbol{K}_c, \qquad \mathrm{tr}[\boldsymbol{K}_c\boldsymbol{P}^{-1}] = n \tag{15.16}$$

From (15.16), it follows that

$$\begin{aligned} \boldsymbol{P} &= \sqrt{\lambda}\,\boldsymbol{W}_o^{-\frac{1}{2}}\left[\boldsymbol{W}_o^{\frac{1}{2}}\boldsymbol{K}_c\boldsymbol{W}_o^{\frac{1}{2}}\right]^{\frac{1}{2}}\boldsymbol{W}_o^{-\frac{1}{2}} \\ \frac{1}{\sqrt{\lambda}}\,\mathrm{tr}[\boldsymbol{K}_c\boldsymbol{W}_o]^{\frac{1}{2}} &= \frac{1}{\sqrt{\lambda}}\Big(\sum_{i=1}^{n}\theta_i\Big) = n \end{aligned} \tag{15.17}$$

where θ_i^2 for $i = 1, 2, \cdots, n$ denote the eigenvalues of $\boldsymbol{K}_c\boldsymbol{W}_o$. Hence

$$\boldsymbol{P} = \frac{1}{n}\Big(\sum_{i=1}^{n}\theta_i\Big)\boldsymbol{W}_o^{-\frac{1}{2}}\left[\boldsymbol{W}_o^{\frac{1}{2}}\boldsymbol{K}_c\boldsymbol{W}_o^{\frac{1}{2}}\right]^{\frac{1}{2}}\boldsymbol{W}_o^{-\frac{1}{2}} \tag{15.18}$$

Substituting (15.18) into (15.14) yields the minimum value of $J(\boldsymbol{P}, \lambda)$ as

$$\min_{\boldsymbol{P},\lambda} J(\boldsymbol{P}, \lambda) = \frac{1}{n}\Big(\sum_{i=1}^{n} \theta_i\Big)^2 \tag{15.19}$$

Referring to (15.18), the optimal coordinate transformation matrix $\boldsymbol{T}$ that minimizes (15.14) can now be obtained in closed form as follows:

$$\boldsymbol{T} = \frac{1}{\sqrt{n}}\Big(\sum_{i=1}^{n} \theta_i\Big)^{\frac{1}{2}} \boldsymbol{W}_o^{-\frac{1}{2}} \Big[\boldsymbol{W}_o^{\frac{1}{2}} \boldsymbol{K}_c \boldsymbol{W}_o^{\frac{1}{2}}\Big]^{\frac{1}{4}} \boldsymbol{U}_o \tag{15.20}$$

where $\boldsymbol{U}_o$ is an arbitrary $n \times n$ orthogonal matrix. From (15.20), it follows that

$$\overline{\boldsymbol{K}}_c = \boldsymbol{T}^{-1}\boldsymbol{K}_c\boldsymbol{T}^{-T} = n\Big(\sum_{i=1}^{n} \theta_i\Big)^{-1} \boldsymbol{U}_o^T \Big[\boldsymbol{W}_o^{\frac{1}{2}} \boldsymbol{K}_c \boldsymbol{W}_o^{\frac{1}{2}}\Big]^{\frac{1}{2}} \boldsymbol{U}_o \tag{15.21}$$

Next, we choose the $n \times n$ orthogonal matrix $\boldsymbol{U}_o$ such that matrix $\overline{\boldsymbol{K}}$ in (15.21) satisfies the l_2-scaling constraints in (15.13). To this end, we perform the eigenvalue-eigenvector decomposition

$$\Big[\boldsymbol{W}_o^{\frac{1}{2}} \boldsymbol{K}_c \boldsymbol{W}_o^{\frac{1}{2}}\Big]^{\frac{1}{2}} = \boldsymbol{Q}\,\mathrm{diag}\{\theta_1, \theta_2, \cdots, \theta_n\}\boldsymbol{Q}^T \tag{15.22}$$

where $\boldsymbol{Q}\boldsymbol{Q}^T = \boldsymbol{I}_n$. As a result, we can write

$$n\Big(\sum_{i=1}^{n} \theta_i\Big)^{-1} \Big[\boldsymbol{W}_o^{\frac{1}{2}} \boldsymbol{K}_c \boldsymbol{W}_o^{\frac{1}{2}}\Big]^{\frac{1}{2}} = \boldsymbol{Q}\boldsymbol{\Lambda}^{-2}\boldsymbol{Q}^T \tag{15.23}$$

where

$$\boldsymbol{\Lambda} = \mathrm{diag}\{\lambda_1, \lambda_2, \cdots, \lambda_n\}$$

$$\lambda_i = \Big(\frac{\theta_1 + \theta_2 + \cdots + \theta_n}{n\theta_i}\Big)^{\frac{1}{2}} \quad \text{for } i = 1, 2, \cdots, n$$

Now, an $n \times n$ orthogonal matrix $\boldsymbol{Z}$ such that

$$\boldsymbol{Z}\boldsymbol{\Lambda}^{-2}\boldsymbol{Z}^T = \begin{bmatrix} 1 & * & \cdots & * \\ * & 1 & \ddots & \vdots \\ \vdots & \ddots & \ddots & * \\ * & \cdots & * & 1 \end{bmatrix} \tag{15.24}$$

can be obtained by numerical manipulations [3, p. 278]. By choosing $U_o = QZ^T$ in (15.20), the optimal coordinate transformation matrix T satisfying (15.13) and (15.19) can be constructed as

$$T = \frac{1}{\sqrt{n}} \Big(\sum_{i=1}^{n} \theta_i \Big)^{\frac{1}{2}} W_o^{-\frac{1}{2}} \Big[W_o^{\frac{1}{2}} K_c W_o^{\frac{1}{2}} \Big]^{\frac{1}{4}} QZ^T \tag{15.25}$$

By substituting (15.25) into (15.10), we can construct the optimal realization $(\overline{A}, \overline{b}, \overline{c}, d)_n$ which minimizes $\mathrm{tr}[T^T W_o T]$ subject to l_2-scaling constraints in (15.13).

15.3 Filters Quantized before Multiplications

15.3.1 State-Space Model with High-Order Error Feedback

Again we consider a stable, controllable and observable state-space digital filter $(A, b, c, d)_n$ of order n described by (15.1). When the quantization is performed before matrix-vector multiplication, an actual state-space digital filter $(A, b, c, d)_n$ can be expressed as

$$\begin{aligned} \tilde{x}(k+1) &= AQ[\tilde{x}(k)] + bu(k) \\ \tilde{y}(k) &= cQ[\tilde{x}(k)] + du(k) \end{aligned} \tag{15.26}$$

where

$$e(k) = \tilde{x}(k) - Q[\tilde{x}(k)]$$

A block diagram of an actual state-space digital filter in (15.26) is shown in Figure 15.5. By taking the quantization performed before matrix-vector multiplication into account, an FWL implementation of (15.1) with error feedforward and high-order error feedback can be obtained as

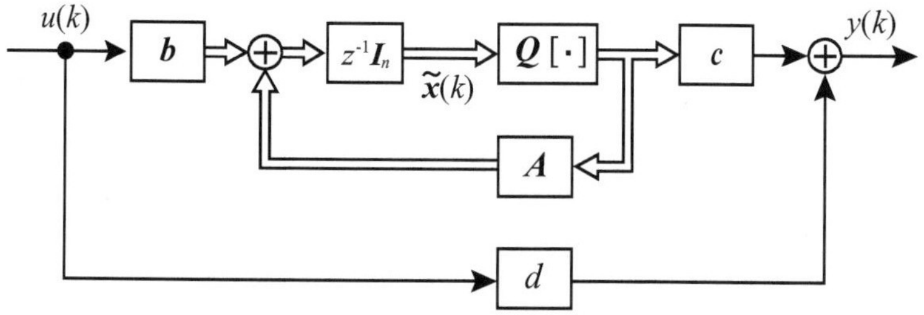

Figure 15.5 Block diagram of an actual state-space digital filter.

$$\tilde{\boldsymbol{x}}(k+1) = \boldsymbol{A}\boldsymbol{Q}[\tilde{\boldsymbol{x}}(k)] + \boldsymbol{b}u(k) + \sum_{i=0}^{N-1} \boldsymbol{D}_i \boldsymbol{e}(k-i)$$
$$\tilde{y}(k) = \boldsymbol{c}\boldsymbol{Q}[\tilde{\boldsymbol{x}}(k)] + du(k) + \boldsymbol{h}\boldsymbol{e}(k) \tag{15.27}$$

where $\boldsymbol{h}$ and $\boldsymbol{D}_0, \boldsymbol{D}_1, \cdots, \boldsymbol{D}_{N-1}$ are referred to as a $1 \times n$ error-feedforward vector and $n \times n$ high-order error feedback matrices, respectively. A block diagram illustrating a state-space model with Nth-order error feedback and an error feedforward path is shown in Figure 15.6.

The coefficient matrices $\boldsymbol{A}, \boldsymbol{b}, \boldsymbol{c}$, and d in (15.27) are assumed to have exact fractional B_c-bit representations. The FWL state-variable vector $\tilde{\boldsymbol{x}}(k)$ and each output $\tilde{y}(k)$ has B-bit fractional representations, while the input $u(k)$ is a $(B - B_c)$-bit fraction. The quantizer $\boldsymbol{Q}[\cdot]$ in (15.27) rounds the B-bit

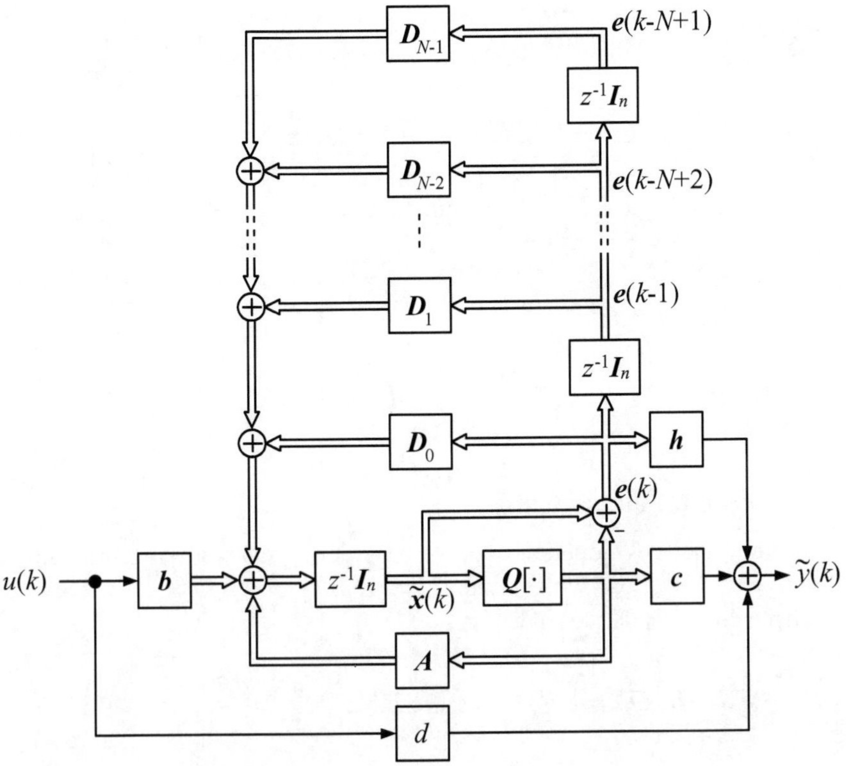

Figure 15.6 State-space model with Nth-order error feedback and error feedforward.

fraction $\tilde{\boldsymbol{x}}(k)$ to $(B - B_c)$-bit after the multiplications and additions, where the sign bit is not counted. It is assumed that the roundoff error $\boldsymbol{e}(k)$ can be modeled as a Gaussian random process with zero mean and covariance $\sigma^2 \boldsymbol{I}_n$. By subtracting (15.27) from (15.1), we obtain

$$\begin{aligned} \Delta\boldsymbol{x}(k+1) &= \boldsymbol{A}\Delta\boldsymbol{x}(k) + \boldsymbol{A}\boldsymbol{e}(k) - \sum_{i=0}^{N-1} \boldsymbol{D}_i \boldsymbol{e}(k-i) \\ \Delta y(k) &= \boldsymbol{c}\Delta\boldsymbol{x}(k) + (\boldsymbol{c} - \boldsymbol{h})\boldsymbol{e}(k) \end{aligned} \tag{15.28}$$

where

$$\Delta\boldsymbol{x}(k) = \boldsymbol{x}(k) - \tilde{\boldsymbol{x}}(k), \qquad \Delta y(k) = y(k) - \tilde{y}(k)$$

By taking the z-transform on both sides of (15.28), we have

$$\begin{aligned} z[\Delta\boldsymbol{X}(z) - \Delta\boldsymbol{x}(0)] &= \boldsymbol{A}\Delta\boldsymbol{X}(z) + \boldsymbol{A}\boldsymbol{E}(z) - \sum_{i=0}^{N-1} \boldsymbol{D}_i z^{-i} \boldsymbol{E}(z) \\ \Delta Y(z) &= \boldsymbol{c}\Delta\boldsymbol{X}(z) + (\boldsymbol{c} - \boldsymbol{h})\boldsymbol{E}(z) \end{aligned} \tag{15.29}$$

where $\Delta\boldsymbol{X}(z)$, $\Delta Y(z)$ and $\boldsymbol{E}(z)$ represent the z-transforms of $\Delta\boldsymbol{x}(k)$, $\Delta y(k)$ and $\boldsymbol{e}(k)$, respectively. Setting $\Delta\boldsymbol{x}(0) = \boldsymbol{0}$ leads (15.29) to

$$\begin{aligned} \Delta Y(z) &= \boldsymbol{H}_e(z)\boldsymbol{E}(z) \\ \boldsymbol{H}_e(z) &= \boldsymbol{c}(z\boldsymbol{I}_n - \boldsymbol{A})^{-1}\Big(\boldsymbol{A} - \sum_{i=0}^{N-1} \boldsymbol{D}_i z^{-i}\Big) + \boldsymbol{c} - \boldsymbol{h} \end{aligned} \tag{15.30}$$

15.3.2 Formula for Noise Gain

Based on the model developed above, we now define the normalized noise gain $J_{e1}(\boldsymbol{h}, \boldsymbol{D}) = \sigma_{out}^2/\sigma^2$ with $\boldsymbol{h}$ and $\boldsymbol{D} = [\boldsymbol{D}_0, \boldsymbol{D}_1, \cdots, \boldsymbol{D}_{N-1}]$ in terms of the transfer function $\boldsymbol{H}_e(z)$ as

$$J_{e1}(\boldsymbol{h}, \boldsymbol{D}) = \mathrm{tr}\Big[\frac{1}{2\pi j} \oint_{|z|=1} \boldsymbol{H}_e^H(z)\boldsymbol{H}_e(z)\frac{dz}{z}\Big] \tag{15.31}$$

In order to derive an easy-to-use formula to evaluate and minimize the noise gain, we write $\boldsymbol{H}_e(z)$ in (15.30) as

$$
\begin{aligned}
\boldsymbol{H}_e(z) &= \boldsymbol{c}\sum_{k=1}^{\infty} \boldsymbol{A}^{k-1} z^{-k}\Big(\boldsymbol{A} - \sum_{i=0}^{N-1} \boldsymbol{D}_i\, z^{-i}\Big) + \boldsymbol{c} - \boldsymbol{h} \\
&= \boldsymbol{c}\left[\sum_{k=1}^{\infty} \boldsymbol{A}^k z^{-k} - \sum_{l=1}^{\infty}\sum_{i=0}^{N-1} \boldsymbol{A}^{l-1}\boldsymbol{D}_i\, z^{-(l+i)}\right] + \boldsymbol{c} - \boldsymbol{h} \\
&= \sum_{k=1}^{\infty} \boldsymbol{c}\Big(\boldsymbol{A}^k - \sum_{i=0}^{N-1} \boldsymbol{A}^{k-i-1}\boldsymbol{D}_i\Big) z^{-k} + \boldsymbol{c} - \boldsymbol{h}
\end{aligned}
\tag{15.32}
$$

where $\boldsymbol{A}^i = \boldsymbol{0}$ for $i < 0$. By substituting (15.32) into (15.31) and making use of the *Cauchy integral theorem*

$$
\frac{1}{2\pi j}\oint_C z^k \frac{dz}{z} = \begin{cases} 1, & k = 0 \\ 0, & k \neq 0 \end{cases}
\tag{15.33}
$$

where C is a counterclockwise contour that encircles the origin, we obtain

$$
\begin{aligned}
J_{e1}(\boldsymbol{h}, \boldsymbol{D}) = \mathrm{tr}\Big[\boldsymbol{W}_o &- \sum_{i=0}^{N-1}\Big\{(\boldsymbol{A}^T)^{i+1}\boldsymbol{W}_o\boldsymbol{D}_i + \boldsymbol{D}_i^T\boldsymbol{W}_o\boldsymbol{A}^{i+1}\Big\} \\
&+ \sum_{i=0}^{N-1}\sum_{j=0}^{N-1} \boldsymbol{D}_i^T\Big\{(\boldsymbol{A}^T)^{j-i}\boldsymbol{W}_o + \boldsymbol{W}_o\boldsymbol{A}^{i-j}\Big\}\boldsymbol{D}_j \\
&- \sum_{i=0}^{N-1} \boldsymbol{D}_i^T\boldsymbol{W}_o\boldsymbol{D}_i - 2\boldsymbol{h}^T\boldsymbol{c} + \boldsymbol{h}^T\boldsymbol{h}\Big]
\end{aligned}
\tag{15.34}
$$

where $\boldsymbol{W}_o$ is the observability Grammian of the filter in (15.1), that can be obtained by solving the Lyapunov equation in (15.7).

It is useful to note that if the high-order error feedback matrices $\boldsymbol{D}_0, \boldsymbol{D}_1, \cdots, \boldsymbol{D}_{N-1}$ are diagonal, then the formula for the noise gain can be considerably simplified to

$$
\begin{aligned}
J_{e1}(\boldsymbol{h}, \boldsymbol{D}) = \mathrm{tr}\Big[\boldsymbol{W}_o &- \boldsymbol{c}^T\boldsymbol{c} - 2\sum_{i=0}^{N-1} \boldsymbol{W}_o\boldsymbol{A}^{i+1}\boldsymbol{D}_i \\
&+ \sum_{i=0}^{N-1}\sum_{j=0}^{N-1} \boldsymbol{W}_o\boldsymbol{A}^{|i-j|}\boldsymbol{D}_i\boldsymbol{D}_j\Big] + (\boldsymbol{c} - \boldsymbol{h})(\boldsymbol{c} - \boldsymbol{h})^T
\end{aligned}
\tag{15.35}
$$

It should also be noted that the l_2-scaling constraints on the state-variable vector $\boldsymbol{x}(k)$ involve the controllability Grammian $\boldsymbol{K}_c$ of the filter in (15.1), which can be computed by solving the Lyapunov equation in (15.8).

15.3.3 Problem Formulation

A different yet equivalent state-space description of (15.1), $(\overline{\boldsymbol{A}}, \overline{\boldsymbol{b}}, \overline{\boldsymbol{c}}, d)_n$, can be obtained via a coordinate transformation in (15.9) as shown in (15.10).

We now choose the error feedforward vector as $\boldsymbol{h} = \overline{\boldsymbol{c}}$ to eliminate the last term in (15.35). With an equivalent state-space realization as specified in (15.10) and assuming the use of high-order diagonal error feedback matrices, the normalized noise gain is then found to be

$$J_{e2}(\boldsymbol{D}, \boldsymbol{T}) = \mathrm{tr}\Big[\boldsymbol{T}^T(\boldsymbol{W}_o - \boldsymbol{c}^T\boldsymbol{c})\boldsymbol{T} - 2\sum_{i=0}^{N-1} \boldsymbol{T}^T\boldsymbol{W}_o\boldsymbol{A}^{i+1}\boldsymbol{T}\boldsymbol{D}_i + \sum_{i=0}^{N-1}\sum_{j=0}^{N-1} \boldsymbol{T}^T\boldsymbol{W}_o\boldsymbol{A}^{|i-j|}\boldsymbol{T}\boldsymbol{D}_i\boldsymbol{D}_j\Big] \tag{15.36}$$

where the noise gain is denoted as $J_{e2}(\boldsymbol{D}, \boldsymbol{T})$ to reflect the fact that the noise gain is now dependent on both high-order error feedback matrices $\boldsymbol{D}_0, \boldsymbol{D}_1, \cdots, \boldsymbol{D}_{N-1}$ as well as state-space coordinate transformation $\boldsymbol{T}$. Formula (15.36) provides an analytic foundation for the minimization of the noise gain by *jointly* optimize the error feedback loop and state-space coordinate transformation. Formally, the problem being considered here can be stated as to jointly deign the high-order error feedback matrices $\boldsymbol{D}_0, \boldsymbol{D}_1, \cdots, \boldsymbol{D}_{N-1}$ and coordinate transformation matrix $\boldsymbol{T}$ to minimize the noise gain $J_{e2}(\boldsymbol{D}, \boldsymbol{T})$ in (15.36) subject to the l_2-scaling constraints in (15.13), assuming the error feedforward vector $\boldsymbol{h} = \overline{\boldsymbol{c}}$ is chosen so as to eliminate the last term in (15.35).

15.3.4 Joint Optimization of Error Feedback and Realization

15.3.4.1 The Use of Quasi-Newton Algorithm

In what follows, it is assumed that the high-order error feedback matrices are all diagonal so that (15.36) is a valid objective function to be minimized. Because the constraints in (15.13) are nonconvex and highly nonlinear with respect to matrix $\boldsymbol{T}$, it is beneficial if these constraints can be eliminated so as to work with an unconstrained problem that admits fast Newton-like algorithms. To this end, we define

$$\hat{\boldsymbol{T}} = \boldsymbol{T}^T \boldsymbol{K}_c^{-\frac{1}{2}} \tag{15.37}$$

which leads (15.13) to

$$(\hat{\boldsymbol{T}}^{-T} \hat{\boldsymbol{T}}^{-1})_{ii} = 1 \ \text{ for } \ i = 1, 2, \cdots, n \tag{15.38}$$

These constraints are always satisfied if matrix $\hat{\boldsymbol{T}}^{-1}$ assumes the form

$$\hat{\boldsymbol{T}}^{-1} = \left[\frac{\boldsymbol{t}_1}{||\boldsymbol{t}_1||}, \frac{\boldsymbol{t}_2}{||\boldsymbol{t}_2||}, \cdots, \frac{\boldsymbol{t}_n}{||\boldsymbol{t}_n||} \right] \tag{15.39}$$

Substituting (15.37) into (15.38), we obtain

$$\begin{aligned} J_{e3}(\boldsymbol{D}, \hat{\boldsymbol{T}}) = \text{tr}\Big[& \hat{\boldsymbol{T}} (\boldsymbol{V}_0 - \hat{\boldsymbol{c}}^T \hat{\boldsymbol{c}}) \hat{\boldsymbol{T}}^T - 2 \sum_{p=0}^{N-1} \hat{\boldsymbol{T}} \boldsymbol{V}_{p+1} \hat{\boldsymbol{T}}^T \boldsymbol{D}_p \\ & + \sum_{p=0}^{N-1} \sum_{q=0}^{N-1} \hat{\boldsymbol{T}} \boldsymbol{V}_{|p-q|} \hat{\boldsymbol{T}}^T \boldsymbol{D}_p \boldsymbol{D}_q \Big] \end{aligned} \tag{15.40}$$

where

$$\boldsymbol{V}_p = \boldsymbol{K}_c^{\frac{1}{2}} \boldsymbol{W}_o \boldsymbol{A}^p \boldsymbol{K}_c^{\frac{1}{2}}, \qquad \hat{\boldsymbol{c}} = \boldsymbol{c} \boldsymbol{K}_c^{\frac{1}{2}}$$

Following the foregoing arguments, the problem of obtaining $\boldsymbol{T}$ and $\boldsymbol{D}_0, \boldsymbol{D}_1, \cdots, \boldsymbol{D}_{N-1}$ that jointly minimize (15.36) subject to the l_2-scaling constraints in (15.13) is now converted into an unconstrained optimization problem of obtaining $\hat{\boldsymbol{T}}$ and $\boldsymbol{D}_0, \boldsymbol{D}_1, \cdots, \boldsymbol{D}_{N-1}$ that jointly minimize $J_{e3}(\boldsymbol{D}, \hat{\boldsymbol{T}})$ in (15.40).

Let $\boldsymbol{x}$ be the column vector that collects the variables in $[\boldsymbol{t}_1, \boldsymbol{t}_2, \cdots, \boldsymbol{t}_n]$ and $\boldsymbol{D}_0, \boldsymbol{D}_1, \cdots, \boldsymbol{D}_{N-1}$. Then $J_{e3}(\boldsymbol{D}, \hat{\boldsymbol{T}})$ in (15.40) is a function of $\boldsymbol{x}$, denoted by $J(\boldsymbol{x})$. The proposed algorithm starts with an initial point $\boldsymbol{x}_0$ obtained from the assignment $\hat{\boldsymbol{T}} = \boldsymbol{D}_0 = \boldsymbol{D}_1 = \cdots = \boldsymbol{D}_{N-1} = \boldsymbol{I}_n$. In the kth iteration, a quasi-Newton algorithm updates the most recent point $\boldsymbol{x}_k$ to point $\boldsymbol{x}_{k+1}$ as [21]

$$\boldsymbol{x}_{k+1} = \boldsymbol{x}_k + \alpha_k \boldsymbol{d}_k \tag{15.41}$$

where

$$\boldsymbol{d}_k = -\boldsymbol{S}_k \nabla J(\boldsymbol{x}_k), \qquad \alpha_k = arg\left[\min_{\alpha} \ J(\boldsymbol{x}_k + \alpha \boldsymbol{d}_k) \right]$$

$$\boldsymbol{S}_{k+1} = \boldsymbol{S}_k + \left(1 + \frac{\boldsymbol{\gamma}_k^T \boldsymbol{S}_k \boldsymbol{\gamma}_k}{\boldsymbol{\gamma}_k^T \boldsymbol{\delta}_k} \right) \frac{\boldsymbol{\delta}_k \boldsymbol{\delta}_k^T}{\boldsymbol{\gamma}_k^T \boldsymbol{\delta}_k} - \frac{\boldsymbol{\delta}_k \boldsymbol{\gamma}_k^T \boldsymbol{S}_k + \boldsymbol{S}_k \boldsymbol{\gamma}_k \boldsymbol{\delta}_k^T}{\boldsymbol{\gamma}_k^T \boldsymbol{\delta}_k}, \qquad \boldsymbol{S}_0 = \boldsymbol{I}$$

$$\boldsymbol{\delta}_k = \boldsymbol{x}_{k+1} - \boldsymbol{x}_k, \qquad \boldsymbol{\gamma}_k = \nabla J(\boldsymbol{x}_{k+1}) - \nabla J(\boldsymbol{x}_k)$$

$\nabla J(\boldsymbol{x})$ denotes the gradient of $J(\boldsymbol{x})$ with respect to $\boldsymbol{x}$, and $\boldsymbol{S}_k$ is a positive-definite approximation of the inverse Hessian matrix of $J(\boldsymbol{x})$.

The iteration process (15.41) continues until

$$|J(\boldsymbol{x}_{k+1}) - J(\boldsymbol{x}_k)| < \varepsilon \tag{15.42}$$

is satisfied where $\varepsilon > 0$ is a prescribed tolerance. If the iteration is terminated at step k, the $\boldsymbol{x}_k$ is viewed as a solution point.

15.3.4.2 Gradient of $J(\boldsymbol{x})$

The implementation efficiency and solution accuracy of the quasi-Newton algorithm greatly depends on how the gradient $\nabla J(\boldsymbol{x})$ is evaluated. With high-order diagonal error feedback matrices, we derive closed-form formulas for computing the partial derivatives of $J(\boldsymbol{x})$ with respect to the elements of $\boldsymbol{T}$ as well as those of $\boldsymbol{D} = [\boldsymbol{D}_0, \boldsymbol{D}_1, \cdots, \boldsymbol{D}_{N-1}]$. The closed-form expressions for the gradient of $J(\boldsymbol{x})$ can be derived below.

Each term of the objective function in (15.40) has the form $J(\boldsymbol{x}) = \mathrm{tr}[\hat{\boldsymbol{T}}\boldsymbol{V}\hat{\boldsymbol{T}}^T]$ which, in the light of (15.39), can be expressed as

$$J(\boldsymbol{x}) = \mathrm{tr}\left[\left[\frac{\boldsymbol{t}_1}{||\boldsymbol{t}_1||}, \frac{\boldsymbol{t}_2}{||\boldsymbol{t}_2||}, \cdots, \frac{\boldsymbol{t}_n}{||\boldsymbol{t}_n||}\right]^{-1} \boldsymbol{V} \left[\frac{\boldsymbol{t}_1}{||\boldsymbol{t}_1||}, \frac{\boldsymbol{t}_2}{||\boldsymbol{t}_2||}, \cdots, \frac{\boldsymbol{t}_n}{||\boldsymbol{t}_n||}\right]^{-T}\right] \tag{15.43}$$

To compute $\partial J(\boldsymbol{x})/\partial t_{ij}$, we perturb the ith component of vector $\boldsymbol{t}_j$ by a small amount, say Δ, and keep the rest of $\hat{\boldsymbol{T}}$ unchanged. If we denote the perturbed jth column of $\hat{\boldsymbol{T}}^{-1}$ by $\tilde{\boldsymbol{t}}_j/||\tilde{\boldsymbol{t}}_j||$, then a linear approximation of $\tilde{\boldsymbol{t}}_j/||\tilde{\boldsymbol{t}}_j||$ can be obtained as

$$\frac{\tilde{\boldsymbol{t}}_j}{||\tilde{\boldsymbol{t}}_j||} \simeq \frac{\boldsymbol{t}_j}{||\boldsymbol{t}_j||} + \Delta\, \partial\left\{\frac{\boldsymbol{t}_j}{||\boldsymbol{t}_j||}\right\}/\partial t_{ij} = \frac{\boldsymbol{t}_j}{||\boldsymbol{t}_j||} - \Delta\, \boldsymbol{g}_{ij} \tag{15.44}$$

where

$$\boldsymbol{g}_{ij} = -\partial\left\{\frac{\boldsymbol{t}_j}{||\boldsymbol{t}_j||}\right\}/\partial t_{ij} = \frac{1}{||\boldsymbol{t}_j||^3}(t_{ij}\boldsymbol{t}_j - ||\boldsymbol{t}_j||^2 \boldsymbol{e}_i)$$

Now let $\hat{\boldsymbol{T}}_{ij}$ be the matrix obtained from $\hat{\boldsymbol{T}}$ with a perturbed (i,j)th component, then we obtain

$$\hat{\boldsymbol{T}}_{ij}^{-1} = \hat{\boldsymbol{T}}^{-1} - \Delta \boldsymbol{g}_{ij}\boldsymbol{e}_j^T \tag{15.45}$$

and up to the first-order, the matrix inversion formula [22, p. 655] gives

$$\hat{\boldsymbol{T}}_{ij} = \hat{\boldsymbol{T}} + \frac{\Delta \hat{\boldsymbol{T}}\boldsymbol{g}_{ij}\boldsymbol{e}_j^T\hat{\boldsymbol{T}}}{1 - \Delta \boldsymbol{e}_j^T\hat{\boldsymbol{T}}\boldsymbol{g}_{ij}} \simeq \hat{\boldsymbol{T}} + \Delta \hat{\boldsymbol{T}}\boldsymbol{g}_{ij}\boldsymbol{e}_j^T\hat{\boldsymbol{T}} \tag{15.46}$$

For convenience, we define $\hat{\boldsymbol{T}}_{ij} = \hat{\boldsymbol{T}} + \Delta \boldsymbol{S}$ with $\boldsymbol{S} = \hat{\boldsymbol{T}} \boldsymbol{g}_{ij} \boldsymbol{e}_j^T \hat{\boldsymbol{T}}$ and write

$$\begin{aligned} \hat{\boldsymbol{T}}_{ij} \boldsymbol{V} \hat{\boldsymbol{T}}_{ij}^T &= (\hat{\boldsymbol{T}} + \Delta \boldsymbol{S}) \boldsymbol{V} (\hat{\boldsymbol{T}} + \Delta \boldsymbol{S})^T \\ &= \hat{\boldsymbol{T}} \boldsymbol{V} \hat{\boldsymbol{T}}^T + \Delta \boldsymbol{S} \boldsymbol{V} \hat{\boldsymbol{T}}^T + \Delta \hat{\boldsymbol{T}} \boldsymbol{V} \boldsymbol{S}^T + \Delta^2 \boldsymbol{S} \boldsymbol{V} \boldsymbol{S}^T \end{aligned} \tag{15.47}$$

which implies that

$$\operatorname{tr}[\hat{\boldsymbol{T}}_{ij} \boldsymbol{V} \hat{\boldsymbol{T}}_{ij}^T] - \operatorname{tr}[\hat{\boldsymbol{T}} \boldsymbol{V} \hat{\boldsymbol{T}}^T] \simeq \Delta \operatorname{tr}[\boldsymbol{S}(\boldsymbol{V} + \boldsymbol{V}^T) \hat{\boldsymbol{T}}^T] \tag{15.48}$$

provided that Δ is sufficiently small. Hence

$$\begin{aligned} \frac{\partial \operatorname{tr}[\hat{\boldsymbol{T}} \boldsymbol{V} \hat{\boldsymbol{T}}^T]}{\partial t_{ij}} &= \lim_{\Delta \to 0} \frac{\operatorname{tr}[\hat{\boldsymbol{T}}_{ij} \boldsymbol{V} \hat{\boldsymbol{T}}_{ij}^T] - \operatorname{tr}[\hat{\boldsymbol{T}} \boldsymbol{V} \hat{\boldsymbol{T}}^T]}{\Delta} \\ &= \operatorname{tr}[\boldsymbol{S}(\boldsymbol{V} + \boldsymbol{V}^T) \hat{\boldsymbol{T}}^T] \end{aligned} \tag{15.49}$$

By substituting $\boldsymbol{S} = \hat{\boldsymbol{T}} \boldsymbol{g}_{ij} \boldsymbol{e}_j^T \hat{\boldsymbol{T}}$ into (15.49), we obtain

$$\begin{aligned} \frac{\partial \operatorname{tr}[\hat{\boldsymbol{T}} \boldsymbol{V} \hat{\boldsymbol{T}}^T]}{\partial t_{ij}} &= \operatorname{tr}[\hat{\boldsymbol{T}} \boldsymbol{g}_{ij} \boldsymbol{e}_j^T \hat{\boldsymbol{T}} (\boldsymbol{V} + \boldsymbol{V}^T) \hat{\boldsymbol{T}}^T] \\ &= \boldsymbol{e}_j^T \hat{\boldsymbol{T}} (\boldsymbol{V} + \boldsymbol{V}^T) \hat{\boldsymbol{T}}^T \hat{\boldsymbol{T}} \boldsymbol{g}_{ij} \end{aligned} \tag{15.50}$$

On comparing (15.40) with (15.50), the components of the gradient of $J(\boldsymbol{x})$ with respect to $\boldsymbol{t}_1, \boldsymbol{t}_2, \cdots, \boldsymbol{t}_n$ are found to be

$$\begin{aligned} \frac{\partial J(\boldsymbol{x})}{\partial t_{ij}} &= 2\boldsymbol{e}_j^T \Big[\hat{\boldsymbol{T}} (\boldsymbol{V}_0 - \hat{\boldsymbol{c}}^T \hat{\boldsymbol{c}}) \hat{\boldsymbol{T}}^T - \sum_{p=0}^{N-1} \hat{\boldsymbol{T}} (\boldsymbol{V}_{p+1} + \boldsymbol{V}_{p+1}^T) \hat{\boldsymbol{T}}^T \boldsymbol{D}_p \\ &\quad + \frac{1}{2} \sum_{p=0}^{N-1} \sum_{q=0}^{N-1} \hat{\boldsymbol{T}} (\boldsymbol{V}_{|p-q|} + \boldsymbol{V}_{|p-q|}^T) \hat{\boldsymbol{T}}^T \boldsymbol{D}_p \boldsymbol{D}_q \Big] \hat{\boldsymbol{T}} \boldsymbol{g}_{ij} \end{aligned} \tag{15.51}$$

for $i, j = 1, 2, \cdots, n$.

Let the high-order error feedback matrices assume the form

$$\boldsymbol{D}_p = \operatorname{diag}\{d_{p1}, d_{p2}, \cdots, d_{pn}\} \quad \text{for} \quad p = 0, 1, \cdots, N-1. \tag{15.52}$$

The gradients of $J(\boldsymbol{x})$ with respect to the diagonal $\boldsymbol{D}_p$ are given by

$$\frac{\partial J(\boldsymbol{x})}{\partial d_{pi}} = -2 \left(\hat{\boldsymbol{T}} \boldsymbol{V}_{p+1} \hat{\boldsymbol{T}}^T \right)_{ii} + 2 \sum_{q=0}^{N-1} d_{qi} \left(\hat{\boldsymbol{T}} \boldsymbol{V}_{|p-q|} \hat{\boldsymbol{T}}^T \right)_{ii} \tag{15.53}$$

for $p = 0, 1, \cdots, N-1$ and $i = 1, 2, \cdots, n$.

15.3.5 Analytical Method for Separate Optimization

Here the term "separate optimization" refers to a procedure where the optimization of transformation matrix $\boldsymbol{T}$ and that of the error feedback matrices $\{\boldsymbol{D}_0, \boldsymbol{D}_1, \cdots, \boldsymbol{D}_{N-1}\}$ are carried out separately as two different steps. First, we fix the error feedback matrices to $\boldsymbol{D}_i = \boldsymbol{0}$ for $i = 0, 1, \cdots, N-1$ in (15.27) so that the objective function in (15.36) is reduced to

$$J_{e2}(\boldsymbol{0}, \boldsymbol{T}) = \mathrm{tr}\big[\boldsymbol{T}^T(\boldsymbol{W}_o - \boldsymbol{c}^T\boldsymbol{c})\boldsymbol{T}\big] \tag{15.54}$$

which is minimized with respect to matrix $\boldsymbol{T}$ subject to the l_2-scaling constraints in (15.13). Second, with $\boldsymbol{T}$ optimized in the first step, (15.36) is minimized under the fixed $\boldsymbol{T}$ with respect to matrices $\boldsymbol{D}_0, \boldsymbol{D}_1, \cdots, \boldsymbol{D}_{N-1}$. To perform the first step, we define the Lagrange function

$$J_o(\boldsymbol{P}, \lambda) = \mathrm{tr}\big[\boldsymbol{X}\boldsymbol{P}\big] + \lambda\left(\mathrm{tr}[\boldsymbol{K}\boldsymbol{P}^{-1}] - n\right) \tag{15.55}$$

where $\boldsymbol{P} = \boldsymbol{T}\boldsymbol{T}^T$ and $\boldsymbol{X} = \boldsymbol{W}_o - \boldsymbol{c}^T\boldsymbol{c}$. By applying the same manner as in Section 15.2.2, we arrive at

$$\boldsymbol{T} = \frac{1}{\sqrt{n}}\Big(\sum_{i=1}^{n}\sigma_i\Big)^{\frac{1}{2}}\boldsymbol{X}^{-\frac{1}{2}}\Big[\boldsymbol{X}^{\frac{1}{2}}\boldsymbol{K}_c\boldsymbol{X}^{\frac{1}{2}}\Big]^{\frac{1}{4}}\boldsymbol{Q}\boldsymbol{Z}^T \tag{15.56}$$

where σ_i^2 for $i = 1, 2, \cdots, n$ denote the eigenvalues of $\boldsymbol{K}_c\boldsymbol{X}$. This is the coordinate transformation matrix $\boldsymbol{T}$ which minimizes (15.54) subject to the l_2-scaling constraints in (15.13).

In the second step, suppose that $\boldsymbol{D}_p$ for $p = 0, 1, \cdots, N-1$ are diagonal matrices. In this case, matrix $\boldsymbol{D}_p$ assumes the form

$$\boldsymbol{D}_p = \mathrm{diag}\{\alpha_{p1}, \alpha_{p2}, \cdots, \alpha_{pn}\} \;\text{ for }\; p = 0, 1, \ldots, N-1 \tag{15.57}$$

It follows that

$$\frac{\partial J_{e2}(\boldsymbol{D}, \boldsymbol{T})}{\partial \alpha_{pi}} = -2\Big(\boldsymbol{T}^T\boldsymbol{W}_o\boldsymbol{A}^{p+1}\boldsymbol{T}\Big)_{ii} + 2\sum_{q=0}^{N-1}\alpha_{qi}\Big(\boldsymbol{T}^T\boldsymbol{W}_o\boldsymbol{A}^{|p-q|}\boldsymbol{T}\Big)_{ii} = 0 \tag{15.58}$$

for $i = 1, 2, \cdots, n$. As a result, matrix $\boldsymbol{D}_p$ can be derived from

$$\begin{bmatrix} \alpha_{0i} \\ \alpha_{1i} \\ \vdots \\ \alpha_{N-1,i} \end{bmatrix} = \boldsymbol{R}_i^{-1}\begin{bmatrix} r_i(1,0) \\ r_i(2,0) \\ \vdots \\ r_i(N,0) \end{bmatrix} \tag{15.59}$$

where

$$\boldsymbol{R}_i = \begin{bmatrix} r_i(0,0) & r_i(0,1) & \cdots & r_i(0,N-1) \\ r_i(1,0) & r_i(1,1) & \cdots & r_i(1,N-1) \\ \vdots & \vdots & \ddots & \vdots \\ r_i(N-1,0) & r_i(N-1,1) & \cdots & r_i(N-1,N-1) \end{bmatrix}$$

with $r_i(p,q) = \left(\boldsymbol{T}^T \boldsymbol{W}_o \boldsymbol{A}^{|p-q|} \boldsymbol{T}\right)_{ii}$ for $i = 1, 2, \cdots, n$.

15.4 Numerical Experiments

15.4.1 Filter Description and Initial Roundoff Noise

We consider a stable 3rd-order lowpass state-space digital filter $(\boldsymbol{A}, \boldsymbol{b}, \boldsymbol{c}, d)_3$ described by

$$\boldsymbol{A} = \begin{bmatrix} 0 & 1 & 0 \\ 0 & 0 & 1 \\ 0.339377 & -1.152652 & 1.520167 \end{bmatrix}, \quad \boldsymbol{b} = \begin{bmatrix} 0 \\ 0 \\ 1 \end{bmatrix}$$

$$\boldsymbol{c} = \begin{bmatrix} 0.093253 & 0.128620 & 0.314713 \end{bmatrix}, \qquad d = 0.065959$$

The controllability and observability Grammians $\boldsymbol{K}_c$ and $\boldsymbol{W}_o$ were computed from (15.7) and (15.8) as

$$\boldsymbol{K}_c = \begin{bmatrix} 5.215397 & 3.869762 & 1.184455 \\ 3.869762 & 5.215397 & 3.869762 \\ 1.184455 & 3.869762 & 5.215397 \end{bmatrix}$$

$$\boldsymbol{W}_o = \begin{bmatrix} 0.138134 & -0.313522 & 0.336218 \\ -0.313522 & 0.872712 & -0.804183 \\ 0.336218 & -0.804183 & 1.123823 \end{bmatrix}$$

The eigenvalues of $\boldsymbol{K}_c \boldsymbol{W}_o$ were as

$$\theta_1^2 = 2.499998, \qquad \theta_2^2 = 0.049748, \qquad \theta_1^3 = 0.729367$$

When a coordinate transformation defined by

$$\boldsymbol{T}_o = \mathrm{diag}\{\, 2.283724,\ 2.283724,\ 2.283724 \,\}$$

was applied to the above filter $(\boldsymbol{A}, \boldsymbol{b}, \boldsymbol{c}, d)_3$, its controllability and observability Grammians $\boldsymbol{K}_c^o = \boldsymbol{T}_o^{-1} \boldsymbol{K}_c \boldsymbol{T}_o^{-T}$ and $\boldsymbol{W}_o^o = \boldsymbol{T}_o^T \boldsymbol{W}_o \boldsymbol{T}_o$ were derived as

$$K_c^o = \begin{bmatrix} 1.000000 & 0.741988 & 0.227107 \\ 0.741988 & 1.000000 & 0.741988 \\ 0.227107 & 0.741988 & 1.000000 \end{bmatrix}$$

$$W_o^o = \begin{bmatrix} 0.720426 & -1.635144 & 1.753511 \\ -1.635144 & 4.551538 & -4.194133 \\ 1.753511 & -4.194133 & 5.861185 \end{bmatrix}$$

The original noise gain subject to l_2-scaling constraints was then computed from (15.14) as

$$J(P,0) = \text{tr}[T_o^T W_o T_o] = \text{tr}[W_o^o] = 11.133150$$

where $P = T_o T_o^T$.

15.4.2 The Use of Analytical Method in Section 15.2.2

The optimal symmetric and positive-definite matrix P which minimizes (15.14) were obtained from (15.18) as

$$P = \begin{bmatrix} 12.205669 & 4.936938 & -0.320033 \\ 4.936938 & 4.593693 & 2.247419 \\ -0.320033 & 2.247419 & 3.190823 \end{bmatrix}$$

and the minimum value of (15.14) was found to be

$$\min_{P,\lambda} J(P,\lambda) = 2.355360$$

As a result, the optimal coordinate transformation matrix T was computed from (15.25) as

$$T = \begin{bmatrix} -1.961239 & 2.849950 & 0.486823 \\ 0.382807 & 2.067061 & -0.417624 \\ 1.519389 & 0.874919 & 0.341756 \end{bmatrix}$$

The optimal realization that minimizes the roundoff noise $\text{tr}[T^T W_o T]$ subject to l_2-scaling constraints in (15.13) was then constructed from (15.10) as

$$\overline{A} = \begin{bmatrix} 0.542471 & -0.340493 & 0.565793 \\ 0.565793 & 0.488848 & 0.159318 \\ -0.340493 & 0.012490 & 0.488848 \end{bmatrix}, \quad \overline{b} = \begin{bmatrix} 0.388814 \\ 0.111998 \\ 0.910741 \end{bmatrix}$$

$$\overline{c} = \begin{bmatrix} 0.344517 & 0.806980 & 0.099238 \end{bmatrix}$$

In this case, the controllability and observability Grammians $\overline{\boldsymbol{K}}_c = \boldsymbol{T}^{-1}\boldsymbol{K}_c\boldsymbol{T}^{-T}$ and $\overline{\boldsymbol{W}}_o = \boldsymbol{T}^T\boldsymbol{W}_o\boldsymbol{T}$ became

$$\overline{\boldsymbol{K}}_c = \begin{bmatrix} 1.000000 & 0.541747 & 0.541747 \\ 0.541747 & 1.000000 & 0.036160 \\ 0.541747 & 0.036160 & 1.000000 \end{bmatrix}$$

$$\overline{\boldsymbol{W}}_o = \begin{bmatrix} 0.785120 & 0.425337 & 0.425337 \\ 0.425337 & 0.785120 & 0.028390 \\ 0.425337 & 0.028390 & 0.785120 \end{bmatrix}$$

and

$$\mathrm{tr}[\overline{\boldsymbol{W}}_o] = 2.355360$$

which coincides with the minimum value of $J(\boldsymbol{P}, \lambda)$ in (15.14).

15.4.3 The Use of Iterative Method in Section 15.3.4

In what follows, the simulation was carried out for the state-space model specified by $(\boldsymbol{T}_o^{-1}\boldsymbol{A}\boldsymbol{T}_o, \boldsymbol{T}_o^{-1}\boldsymbol{b}, \boldsymbol{c}\boldsymbol{T}_o, d)_3$. Simulation results are shown for two system set-up's—the first employed a static error feedback (i.e. $N = 1$) while the second used a dynamic error feedback with $N = 2$. In both cases the error feedback matrices involved were all diagonal.

In the case of $N = 1$, the quasi-Newton algorithm in (15.41) was applied to minimize (15.40) with tolerance $\varepsilon = 10^{-8}$ in (15.42). It took the algorithm 64 iterations to converge to the solution

$$\hat{\boldsymbol{T}} = \begin{bmatrix} 5.899503 & -0.516441 & 1.228943 \\ 5.935833 & -0.493021 & 2.377971 \\ 4.442394 & -1.380516 & 1.478971 \end{bmatrix}$$

$$\boldsymbol{D}_0 = \mathrm{diag}\{\ 0.390401 \quad 0.574984 \quad 0.659618\ \}$$

The minimized noise gain was found to be

$$J_{e3}(\boldsymbol{D}, \hat{\boldsymbol{T}}) = 0.273955$$

The profile of $J_{e3}(\boldsymbol{D}, \hat{\boldsymbol{T}})$ during the first 64 iterations of the algorithm is depicted in Figure 15.7. The values of function $J(\boldsymbol{x})$ at the initial point and seven subsequent iterates were found to be

$$\begin{bmatrix} J(\boldsymbol{x}_0) & J(\boldsymbol{x}_1) & J(\boldsymbol{x}_2) & J(\boldsymbol{x}_3) \\ J(\boldsymbol{x}_4) & J(\boldsymbol{x}_5) & J(\boldsymbol{x}_6) & J(\boldsymbol{x}_7) \end{bmatrix} = \begin{bmatrix} 1.2156 & 0.7980 & 0.6502 & 0.5258 \\ 0.4529 & 0.4132 & 0.3847 & 0.3564 \end{bmatrix}$$

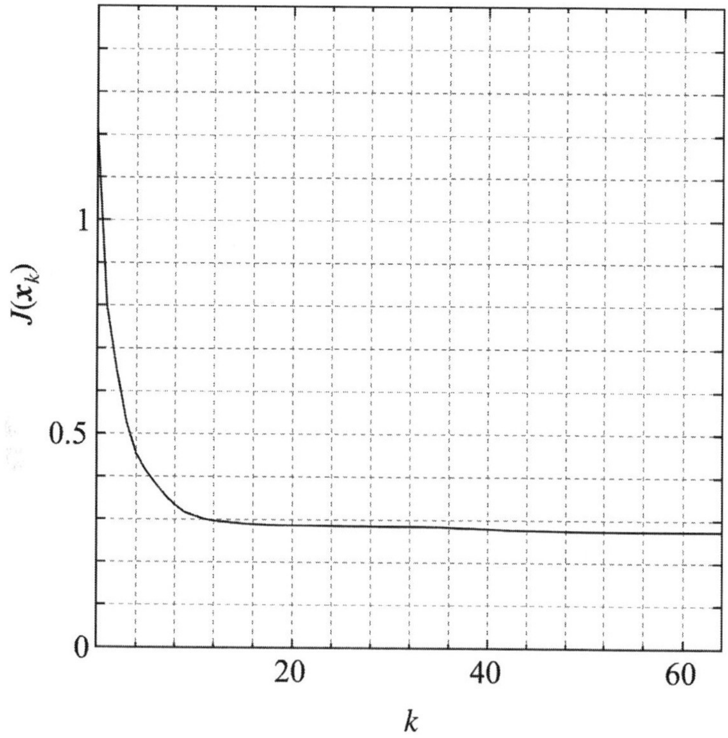

Figure 15.7 Profile of $J_{e3}(\boldsymbol{D}, \hat{\boldsymbol{T}})$ during the first 64 iterations.

The coordinate transformation matrix $\boldsymbol{T}$ was then computed from (15.37) as

$$\boldsymbol{T} = \begin{bmatrix} 5.119779 & 5.187942 & 3.438106 \\ 2.677953 & 3.209672 & 1.472176 \\ 1.013689 & 2.060072 & 0.832834 \end{bmatrix}$$

which yields an equivalent realization in (15.10) with

$$\overline{\boldsymbol{A}} = \begin{bmatrix} 0.561391 & 0.228938 & -0.345628 \\ -0.410493 & 0.581361 & 0.374739 \\ 0.562335 & -0.284606 & 0.377416 \end{bmatrix}, \quad \overline{\boldsymbol{b}} = \begin{bmatrix} -0.704730 \\ 0.346362 \\ 0.526789 \end{bmatrix}$$

$$\overline{\boldsymbol{c}} = \begin{bmatrix} 2.605487 & 3.528241 & 1.763191 \end{bmatrix}$$

From (15.11), it follows that the corresponding controllability and observability Grammians $\overline{\boldsymbol{K}}_c$ and $\overline{\boldsymbol{W}}_o$ became

$$\overline{\boldsymbol{K}}_c = \begin{bmatrix} 1.000000 & -0.692295 & -0.317993 \\ -0.692295 & 1.000000 & -0.447851 \\ -0.317993 & -0.447851 & 1.000000 \end{bmatrix}$$

$$\overline{\boldsymbol{W}}_o = \begin{bmatrix} 8.140416 & 11.841925 & 6.169088 \\ 11.841925 & 18.715668 & 9.945036 \\ 6.169088 & 9.945036 & 5.650477 \end{bmatrix}$$

In the case of $N = 2$, the quasi-Newton algorithm in (15.41) was applied to minimize (15.40) with tolerance $\varepsilon = 10^{-8}$ in (15.42). It took the algorithm 36 iterations to converge to the solution

$$\hat{\boldsymbol{T}} = \begin{bmatrix} 1.367924 & -0.014714 & 0.386079 \\ -0.849858 & 1.032268 & 0.308031 \\ 0.078696 & 0.047338 & 1.218360 \end{bmatrix}$$

$$\boldsymbol{D}_0 = \mathrm{diag}\{\ 0.478043 \quad 0.923417 \quad 1.187428\ \}$$

$$\boldsymbol{D}_1 = \mathrm{diag}\{\ -0.060347 \quad -0.622454 \quad -0.584187\ \}$$

The minimized noise gain in this case was found to be

$$J_{e3}(\boldsymbol{D}, \hat{\boldsymbol{T}}) = 0.031801$$

which is approximately nine times smaller than what the best static error-feedback can achieve. The profile of $J_{e3}(\boldsymbol{D}, \hat{\boldsymbol{T}})$ during the first 36 iterations of the algorithm is depicted in Figure 15.8. The values of function $J(\boldsymbol{x})$ at the initial point and seven subsequent iterates were found to be

$$\begin{bmatrix} J(\boldsymbol{x}_0) & J(\boldsymbol{x}_1) & J(\boldsymbol{x}_2) & J(\boldsymbol{x}_3) \\ J(\boldsymbol{x}_4) & J(\boldsymbol{x}_5) & J(\boldsymbol{x}_6) & J(\boldsymbol{x}_7) \end{bmatrix} = \begin{bmatrix} 7.7554 & 2.5386 & 1.3188 & 0.4822 \\ 0.3056 & 0.1281 & 0.0721 & 0.0575 \end{bmatrix}$$

The coordinate transformation matrix $\boldsymbol{T}$ was then computed from (15.37) as

$$\boldsymbol{T} = \begin{bmatrix} 1.234828 & -0.312148 & 0.118222 \\ 0.747764 & 0.581564 & 0.598990 \\ 0.371646 & 0.705802 & 1.120151 \end{bmatrix}$$

which yields an equivalent realization in (15.10) with

$$\overline{\boldsymbol{A}} = \begin{bmatrix} 0.588282 & 0.709731 & 0.690452 \\ -0.081529 & 0.771527 & 0.873168 \\ -0.034786 & -0.456772 & 0.160358 \end{bmatrix}, \qquad \overline{\boldsymbol{b}} = \begin{bmatrix} -0.219057 \\ -0.557861 \\ 0.815097 \end{bmatrix}$$

$$\overline{\boldsymbol{c}} = \begin{bmatrix} 0.749725 & 0.611620 & 1.006192 \end{bmatrix}$$

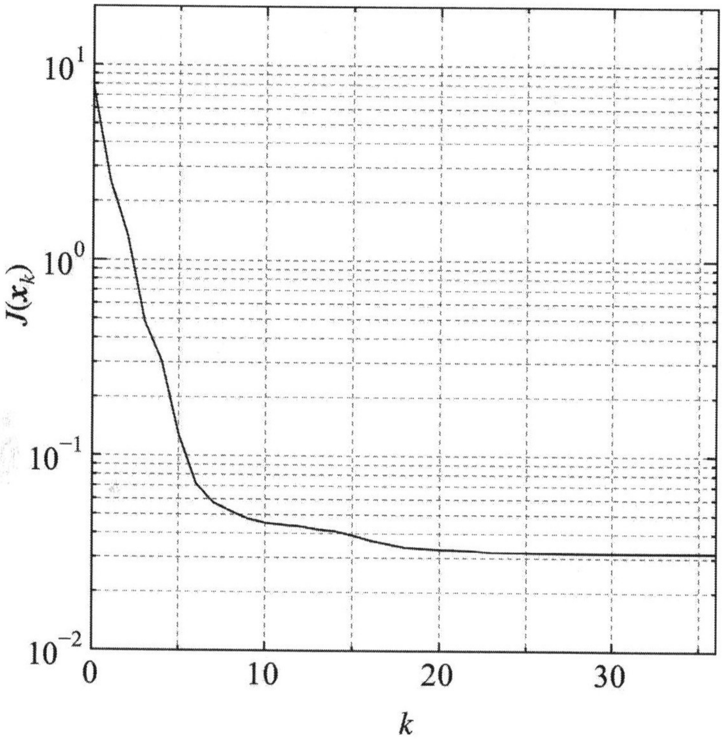

Figure 15.8 Profile of $J_{e3}(\boldsymbol{D}, \hat{\boldsymbol{T}})$ during the first 36 iterations.

From (15.11), it follows that the corresponding controllability and observability Grammians $\overline{\boldsymbol{K}}_c$ and $\overline{\boldsymbol{W}}_o$ became

$$\overline{\boldsymbol{K}}_c = \begin{bmatrix} 1.000000 & 0.666020 & -0.546135 \\ 0.666020 & 1.000000 & -0.496903 \\ -0.546135 & -0.496903 & 1.000000 \end{bmatrix}$$

$$\overline{\boldsymbol{W}}_o = \begin{bmatrix} 0.711711 & 0.651290 & 1.285624 \\ 0.651290 & 0.907285 & 1.413957 \\ 1.285624 & 1.413957 & 3.602021 \end{bmatrix}$$

The above and several other simulation results regarding the noise gain $J_{e3}(\boldsymbol{D}, \hat{\boldsymbol{T}})$ in (15.40) plus $(\hat{\boldsymbol{c}} - \boldsymbol{h})(\hat{\boldsymbol{c}} - \boldsymbol{h})^T$ were summarized in Table 15.1 where the column with "Infinite Precision" shows the value of $J_{e3}(\boldsymbol{D}, \hat{\boldsymbol{T}})$ derived from the optimal $\hat{\boldsymbol{T}}$ and $\boldsymbol{D}$. The column with "3-Bit Quantization" means that of $J_{e3}(\boldsymbol{D}, \hat{\boldsymbol{T}}) + (\hat{\boldsymbol{c}} - \boldsymbol{h})(\hat{\boldsymbol{c}} - \boldsymbol{h})^T$ where each entry of the optimal

Table 15.1 Performance comparison

N	Optimization	Infinite Precision	3-Bit Quantization
1	Separate	0.644452	0.659679
	Joint	0.273955	0.295692
2	Separate	0.353326	0.373479
	Joint	0.031801	0.046057

$\boldsymbol{D}$ was rounded to a power-of-two representation with 3 bits after the binary point, matrix $\hat{\boldsymbol{T}}$ was updated so as to minimize $J_{e3}(\boldsymbol{D}, \hat{\boldsymbol{T}})$ with such a $\boldsymbol{D}$ fixed, and each entry of $\hat{c}$ derived from the updated $\hat{\boldsymbol{T}}$ was rounded to a power-of-two representation with 3 bits after the binary point to set a new vector $\boldsymbol{h}$.

The proposed joint optimization strategy was evaluated in comparison with the separate optimization technique shown in Section 15.3.5. The separate optimization technique finds an optimal coordinate transformation $\boldsymbol{T}$ first in absence of error feedback. With T fixed, it then finds an optimal error-feedback matrix $\boldsymbol{D}$ as well as a feedforward vector $\boldsymbol{h}$. Our simulations considered two cases, namely the case where the system parameters were implemented with infinite precision and the case where the system parameters were implemented using 3-bit quantization. Table 15.1 includes numerical results obtained using the separate optimization and joint optimization, respectively, where each optimization procedure was applied with static ($N = 1$) and dynamic ($N = 2$) error-feedback. From Table 15.1, it is observed that (i) the use of a dynamic error-feedback (i.e. with $N > 1$) led to considerable improvement in roundoff noise reduction relative to static error-feedback (i.e. with $N = 1$) for both separate and joint optimization; (ii) in each of the four scenarios, joint optimization outperforms its separate-optimization counterpart in a significant manner; and (iii) for both infinite precision and 3-bit quantization cases, the best performance was achieved when the joint optimization technique was employed in conjunction with a dynamic error-feedback.

15.5 Summary

For state-space digital filters, two techniques for minimizing the roundoff noise subject to l_2-scaling constraints have been presented. One has relied on the lines studied by Mullis-Roberts and Hwang with the relaxation of l_2-scaling constraints into a single constraint on matrix trace, and the optimal matrix solution has been analytically obtained in closed form. The other has been a joint optimization technique of high-order error feedback and state-space realization for minimizing the effects of roundoff noise at the filter output subject to l_2-scaling constraints, and an efficient quasi-Newton algorithm has been employed to minimize the objective function iteratively. The simulation results in numerical experiments have demonstrated the validity and effectiveness of the present techniques.

References

[1] S. Y. Hwang, "Roundoff noise in state-space digital filtering: A general analysis," *IEEE Trans. Acoust., Speech, Signal Process.*, vol. ASSP-24, no. 3, pp. 256–262, June 1976.

[2] C. T. Mullis and R. A. Roberts, "Synthesis of minimum roundoff noise fixed point digital filters," *IEEE Trans. Circuits Syst.*, vol. CAS-23, no. 9, pp. 551–562, Sept. 1976.

[3] S. Y. Hwang, "Minimum uncorrelated unit noise in state-space digital filtering," *IEEE Trans. Acoust., Speech, Signal Process.*, vol. ASSP-25, no. 4, pp. 273–281, Aug. 1977.

[4] L. B. Jackson, A. G. Lindgren and Y. Kim, "Optimal synthesis of second-order state-space structures for digital filters," *IEEE Trans. Circuits Syst.*, vol. CAS-26, no. 3, pp. 149–153, Mar. 1979.

[5] H. A. Spang, III and P. M. Shultheiss, "Reduction of quantizing noise by use of feedback," *IRE Trans. Commun. Syst.*, vol. CS-10, no. 4, pp. 373–380, Dec. 1962.

[6] T. Thong and B. Liu, "Error spectrum shaping in narrowband recursive digital filters," *IEEE Trans. Acoust., Speech, Signal Process.*, vol. ASSP-25, no. 2, pp. 200–203, Apr. 1977.

[7] T.-L. Chang and S. A. White, "An error cancellation digital-filter structure and its distributed-arithmetic implementation," *IEEE Trans. Circuits Syst.*, vol. CAS-28, no. 4, pp. 339–342, Apr. 1981.

[8] D. C. Munson and D. Liu, "Narrow-band recursive filters with error spectrum shaping," *IEEE Trans. Circuits Syst.*, vol. CAS-28, no. 2, pp. 160–163, Feb. 1981.

[9] W. E. Higgins and D. C. Munson, "Noise reduction strategies for digital filters: Error spectrum shaping versus the optimal linear state-space formulation," *IEEE Trans. Acoust., Speech, Signal Process.*, vol. ASSP-30, no. 6, pp. 963–973, Dec. 1982.

[10] M. Renfors, "Roundoff noise in error-feedback state-space filters," in *Pro (c. Int. Conf. Acoust., Speech, Signal Process. (ICASSP'83)*, Apr. 1983, pp. 619–622.

[11] W. E. Higgins and D. C. Munson, "Optimal and suboptimal error spectrum shaping for cascade-form digital filters," *IEEE Trans. Circuits Syst.*, vol. CAS-31, no. 5, pp. 429–437, May 1984.

[12] T. I. Laakso and I. O. Hartimo, "Noise reduction in recursive digital filters using high-order error feedback," *IEEE Trans. Signal Process.*, vol. 40, no. 5, pp. 1096–1107, May 1992.

[13] P. P. Vaidyanathan, "On error-spectrum shaping in state-space digital filters," *IEEE Trans. Circuits Syst.*, vol. CAS-32, no. 1, pp. 88–92, Jan. 1985.

[14] D. Williamson, "Roundoff noise minimization and pole-zero sensitivity in fixed-point digital filters using residue feedback," *IEEE Trans. Acoust., Speech, Signal Process.*, vol. ASSP-34, no. 5, pp. 1210–1220, Oct. 1986.

[15] D. Williamson, "Delay replacement in direct form structures", *IEEE Trans. Acoust., Speech, Signal Process.*, vol. ASSP-36, no. 4, pp. 453–460, Apr. 1988.

[16] G. Li and M. Gevers, "Roundoff noise minimization using delta-operator realizations," *IEEE Trans. Signal Process.*, vol. 41, no. 2, pp. 629–637, Feb. 1993.

[17] G. Li and Z. Zhao, "On the generalized DFIIt structure and its state-space realization in digital filter implementation," *IEEE Trans. Circuits Syst. I*, vol. 51, no. 4, pp. 769–778, Apr. 2004.

[18] T. Hinamoto, H. Ohnishi and W.-S. Lu, "Roundoff noise minimization of state-space digital filters using separate and joint error feedback/ coordinate transformation," *IEEE Trans. Circuits Syst. I*, vol. 50, no. 1, pp. 23–33, Jan. 2003.

[19] W.-S. Lu and T. Hinamoto, "Jointly optimized error-feedback and realization for roundoff noise minimization in state-space digital filters," *IEEE Trans. Signal Process.*, vol. 53, no. 6, pp. 2135–2145, June 2005.

[20] T. Hinamoto, A. Doi and W.-S. Lu, "Jointly optimal high-order error-feedback and realization for roundoff noise minimization in 1-D and 2-D state-space digital filters," *IEEE Trans. Signal Process.*, vol. 61, no. 23, pp. 5893–5904, Dec. 1, 2013.

[21] R. Fletcher, *Practical Methods of Optimization*, 2nd ed. New York, NY: Wiley, 1987.

[22] T. Kailath, *Linear Systems*, Engle Cliffs, NJ: Prentice Hall, 1980.

16

Generalized Transposed Direct-Form II Realization

16.1 Preview

Delta operator is often applied to the implementation of IIR digital filters due to its good numerical properties subject to fast sampling. When the sampling rate is much higher than the underlying signal bandwidth, then finite-word-length (FWL) effects become worse. If the poles of a narrow-band lowpass filter cluster near the point $z = 1$ in the z plane, the filter becomes very noisy and sensitive to coefficient quantization. For example, notch filters have pole(s) close to the point $z = 1$ if the rejected signal component exists at the normalized frequency near zero. To achieve good finite-word-length performance subject to fast sampling, the forward shift operator z is replaced by the delta operator, defined by

$$\delta(z) = \frac{z-1}{\Delta}$$

where Δ is originally the sampling interval [1]. Later on, the parameter Δ is often used to improve the finite-word-length effects as a free parameter by several researchers. An implementation of $\delta^{-1}(z)$ is shown in Figure 16.1.

It is known that delta operator realizations generally yield better roundoff noise performance and more robust coefficient sensitivity [10]. For example, a high-order transfer function may be implemented as a series of second-order

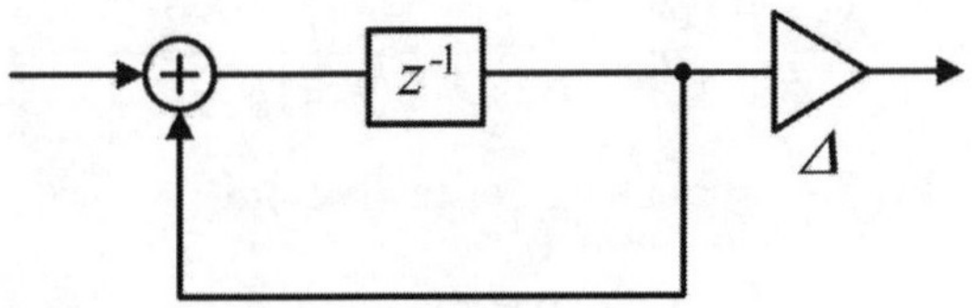

Figure 16.1 Implementation of $\delta^{-1}(z)$.

sections that are connected in cascade with each section implemented by a direct form in delta operator. The roundoff noise of such an implementation was analyzed in [2] where the transposed direct-form II is found to yield the lowest roundoff noise gain at the output among all the direct forms. A modified delta transposed direct-form II second-order section where the Δs and filter coefficients at different branches are separately scaled to achieve improved roundoff noise gain minimization was proposed [3]. Alternatively, an nth-order delta transposed direct-form II IIR filter with minimum roundoff noise gain and sensitivity has been derived by utilizing different *coupling coefficients* at different branch nodes for better noise gain suppression [4]. Moreover, a set of special operators was employed to obtain the generalized transposed direct-form II structure of a general order, say pth-order, IIR digital filter and its equivalent state-space realization subject to l_2-scaling where p free parameters in the operators are appropriately chosen to minimize the roundoff noise gain or the coefficient sensitivity measure of each structure [5].

This chapter is written mainly along the line of the paper in [5], but we elaborate the subject matter with further details. To begin with, the transposed direct-form II structure of an nth-order IIR digital filter using a set of special operators is introduced, and its equivalent state-space realization is constructed. These are followed by analyzing the roundoff noise and l_2-sensitivity for the generalized transposed direct-form II structure and its equivalent state-space realization. Then, given a transfer function and n free parameters in the operators, a concrete procedure for evaluating the roundoff noise gain and the l_2-sensitivity measure of each structure is summarized. Finally, numerical experiments are presented to demonstrate the validity and effectiveness of the techniques addressed in this chapter.

16.2 Structural Transformation

Consider an SISO (single-input/single-output) time-invariant linear digital filter described by

$$H(z) = \frac{b_0 z^n + b_1 z^{n-1} + \cdots + b_{n-1} z + b_n}{z^n + a_1 z^{n-1} + \cdots + a_{n-1} z + a_n} \tag{16.1}$$

Let $\boldsymbol{P}$ be an $(n+1) \times (n+1)$ nonsingular matrix and

$$\boldsymbol{q}(z) = \boldsymbol{P}\boldsymbol{z} \tag{16.2}$$

where

$$\boldsymbol{q}(z) = \begin{bmatrix} q_0(z) & q_1(z) & \cdots & q_n(z) \end{bmatrix}^T, \qquad \boldsymbol{z} = \begin{bmatrix} z^n & \cdots & z & 1 \end{bmatrix}^T$$

By defining scalars $\{\alpha_i | \, i = 1, 2, \cdots, n\}$ and $\{\beta_i | \, i = 0, 1, \cdots, n\}$ such that

$$\begin{aligned} \kappa \begin{bmatrix} 1 & \alpha_1 & \cdots & \alpha_n \end{bmatrix} \boldsymbol{P} &= \begin{bmatrix} 1 & a_1 & \cdots & a_n \end{bmatrix} \\ \kappa \begin{bmatrix} \beta_0 & \beta_1 & \cdots & \beta_n \end{bmatrix} \boldsymbol{P} &= \begin{bmatrix} b_0 & b_1 & \cdots & b_n \end{bmatrix} \end{aligned} \tag{16.3}$$

the transfer function in (16.1) can be written as

$$\begin{aligned} H(z) &= \frac{\beta_0 q_0(z) + \beta_1 q_1(z) + \cdots + \beta_n q_n(z)}{q_0(z) + \alpha_1 q_1(z) + \cdots + \alpha_n q_n(z)} \\ &= \frac{\beta_0 + \displaystyle\sum_{i=1}^{n} \beta_i \frac{q_i(z)}{q_0(z)}}{1 + \displaystyle\sum_{i=1}^{n} \alpha_i \frac{q_i(z)}{q_0(z)}} \end{aligned} \tag{16.4}$$

where κ is a scaling factor such that $\alpha_0 = 1$. From (16.4), it is obvious that the filter is characterized by scalars $\{\alpha_i | \, i = 1, 2, \cdots, n\}$ and $\{\beta_i | \, i = 0, 1, \cdots, n\}$ under the polynomial operators $\{q_i(z) | \, i = 0, 1, \cdots, n\}$. We now consider a special set of polynomial operators, that leads to an interesting structure for filter implementation.

Define

$$\rho_i(z) = \frac{z - \gamma_i}{\Delta_i} \quad \text{for } i = 1, 2, \cdots, n \tag{16.5}$$

where $\{\gamma_i\}$ and $\{\Delta_i > 0\}$ are two sets of constants to be discussed later. Let the polynomial operators be chosen as

$$\begin{aligned} q_i(z) &= \rho_{i+1}(z)\rho_{i+2}(z) \cdots \rho_n(z) \quad \text{for } i = 0, 1, \cdots, n-1 \\ q_n(z) &= 1 \end{aligned} \tag{16.6a}$$

which are equivalent to

$$\begin{bmatrix} q_0(z) \\ q_1(z) \\ \vdots \\ q_{n-1}(z) \\ q_n(z) \end{bmatrix} = \begin{bmatrix} \rho_1(z)\rho_2(z) \cdots \rho_n(z) \\ \rho_2(z)\rho_3(z) \cdots \rho_n(z) \\ \vdots \\ \rho_n(z) \\ 1 \end{bmatrix} = \boldsymbol{P}\boldsymbol{z} \tag{16.6b}$$

As an example, consider the case where $n = 3$. Then we have

$$\begin{aligned} q_0(z) &= \frac{z^3 - (\gamma_1 + \gamma_2 + \gamma_3)z^2 + (\gamma_1\gamma_2 + \gamma_2\gamma_3 + \gamma_3\gamma_1)z - \gamma_1\gamma_2\gamma_3}{\Delta_1\Delta_2\Delta_3} \\ q_1(z) &= \frac{z^2 - (\gamma_2 + \gamma_3)z + \gamma_2\gamma_3}{\Delta_2\Delta_3} \\ q_2(z) &= \frac{z - \gamma_3}{\Delta_3} \\ q_3(z) &= 1 \end{aligned} \tag{16.7a}$$

which are equivalent to

$$\begin{bmatrix} q_0(z) \\ q_1(z) \\ q_2(z) \\ q_3(z) \end{bmatrix} = \frac{1}{\Delta_1\Delta_2\Delta_3} \cdot \begin{bmatrix} 1 & -(\gamma_1 + \gamma_2 + \gamma_3) & \gamma_1\gamma_2 + \gamma_2\gamma_3 + \gamma_3\gamma_1 & -\gamma_1\gamma_2\gamma_3 \\ 0 & \Delta_1 & -(\gamma_2 + \gamma_3)\Delta_1 & \gamma_2\gamma_3\Delta_1 \\ 0 & 0 & \Delta_1\Delta_2 & -\gamma_3\Delta_1\Delta_2 \\ 0 & 0 & 0 & \Delta_1\Delta_2\Delta_3 \end{bmatrix} \begin{bmatrix} z^3 \\ z^2 \\ z \\ 1 \end{bmatrix} \tag{16.7b}$$

hence

$$\boldsymbol{P} = \frac{1}{\Delta_1\Delta_2\Delta_3} \begin{bmatrix} 1 & -(\gamma_1 + \gamma_2 + \gamma_3) & \gamma_1\gamma_2 + \gamma_2\gamma_3 + \gamma_3\gamma_1 & -\gamma_1\gamma_2\gamma_3 \\ 0 & \Delta_1 & -(\gamma_2 + \gamma_3)\Delta_1 & \gamma_2\gamma_3\Delta_1 \\ 0 & 0 & \Delta_1\Delta_2 & -\gamma_3\Delta_1\Delta_2 \\ 0 & 0 & 0 & \Delta_1\Delta_2\Delta_3 \end{bmatrix} \tag{16.7c}$$

Referring to (16.7b), with the choice of $\{\rho_i(z) | \, i = 1, 2, \cdots, n\}$ in (16.5) it is possible to specify the corresponding transformation matrix $\boldsymbol{P}$ and scalar $\kappa = \Delta_1\Delta_2 \cdots \Delta_n$.

From (16.6a), it follows that

$$\frac{q_i(z)}{q_0(z)} = \frac{\rho_{i+1}(z)\rho_{i+2}(z) \cdots \rho_n(z)}{\rho_1(z)\rho_2(z) \cdots \rho_n(z)} = \rho_1^{-1}(z)\rho_2^{-1}(z) \cdots \rho_i^{-1}(z) \tag{16.8}$$

Hence the transfer function in (16.4) can be written as

$$H(z) = \frac{\beta_0 + \sum_{i=1}^{n} \beta_i \, \rho_1^{-1}(z)\rho_2^{-1}(z)\cdots\rho_i^{-1}(z)}{1 + \sum_{i=1}^{n} \alpha_i \, \rho_1^{-1}(z)\rho_2^{-1}(z)\cdots\rho_i^{-1}(z)} \tag{16.9}$$

which can be expressed in terms of its difference equation as

$$\begin{aligned} y(k) = &\left[\beta_0 + \sum_{i=1}^{n} \beta_i \, \rho_1^{-1}(z)\rho_2^{-1}(z)\cdots\rho_i^{-1}(z)\right] u(k) \\ &- \left[\sum_{i=1}^{n} \alpha_i \, \rho_1^{-1}(z)\rho_2^{-1}(z)\cdots\rho_i^{-1}(z)\right] y(k) \end{aligned} \tag{16.10}$$

The transposed direct-form II structure of a ρ operator-based IIR digital filter specified by (16.10) is illustrated in Figure 16.2, and an implementation of $\rho_i^{-1}(z)$ is shown in Figure 16.3.

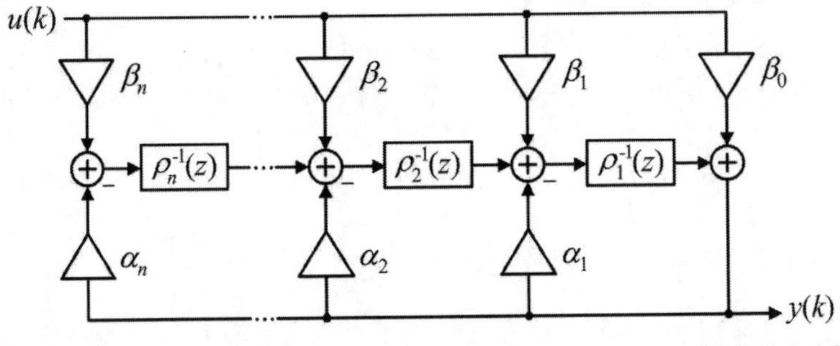

Figure 16.2 Transposed direct-form II structure of a ρ operator-based IIR digital filter.

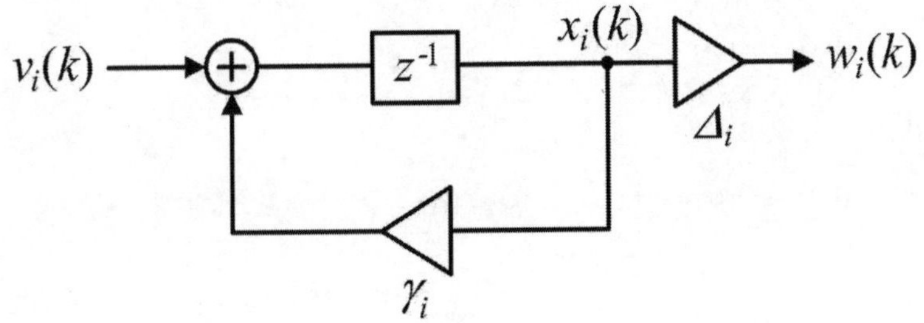

Figure 16.3 Implementation of $\rho_i^{-1}(z)$.

From Figures 16.2 and 16.3, we obtain

$$\begin{aligned} y(k) &= w_1(k) + \beta_0 u(k) \\ w_i(k) &= \rho_i^{-1}(z)\left[w_{i+1}(k) + \beta_i u(k) - \alpha_i y(k)\right] \\ &\qquad\qquad\qquad \text{for } i = 1, 2, \cdots, n-1 \\ w_n(k) &= \rho_n^{-1}(z)\left[\beta_n u(k) - \alpha_n y(k)\right], \qquad w_{n+1}(k) = 0 \end{aligned} \tag{16.11}$$

where $w_i(k)$ is the output of $\rho_i^{-1}(z)$.

It is seen that this structure has $3n+1$ nontrivial parameters $\{\alpha_i\}$, $\{\beta_i\}$ and $\{\Delta_i\}$ plus n free parameters $\{\gamma_i\}$. If $\gamma_i = 0$ and $\Delta_i = 1$ for $i = 1, 2, \cdots, n$, then Figure 16.2 becomes identical to the transposed direct form II structure [9, p. 155], while if $\gamma_i = 1$ for $i = 1, 2, \cdots, n$, then Figure 16.2 is identical to the direct-form delta operator-based filter structure [4]. With the n free parameters $\{\gamma_i\}$ one can enjoy more degrees of freedom to minimize the FWL effects.

16.3 Equivalent State-Space Realization

16.3.1 State-Space Realization I

We now derive an equivalent state-space model from (16.11). Since the operator $\rho_i^{-1}(z)$ in Figure 16.3 is expressed in terms of its transfer function as

$$\rho_i^{-1}(z) = \frac{\Delta_i}{z - \gamma_i} \quad \text{for } i = 1, 2, \cdots, n \tag{16.12}$$

we can write (16.11) as

$$\begin{aligned} y(k) &= \Delta_1 x_1(k) + \beta_0 u(k) \\ w_i(k) &= \frac{\Delta_i}{z - \gamma_i}\left[w_{i+1}(k) + \beta_i u(k) - \alpha_i\{\Delta_1 x_1(k) + \beta_0 u(k)\}\right] \\ &\qquad i = 1, 2, \cdots, n \end{aligned} \tag{16.13}$$

which leads to

$$\begin{aligned} (z - \gamma_i)x_i(k) &= -\alpha_i \Delta_1 x_1(k) + \Delta_{i+1} x_{i+1}(k) + (\beta_i - \alpha_i \beta_0)u(k) \\ &\qquad i = 1, 2, \cdots, n \end{aligned} \tag{16.14a}$$

or, equivalently,

$$x_i(k+1) = -\alpha_i \Delta_1 x_1(k) + \gamma_i x_i(k) + \Delta_{i+1} x_{i+1}(k) + (\beta_i - \alpha_i \beta_0) u(k)$$
$$i = 1, 2, \cdots, n \tag{16.14b}$$

where $x_{n+1}(k) = 0$. Using (16.14b) and the first equation in (16.13), an equivalent state-space realization $(\boldsymbol{A}_\rho, \boldsymbol{b}_\rho, \boldsymbol{c}_\rho, \beta_0)_n$ can be constructed as

$$\begin{aligned} \boldsymbol{x}(k+1) &= \boldsymbol{A}_\rho \boldsymbol{x}(k) + \boldsymbol{b}_\rho u(k) \\ y(k) &= \boldsymbol{c}_\rho \boldsymbol{x}(k) + \beta_0 u(k) \end{aligned} \tag{16.15}$$

where

$$\boldsymbol{A}_\rho = \begin{bmatrix} -\alpha_1\Delta_1 & \Delta_2 & \cdots & 0 \\ -\alpha_2\Delta_1 & 0 & \ddots & \vdots \\ \vdots & \vdots & \ddots & \Delta_n \\ -\alpha_n\Delta_1 & 0 & \cdots & 0 \end{bmatrix} + \begin{bmatrix} \gamma_1 & 0 & \cdots & 0 \\ 0 & \gamma_2 & \ddots & \vdots \\ \vdots & \ddots & \ddots & 0 \\ 0 & \cdots & 0 & \gamma_n \end{bmatrix}$$

$$= \begin{bmatrix} -\alpha_1 & 1 & \cdots & 0 \\ -\alpha_2 & 0 & \ddots & \vdots \\ \vdots & \vdots & \ddots & 1 \\ -\alpha_n & 0 & \cdots & 0 \end{bmatrix} \begin{bmatrix} \Delta_1 & 0 & \cdots & 0 \\ 0 & \Delta_2 & \ddots & \vdots \\ \vdots & \ddots & \ddots & 0 \\ 0 & \cdots & 0 & \Delta_n \end{bmatrix} + \begin{bmatrix} \gamma_1 & 0 & \cdots & 0 \\ 0 & \gamma_2 & \ddots & \vdots \\ \vdots & \ddots & \ddots & 0 \\ 0 & \cdots & 0 & \gamma_n \end{bmatrix}$$

$$\boldsymbol{b}_\rho = \boldsymbol{\beta} - \beta_0\,\boldsymbol{\alpha}, \qquad \boldsymbol{c}_\rho = \begin{bmatrix} \Delta_1 & 0 & \cdots & 0 \end{bmatrix}$$

$$\boldsymbol{x}(k) = \begin{bmatrix} x_1(k) \\ x_2(k) \\ \vdots \\ x_n(k) \end{bmatrix}, \qquad \boldsymbol{\alpha} = \begin{bmatrix} \alpha_1 \\ \alpha_2 \\ \vdots \\ \alpha_n \end{bmatrix}, \qquad \boldsymbol{\beta} = \begin{bmatrix} \beta_1 \\ \beta_2 \\ \vdots \\ \beta_n \end{bmatrix}$$

From (16.15), the transfer function from the input $u(k)$ to the output $y(k)$ is given by

$$H(z) = \boldsymbol{c}_\rho (z\boldsymbol{I}_n - \boldsymbol{A}_\rho)^{-1} \boldsymbol{b}_\rho + \beta_0 \tag{16.16}$$

16.3.2 State-Space Realization II

We now consider constructing an equivalent state-space realization in the case where $\Delta_i = 1$ for $i = 1, 2, \cdots, n$. Suppose that (16.10) is expressed with $\Delta_i = 1$ for $i = 1, 2, \cdots, n$ as

$$\begin{aligned} y(k) = & \left[\beta_0 + \sum_{i=1}^{n} \overline{\beta}_i \, \rho_1^{-1}(z)\rho_2^{-1}(z)\cdots\rho_i^{-1}(z)\right] u(k) \\ & - \left[\sum_{i=1}^{n} \overline{\alpha}_i \, \rho_1^{-1}(z)\rho_2^{-1}(z)\cdots\rho_i^{-1}(z)\right] y(k) \end{aligned} \tag{16.17a}$$

where

$$\rho_i^{-1}(z) = \frac{1}{z - \gamma_i} \quad \text{for } i = 1, 2, \cdots, n \tag{16.17b}$$

The transposed direct-form II structure of a ρ operator-based IIR digital filter in (16.17a) is depicted in Figure 16.4, and an implementation of $\rho_i^{-1}(z)$ in case $\Delta_i = 1$ is drawn in Figure 16.5.

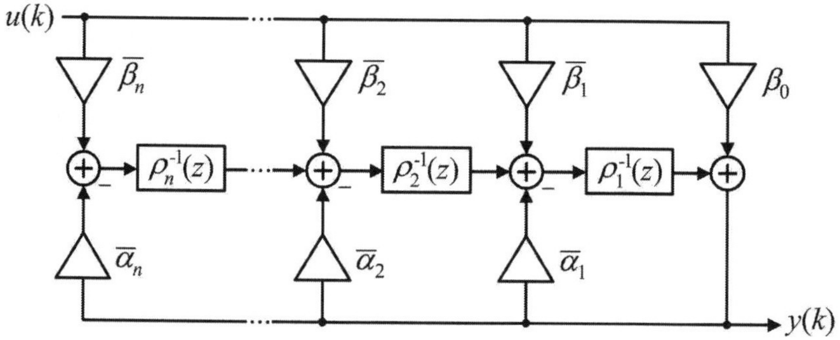

Figure 16.4 Transposed direct-form II structure of a ρ operator-based IIR digital filter.

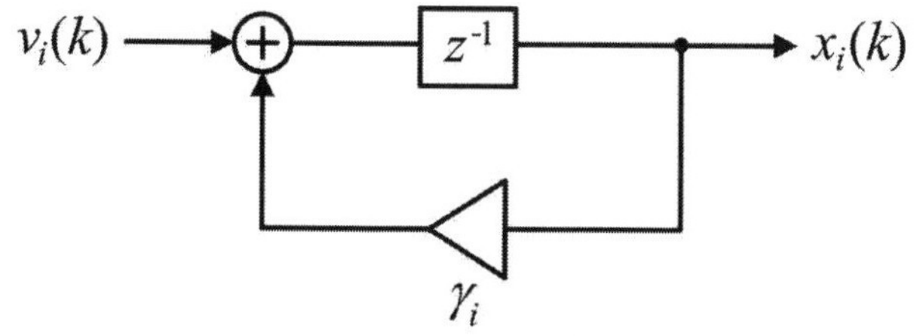

Figure 16.5 Implementation of $\rho_i^{-1}(z)$ in case $\Delta_i = 1$.

In this case, (16.13) is written as

$$
\begin{aligned}
y(k) &= x_1(k) + \beta_0 u(k) \\
x_i(k) &= \frac{1}{z-\gamma_i}\left[x_{i+1}(k) + \overline{\beta}_i u(k) - \overline{\alpha}_i\{x_1(k) + \beta_0 u(k)\}\right] \\
&\quad i = 1, 2, \cdots, n
\end{aligned}
\tag{16.18}
$$

which is equivalent to

$$
\begin{aligned}
y(k) &= x_1(k) + \beta_0 u(k) \\
x_i(k+1) &= -\overline{\alpha}_i x_1(k) + \gamma_i x_i(k) + x_{i+1}(k) + (\overline{\beta}_i - \overline{\alpha}_i \beta_0) u(k) \\
&\quad i = 1, 2, \cdots, n
\end{aligned}
\tag{16.19}
$$

where $x_{n+1}(k) = 0$. By virtue of (16.19), the corresponding equivalent state-space realization $(\overline{\boldsymbol{A}}_\rho, \overline{\boldsymbol{b}}_\rho, \overline{\boldsymbol{c}}_\rho, \beta_0)_n$ can be constructed as

$$
\begin{aligned}
\overline{\boldsymbol{x}}(k+1) &= \overline{\boldsymbol{A}}_\rho \overline{\boldsymbol{x}}(k) + \overline{\boldsymbol{b}}_\rho u(k) \\
y(k) &= \overline{\boldsymbol{c}}_\rho \overline{\boldsymbol{x}}(k) + \beta_0 u(k)
\end{aligned}
\tag{16.20}
$$

where

$$
\overline{\boldsymbol{A}}_\rho = \begin{bmatrix} -\overline{\alpha}_1 & 1 & \cdots & 0 \\ -\overline{\alpha}_2 & 0 & \ddots & \vdots \\ \vdots & \vdots & \ddots & 1 \\ -\overline{\alpha}_n & 0 & \cdots & 0 \end{bmatrix} + \begin{bmatrix} \gamma_1 & 0 & \cdots & 0 \\ 0 & \gamma_2 & \ddots & \vdots \\ \vdots & \ddots & \ddots & 0 \\ 0 & \cdots & 0 & \gamma_n \end{bmatrix}
$$

$$
\overline{\boldsymbol{b}}_\rho = \overline{\boldsymbol{\beta}} - \beta_0 \overline{\boldsymbol{\alpha}}, \qquad \overline{\boldsymbol{c}}_\rho = \begin{bmatrix} 1 & 0 & \cdots & 0 \end{bmatrix}
$$

$$
\overline{\boldsymbol{x}}(k) = \begin{bmatrix} x_1(k) \\ x_2(k) \\ \vdots \\ x_n(k) \end{bmatrix}, \qquad \overline{\boldsymbol{\alpha}} = \begin{bmatrix} \overline{\alpha}_1 \\ \overline{\alpha}_2 \\ \vdots \\ \overline{\alpha}_n \end{bmatrix}, \qquad \overline{\boldsymbol{\beta}} = \begin{bmatrix} \overline{\beta}_1 \\ \overline{\beta}_2 \\ \vdots \\ \overline{\beta}_n \end{bmatrix}
$$

If the state-space model in (16.15) is related to that in (16.20) by

$$
\overline{\boldsymbol{x}}(k) = \boldsymbol{T}^{-1}\boldsymbol{x}(k), \qquad \boldsymbol{T} = \mathrm{diag}\{t_1\, t_2\, \cdots\, t_n\}
\tag{16.21}
$$

then we obtain

$$
\overline{\boldsymbol{A}}_\rho = \boldsymbol{T}^{-1}\boldsymbol{A}_\rho\boldsymbol{T}, \qquad \overline{\boldsymbol{b}}_\rho = \boldsymbol{T}^{-1}\boldsymbol{b}_\rho, \qquad \overline{\boldsymbol{c}}_\rho = \boldsymbol{c}_\rho\boldsymbol{T}
\tag{16.22}
$$

By making use of $\overline{c}_\rho = c_\rho T$ and $T\overline{A}_\rho = A_\rho T$, it follows that

$$\begin{bmatrix} 1 & 0 & \cdots & 0 \end{bmatrix} = \begin{bmatrix} \Delta_1 t_1 & 0 & \cdots & 0 \end{bmatrix} \;:\; t_1 = \Delta_1^{-1}$$

$$\begin{bmatrix} -\overline{\alpha}_1\Delta_1^{-1} & \Delta_1^{-1} & 0 & \cdots & 0 \\ -\overline{\alpha}_2 t_2 & 0 & t_2 & \ddots & \vdots \\ \vdots & \vdots & \ddots & \ddots & 0 \\ -\overline{\alpha}_{n-1}t_{n-1} & 0 & \cdots & 0 & t_{n-1} \\ -\overline{\alpha}_n t_n & 0 & \cdots & 0 & 0 \end{bmatrix} \qquad (16.23)$$

$$= \begin{bmatrix} -\alpha_1 & \Delta_2 t_2 & 0 & \cdots & 0 \\ -\alpha_2 & 0 & \Delta_3 t_3 & \ddots & \vdots \\ \vdots & \vdots & \ddots & \ddots & 0 \\ -\alpha_{n-1} & 0 & \cdots & 0 & \Delta_n t_n \\ -\alpha_n & 0 & \cdots & 0 & 0 \end{bmatrix}$$

which yields

$$\begin{aligned} &T = \text{diag}\{\Delta_1^{-1},\ \Delta_1^{-1}\Delta_2^{-1},\ \cdots,\ \Delta_1^{-1}\Delta_2^{-1}\cdots\Delta_n^{-1}\} \\ &\overline{\alpha} = T^{-1}\alpha, \qquad \overline{\beta} = T^{-1}\beta \end{aligned} \qquad (16.24)$$

because $\overline{b}_\rho = T^{-1}b_\rho$.

16.3.3 Choice of $\{\Delta_i\}$ Satisfying l_2-Scaling Constraints

The controllability Grammian $\overline{K}_\rho$ of the state-space model in (16.20) plays an important role in the dynamic-range scaling of the state-variable vector $x(k)$ in (16.15), and matrix $\overline{K}_\rho$ can be obtained by solving the Lyapunov equation

$$\overline{K}_\rho = \overline{A}_\rho\overline{K}_\rho\overline{A}_\rho^T + \overline{b}_\rho\overline{b}_\rho^T \qquad (16.25)$$

With an equivalent state-space realization as specified in (16.22), the controllability Grammian K_ρ of the state-space model in (16.15) assumes the form

$$K_\rho = T\overline{K}_\rho T^T, \qquad T = \text{diag}\{t_1,\ t_2,\ \cdots,\ t_n\} \qquad (16.26)$$

If l_2-scaling constraints are imposed on the state-variable vector $x(k)$ in (16.15), it is required that the controllability Grammian K_ρ of the state-space model in (16.15) is subject to the constraints

$$e_i^T K_\rho e_i = e_i^T T\overline{K}_\rho T^T e_i = t_i^2 e_i^T \overline{K}_\rho e_i = 1 \ \text{ for } i = 1, 2, \cdots, n \qquad (16.27)$$

where $\boldsymbol{e}_i$ indicates an $n \times 1$ unit vector whose ith element equals unity. Consequently, noting that $\boldsymbol{T} = \text{diag}\{\Delta_1^{-1}, \Delta_1^{-1}\Delta_2^{-1}, \ldots, \Delta_1^{-1}\Delta_2^{-1}\cdots\Delta_n^{-1}\}$ in (16.24), *i.e.*, $t_i = \Delta_1^{-1}\Delta_2^{-1}\cdots\Delta_i^{-1}$ for $i = 1, 2, \cdots, n$, we obtain

$$\Delta_1 = \sqrt{\boldsymbol{e}_1^T \overline{\boldsymbol{K}}_\rho \boldsymbol{e}_1}, \quad \Delta_2 = \sqrt{\frac{\boldsymbol{e}_2^T \overline{\boldsymbol{K}}_\rho \boldsymbol{e}_2}{\boldsymbol{e}_1^T \overline{\boldsymbol{K}}_\rho \boldsymbol{e}_1}}, \ldots,$$
$$\Delta_n = \sqrt{\frac{\boldsymbol{e}_n^T \overline{\boldsymbol{K}}_\rho \boldsymbol{e}_n}{\boldsymbol{e}_{n-1}^T \overline{\boldsymbol{K}}_\rho \boldsymbol{e}_{n-1}}} \tag{16.28}$$

Therefore, if the filter under discussion is l_2-scaled, then (16.5) and (16.6a) imply that

$$q_0(z) = \prod_{l=1}^{n} \rho_l(z) = \frac{1}{\Delta_1\Delta_2\cdots\Delta_n}\prod_{l=1}^{n}(z - \gamma_l) = \frac{1}{\sqrt{\boldsymbol{e}_n^T \overline{\boldsymbol{K}}_\rho \boldsymbol{e}_n}}\prod_{l=1}^{n}(z - \gamma_l)$$
$$q_i(z) = \prod_{l=i+1}^{n} \rho_l(z) = \frac{1}{\Delta_{i+1}\Delta_{i+2}\cdots\Delta_n}\prod_{l=i+1}^{n}(z - \gamma_l) \tag{16.29}$$
$$= \sqrt{\frac{\boldsymbol{e}_i^T \overline{\boldsymbol{K}}_\rho \boldsymbol{e}_i}{\boldsymbol{e}_n^T \overline{\boldsymbol{K}}_\rho \boldsymbol{e}_n}}\prod_{l=i+1}^{n}(z - \gamma_l)$$
$$q_n(z) = 1 \quad \text{for } i = 1, 2, \cdots, n-1$$

and we have

$$\kappa = \prod_{i=1}^{n} \Delta_i = \sqrt{\boldsymbol{e}_n^T \overline{\boldsymbol{K}}_\rho \boldsymbol{e}_n} \tag{16.30}$$

The parameters $\{\Delta_i\}$ is called the *coupling coefficients* [4]. These parameters affect the dynamic range of the signals at the branch nodes, which are actually the state variables $\{x_i(k)\}$. The analysis presented above suggests that one can avoid overflow oscillations if these coupling coefficients are used to scale the state variables.

16.4 Analysis of Roundoff Noise

16.4.1 Roundoff Noise of ρ-Operator Transposed Direct-Form II Structure

In this section, we first investigate the effect of roundoff noise produced by the term $\alpha_i y(k)$ at the output. Due to the product quantization, for the actual filter implemented by a FWL machine, (16.11) is written as

$$
\begin{aligned}
\tilde{y}(k) &= \tilde{w}_1(k) + \beta_0 u(k) \\
&\vdots \\
\tilde{w}_{i-1}(k) &= \rho_{i-1}^{-1}(z)\left[\tilde{w}_i(k) + \beta_{i-1}u(k) - \alpha_{i-1}\tilde{y}(k)\right] \\
\tilde{w}_i(k) &= \rho_i^{-1}(z)\left[\tilde{w}_{i+1}(k) + \beta_i u(k) - \{\alpha_i\tilde{y}(k) + \varepsilon_i(k)\}\right] \\
\tilde{w}_{i+1}(k) &= \rho_{i+1}^{-1}(z)\left[\tilde{w}_{i+2}(k) + \beta_{i+1}u(k) - \alpha_{i+1}\tilde{y}(k)\right] \\
&\vdots
\end{aligned}
\tag{16.31}
$$

where $\tilde{y}(k)$ is the actual output, $\tilde{w}_i(k)$ is the actual signal of $w_i(k)$, and

$$\varepsilon_i(k) \stackrel{\triangle}{=} Q[\alpha_i\tilde{y}(k)] - \alpha_i\tilde{y}(k)$$

is the roundoff noise due to the quantizer $Q[\cdot]$. Subtracting (16.11) from (16.31) yields

$$
\begin{aligned}
\Delta y(k) &= \Delta w_1(k) \\
&\vdots \\
\Delta w_{i-1}(k) &= \rho_{i-1}^{-1}(z)\left[\Delta w_i(k) - \alpha_{i-1}\Delta y(k)\right] \\
\Delta w_i(k) &= \rho_i^{-1}(z)\left[\Delta w_{i+1}(k) - \alpha_i\Delta y(k) - \varepsilon_i(k)\right] \\
\Delta w_{i+1}(k) &= \rho_{i+1}^{-1}(z)\left[\Delta w_{i+2}(k) - \alpha_{i+1}\Delta y(k)\right] \\
&\vdots
\end{aligned}
\tag{16.32}
$$

where

$$\Delta y(k) = \tilde{y}(k) - y(k), \qquad \Delta w_i(k) = \tilde{w}_i(k) - w_i(k)$$

By comparing (16.32) with (16.11), it is seen that the transfer function $H_i(z)$ from $-\varepsilon_i(k)$ to $\Delta y(k)$ is given by (16.16) with $\beta_0 = 0$ and $\boldsymbol{\beta}$ replaced by $\boldsymbol{e}_i$, that is,

$$H_i(z) = \boldsymbol{c}_\rho(z\boldsymbol{I}_n - \boldsymbol{A}_\rho)^{-1}\boldsymbol{e}_i \ \text{ for } \ i = 1, 2, \cdots, n \tag{16.33}$$

Based on the model developed above, we now define the normalized noise gain in terms of $H_i(z)$ as

$$J_1(\alpha_i) = \frac{E[y(k)^2]}{E[\varepsilon_i(k)^2]} = \frac{1}{2\pi j}\oint_{|z|=1} H_i^H(z)\, H_i(z)\frac{dz}{z} \tag{16.34}$$

where $\boldsymbol{A}^H$ denotes the conjugate transpose of matrix $\boldsymbol{A}$. Substituting (16.33) into (16.34) yields

$$J_1(\alpha_i) = \boldsymbol{e}_i^T \boldsymbol{W}_\rho \boldsymbol{e}_i \ \text{ for } i = 1, 2, \cdots, n \tag{16.35}$$

where $\boldsymbol{W}_\rho$ is the observability Grammian of the state-space model in (16.15) which can be obtained by solving the Lyapunov equation

$$\boldsymbol{W}_\rho = \boldsymbol{A}_\rho^T \boldsymbol{W}_\rho \boldsymbol{A}_\rho + \boldsymbol{c}_\rho^T \boldsymbol{c}_\rho \tag{16.36}$$

Similarly, the roundoff noise gain produced by β_i can be expressed as

$$J_2(\beta_i) = \boldsymbol{e}_i^T \boldsymbol{W}_\rho \boldsymbol{e}_i \ \text{ for } i = 1, 2, \cdots, n \tag{16.37}$$

Regarding the term β_0, the first equation in (16.32) is changed to

$$\Delta y(k) = \Delta w_1(k) + \varepsilon_0(k) \tag{16.38}$$

where

$$\varepsilon_0(k) \triangleq Q[\beta_0 u(k)] - \beta_0 u(k)$$

By comparing (16.32) whose first equation is replaced by (16.38) with (16.11), it is seen that the transfer function $H_0(z)$ from $\varepsilon_0(k)$ to $\Delta y(k)$ is described by (16.16) with $\beta_0 = 1$ and $\boldsymbol{\beta} = \boldsymbol{0}$, that is,

$$H_0(z) = -\boldsymbol{c}_\rho(z\boldsymbol{I}_n - \boldsymbol{A}_\rho)^{-1}\boldsymbol{\alpha} + 1 \tag{16.39}$$

Hence the roundoff noise gain caused by β_0 is given by

$$J_3(\beta_0) = \boldsymbol{\alpha}^T \boldsymbol{W}_\rho \boldsymbol{\alpha} + 1 \tag{16.40}$$

With $w_{i+1}(k)$ replaced by $\Delta_{i+1} x_{i+1}(k)$ in (16.11), the roundoff noise gain due to Δ_{i+1} can be viewed as a function of β_i, *i.e.*,

$$J_2(\Delta_{i+1}) = J_2(\beta_i) = \boldsymbol{e}_i^T \boldsymbol{W}_\rho \boldsymbol{e}_i \ \text{ for } i = 1, 2, \cdots, n-1 \tag{16.41}$$

Similarly, $w_1(k)$ replaced by $\Delta_1 x_1(k)$ in (16.11), the roundoff noise gain due to Δ_1 can be considered as that produced by β_0, *i.e.*,

$$J_3(\Delta_1) = J_3(\beta_0) = \boldsymbol{\alpha}^T \boldsymbol{W}_\rho \boldsymbol{\alpha} + 1 \tag{16.42}$$

As shown in Figure 16.3, parameter γ_i yields a multiplication $\gamma_i x_i(k)$. This multiplication produces no roundoff noise if $\gamma_i = 0, \pm 1$. Let $\psi(\gamma_i)\omega_i(k)$ denote the roundoff noise due to γ_i, where $\psi(\gamma_i) = 1$ for all γ_i except $\gamma_i = 0, \pm 1$, for which $\psi(\gamma_i) = 0$, and let $\Delta y(k)$ be the corresponding output deviation. Then the transfer function from $\psi(\gamma_i)\omega_i(k)$ to $\Delta y(k)$ becomes $H_i(z)$ defined above. Actually, this roundoff noise can be viewed as the one generated by $\beta_i u(k)$. Hence

$$J_4(\gamma_i) = \psi(\gamma_i)J_2(\beta_i) = \psi(\gamma_i)\boldsymbol{e}_i^T\boldsymbol{W}_\rho\boldsymbol{e}_i \ \text{ for } i = 1, 2, \cdots, n \tag{16.43}$$

Based on the above analysis, the total roundoff noise gain of the filter structure in Figure 16.2 is defined as

$$\begin{aligned} J_\rho = &\sum_{i=1}^{n} J_1(\alpha_i) + \sum_{i=1}^{n} J_2(\beta_i) + J_3(\beta_0) \\ &+ \sum_{i=1}^{n-1} J_2(\Delta_{i+1}) + J_3(\Delta_1) + \sum_{i=1}^{n} J_4(\gamma_i) \end{aligned} \tag{16.44}$$

which can be written as

$$\begin{aligned} J_\rho &= \sum_{i=1}^{n} \left[3 + \psi(\gamma_i)\right]\boldsymbol{e}_i^T\boldsymbol{W}_\rho\boldsymbol{e}_i - \boldsymbol{e}_n^T\boldsymbol{W}_\rho\boldsymbol{e}_n + 2\left(\boldsymbol{\alpha}^T\boldsymbol{W}_\rho\boldsymbol{\alpha} + 1\right) \\ &= 3\,\mathrm{tr}\left[\boldsymbol{W}_\rho\right] + \mathrm{tr}\left[\boldsymbol{\Psi}\boldsymbol{W}_\rho\right] - \boldsymbol{e}_n^T\boldsymbol{W}_\rho\boldsymbol{e}_n + 2\left(\boldsymbol{\alpha}^T\boldsymbol{W}_\rho\boldsymbol{\alpha} + 1\right) \end{aligned} \tag{16.45}$$

where

$$\boldsymbol{\Psi} = \mathrm{diag}\left\{\psi(\gamma_1),\ \psi(\gamma_2),\ \cdots,\ \psi(\gamma_n)\right\}$$

16.4.2 Roundoff Noise of Equivalent State-Space Realization

Consider a stable, controllable and observable state-space model $(\boldsymbol{A}, \boldsymbol{b}, \boldsymbol{c}, d)_n$ described by

$$\begin{aligned} \boldsymbol{x}(k+1) &= \boldsymbol{A}\boldsymbol{x}(k) + \boldsymbol{b}u(k) \\ y(k) &= \boldsymbol{c}\boldsymbol{x}(k) + du(k) \end{aligned} \tag{16.46}$$

where $\boldsymbol{x}(k)$ is an $n \times 1$ state-variable vector, $u(k)$ is a scalar input, $y(k)$ is a scalar output, and $\boldsymbol{A}$, $\boldsymbol{b}$, $\boldsymbol{c}$, and d are $n \times n$, $n \times 1$, $1 \times n$, and 1×1 real constant matrices, respectively. Section 15.2.1 states that the roundoff noise gain due to product quantization associated with the $\boldsymbol{A}$, $\boldsymbol{b}$, $\boldsymbol{c}$, and d matrices in (16.46) can be expressed as

$$I_S = \mathrm{tr}[\boldsymbol{Q}\boldsymbol{W}_o] + \mu + \nu \tag{16.47a}$$

where $\boldsymbol{W}_o$ is the observability Grammian of the state-space model in (16.46), which can be obtained by solving the Lyapunov equation

$$\boldsymbol{W}_o = \boldsymbol{A}^T \boldsymbol{W}_o \boldsymbol{A} + \boldsymbol{c}^T \boldsymbol{c} \tag{16.47b}$$

$\boldsymbol{Q}$ is a diagonal matrix whose ith diagonal element q_i is the number of coefficients in the ith rows of $\boldsymbol{A}$ and $\boldsymbol{b}$ that are neither 0 nor ± 1, and $\mu + \nu$ is the number of neither 0 nor ± 1 constants in $\boldsymbol{c}$ and d.

If (16.47a) is applied to the equivalent state-space realization in (16.15), the corresponding roundoff noise gain, say $I_{S\rho}$, is given by

$$I_{S\rho} = \mathrm{tr}[\boldsymbol{Q}\boldsymbol{W}_\rho] + 2 \tag{16.48}$$

where

$$q_1 = 3, \qquad q_i = 3 + \psi(\gamma_i) \text{ for } i = 2, 3, \cdots, n-1, \qquad q_n = 2 + \psi(\gamma_n)$$

and $d = \beta_0$. Hence

$$I_{S\rho} = 3\,\mathrm{tr}\big[\boldsymbol{W}_\rho\big] + \mathrm{tr}\big[\boldsymbol{\Psi}\boldsymbol{W}_\rho\big] - \boldsymbol{e}_n^T \boldsymbol{W}_\rho \boldsymbol{e}_n + 2 - \psi(\gamma_1)\, \boldsymbol{e}_1^T \boldsymbol{W}_\rho \boldsymbol{e}_1 \tag{16.49}$$

Notice that

$$J_\rho - I_{S\rho} = 2\,\boldsymbol{\alpha}^T \boldsymbol{W}_\rho \boldsymbol{\alpha} + \psi(\gamma_1)\, \boldsymbol{e}_1^T \boldsymbol{W}_\rho \boldsymbol{e}_1 > 0 \tag{16.50}$$

This reveals that for a given set of $\{\gamma_i\}$, the equivalent state-space realization always has lower roundoff noise gain than the corresponding ρ-operator transposed direct form II structure in Figure 16.2.

16.5 Analysis of l_2-Sensitivity

In this section, the l_2-sensitivity of the transfer function $H(z)$ in (16.16) with respect to non-zero parameters is analyzed for both ρ-operator transposed direct form II structure and its equivalent state-space realization. It is noted that a mixture of l_1 and l_2 norms was employed in [4] to analyze the sensitivity, whereas only a pure l_2 norm was used in [5].

16.5.1 l_2-Sensitivity of ρ-Operator Transposed Direct-Form II Structure

Definition 16.1

Let $\boldsymbol{X}$ be an $m \times n$ real matrix and let $f(\boldsymbol{X})$ be a scalar complex function of $\boldsymbol{X}$, differentiable with respect to all entries of $\boldsymbol{X}$. The sensitivity function of $f(\boldsymbol{X})$ with respect to $\boldsymbol{X}$ is then defined as

$$\frac{\partial f(\boldsymbol{X})}{\partial \boldsymbol{X}} = \begin{bmatrix} \frac{\partial f(\boldsymbol{X})}{\partial x_{11}} & \frac{\partial f(\boldsymbol{X})}{\partial x_{12}} & \cdots & \frac{\partial f(\boldsymbol{X})}{\partial x_{1n}} \\ \frac{\partial f(\boldsymbol{X})}{\partial x_{21}} & \frac{\partial f(\boldsymbol{X})}{\partial x_{22}} & \cdots & \frac{\partial f(\boldsymbol{X})}{\partial x_{2n}} \\ \vdots & \vdots & \ddots & \vdots \\ \frac{\partial f(\boldsymbol{X})}{\partial x_{m1}} & \frac{\partial f(\boldsymbol{X})}{\partial x_{m2}} & \cdots & \frac{\partial f(\boldsymbol{X})}{\partial x_{mn}} \end{bmatrix} \tag{16.51}$$

where x_{ij} denotes the (i, j)th entry of matrix $\boldsymbol{X}$.

Definition 16.2

Let $\boldsymbol{X}(z)$ be an $m \times n$ complex matrix-valued function of a complex variable z and let $x_{pq}(z)$ be the (p, q)th entry of $\boldsymbol{X}(z)$. The l_2-norm of $\boldsymbol{X}(z)$ is then defined as

$$\begin{aligned} ||\boldsymbol{X}(z)||_2 &= \left[\frac{1}{2\pi}\int_0^{2\pi}\left(\sum_{p=1}^{m}\sum_{q=1}^{n}|x_{pq}(e^{j\omega})|^2\right)d\omega\right]^{\frac{1}{2}} \\ &= \left(\mathrm{tr}\left[\frac{1}{2\pi j}\oint_{|z|=1}\boldsymbol{X}(z)\boldsymbol{X}^H(z)\frac{dz}{z}\right]\right)^{\frac{1}{2}} \end{aligned} \tag{16.52}$$

Based on Definitions 16.1 and 16.2, the overall l_2-sensitivity measure for ρ-operator transposed direct form II structure in Figure 16.2 is defined as

$$\begin{aligned} M_\rho = &\left\|\frac{\partial H(z)}{\partial \boldsymbol{\alpha}}\right\|_2^2 + \left\|\frac{\partial H(z)}{\partial \boldsymbol{\beta}}\right\|_2^2 + \left\|\frac{\partial H(z)}{\partial \beta_0}\right\|_2^2 \\ &+ \sum_{i=1}^{n}\left\|\frac{\partial H(z)}{\partial \Delta_i}\right\|_2^2 + \sum_{i=1}^{n}\psi(\gamma_i)\left\|\frac{\partial H(z)}{\partial \gamma_i}\right\|_2^2 \end{aligned} \tag{16.53}$$

From (16.15) and (16.16), it follows that

$$\frac{\partial H(z)}{\partial \boldsymbol{\alpha}} = -H(z)(z\boldsymbol{I}_n - \boldsymbol{A}_\rho^T)^{-1}\boldsymbol{c}_\rho^T$$

$$\frac{\partial H(z)}{\partial \boldsymbol{\beta}} = (z\boldsymbol{I}_n - \boldsymbol{A}_\rho^T)^{-1}\boldsymbol{c}_\rho^T$$

$$\frac{\partial H(z)}{\partial \beta_0} = 1 - \boldsymbol{c}_\rho(z\boldsymbol{I}_n - \boldsymbol{A}_\rho)^{-1}\boldsymbol{\alpha}$$

$$\frac{\partial H(z)}{\partial \Delta_1} = \big[1 - \boldsymbol{c}_\rho (z\boldsymbol{I}_n - \boldsymbol{A}_\rho)^{-1}\boldsymbol{\alpha}\big]\boldsymbol{e}_1^T (z\boldsymbol{I}_n - \boldsymbol{A}_\rho)^{-1}\boldsymbol{b}_\rho$$

$$\frac{\partial H(z)}{\partial \Delta_i} = \boldsymbol{c}_\rho (z\boldsymbol{I}_n - \boldsymbol{A}_\rho)^{-1}\boldsymbol{e}_{i-1}\boldsymbol{e}_i^T (z\boldsymbol{I}_n - \boldsymbol{A}_\rho)^{-1}\boldsymbol{b}_\rho \ \text{ for } i = 2, 3, \cdots, n$$

$$\frac{\partial H(z)}{\partial \gamma_i} = \boldsymbol{c}_\rho (z\boldsymbol{I}_n - \boldsymbol{A}_\rho)^{-1}\boldsymbol{e}_i\boldsymbol{e}_i^T (z\boldsymbol{I}_n - \boldsymbol{A}_\rho)^{-1}\boldsymbol{b}_\rho \ \text{ for } i = 1, 2, \cdots, n \tag{16.54}$$

Since $\partial H(z)/\partial \boldsymbol{\alpha}$, $\partial H(z)/\partial \Delta_i$, and $\partial H(z)/\partial \gamma_i$ can be expressed as

$$\frac{\partial H(z)}{\partial \boldsymbol{\alpha}} = -\begin{bmatrix}\boldsymbol{I}_n & \boldsymbol{0}\end{bmatrix}\left(z\boldsymbol{I}_{2n} - \begin{bmatrix}\boldsymbol{A}_\rho^T & \boldsymbol{c}_\rho^T\boldsymbol{c}_\rho \\ \boldsymbol{0} & \boldsymbol{A}_\rho\end{bmatrix}\right)^{-1}\begin{bmatrix}\beta_0\boldsymbol{c}_\rho^T \\ \boldsymbol{b}_\rho\end{bmatrix}$$

$$\frac{\partial H(z)}{\partial \Delta_1} = \begin{bmatrix}\boldsymbol{e}_1^T & \boldsymbol{0}\end{bmatrix}\left(z\boldsymbol{I}_{2n} - \begin{bmatrix}\boldsymbol{A}_\rho & \boldsymbol{b}_\rho\boldsymbol{c}_\rho \\ \boldsymbol{0} & \boldsymbol{A}_\rho\end{bmatrix}\right)^{-1}\begin{bmatrix}\boldsymbol{b}_\rho \\ -\boldsymbol{\alpha}\end{bmatrix}$$

$$\frac{\partial H(z)}{\partial \Delta_i} = \begin{bmatrix}\boldsymbol{c}_\rho & \boldsymbol{0}\end{bmatrix}\left(z\boldsymbol{I}_{2n} - \begin{bmatrix}\boldsymbol{A}_\rho & \boldsymbol{e}_{i-1}\boldsymbol{e}_i^T \\ \boldsymbol{0} & \boldsymbol{A}_\rho\end{bmatrix}\right)^{-1}\begin{bmatrix}\boldsymbol{0} \\ \boldsymbol{b}_\rho\end{bmatrix} \ \text{ for } i = 2, 3, \cdots, n$$

$$\frac{\partial H(z)}{\partial \gamma_i} = \begin{bmatrix}\boldsymbol{e}_i^T & \boldsymbol{0}\end{bmatrix}\left(z\boldsymbol{I}_{2n} - \begin{bmatrix}\boldsymbol{A}_\rho & \boldsymbol{b}_\rho\boldsymbol{c}_\rho \\ \boldsymbol{0} & \boldsymbol{A}_\rho\end{bmatrix}\right)^{-1}\begin{bmatrix}\boldsymbol{0} \\ \boldsymbol{e}_i\end{bmatrix} \ \text{ for } i = 1, 2, \cdots, n \tag{16.55}$$

substituting (16.54) and (16.55) into (16.53) leads to

$$\begin{aligned} M_\rho &= \begin{bmatrix}\beta_0\boldsymbol{c}_\rho & \boldsymbol{b}_\rho^T\end{bmatrix}\boldsymbol{P}\begin{bmatrix}\beta_0\boldsymbol{c}_\rho^T \\ \boldsymbol{b}_\rho\end{bmatrix} + \mathrm{tr}[\boldsymbol{W}_\rho] + 1 + \boldsymbol{\alpha}^T\boldsymbol{W}_\rho\boldsymbol{\alpha} \\ &\quad + \begin{bmatrix}\boldsymbol{b}_\rho^T & -\boldsymbol{\alpha}^T\end{bmatrix}\boldsymbol{N}_1\begin{bmatrix}\boldsymbol{b}_\rho \\ -\boldsymbol{\alpha}\end{bmatrix} + \sum_{i=2}^{n}\begin{bmatrix}\boldsymbol{0} & \boldsymbol{b}_\rho^T\end{bmatrix}\boldsymbol{M}_i\begin{bmatrix}\boldsymbol{0} \\ \boldsymbol{b}_\rho\end{bmatrix} \\ &\quad + \sum_{i=1}^{n}\psi(\gamma_i)\begin{bmatrix}\boldsymbol{0} & \boldsymbol{e}_i^T\end{bmatrix}\boldsymbol{N}_i\begin{bmatrix}\boldsymbol{0} \\ \boldsymbol{e}_i\end{bmatrix} \end{aligned} \tag{16.56}$$

where $\boldsymbol{W}_\rho$ is the observability Grammian of the state-space model in (16.15), which can be obtained by solving the Lyapunov equation in (16.36). In addition, matrices $\boldsymbol{P}$, $\boldsymbol{M}_i$ and $\boldsymbol{N}_i$ can be obtained by solving the Lyapunov equations

$$
\begin{aligned}
\boldsymbol{P} &= \begin{bmatrix} \boldsymbol{A}_\rho^T & \boldsymbol{c}_\rho^T \boldsymbol{c}_\rho \\ \boldsymbol{0} & \boldsymbol{A}_\rho \end{bmatrix}^T \boldsymbol{P} \begin{bmatrix} \boldsymbol{A}_\rho^T & \boldsymbol{c}_\rho^T \boldsymbol{c}_\rho \\ \boldsymbol{0} & \boldsymbol{A}_\rho \end{bmatrix} + \begin{bmatrix} \boldsymbol{I}_n & \boldsymbol{0} \\ \boldsymbol{0} & \boldsymbol{0} \end{bmatrix} \\
\boldsymbol{M}_i &= \begin{bmatrix} \boldsymbol{A}_\rho & \boldsymbol{e}_{i-1} \boldsymbol{e}_i^T \\ \boldsymbol{0} & \boldsymbol{A}_\rho \end{bmatrix}^T \boldsymbol{M}_i \begin{bmatrix} \boldsymbol{A}_\rho & \boldsymbol{e}_{i-1} \boldsymbol{e}_i^T \\ \boldsymbol{0} & \boldsymbol{A}_\rho \end{bmatrix} + \begin{bmatrix} \boldsymbol{c}_\rho^T \boldsymbol{c}_\rho & \boldsymbol{0} \\ \boldsymbol{0} & \boldsymbol{0} \end{bmatrix} \\
&\qquad \text{for} \quad i = 2, 3, \cdots, n \\
\boldsymbol{N}_i &= \begin{bmatrix} \boldsymbol{A}_\rho & \boldsymbol{b}_\rho \boldsymbol{c}_\rho \\ \boldsymbol{0} & \boldsymbol{A}_\rho \end{bmatrix}^T \boldsymbol{N}_i \begin{bmatrix} \boldsymbol{A}_\rho & \boldsymbol{b}_\rho \boldsymbol{c}_\rho \\ \boldsymbol{0} & \boldsymbol{A}_\rho \end{bmatrix} + \begin{bmatrix} \boldsymbol{e}_i \boldsymbol{e}_i^T & \boldsymbol{0} \\ \boldsymbol{0} & \boldsymbol{0} \end{bmatrix} \\
&\qquad \text{for} \quad i = 1, 2, \cdots, n
\end{aligned}
\tag{16.57}
$$

16.5.2 l_2-Sensitivity of Equivalent State-Space Realization

(1) Improved l_2-Sensitivity Measure for General State-Space Models

The transfer function of the state-space model in (16.46) can be expressed as

$$
H(z) = \boldsymbol{c}(z\boldsymbol{I}_n - \boldsymbol{A})^{-1}\boldsymbol{b} + d \tag{16.58}
$$

where

$$
\boldsymbol{A} = \begin{bmatrix} a_{11} & a_{12} & \cdots & a_{1n} \\ a_{21} & a_{22} & \cdots & a_{2n} \\ \vdots & \vdots & \ddots & \vdots \\ a_{n1} & a_{n2} & \cdots & a_{nn} \end{bmatrix}, \qquad \boldsymbol{b} = \begin{bmatrix} b_1 \\ b_2 \\ \vdots \\ b_n \end{bmatrix}, \qquad \boldsymbol{c} = \begin{bmatrix} c_1 & c_2 & \cdots & c_n \end{bmatrix}
$$

The l_2-sensitivity measure for the state-space model in (16.46) is defined as [11]

$$
\begin{aligned}
S = &\sum_{i=1}^{n} \sum_{l=1}^{n} \frac{1}{2\pi j} \oint_{|z|=1} \left| \frac{\partial H(z)}{\partial a_{il}} \right|^2 \frac{dz}{z} + \sum_{i=1}^{n} \frac{1}{2\pi j} \oint_{|z|=1} \left| \frac{\partial H(z)}{\partial b_i} \right|^2 \frac{dz}{z} \\
&+ \sum_{l=1}^{n} \frac{1}{2\pi j} \oint_{|z|=1} \left| \frac{\partial H(z)}{\partial c_l} \right|^2 \frac{dz}{z} + \frac{1}{2\pi j} \oint_{|z|=1} \left| \frac{\partial H(z)}{\partial d} \right|^2 \frac{dz}{z}
\end{aligned}
\tag{16.59}
$$

where

$$\frac{\partial H(z)}{\partial a_{il}} = \boldsymbol{g}(z)\boldsymbol{e}_i\boldsymbol{e}_l^T\boldsymbol{f}(z), \qquad \frac{\partial H(z)}{\partial b_i} = \boldsymbol{g}(z)\boldsymbol{e}_i$$

$$\frac{\partial H(z)}{\partial c_l} = \boldsymbol{e}_l^T\boldsymbol{f}(z), \qquad \frac{\partial H(z)}{\partial d} = 1$$

with

$$\boldsymbol{f}(z) = (z\boldsymbol{I}_n - \boldsymbol{A})^{-1}\boldsymbol{b}, \qquad \boldsymbol{g}(z) = \boldsymbol{c}(z\boldsymbol{I}_n - \boldsymbol{A})^{-1}$$

Since coefficients 0 and ± 1 can be realized precisely in the implementation of FWL digital filters, the l_2-sensitivity is not affected by these coefficients. Consequently, the sensitivity of individual elements of coefficient matrices $\boldsymbol{A}$, $\boldsymbol{b}$ and $\boldsymbol{c}$ should be changed to [7]

$$\begin{aligned} &\frac{\partial H(z)}{\partial a_{il}} = \psi(a_{il})\boldsymbol{g}(z)\boldsymbol{e}_i\boldsymbol{e}_l^T\boldsymbol{f}(z), \qquad \frac{\partial H(z)}{\partial b_i} = \psi(b_i)\boldsymbol{g}(z)\boldsymbol{e}_i \\ &\frac{\partial H(z)}{\partial c_l} = \psi(c_l)\boldsymbol{e}_l^T\boldsymbol{f}(z), \qquad \frac{\partial H(z)}{\partial d} = \psi(d) \end{aligned} \tag{16.60}$$

where

$$\psi(a_{il}) = \begin{cases} 1 & \text{for} \quad a_{il} \neq 0, \pm 1 \\ 0 & \text{for} \quad a_{il} = 0, \pm 1 \end{cases}, \qquad \psi(b_i) = \begin{cases} 1 & \text{for} \quad b_i \neq 0, \pm 1 \\ 0 & \text{for} \quad b_i = 0, \pm 1 \end{cases}$$

$$\psi(c_l) = \begin{cases} 1 & \text{for} \quad c_l \neq 0, \pm 1 \\ 0 & \text{for} \quad c_l = 0, \pm 1 \end{cases}, \qquad \psi(d) = \begin{cases} 1 & \text{for} \quad d \neq 0, \pm 1 \\ 0 & \text{for} \quad d = 0, \pm 1 \end{cases}$$

Lemma 16.1
The improved l_2-sensitivity measure for the state-space model $(\boldsymbol{A}, \boldsymbol{b}, \boldsymbol{c}, d)_n$ in (16.46) is presented by [7]

$$\begin{aligned} S_I = &\sum_{i=1}^{n}\sum_{l=1}^{n} \psi(a_{il}) \left[\, \boldsymbol{c} \quad \boldsymbol{0} \,\right] \boldsymbol{R}(i,l) \begin{bmatrix} \boldsymbol{c}^T \\ \boldsymbol{0} \end{bmatrix} + \sum_{i=1}^{n} \psi(b_i) W_{ii} \\ &+ \sum_{l=1}^{n} \psi(c_l) K_{ll} + \psi(d) \end{aligned} \tag{16.61}$$

where K_{ll} for $l = 1, 2, \cdots, n$ is the (l,l)th entry of the controllability Grammian $\boldsymbol{K}_c$, W_{ii} for $i = 1, 2, \cdots, n$ is the (i,i)th entry of the observability Grammian $\boldsymbol{W}_o$, and matrices $\boldsymbol{R}(i,l)$, $\boldsymbol{K}_c$ and $\boldsymbol{W}_o$ are obtained by solving the Lyapunov equations

$$\boldsymbol{R}(i,l) = \begin{bmatrix} \boldsymbol{A} & \boldsymbol{e}_i\boldsymbol{e}_l^T \\ \boldsymbol{0} & \boldsymbol{A} \end{bmatrix} \boldsymbol{R}(i,l) \begin{bmatrix} \boldsymbol{A} & \boldsymbol{e}_i\boldsymbol{e}_l^T \\ \boldsymbol{0} & \boldsymbol{A} \end{bmatrix}^T + \begin{bmatrix} \boldsymbol{0} & \boldsymbol{0} \\ \boldsymbol{0} & \boldsymbol{b}\boldsymbol{b}^T \end{bmatrix}$$

$$\text{for } i,l = 1,2,\cdots,n$$

$$\boldsymbol{K}_c = \boldsymbol{A}\boldsymbol{K}_c\boldsymbol{A}^T + \boldsymbol{b}\boldsymbol{b}^T, \qquad \boldsymbol{W}_o = \boldsymbol{A}^T\boldsymbol{W}_o\boldsymbol{A} + \boldsymbol{c}^T\boldsymbol{c}$$

The improved l_2-sensitivity measure given in (16.61) can be modified to two novel forms so that the number of the Lyapunov equations to be solved is reduced from $n^2 + 2$ to $n + 2$. [8]

Theorem 16.1

The improved l_2-sensitivity measure in (16.61) can be expressed in the form

$$\begin{aligned} S_I = &\sum_{i=1}^{n}\sum_{l=1}^{n} \psi(a_{il}) \begin{bmatrix} \boldsymbol{e}_l^T & \boldsymbol{0} \end{bmatrix} \boldsymbol{M}(i) \begin{bmatrix} \boldsymbol{e}_l \\ \boldsymbol{0} \end{bmatrix} + \sum_{i=1}^{n} \psi(b_i) W_{ii} \\ &+ \sum_{l=1}^{n} \psi(c_l) K_{ll} + \psi(d) \end{aligned} \tag{16.62a}$$

where $\boldsymbol{M}(i)$ is obtained by solving the Lyapunov equations

$$\boldsymbol{M}(i) = \begin{bmatrix} \boldsymbol{A} & \boldsymbol{b}\boldsymbol{c} \\ \boldsymbol{0} & \boldsymbol{A} \end{bmatrix} \boldsymbol{M}(i) \begin{bmatrix} \boldsymbol{A} & \boldsymbol{b}\boldsymbol{c} \\ \boldsymbol{0} & \boldsymbol{A} \end{bmatrix}^T + \begin{bmatrix} \boldsymbol{0} & \boldsymbol{0} \\ \boldsymbol{0} & \boldsymbol{e}_i\boldsymbol{e}_i^T \end{bmatrix} \tag{16.62b}$$

$$\text{for } i = 1,2,\cdots,n$$

Proof

Noting that

$$\begin{aligned} \boldsymbol{f}(z)\boldsymbol{g}(z) &= (z\boldsymbol{I}_n - \boldsymbol{A})^{-1}\boldsymbol{b}\boldsymbol{c}(z\boldsymbol{I}_n - \boldsymbol{A})^{-1} \\ &= \begin{bmatrix} \boldsymbol{I}_n & \boldsymbol{0} \end{bmatrix} \left(z\boldsymbol{I}_{2n} - \begin{bmatrix} \boldsymbol{A} & \boldsymbol{b}\boldsymbol{c} \\ \boldsymbol{0} & \boldsymbol{A} \end{bmatrix} \right)^{-1} \begin{bmatrix} \boldsymbol{0} \\ \boldsymbol{I}_n \end{bmatrix} \end{aligned} \tag{16.63}$$

and defining $\Phi(z) = \boldsymbol{f}(z)\boldsymbol{g}(z)$, (16.60) clearly implies that

$$\begin{aligned} \frac{1}{2\pi j}\oint_{|z|=1} \left| \frac{\partial H(z)}{\partial a_{il}} \right|^2 \frac{dz}{z} &= \psi(a_{il})\, \boldsymbol{e}_l^T \left[\frac{1}{2\pi j}\oint_{|z|=1} \Phi(z)\boldsymbol{e}_i\boldsymbol{e}_i^T\Phi^T(z^{-1})\frac{dz}{z} \right] \boldsymbol{e}_l \\ &= \psi(a_{il}) \begin{bmatrix} \boldsymbol{e}_l^T & \boldsymbol{0} \end{bmatrix} \boldsymbol{M}(i) \begin{bmatrix} \boldsymbol{e}_l \\ \boldsymbol{0} \end{bmatrix} \end{aligned} \tag{16.64}$$

where

$$\boldsymbol{M}(i) = \sum_{k=0}^{\infty} \begin{bmatrix} \boldsymbol{A} & \boldsymbol{bc} \\ \boldsymbol{0} & \boldsymbol{A} \end{bmatrix}^k \begin{bmatrix} \boldsymbol{0} \\ \boldsymbol{e}_i \end{bmatrix} \begin{bmatrix} \boldsymbol{0} \\ \boldsymbol{e}_i \end{bmatrix}^T \begin{bmatrix} \boldsymbol{A}^T & \boldsymbol{0} \\ (\boldsymbol{bc})^T & \boldsymbol{A}^T \end{bmatrix}^k$$

which yields the Lyapunov equations in (16.62b), hence the proof is complete. ■

Theorem 16.2
The improved l_2-sensitivity measure in (16.61) can be modified to

$$S_I = \sum_{i=1}^{n} \sum_{l=1}^{n} \psi(a_{il}) \, [\boldsymbol{0} \;\; \boldsymbol{e}_i^T] \, \boldsymbol{N}(l) \begin{bmatrix} \boldsymbol{0} \\ \boldsymbol{e}_i \end{bmatrix} + \sum_{i=1}^{n} \psi(b_i) W_{ii} + \sum_{l=1}^{n} \psi(c_l) K_{ll} + \psi(d) \tag{16.65a}$$

where $\boldsymbol{N}(l)$ is obtained by solving the Lyapunov equations

$$\boldsymbol{N}(l) = \begin{bmatrix} \boldsymbol{A} & \boldsymbol{bc} \\ \boldsymbol{0} & \boldsymbol{A} \end{bmatrix}^T \boldsymbol{N}(l) \begin{bmatrix} \boldsymbol{A} & \boldsymbol{bc} \\ \boldsymbol{0} & \boldsymbol{A} \end{bmatrix} + \begin{bmatrix} \boldsymbol{e}_l \boldsymbol{e}_l^T & \boldsymbol{0} \\ \boldsymbol{0} & \boldsymbol{0} \end{bmatrix} \quad \text{for } l = 1, 2, \cdots, n \tag{16.65b}$$

Proof
From (16.60) and (16.63), it follows that

$$\begin{aligned} \frac{1}{2\pi j} \oint_{|z|=1} \left| \frac{\partial H(z)}{\partial a_{il}} \right|^2 \frac{dz}{z} &= \psi(a_{il}) \, \boldsymbol{e}_i^T \left[\frac{1}{2\pi j} \oint_{|z|=1} \Phi^T(z^{-1}) \boldsymbol{e}_l \boldsymbol{e}_l^T \Phi(z) \frac{dz}{z} \right] \boldsymbol{e}_i \\ &= \psi(a_{il}) \, [\boldsymbol{0} \;\; \boldsymbol{e}_i^T] \, \boldsymbol{N}(l) \begin{bmatrix} \boldsymbol{0} \\ \boldsymbol{e}_i \end{bmatrix} \end{aligned} \tag{16.66}$$

where

$$\boldsymbol{N}(l) = \sum_{k=0}^{\infty} \begin{bmatrix} \boldsymbol{A}^T & \boldsymbol{0} \\ (\boldsymbol{bc})^T & \boldsymbol{A}^T \end{bmatrix}^k \begin{bmatrix} \boldsymbol{e}_l \\ \boldsymbol{0} \end{bmatrix} \begin{bmatrix} \boldsymbol{e}_l \\ \boldsymbol{0} \end{bmatrix}^T \begin{bmatrix} \boldsymbol{A} & \boldsymbol{bc} \\ \boldsymbol{0} & \boldsymbol{A} \end{bmatrix}^k$$

which yields the Lyapunov equations in (16.65b), hence the proof is complete. ■

(2) l_2-Sensitivity Measure for Equivalent State-Space Realization

As for the implementation of an equivalent state-space realization $(\boldsymbol{A}_\rho, \boldsymbol{b}_\rho, \boldsymbol{c}_\rho, \beta_0)_n$ in (16.15), the corresponding l_2-sensitivity measure, say $S_{I\rho}$, can be evaluated directly by employing either of (16.61), (16.62a), and (16.65a) with

$$\psi(a_{il}^\rho) = \begin{cases} 1 & \text{for} \quad (i,l) = (k,1),\ k = 1,2,\cdots,n \\ 1 & \text{for} \quad (i,l) = (k,k+1),\ k = 1,2,\cdots,n-1 \\ \psi(\gamma_i) & \text{for} \quad i = l = k,\ k = 2,3,\cdots,n \\ 0 & \text{otherwise} \end{cases}$$
$$\psi(b_i^\rho) = 1 \ \text{ for } i = 1,2,\cdots,n$$
$$\psi(c_l^\rho) = \begin{cases} 1 & \text{for} \quad l = 1 \\ 0 & \text{otherwise} \end{cases} \tag{16.67}$$

16.6 Filter Synthesis

16.6.1 Computation of Roundoff Noise and l_2-Sensitivity

For given parameters $\{\gamma_i | \, i = 1,2,\cdots,n\}$, a concrete procedure for evaluating the roundoff noise gain and the l_2-sensitivity measure for both a ρ-operator transposed direct form II structure and its equivalent state-space realization is summarized as the following steps:

1. Compute $\overline{\boldsymbol{\alpha}} = [\overline{\alpha}_1,\ \overline{\alpha}_2,\ \cdots,\ \overline{\alpha}_n]^T$ and $\overline{\boldsymbol{\beta}} = [\overline{\beta}_0,\ \overline{\beta}_1,\ \cdots,\ \overline{\beta}_n]^T$ using
$$\begin{bmatrix}1 & \overline{\alpha}_1 & \cdots & \overline{\alpha}_n\end{bmatrix} \boldsymbol{P} = \begin{bmatrix}1 & a_1 & \cdots & a_n\end{bmatrix}$$
$$\begin{bmatrix}\overline{\beta}_0 & \overline{\beta}_1 & \cdots & \overline{\beta}_n\end{bmatrix} \boldsymbol{P} = \begin{bmatrix}b_0 & b_1 & \cdots & b_n\end{bmatrix} \tag{16.68}$$
for given $\{\gamma_i | \, i = 1,2,\cdots,n\}$, $\{a_i | \, i = 1,2,\cdots,n\}$ and $\{b_i | \, i = 0,1,\cdots,n\}$ where $\Delta_i = 1$ for $i = 1,2,\cdots,n$.
2. Construct the corresponding equivalent state-space realization $(\overline{\boldsymbol{A}}_\rho, \overline{\boldsymbol{b}}_\rho, \overline{\boldsymbol{c}}_\rho, \beta_0)_n$ in (16.20).
3. Obtain the controllability Grammian $\overline{\boldsymbol{K}}_\rho$ by solving the Lyapunov equation in (16.25).
4. Compute the l_2-scaling factors Δ_i for $i = 1,2,\cdots,n$ via (16.28).
5. Find the l_2-scaled $\boldsymbol{\alpha} = [\alpha_1,\ \alpha_2,\cdots,\alpha_n]^T$ and $\boldsymbol{\beta} = [\beta_0,\ \beta_1,\cdots,\beta_n]^T$ by using (16.24) and construct the l_2-scaled equivalent state-space realization $(\boldsymbol{A}_\rho, \boldsymbol{b}_\rho, \boldsymbol{c}_\rho, \beta_0)_n$ in (16.15).

6. Obtain the observability Grammian $\boldsymbol{W}_\rho$ by solving the Lyapunov equation in (16.36).
7. Compute the roundoff noise gains J_ρ and $I_{S\rho}$ via (16.45) and (16.49), respectively.
8. Calculate the l_2-sensitivity measures M_ρ and S_I (say $S_{I\rho}$) from (16.56) and (16.61) with (16.67), respectively.

16.6.2 Choice of Parameters $\{\gamma_i | \, i = 1, 2, \cdots, n\}$

It is assumed here that $|\gamma_i| \le 1$ for $i = 1, 2, \cdots, n$. For a fixed-point implementation of B_c bits, every γ_i must be truncated or rounded into a B_γ-bit number ($B_\gamma \le B_c$) of the form

$$v = \pm \sum_{p=1}^{B_\gamma} b_p 2^{-p}, \quad b_p = 0, 1 \ \forall \, p \tag{16.69}$$

unless $v = \pm 1$. Hence γ_i for $i = 1, 2, \cdots, n$ should take values within a *discrete space* defined by

$$S_\gamma = \{-1, 1\} \bigcup \left\{ \pm \sum_{p=1}^{B_\gamma} b_p 2^{-p}, \quad b_p = 0, 1 \ \forall \, p \right\} \tag{16.70}$$

which contains $(2^{B_\gamma + 1} + 1)$ elements. Hereafter, it is assumed that all parameters γ_i for $i = 1, 2, \cdots, n$ belong to S_γ, i.e., $\gamma_i \in S_\gamma \ \ \forall \, i$. As a result, these parameters do no contribute to the overall l_2-sensitivity measure. However, they do cause roundoff noise unless they are trivial, i.e., if the quantization error is generated after the product quantization for some γ_i.

16.6.3 Search of Optimal Vector $\boldsymbol{\gamma} = [\gamma_1, \ \gamma_2, \cdots, \gamma_n]^T$

Let $\overline{S}_{\boldsymbol{\gamma}} \in \boldsymbol{R}^{n \times 1}$ be the space in which $\boldsymbol{\gamma}$ takes values, that is,

$$\overline{S}_{\boldsymbol{\gamma}} = \{\boldsymbol{\gamma} \, | \, \gamma_i \in S_\gamma \ \ \forall \, i\} \tag{16.71}$$

It is obvious that J_ρ and $I_{S\rho}$ are functions of $\boldsymbol{\gamma}$, respectively. Therefore, we can consider the problem of minimizing either J_ρ or $I_{S\rho}$ with respect to vector $\boldsymbol{\gamma}$. That is

$$\begin{aligned} \boldsymbol{\gamma}\big(J_\rho^{opt}\big) &= \arg \min_{\boldsymbol{\gamma} \in \overline{S}_{\boldsymbol{\gamma}}} J_\rho \\ \boldsymbol{\gamma}\big(I_{S\rho}^{opt}\big) &= \arg \min_{\boldsymbol{\gamma} \in \overline{S}_{\boldsymbol{\gamma}}} I_{S\rho} \end{aligned} \tag{16.72}$$

Since J_ρ and $I_{S\rho}$ are highly nonlinear and nonconvex functions with respect to γ, it is very difficult to find the optimal solutions. However, the problems can be solved easily using exhaustive searching since the space $\overline{S}_\gamma$ includes $(2^{B_\gamma+1}+1)^n$ elements where n is the filter order, and B_γ is the number of bits for implementing $\{\gamma_i\}$ with 4 to 8 bits typically. Hence by repeating the procedure summarized in Section 16.6.1 a finite number of times, we can find the $\gamma(J_\rho^{opt})$ and $\gamma(I_{S\rho}^{opt})$ eventually. For comparison purposes, we define

$$\gamma_\delta = [1,\ 1,\ \cdots,\ 1]^T, \qquad \gamma_z = [0,\ 0,\ \cdots,\ 0]^T \tag{16.73}$$

Finally, we should mention that just like the roundoff noise gains J_ρ and $I_{S\rho}$, the l_2-sensitivity measures M_ρ and $S_{I\rho}$ are all function of γ, and therefore the problem of minimizing either M_ρ or $S_{I\rho}$ with respect to γ is also considered. That is

$$\begin{aligned} \gamma(M_\rho^{opt}) &= \arg \min_{\gamma \in \overline{S}_\gamma} M_\rho \\ \gamma(S_{I\rho}^{opt}) &= \arg \min_{\gamma \in \overline{S}_\gamma} S_{I\rho} \end{aligned} \tag{16.74}$$

We can solve these problems by repeating the same procedure a finite number of times and find the $\gamma(M_\rho^{opt})$ and $\gamma(S_{I\rho}^{opt})$ eventually.

16.7 Numerical Experiments

Consider a fourth Butterworth lowpass filter, with very narrow bandwidth, described by

$$H(z) = 10^{-3}\,\frac{\begin{array}{c}0.031239z^4 + 0.124956z^3 + 0.187434z^2 + 0.124956z \\ +0.031239\end{array}}{z^4 - 3.589734z^3 + 4.851276z^2 - 2.924053z + 0.663010}$$

which has a normalized bandwidth 0.025, poles $0.931900 \pm j0.136363$, $0.862967 \pm j0.052305$ and zeros -1.000226, $-1.000000 \pm j0.000226$, -0.999774. Note that the poles are clustered around $z = 1$. It is assumed here that $B_\gamma = 4$. Hence the optimal solutions of γ are found through exhaustive searching within the set $\overline{S}_\gamma$ in (16.71). Since $\gamma_i \in S_\gamma\ \forall\ i$, we set $\psi(\gamma_i) = 0$ for $i = 1,2,3,4$ to compute the l_2-sensitivity measures M_ρ and $S_{I\rho}$ from (16.56) and (16.61) with (16.67), respectively, in the simulation [5].

Suppose that $\gamma_\delta = [1,\ 1\ ,1\ ,\ 1]^T$. In this case, matrix $\boldsymbol{P}$ in (16.6b) was found to be

$$\boldsymbol{P} = \begin{bmatrix} 1 & -4 & 6 & -4 & 1 \\ 0 & 1 & -3 & 3 & -1 \\ 0 & 0 & 1 & -2 & 1 \\ 0 & 0 & 0 & 1 & -1 \\ 0 & 0 & 0 & 0 & 1 \end{bmatrix}$$

By using (16.68), $\overline{\boldsymbol{\alpha}}$ and $\overline{\boldsymbol{\beta}}$ were computed as

$$\overline{\boldsymbol{\alpha}} = \begin{bmatrix}0.410266, & 0.082074, & 0.009297, & 0.000500\end{bmatrix}^T$$

$$\overline{\boldsymbol{\beta}} = 10^{-3}\begin{bmatrix}0.031239, & 0.249912, & 0.749735, & 0.999647, & 0.499824\end{bmatrix}^T$$

where $\Delta_i = 1$ for $i = 1, 2, 3, 4$ and the equivalent state-space realization $(\overline{\boldsymbol{A}}_\rho, \overline{\boldsymbol{b}}_\rho, \overline{\boldsymbol{c}}_\rho, \beta_0)_n$ was constructed as

$$\overline{\boldsymbol{A}}_\rho = \begin{bmatrix} 0.589734 & 1 & 0 & 0 \\ -0.082074 & 1 & 1 & 0 \\ -0.009297 & 0 & 1 & 1 \\ -0.000500 & 0 & 0 & 1 \end{bmatrix}, \qquad \overline{\boldsymbol{b}}_\rho = 10^{-3}\begin{bmatrix} 0.237096 \\ 0.747172 \\ 0.999357 \\ 0.499808 \end{bmatrix}$$

$$\overline{\boldsymbol{c}}_\rho = \begin{bmatrix} 1 & 0 & 0 & 0 \end{bmatrix}, \qquad \beta_0 = 3.123898 \times 10^{-5}$$

The controllability Grammian $\overline{\boldsymbol{K}}_\rho$ was obtained by solving the Lyapunov equation in (16.25) as

$$\overline{\boldsymbol{K}}_\rho = 10^{-2}\begin{bmatrix} 5.128326 & 2.077916 & 0.358716 & 0.026271 \\ 2.077916 & 0.893925 & 0.165774 & 0.013317 \\ 0.358716 & 0.165774 & 0.033821 & 0.003126 \\ 0.026271 & 0.013317 & 0.003126 & 0.000363 \end{bmatrix}$$

and the l_2-scaling factors Δ_i for $i = 1, 2, 3, 4$ were computed via (16.28) as

$$\begin{bmatrix} \Delta_1 & \Delta_2 & \Delta_3 & \Delta_4 \end{bmatrix} = \begin{bmatrix} 0.226458, & 0.417506, & 0.194510, & 0.103586 \end{bmatrix}$$

The coordinate transformation matrix $\boldsymbol{T}$, the l_2-scaled $\boldsymbol{\alpha}$ and $\boldsymbol{\beta}$ were then found using (16.24) as

$$\boldsymbol{T} = 10^2 \,\mathrm{diag}\{0.044158,\ 0.105767,\ 0.543759,\ 5.249357\}$$

$$\boldsymbol{\alpha} = \begin{bmatrix} 1.811665 \\ 0.868073 \\ 0.505557 \\ 0.262375 \end{bmatrix}, \qquad \boldsymbol{\beta} = 10^{-1}\begin{bmatrix} 0.011036 \\ 0.079297 \\ 0.543567 \\ 2.623753 \end{bmatrix}$$

and the l_2-scaled equivalent state-space realization $(\boldsymbol{A}_\rho, \boldsymbol{b}_\rho, \boldsymbol{c}_\rho, \beta_0)_n$ was constructed from (16.15) as

$$\boldsymbol{A}_\rho = \begin{bmatrix} 0.589734 & 0.417506 & 0 & 0 \\ -0.196582 & 1 & 0.194510 & 0 \\ -0.114488 & 0 & 1 & 0.103586 \\ -0.059417 & 0 & 0 & 1 \end{bmatrix}$$

$$\boldsymbol{b}_\rho = 10^{-1} \begin{bmatrix} 0.010470 \\ 0.079026 \\ 0.543409 \\ 2.623671 \end{bmatrix}$$

$$\boldsymbol{c}_\rho = 10^{-1} \begin{bmatrix} 2.264581 & 0 & 0 & 0 \end{bmatrix}, \qquad \beta_0 = 3.123898 \times 10^{-5}$$

The controllability Grammian $\boldsymbol{K}_\rho$ was derived from (16.26) as

$$\boldsymbol{K}_\rho = \begin{bmatrix} 1.000000 & 0.970487 & 0.861329 & 0.608974 \\ 0.970487 & 1.000000 & 0.953392 & 0.739347 \\ 0.861329 & 0.953392 & 1.000000 & 0.892383 \\ 0.608974 & 0.739347 & 0.892383 & 1.000000 \end{bmatrix}$$

and the observability Grammian $\boldsymbol{W}_\rho$ was obtained by solving the Lyapunov equation in (16.36) as

$$\boldsymbol{W}_\rho = 10^{-1} \begin{bmatrix} 1.605135 & -0.335077 & -2.146692 & 0.336926 \\ -0.335077 & 4.747654 & -0.461734 & -3.803713 \\ -2.146692 & -0.461734 & 7.232309 & -0.374583 \\ 0.336926 & -3.803713 & -0.374583 & 7.526435 \end{bmatrix}$$

Consequently, the roundoff noise gains J_ρ and $I_{S\rho}$ were computed from (16.45) and (16.49) as

$$J_\rho = 8.442661, \qquad I_{S\rho} = 7.580816$$

respectively, where $\boldsymbol{\Psi}$ in (16.45) was set to $\boldsymbol{\Psi} = \mathrm{diag}\{0,\ 0,\ 0,\ 0\}$.

The l_2-sensitivity measures M_ρ and S_I (say $S_{I\rho}$) were calculated from (16.56) and (16.61) with (16.67) as

$$M_\rho = 16.528854, \qquad S_{I\rho} = 41.019991$$

respectively.

Table 16.1 Performance comparison among verious γ

γ	J_ρ^{opt}	$I_{S\rho}^{opt}$	M_ρ^{opt}	$S_{I\rho}^{opt}$
γ_δ	8.442661	7.580816	16.528854	41.019991
γ_z	8.349654×10^5	4.201743×10^5	6.333195×10^6	4.882296×10^6
$\gamma(J_\rho^{opt})$	4.512680	4.286420	8.958807	16.776958
$\gamma(I_{S\rho}^{opt})$	4.512680	4.286420	8.958807	16.776958
$\gamma(M_\rho^{opt})$	5.101317	4.681570	7.149017	16.590286
$\gamma(S_{I\rho}^{opt})$	4.567436	4.294351	7.735488	15.470568

After repeating the procedure summarized in Section 16.6.1 plenty of times, we arrived at

$$\gamma(J_\rho^{opt}) = \begin{bmatrix} 1.0000 \\ 0.8750 \\ 0.8750 \\ 0.8750 \end{bmatrix}, \quad \gamma(I_{S\rho}^{opt}) = \begin{bmatrix} 1.0000 \\ 0.8750 \\ 0.8750 \\ 0.8750 \end{bmatrix}$$

$$\gamma(M_\rho^{opt}) = \begin{bmatrix} 0.9375 \\ 0.9375 \\ 0.8750 \\ 0.9375 \end{bmatrix}, \quad \gamma(S_{I\rho}^{opt}) = \begin{bmatrix} 1.0000 \\ 0.9375 \\ 0.8750 \\ 0.8750 \end{bmatrix}$$

Detailed numerical results of applying the present technique to this example are summarized in comparison with $\boldsymbol{\gamma}_\delta = [1, 1, 1, 1]^T$ and $\boldsymbol{\gamma}_z = [0, 0, 0, 0]^T$ in Table 16.1.

16.8 Summary

In this chapter, the transposed direct-form II structure of an nth-order IIR digital filter using a set of special operators has been introduced, and its equivalent state-space realization has been constructed. Moreover, the roundoff noise and l_2-sensitivity have been analyzed for the generalized transposed direct-form II structure and its equivalent state-space realization. Given a transfer function and a set of n free parameters, a concrete procedure for evaluating the roundoff noise gain has been presented and the l_2-sensitivity measure of each structure has been addressed. Numerical experiments are presented to illustrate the validity and effectiveness of the present techniques.

References

[1] R. H. Middleton and G. C. Goodwin, "Improved finite word length characteristics in digital control using delta operators," *IEEE Trans. Autom. Control*, vol. AC-31, no. 11, pp. 1015–1021, Nov. 1986.

[2] J. Kauraniemi, T. I. Laakso, I. Hartimo and S. J. Ovaska, "Delta operator realizations of direct-form IIR filters," *IEEE Trans. Circuits Syst.-II*, vol. 45, no. 1, pp. 41–52, Jan. 1998.

[3] N. Wong and T.-S. Ng, "Roundoff noise minimization in a modified direct-form delta operator IIR structure," *IEEE Trans. Circuits Syst.-II*, vol. 47, no. 12, pp. 1533–1536, Dec. 2000.

[4] N. Wong and T.-S. Ng, "A generalized direct-form delta operator-based IIR filter with minimum noise gain and sensitivity," *IEEE Trans. Circuits Syst.-II*, vol. 48, no. 4, pp. 425–431, Apr. 2001.

[5] G. Li and Z. Zhao, "On the generalized DFIIt structure and its state-space realization in digital filter implementation," *IEEE Trans. Circuits Syst.-I*, vol. 51, no. 4, pp. 769–778, Apr. 2004.

[6] S. Y. Hwang, "Minimum uncorrelated unit noise in state-space digital filtering," *IEEE Trans. Acoust., Speech, Signal Process.*, vol. 25, no. 4, pp. 273–281, Aug. 1977.

[7] C. Xiao, "Improved L_2-sensitivity for state-space digital system," *IEEE Trans. Signal Process.*, vol. 45, no. 4, pp. 837–840, Apr. 1997.

[8] Y. Hinamoto and A. Doi, "Simplified computation of l_2-sensitivity for 1-D and a class of 2-D state-space digital filters considering 0 and ± 1 elements," Proc. *SIGMAP 2013 – Int. Conf. Signal Process. and Multimedia Appl.*, pp. 53–58, Reykjavik, Iceland, Jul. 2013.

[9] A. V. Oppenheim and R. W. Schafer, *Digital Signal Processing*, Englewood Cliffs, NJ: Prentice Hall, 1975.

[10] R. H. Middleton and G. C. Goodwin, *Digital Control and Estimation: A Unified Approach*, Englewood Cliffs, NJ: Prentice Hall, 1990.

[11] W.-Y. Yan and J. B. Moore, "On L^2-sensitivity minimization of linear state-space systems," *IEEE Trans. Circuits Syst. I*, vol. 39, no. 8, pp. 641–648, Aug. 1992.

17

Block-State Realization of IIR Digital Filters

17.1 Preview

An important issue involved in the implementation of IIR digital filters using fixed-point arithmetic is reducing roundoff noise at the filter's output. As is well known, roundoff noise is critically dependent on the internal structure of IIR digital filters. In this regard, state-space models for digital filters provide a suitable platform in which the internal structure of a filter can be explored so as to minimize its roundoff noise by choosing an appropriate linear transformation for state-space coordinates without altering the input-output characteristic of the filter. Effective techniques for constructing an optimal filter structure that minimizes the roundoff noise at the filter output subject to l_2-scaling constraints have been explored [1–3]. However, such a realization requires $(n+1)^2$ multiplications to compute each output sample for an nth-order filter, an increase of n^2 multiplications over canonical direct-form structures. This is a major disadvantage especially for high-order digital filters as its effect on data throughput rate becomes much greater when n is large. Alternatively, block-state realization of IIR digital filters has been proposed as a method of increasing data throughput rate and reducing finite-word-length (FWL) effects [4–6]. In the block-state realization, we implement single-input/single-output (SISO) state-space model by dividing the scalar input data stream into data vectors of length L, process the data vectors with an L-input/L-output state-space model, say (L, L) system, and then reconstruct the scalar output stream from the processed data vectors [5]. At the input of an (L, L) system, a serial-in/parallel-out register converts the scalar input into a vector input. At the output of the (L, L) system, a parallel-in/serial-out register converts the vector output into a scalar output. Obviously, the scalar sample throughput rate is L times the fundamental clock rate of the (L, L) system. It has been shown [6] that for an nth-order filter, the optimal blocklength is $L = \sqrt{2n}$ which is noninteger and requires rounding to the closest integer. This optimal block length minimizes average number of multiplications needed to compute each output sample to $(2+\sqrt{2})n+1/2$ [6].

Another important and related issue is reducing the effects of quatizing filter coefficients. It is of importance to note that coefficient quantization usually alters filter characteristics. For instance, a stable filter designed under the assumption of infinite precision may become unstable after coefficient quantization. This has motivated the study of coefficient sensitivity and its minimization for digital filters. Several techniques has also been explored to analyze l_2-sensitivity and to synthesize the state-space filter structure that minimizes l_2-sensitivity for digital filters [7, 8]. In addition, the minimization problem of l_2-sensitivity subject to l_2-scaling constraints has been treated for state-space digital filters [9–11]. It is known that the use of scaling constraints can be beneficial for suppressing overflow oscillations [2, 3].

In this chapter, we shall consider the block-state realization for a given SISO state-space model, and examine some of the properties of the block-state realization. Second, we analyze the roundoff noise in the block-state realization and minimize the average roundoff noise gain subject to l_2-scaling constraints. Third, we present a quantitative analysis on l_2-sensitivity for the block-state realization of state-space digital filters. Following the analysis, we study two techniques for minimizing a sensitivity measure known as average l_2-sensitivity subject to l_2-scaling constraints for the block-state realization of state-space digital filters. One of the techniques is based on a Lagrange function, while the other relies on an efficient quasi-Newton algorithm. Finally, numerical experiments are presented to demonstrate the validity and effectiveness of the techniques addressed in this chapter.

17.2 Block-State Realization

Consider a single-input/single-output (SISO) state-space model $(\boldsymbol{A}, \boldsymbol{b}, \boldsymbol{c}, d)_n$ described by

$$\begin{aligned} \boldsymbol{x}(k+1) &= \boldsymbol{A}\boldsymbol{x}(k) + \boldsymbol{b}u(k) \\ y(k) &= \boldsymbol{c}\boldsymbol{x}(k) + du(k) \end{aligned} \tag{17.1}$$

where $\boldsymbol{x}(k)$ is an $n \times 1$ state-variable vector, $u(k)$ is a scalar input, $y(k)$ is a scalar output, and $\boldsymbol{A}$, $\boldsymbol{b}$, $\boldsymbol{c}$, and d are $n \times n$, $n \times 1$, $1 \times n$, and 1×1 real constant matrices, respectively. The SISO state-space model in (17.1) is assumed to be stable, controllable and observable. A block-diagram of the state-space model in (17.1) is depicted in Figure 17.1. Note that in the most pessimistic case where all elements of coefficient matrices in (17.1) are nontrivial, $(n+1)^2$

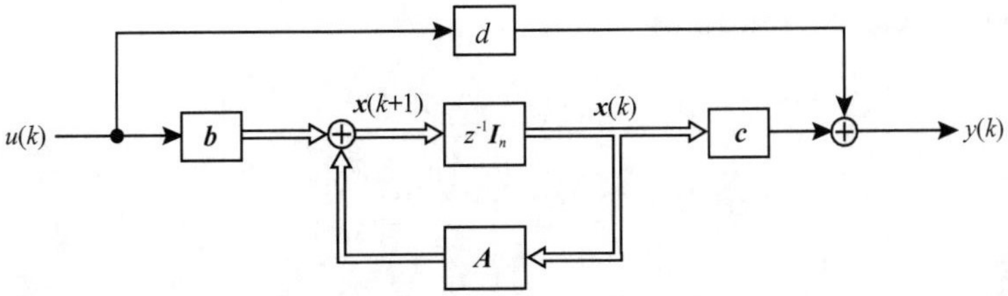

Figure 17.1 A state-space model.

multiplications are required to compute the output $y(k)$ during each sample interval.

From (17.1), it follows that for some integer $L > 0$,

$$
\begin{aligned}
\boldsymbol{x}(kL+i) &= \boldsymbol{A}^i\boldsymbol{x}(kL) + \boldsymbol{A}^{i-1}\boldsymbol{b}u(kL) + \cdots \\
&\quad + \boldsymbol{A}\boldsymbol{b}u(kL+i-2) + \boldsymbol{b}u(kL+i-1) \\
y(kL+i) &= \boldsymbol{c}\boldsymbol{A}^i\boldsymbol{x}(kL) + \boldsymbol{c}\boldsymbol{A}^{i-1}\boldsymbol{b}u(kL) + \cdots \\
&\quad + \boldsymbol{c}\boldsymbol{b}u(kL+i-1) + du(kL+i) \\
&\qquad i = 0, 1, \cdots, L-1
\end{aligned}
\tag{17.2}
$$

which leads to

$$
\begin{aligned}
\boldsymbol{x}(kL+L) &= \boldsymbol{A}^L\boldsymbol{x}(kL) \\
&\quad + \begin{bmatrix} \boldsymbol{A}^{L-1}\boldsymbol{b} & \boldsymbol{A}^{L-2}\boldsymbol{b} & \cdots & \boldsymbol{b} \end{bmatrix} \begin{bmatrix} u(kL) \\ u(kL+1) \\ \vdots \\ u(kL+L-1) \end{bmatrix}
\end{aligned}
$$

$$
\begin{aligned}
\begin{bmatrix} y(kL) \\ y(kL+1) \\ \vdots \\ y(kL+L-1) \end{bmatrix} &= \begin{bmatrix} \boldsymbol{c} \\ \boldsymbol{c}\boldsymbol{A} \\ \vdots \\ \boldsymbol{c}\boldsymbol{A}^{L-1} \end{bmatrix} \boldsymbol{x}(kL) \\
&\quad + \begin{bmatrix} d & 0 & \cdots & 0 \\ \boldsymbol{c}\boldsymbol{b} & d & \ddots & \vdots \\ \vdots & \ddots & \ddots & 0 \\ \boldsymbol{c}\boldsymbol{A}^{L-2}\boldsymbol{b} & \cdots & \boldsymbol{c}\boldsymbol{b} & d \end{bmatrix} \begin{bmatrix} u(kL) \\ u(kL+1) \\ \vdots \\ u(kL+L-1) \end{bmatrix}
\end{aligned}
\tag{17.3}
$$

Equation (17.3) can be written as

$$\begin{aligned}\hat{\boldsymbol{x}}(k+1) &= \hat{\boldsymbol{A}}\hat{\boldsymbol{x}}(k) + \hat{\boldsymbol{B}}\boldsymbol{u}(k)\\ \hat{\boldsymbol{y}}(k) &= \hat{\boldsymbol{C}}\hat{\boldsymbol{x}}(k) + \hat{\boldsymbol{D}}\boldsymbol{u}(k)\end{aligned} \tag{17.4a}$$

where

$$\hat{\boldsymbol{x}}(k) = \boldsymbol{x}(kL), \quad \boldsymbol{u}(k) = \begin{bmatrix} u(kL) \\ u(kL+1) \\ \vdots \\ u(kL+L-1) \end{bmatrix}, \quad \hat{\boldsymbol{y}}(k) = \begin{bmatrix} y(kL) \\ y(kL+1) \\ \vdots \\ y(kL+L-1) \end{bmatrix}$$

$$\hat{\boldsymbol{A}} = \boldsymbol{A}^L, \qquad \hat{\boldsymbol{B}} = \begin{bmatrix} \boldsymbol{A}^{L-1}\boldsymbol{b} & \boldsymbol{A}^{L-2}\boldsymbol{b} & \cdots & \boldsymbol{b} \end{bmatrix}$$

$$\hat{\boldsymbol{C}} = \begin{bmatrix} \boldsymbol{c} \\ \boldsymbol{c}\boldsymbol{A} \\ \vdots \\ \boldsymbol{c}\boldsymbol{A}^{L-1} \end{bmatrix}, \qquad \hat{\boldsymbol{D}} = \begin{bmatrix} d & 0 & \cdots & 0 \\ \boldsymbol{c}\boldsymbol{b} & d & \ddots & \vdots \\ \vdots & \ddots & \ddots & 0 \\ \boldsymbol{c}\boldsymbol{A}^{L-2}\boldsymbol{b} & \cdots & \boldsymbol{c}\boldsymbol{b} & d \end{bmatrix} \tag{17.4b}$$

From (17.4a), it follows that a total of

$$m = n^2 + 2nL + \frac{L(L+1)}{2} \tag{17.5}$$

multiplications are required in each block of L output samples or, equivalently, an average of

$$\frac{m}{L} = \frac{n^2}{L} + 2n + \frac{L+1}{2} \tag{17.6}$$

multiplications for each output sample. If the average measure, m/L, in (17.6) is minimized with respect to L, then the optimal blocklength is obtained as

$$L = \sqrt{2}n \tag{17.7}$$

which is noninteger and requires rounding to the closest integer. By substituting (17.7) into (17.6), the minimum value of m/L becomes

$$\left(\frac{m}{L}\right)_{min} = (2+\sqrt{2})n + \frac{1}{2} \tag{17.8}$$

This result compares favorably with the processing complexity for the canonical forms of the state-space model in (17.1), which require $(2n + 1)$ multiplications per output sample, and reveals that the system in (17.4a) enables us to perform fast processing for high-order digital filters.

In the rest of this chapter, the L-input/L-output state-space model described by (17.4a) is referred to as *block-state realization*, $(\hat{\boldsymbol{A}}, \hat{\boldsymbol{B}}, \hat{\boldsymbol{C}}, \hat{\boldsymbol{D}})_n$, that is generated by the SISO state-space model $(\boldsymbol{A}, \boldsymbol{b}, \boldsymbol{c}, d)_n$ in (17.1). In this way, (17.4b) defines a mapping $(\boldsymbol{A}, \boldsymbol{b}, \boldsymbol{c}, d)_n \longrightarrow (\hat{\boldsymbol{A}}, \hat{\boldsymbol{B}}, \hat{\boldsymbol{C}}, \hat{\boldsymbol{D}})_n$ that transforms the state-space model in (17.1) to the block-state realization in (17.4a). This block-state realization corresponds to the time-invariant case for the block-state realization of periodically time-varying digital filter, that was derived by Meyer and Burrus [12].

The system in (17.4a) can be realized by an L-input/L-output IIR digital filter implemented with serial-in/parallel-out and parallel-in/serial-out registers, as shown in Figure 17.2. It is obvious that the serial input and output sample rates are L times the fundamental clock rate of the L-input/L-output system. The internal structure of the block-state realization for block length of $L = 3$ is illustrated by the flow graph in Figure 17.3.

We now examine some of the properties of the mapping $(\boldsymbol{A}, \boldsymbol{b}, \boldsymbol{c}, d)_n \rightarrow (\hat{\boldsymbol{A}}, \hat{\boldsymbol{B}}, \hat{\boldsymbol{C}}, \hat{\boldsymbol{D}})_n$, which is defined by (17.4b).

Lemma 17.1

Since $\hat{\boldsymbol{A}} = \boldsymbol{A}^L$, the eigenvalues of matrix $\hat{\boldsymbol{A}}$ are the Lth power of those of matrix $\boldsymbol{A}$, i.e., the poles of the block-state realization are the Lth power of the poles of the associated SISO state-space model.

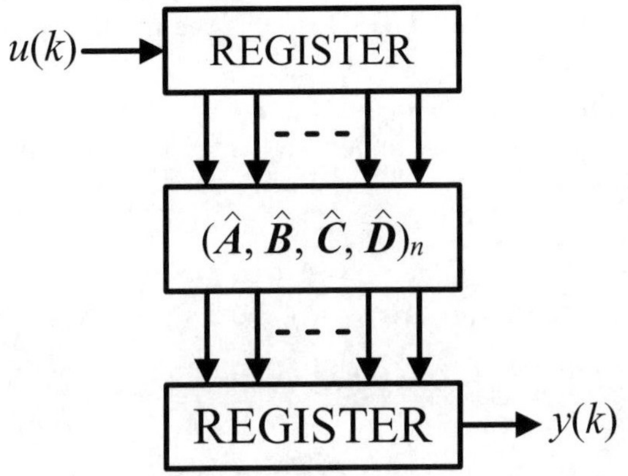

Figure 17.2 Block-state realization using serial-in/parallel-out and parallel-in/serial-out registers.

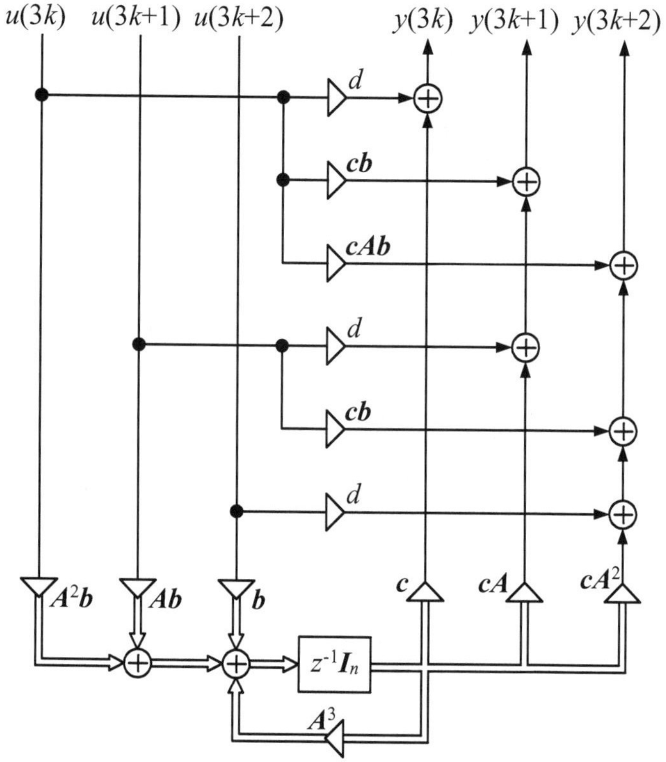

Figure 17.3 Flow graph structure of a block-state realization for block length of three.

Lemma 17.2
Since $\hat{\boldsymbol{A}} = \boldsymbol{A}^L$, the dimension of the state space of the block-state realization is the same as that of the associated SISO state-space model.

Theorem 17.1: *Controllability Invariance*
The controllability is invariant under the mapping $(\boldsymbol{A}, \boldsymbol{b}, \boldsymbol{c}, d)_n \rightarrow (\hat{\boldsymbol{A}}, \hat{\boldsymbol{B}}, \hat{\boldsymbol{C}}, \hat{\boldsymbol{D}})_n$.

Proof
Let the controllability matrix for $(\boldsymbol{A}, \boldsymbol{b}, \boldsymbol{c}, d)_n$ be denoted by

$$\boldsymbol{V}_n = \begin{bmatrix} \boldsymbol{b} & \boldsymbol{A}\boldsymbol{b} & \cdots & \boldsymbol{A}^{n-1}\boldsymbol{b} \end{bmatrix}$$

and let the controllability matrix for $(\hat{A}, \hat{B}, \hat{C}, \hat{D})_n$ be denoted by

$$\hat{V}_n = \begin{bmatrix} \hat{B} & \hat{A}\hat{B} & \cdots & \hat{A}^{n-1}\hat{B} \end{bmatrix}$$

Applying the Cayley-Hamilton theorem to matrix A, all columns of $\hat{V}_n$ can be expressed as linear combinations of the columns of V_n, i.e.,

$$\text{rank}\,[\hat{V}_n] \leq \text{rank}\,[V_n] \tag{17.9}$$

Alternatively, all the columns of V_n are included among the columns of $\hat{V}_n$, i.e.,

$$\text{rank}\,[V_n] \leq \text{rank}\,[\hat{V}_n] \tag{17.10}$$

Hence

$$\text{rank}\,[V_n] = \text{rank}\,[\hat{V}_n] \tag{17.11}$$

that completes the proof of the theorem. ■

Theorem 17.2: *Observability Invariance*
The observability is invariant under the mapping $(A, b, c, d)_n \rightarrow (\hat{A}, \hat{B}, \hat{C}, \hat{D})_n$.

Proof
Due to the duality, the proof of this theorem is essentially the same as that of Theorem 17.1. ■

Corollary 17.1: *Irreducibility Invariance*
$(\hat{A}, \hat{B}, \hat{C}, \hat{D})_n$ is the minimal realization if and only if the associated SISO state-space model $(A, b, c, d)_n$ is minimal, i.e., controllable and observable.

Theorem 17.3: *Equivalence Invariance*
$(T^{-1}AT, T^{-1}b, cT, d)_n \rightarrow (T^{-1}\hat{A}T, T^{-1}\hat{B}, \hat{C}T, \hat{D})_n$ if and only if $(A, b, c, d)_n \rightarrow (\hat{A}, \hat{B}, \hat{C}, \hat{D})_n$ where T is an $n \times n$ nonsingular matrix.

Proof
This proof follows directly from the definition of the mapping $(A, b, c, d)_n \rightarrow (\hat{A}, \hat{B}, \hat{C}, \hat{D})_n$ given by (17.4b). That is, $\hat{A} = A^L$ if and only if $T^{-1}\hat{A}T = T^{-1}A^LT = (T^{-1}AT)^L$, etc. ■

Theorem 17.4: *Controllability Grammian Invariance*
The controllability Grammian is invariant under the mapping $(\boldsymbol{A}, \boldsymbol{b}, \boldsymbol{c}, d)_n \rightarrow (\hat{\boldsymbol{A}}, \hat{\boldsymbol{B}}, \hat{\boldsymbol{C}}, \hat{\boldsymbol{D}})_n$.

Proof
Let the controllability Grammians for $(\boldsymbol{A}, \boldsymbol{b}, \boldsymbol{c}, d)_n$ and $(\hat{\boldsymbol{A}}, \hat{\boldsymbol{B}}, \hat{\boldsymbol{C}}, \hat{\boldsymbol{D}})_n$ be denoted by

$$\boldsymbol{K}_c = \sum_{k=0}^{\infty} \boldsymbol{A}^k \boldsymbol{b} (\boldsymbol{A}^k \boldsymbol{b})^T \quad \text{and} \quad \hat{\boldsymbol{K}}_c = \sum_{k=0}^{\infty} \hat{\boldsymbol{A}}^k \hat{\boldsymbol{B}} (\hat{\boldsymbol{A}}^k \hat{\boldsymbol{B}})^T$$

respectively. Then it follows that

$$\begin{aligned} \hat{\boldsymbol{K}}_c &= \hat{\boldsymbol{B}} \hat{\boldsymbol{B}}^T + \hat{\boldsymbol{A}} \hat{\boldsymbol{B}} (\hat{\boldsymbol{A}} \hat{\boldsymbol{B}})^T + \hat{\boldsymbol{A}}^2 \hat{\boldsymbol{B}} (\hat{\boldsymbol{A}}^2 \hat{\boldsymbol{B}})^T + \cdots \\ &= \sum_{k=0}^{\infty} \boldsymbol{A}^k \boldsymbol{b} (\boldsymbol{A}^k \boldsymbol{b})^T = \boldsymbol{K}_c \end{aligned} \tag{17.12}$$

where

$$\begin{aligned} \hat{\boldsymbol{B}} &= \begin{bmatrix} \boldsymbol{A}^{L-1}\boldsymbol{b} & \boldsymbol{A}^{L-2}\boldsymbol{b} & \cdots & \boldsymbol{A}\boldsymbol{b} & \boldsymbol{b} \end{bmatrix} \\ \hat{\boldsymbol{A}}\hat{\boldsymbol{B}} &= \begin{bmatrix} \boldsymbol{A}^{2L-1}\boldsymbol{b} & \boldsymbol{A}^{2L-2}\boldsymbol{b} & \cdots & \boldsymbol{A}^{L+1}\boldsymbol{b} & \boldsymbol{A}^{L}\boldsymbol{b} \end{bmatrix} \\ \hat{\boldsymbol{A}}^2\hat{\boldsymbol{B}} &= \begin{bmatrix} \boldsymbol{A}^{3L-1}\boldsymbol{b} & \boldsymbol{A}^{3L-2}\boldsymbol{b} & \cdots & \boldsymbol{A}^{2L+1}\boldsymbol{b} & \boldsymbol{A}^{2L}\boldsymbol{b} \end{bmatrix} \\ &\vdots \end{aligned}$$

This completes the proof of the theorem. ■

Theorem 17.5: *Observability Grammian Invariance*
The observability Grammian is invariant under the mapping $(\boldsymbol{A}, \boldsymbol{b}, \boldsymbol{c}, d)_n \rightarrow (\hat{\boldsymbol{A}}, \hat{\boldsymbol{B}}, \hat{\boldsymbol{C}}, \hat{\boldsymbol{D}})_n$.

Proof
Suppose that the observability Grammians for $(\boldsymbol{A}, \boldsymbol{b}, \boldsymbol{c}, d)_n$ and $(\hat{\boldsymbol{A}}, \hat{\boldsymbol{B}}, \hat{\boldsymbol{C}}, \hat{\boldsymbol{D}})_n$ are denoted by

$$\boldsymbol{W}_o = \sum_{k=0}^{\infty} (\boldsymbol{c}\boldsymbol{A}^k)^T \boldsymbol{c}\boldsymbol{A}^k \quad \text{and} \quad \hat{\boldsymbol{W}}_o = \sum_{k=0}^{\infty} (\hat{\boldsymbol{C}}\hat{\boldsymbol{A}}^k)^T \hat{\boldsymbol{C}}\hat{\boldsymbol{A}}^k$$

respectively. Then it follows that

$$\begin{aligned}\hat{\boldsymbol{W}}_o &= \hat{\boldsymbol{C}}^T\hat{\boldsymbol{C}} + (\hat{\boldsymbol{C}}\hat{\boldsymbol{A}})^T\hat{\boldsymbol{C}}\hat{\boldsymbol{A}} + (\hat{\boldsymbol{C}}\hat{\boldsymbol{A}}^2)^T\hat{\boldsymbol{C}}\hat{\boldsymbol{A}}^2 + \cdots \\ &= \sum_{k=0}^{\infty}(\boldsymbol{c}\boldsymbol{A}^k)^T\boldsymbol{c}\boldsymbol{A}^k = \boldsymbol{W}_o\end{aligned} \tag{17.13}$$

where

$$\hat{\boldsymbol{C}} = \begin{bmatrix}\boldsymbol{c}\\ \boldsymbol{c}\boldsymbol{A}\\ \vdots\\ \boldsymbol{c}\boldsymbol{A}^{L-1}\end{bmatrix}, \quad \hat{\boldsymbol{C}}\hat{\boldsymbol{A}} = \begin{bmatrix}\boldsymbol{c}\boldsymbol{A}^{L}\\ \boldsymbol{c}\boldsymbol{A}^{L+1}\\ \vdots\\ \boldsymbol{c}\boldsymbol{A}^{2L-1}\end{bmatrix}, \quad \hat{\boldsymbol{C}}\hat{\boldsymbol{A}}^2 = \begin{bmatrix}\boldsymbol{c}\boldsymbol{A}^{2L}\\ \boldsymbol{c}\boldsymbol{A}^{2L+1}\\ \vdots\\ \boldsymbol{c}\boldsymbol{A}^{3L-1}\end{bmatrix}, \quad \cdots$$

This completes the proof of the theorem. ■

17.3 Roundoff Noise Analysis and Minimization

17.3.1 Roundoff Noise Analysis

In what follows, we use the following symbols.

- σ^2 : Variance of the noise generated by a single scalar roundoff operation.
- σ_y^2 : Variance of the roundoff noise in the output of a SISO state-space model.
- $\hat{\sigma}_y^2$: Variance of the roundoff noise in the single output of a block-state realization.
- $\sigma_{\hat{y}_i}^2$: Variance of the roundoff noise in the ith output of the L-input/L-output system in the block-state realization.

It is assumed that roundoff is carried out only at the outputs of state variable summing nodes, and at the outputs of the summing nodes at the filter's output. The effects of roundoff noise is modelled as stationary white noises $\boldsymbol{w}(k)$ and $\boldsymbol{v}(k)$ with zero mean and covariance matrices $\sigma^2\boldsymbol{I}_n$ and $\sigma^2\boldsymbol{I}_L$, respectively, and these noise sources are introduced into the state and output equations as

$$\begin{aligned}\hat{\boldsymbol{x}}(k+1) &= \hat{\boldsymbol{A}}\hat{\boldsymbol{x}}(k) + \hat{\boldsymbol{B}}\boldsymbol{u}(k) + \boldsymbol{w}(k)\\ \hat{\boldsymbol{y}}(k) &= \hat{\boldsymbol{C}}\hat{\boldsymbol{x}}(k) + \hat{\boldsymbol{D}}\boldsymbol{u}(k) + \boldsymbol{v}(k)\end{aligned} \tag{17.14}$$

As a result, the autocorrelation matrix of the vector output $\hat{\boldsymbol{y}}(k)$ can be written as

$$\begin{aligned}\boldsymbol{R}_{\hat{\boldsymbol{y}}}(l) &= E\{\hat{\boldsymbol{y}}(k+l)\hat{\boldsymbol{y}}(k)^T\}\\ &= \left\{\sum_{k=0}^{\infty} \hat{\boldsymbol{C}}\hat{\boldsymbol{A}}^{k+l}(\hat{\boldsymbol{C}}\hat{\boldsymbol{A}}^{k})^T + \boldsymbol{I}_L\delta(l)\right\}\sigma^2\end{aligned} \tag{17.15}$$

whose elements are given by

$$\left[\boldsymbol{R}_{\hat{\boldsymbol{y}}}(l)\right]_{i,j} = \left\{\sum_{k=0}^{\infty} \boldsymbol{c}\boldsymbol{A}^{(k+l)L+i-1}(\boldsymbol{c}\boldsymbol{A}^{kL+j-1})^T + \delta(i-j)\right\}\sigma^2 \tag{17.16}$$

for $i, j = 1, 2, \cdots, L$ where $l \geq 0$. Hence, the variances associated with individual outputs can be expressed as

$$\sigma_{\hat{y}i}^2 = \left[\boldsymbol{R}_{\hat{\boldsymbol{y}}}(0)\right]_{i,i} = \left\{\sum_{k=0}^{\infty} \boldsymbol{c}\boldsymbol{A}^{kL+i-1}(\boldsymbol{c}\boldsymbol{A}^{kL+i-1})^T + 1\right\}\sigma^2 \tag{17.17}$$

for $i = 1, 2, \cdots, L$.

When the individual outputs are combined by a parallel-in/serial-output register to form a single output, the resulting noise will no longer be stationary. The autocorrelation function of the single output can be written as

$$R_y(kL+i-1, lL+j-1) = E\{y(kL+i-1)y(lL+j-1)\} = \left[\boldsymbol{R}_{\hat{\boldsymbol{y}}}(k-l)\right]_{i,j} \tag{17.18}$$

for $i = 1, 2, \cdots, L$. Hence the variance of the nonstationary output noise is given by

$$\hat{\sigma}_y^2(kL+i-1) = \sigma_{\hat{y}i}^2 \tag{17.19}$$

for $i = 1, 2, \cdots, L$. From (17.19), it is observed that the noise variance is periodic with period L.

We now examine the relationship between the roundoff noise in the block-state realization and the roundoff noise in the associated SISO state-space model.

For the SISO state-space model $(\boldsymbol{A}, \boldsymbol{b}, \boldsymbol{c}, d)_n$ in (17.1), the roundoff noise is given by [1]

$$\begin{aligned}\sigma_y^2 &= \left\{\sum_{k=0}^{\infty} \boldsymbol{c}\boldsymbol{A}^k(\boldsymbol{c}\boldsymbol{A}^k)^T + 1\right\}\sigma^2\\ &= \left\{\mathrm{tr}[\boldsymbol{W}_o] + 1\right\}\sigma^2\end{aligned} \tag{17.20}$$

where $\boldsymbol{W}_o$ is the observability Grammian of the SISO state-space model in (17.1), which can be obtained by solving the Lyapunov equation

$$\boldsymbol{W}_o = \boldsymbol{A}^T \boldsymbol{W}_o \boldsymbol{A} + \boldsymbol{c}^T \boldsymbol{c}$$

We remark that the formula in (17.20) can also be obtained from (17.17) by setting $L = 1$. Noting that

$$\boldsymbol{c}\boldsymbol{A}^k (\boldsymbol{c}\boldsymbol{A}^k)^T \geq 0$$

$$\sum_{k=0}^{\infty} \boldsymbol{c}\boldsymbol{A}^{kL+i-1} (\boldsymbol{c}\boldsymbol{A}^{kL+i-1})^T = \boldsymbol{c}\boldsymbol{A}^{i-1} (\boldsymbol{c}\boldsymbol{A}^{i-1})^T + \boldsymbol{c}\boldsymbol{A}^{L+i-1} (\boldsymbol{c}\boldsymbol{A}^{L+i-1})^T + \boldsymbol{c}\boldsymbol{A}^{2L+i-1} (\boldsymbol{c}\boldsymbol{A}^{2L+i-1})^T + \cdots \tag{17.21}$$

it follows that

$$\sum_{k=0}^{\infty} \boldsymbol{c}\boldsymbol{A}^{kL+i-1} (\boldsymbol{c}\boldsymbol{A}^{kL+i-1})^T \leq \sum_{k=0}^{\infty} \boldsymbol{c}\boldsymbol{A}^k (\boldsymbol{c}\boldsymbol{A}^k)^T \tag{17.22}$$

for $i = 1, 2, \cdots, L$. This in conjunction with (17.17), (17.19), (17.20) and (17.22) implies that

$$\hat{\sigma}_y^2(kL + i - 1) \leq \sigma_y^2 \tag{17.23}$$

holds for any k and $i = 1, 2, \cdots, L$, or equivalently,

$$\hat{\sigma}_y^2(k) \leq \sigma_y^2 \tag{17.24}$$

holds for all k. This shows that the variance of the roundoff noise for block-state realization never exceeds that of the associated SISO state-space model.

Let the average roundoff noise (averaged over one block period) in the output of a block-state realization be defined from (17.19) by

$$(\hat{\sigma}_y^2)_{ave} = \frac{1}{L} \sum_{i=1}^{L} \hat{\sigma}_y^2(kL + i - 1) = \frac{1}{L} \sum_{i=1}^{L} \sigma_{\hat{y}_i}^2 \tag{17.25}$$

Since

$$\sum_{i=1}^{L} \sum_{k=0}^{\infty} \boldsymbol{c}\boldsymbol{A}^{kL+i-1} (\boldsymbol{c}\boldsymbol{A}^{kL+i-1})^T = \sum_{k=0}^{\infty} \boldsymbol{c}\boldsymbol{A}^k (\boldsymbol{c}\boldsymbol{A}^k)^T \tag{17.26}$$

we can deduce from (17.17) that

$$(\hat{\sigma}_y^2)_{ave} = \left\{ \frac{1}{L} \sum_{k=0}^{\infty} \boldsymbol{c}\boldsymbol{A}^k (\boldsymbol{c}\boldsymbol{A}^k)^T + 1 \right\} \sigma^2$$
$$= \left\{ \frac{1}{L} \mathrm{tr}[\boldsymbol{W}_o] + 1 \right\} \sigma^2 \tag{17.27}$$

On comparing (17.27) with (17.20), it can be seen that in block-state realization the average roundoff noise variance from internal roundoff noise sources is reduced by a factor of the block length L.

17.3.2 Roundoff Noise Minimization Subject to l_2-Scaling Constraints

If a coordinate transformation defined by

$$\overline{\boldsymbol{x}}(k) = \boldsymbol{T}^{-1}\boldsymbol{x}(k) \tag{17.28}$$

is applied to the SISO state-space model in (17.1), the new realization $(\overline{\boldsymbol{A}}, \overline{\boldsymbol{b}}, \overline{\boldsymbol{c}}, d)_n$ can be characterized by

$$\overline{\boldsymbol{A}} = \boldsymbol{T}^{-1}\boldsymbol{A}\boldsymbol{T}, \qquad \overline{\boldsymbol{b}} = \boldsymbol{T}^{-1}\boldsymbol{b}, \qquad \overline{\boldsymbol{c}} = \boldsymbol{c}\boldsymbol{T} \tag{17.29}$$

For this realization, the average roundoff noise variance in (17.27) is changed to

$$\hat{\sigma}_y^2(\boldsymbol{T})_{ave} = \left\{ \frac{1}{L} \mathrm{tr}[\boldsymbol{T}^T \boldsymbol{W}_o \boldsymbol{T}] + 1 \right\} \sigma^2 \tag{17.30}$$

The controllability Grammian $\hat{\boldsymbol{K}}_c$ for $(\hat{\boldsymbol{A}}, \hat{\boldsymbol{B}}, \hat{\boldsymbol{C}}, \hat{\boldsymbol{D}})_n$ described by (17.4a) plays an important role in the dynamic-range scaling of the state-variable vector $\hat{\boldsymbol{x}}(k)$. Theorem 17.4 states that the controllability Grammian is invariant under the mapping $(\boldsymbol{A}, \boldsymbol{b}, \boldsymbol{c}, d)_n \to (\hat{\boldsymbol{A}}, \hat{\boldsymbol{B}}, \hat{\boldsymbol{C}}, \hat{\boldsymbol{D}})_n$, i.e., $\boldsymbol{K}_c = \hat{\boldsymbol{K}}_c$ where $\boldsymbol{K}_c$ can be obtained by solving the Lyapunov equation

$$\boldsymbol{K}_c = \boldsymbol{A}\boldsymbol{K}_c\boldsymbol{A}^T + \boldsymbol{b}\boldsymbol{b}^T$$

Thus, if the SISO state-space model $(\boldsymbol{A}, \boldsymbol{b}, \boldsymbol{c}, d)_n$ in (17.1) is scaled, the resulting block-state realization $(\hat{\boldsymbol{A}}, \hat{\boldsymbol{B}}, \hat{\boldsymbol{C}}, \hat{\boldsymbol{D}})_n$ will automatically be scaled as well.

With an equivalent realization as specified in (17.29), the controllability Grammian assumes the form

$$\overline{\boldsymbol{K}}_c = \boldsymbol{T}^{-1}\boldsymbol{K}_c\boldsymbol{T}^{-T} \tag{17.31}$$

If l_2-scaling constraints are imposed on the new state-variable vector $\overline{\boldsymbol{x}}(k)$ defined by (17.28), it is required that

$$(\boldsymbol{T}^{-1}\hat{\boldsymbol{K}}_c\boldsymbol{T}^{-T})_{ii} = (\boldsymbol{T}^{-1}\boldsymbol{K}_c\boldsymbol{T}^{-T})_{ii} = 1 \ \text{ for } i = 1, 2, \cdots, n \qquad (17.32)$$

The problem being considered here is to obtain an $n \times n$ coordinate transformation matrix $\boldsymbol{T}$ that minimizes tr$[\boldsymbol{T}^T\boldsymbol{W}_o\boldsymbol{T}]$ in (17.30) subject to the l_2-scaling constraints in (17.32). This problem can readily be solved by applying the technique in Section 15.2.2. In short, the solution of this problem is given by

$$\boldsymbol{T} = \frac{1}{\sqrt{n}}\Big(\sum_{i=1}^{n}\theta_i\Big)^{\frac{1}{2}}\boldsymbol{W}_o^{-\frac{1}{2}}\Big[\boldsymbol{W}_o^{\frac{1}{2}}\boldsymbol{K}_c\boldsymbol{W}_o^{\frac{1}{2}}\Big]^{\frac{1}{4}}\boldsymbol{Q}\boldsymbol{Z}^T \qquad (17.33)$$

where θ_i^2 for $i = 1, 2, \cdots, n$ are the eigenvalues of $\boldsymbol{K}_c\boldsymbol{W}_o$, matrix $\boldsymbol{Q}$ is derived from the eigenvalue-eigenvector decomposition

$$\Big[\boldsymbol{W}_o^{\frac{1}{2}}\boldsymbol{K}_c\boldsymbol{W}_o^{\frac{1}{2}}\Big]^{\frac{1}{2}} = \boldsymbol{Q}\,\text{diag}\{\theta_1, \theta_2, \cdots, \theta_n\}\boldsymbol{Q}^T$$

and matrix $\boldsymbol{Z}$ is an $n \times n$ orthogonal matrix such that

$$\left(\boldsymbol{Z}\boldsymbol{\Lambda}^{-2}\boldsymbol{Z}^T\right)_{ii} = 1 \ \text{ for } i = 1, 2, \cdots, n$$

which can be obtained by numerical manipulation [3, p. 278] where

$$\boldsymbol{\Lambda} = \text{diag}\{\lambda_1, \lambda_2, \cdots, \lambda_n\}$$

$$\lambda_i = \left(\frac{\theta_1 + \theta_2 + \cdots + \theta_n}{n\theta_i}\right)^{\frac{1}{2}} \ \text{ for } i = 1, 2, \cdots, n$$

17.4 l_2-Sensitivity Analysis and Minimization

17.4.1 l_2-Sensitivity Analysis

The transfer function of the block-state realization in (17.4a) is given by

$$\boldsymbol{H}(z) = \hat{\boldsymbol{C}}(z\boldsymbol{I}_n - \hat{\boldsymbol{A}})^{-1}\hat{\boldsymbol{B}} + \hat{\boldsymbol{D}} \qquad (17.34)$$

whose (i, j)th element is described by

$$H_{ij}(z) = \boldsymbol{c}_i(z\boldsymbol{I}_n - \hat{\boldsymbol{A}})^{-1}\boldsymbol{b}_j + d_{ij} \qquad (17.35)$$

where

$$\hat{\boldsymbol{B}} = \begin{bmatrix} \boldsymbol{b}_1 & \boldsymbol{b}_2 & \cdots & \boldsymbol{b}_L \end{bmatrix}$$

$$\hat{\boldsymbol{C}} = \begin{bmatrix} \boldsymbol{c}_1 \\ \boldsymbol{c}_2 \\ \vdots \\ \boldsymbol{c}_L \end{bmatrix}, \qquad d_{ij} = \begin{cases} 0 & \text{for} \quad i < j \\ d & \text{for} \quad i = j \\ \boldsymbol{c}\boldsymbol{A}^{i-j-1}\boldsymbol{b} & \text{for} \quad i > j \end{cases}$$

We are now in a position to define the l_2-sensitivity of the block-state realization in (17.4a).

Definition 17.1
Let $\boldsymbol{X}$ be an $m \times n$ real matrix and let $f(\boldsymbol{X})$ be a scalar complex function of $\boldsymbol{X}$, differentiable with respect to all entries of $\boldsymbol{X}$. The sensitivity function of $f(\boldsymbol{X})$ with respect to $\boldsymbol{X}$ is then defined as

$$\boldsymbol{S}_{\boldsymbol{X}} = \frac{\partial f(\boldsymbol{X})}{\partial \boldsymbol{X}}, \qquad (\boldsymbol{S}_{\boldsymbol{X}})_{ij} = \frac{\partial f(\boldsymbol{X})}{\partial x_{ij}} \tag{17.36}$$

where x_{ij} denotes the (i, j)th entry of matrix $\boldsymbol{X}$.

Definition 17.2
Let $\boldsymbol{X}(z)$ be an $m \times n$ complex matrix-valued function of a complex variable z and let $x_{pq}(z)$ be the (p, q)th entry of $\boldsymbol{X}(z)$. The l_2-norm of $\boldsymbol{X}(z)$ is then defined as

$$\begin{aligned} \|\boldsymbol{X}(z)\|_2 &= \left[\frac{1}{2\pi} \int_0^{2\pi} \sum_{p=1}^{m} \sum_{q=1}^{n} \left| x_{pq}(e^{j\omega}) \right|^2 d\omega \right]^{\frac{1}{2}} \\ &= \left(\operatorname{tr} \left[\frac{1}{2\pi j} \oint_{|z|=1} \boldsymbol{X}(z) \boldsymbol{X}^H(z) \frac{dz}{z} \right] \right)^{\frac{1}{2}} \end{aligned} \tag{17.37}$$

From Definitions 17.1 and 17.2, the l_2-sensitivity measure for the subsystem in (17.35) is defined by

$$\begin{aligned} S_{ij} &= \left\| \frac{\partial H_{ij}(z)}{\partial \hat{\boldsymbol{A}}} \right\|_2^2 + \left\| \frac{\partial H_{ij}(z)}{\partial \boldsymbol{b}_j} \right\|_2^2 + \left\| \frac{\partial H_{ij}(z)}{\partial \boldsymbol{c}_i^T} \right\|_2^2 + \left\| \frac{\partial H_{ij}(z)}{\partial d_{ij}} \right\|_2^2 \\ &= \left\| [\boldsymbol{f}_j(z) \boldsymbol{g}_i(z)]^T \right\|_2^2 + \left\| \boldsymbol{g}_i^T(z) \right\|_2^2 + \left\| \boldsymbol{f}_j(z) \right\|_2^2 + u_o(i - j) \end{aligned} \tag{17.38}$$

where

$$\boldsymbol{f}_j(z) = (z\boldsymbol{I}_n - \hat{\boldsymbol{A}})^{-1}\boldsymbol{b}_j = (z\boldsymbol{I}_n - \boldsymbol{A}^L)^{-1}\boldsymbol{A}^{L-j}\boldsymbol{b}$$

$$\boldsymbol{g}_i(z) = \boldsymbol{c}_i(z\boldsymbol{I}_n - \hat{\boldsymbol{A}})^{-1} = \boldsymbol{c}\boldsymbol{A}^{i-1}(z\boldsymbol{I}_n - \boldsymbol{A}^L)^{-1}$$

$$u_o(i) = \begin{cases} 1 & \text{for} \quad i \geq 0 \\ 0 & \text{for} \quad i < 0 \end{cases}$$

In the rest of this section, $\boldsymbol{f}_j(z)$ and $\boldsymbol{g}_i(z)$ are referred to as *intermediate functions*.

Using simple algebraic manipulations, the l_2-sensitivity measure in (17.38) can be expressed as

$$\begin{aligned} S_{ij} = \mathrm{tr}\big[\boldsymbol{N}_{ij}(\boldsymbol{I}_n)\big] + \sum_{k=0}^{\infty} \boldsymbol{c}\boldsymbol{A}^{kL+i-1}\big(\boldsymbol{c}\boldsymbol{A}^{kL+i-1}\big)^T \\ + \sum_{k=0}^{\infty}\big(\boldsymbol{A}^{(k+1)L-j}\boldsymbol{b}\big)^T\boldsymbol{A}^{(k+1)L-j}\boldsymbol{b} + u_o(i-j) \end{aligned} \tag{17.39}$$

where

$$\boldsymbol{N}_{ij}(\boldsymbol{I}_n) = \frac{1}{2\pi j}\oint_{|z|=1}[\boldsymbol{f}_j(z)\boldsymbol{g}_i(z)]^T\boldsymbol{f}_j(z^{-1})\boldsymbol{g}_i(z^{-1})\frac{dz}{z}$$

A closed-form solution for evaluating $\boldsymbol{N}_{ij}(\boldsymbol{I}_n)$ will be deduced shortly.

We note that each single output $y(kL+i-1)$ for $i = 1, 2, \cdots, L$ in the output vector $\hat{\boldsymbol{y}}(k)$ is generated by the subsystem

$$\begin{aligned} \boldsymbol{H}_i(z) &= \big[H_{i1}(z), H_{i2}(z), \cdots, H_{iL}(z)\big] \\ &= \boldsymbol{c}_i\big(z\boldsymbol{I} - \hat{\boldsymbol{A}}\big)^{-1}\hat{\boldsymbol{B}} + \boldsymbol{d}_i \end{aligned} \tag{17.40}$$

where $\boldsymbol{d}_i = \big[d_{i1}, d_{i2}, \cdots, d_{iL}\big]$. From this in conjunction with (17.39), the l_2-sensitivity measure for the subsystem in (17.40) is found to be

$$\begin{aligned} S_i = \sum_{j=1}^{L} S_{ij} = \sum_{j=1}^{L}\mathrm{tr}\big[\boldsymbol{N}_{ij}(\boldsymbol{I}_n)\big] + \sum_{j=1}^{L}\sum_{k=0}^{\infty}\boldsymbol{c}\boldsymbol{A}^{kL+i-1}\big(\boldsymbol{c}\boldsymbol{A}^{kL+i-1}\big)^T \\ + \sum_{k=0}^{\infty}\big(\boldsymbol{A}^k\boldsymbol{b}\big)^T\boldsymbol{A}^k\boldsymbol{b} + i \end{aligned} \tag{17.41}$$

for $i = 1, 2, \cdots, L$. As a result, the overall l_2-sensitivity for the block-state realization in (17.34) can be expressed as

$$S = \sum_{i=1}^{L} S_i = \sum_{i=1}^{L}\sum_{j=1}^{L} \mathrm{tr}\big[\boldsymbol{N}_{ij}(\boldsymbol{I}_n)\big] + \sum_{j=1}^{L}\sum_{k=0}^{\infty} \boldsymbol{c}\boldsymbol{A}^k(\boldsymbol{c}\boldsymbol{A}^k)^T + \sum_{i=1}^{L}\sum_{k=0}^{\infty} (\boldsymbol{A}^k\boldsymbol{b})^T\boldsymbol{A}^k\boldsymbol{b} + \frac{L(L+1)}{2} \tag{17.42}$$

which is equivalent to

$$S = \sum_{i=1}^{L}\sum_{j=1}^{L} \mathrm{tr}\big[\boldsymbol{N}_{ij}(\boldsymbol{I}_n)\big] + L\,\mathrm{tr}\big[\boldsymbol{W}_o\big] + L\,\mathrm{tr}\big[\boldsymbol{K}_c\big] + \frac{L(L+1)}{2} \tag{17.43}$$

where $\boldsymbol{K}_c$ and $\boldsymbol{W}_o$ are the controllability and observability Grammians of the system in (17.1), respectively.

We now define the average l_2-sensitivity (over one block period of length L) in the output of a block-state realization as

$$(S_i)_{ave} = \frac{1}{L}\sum_{i=1}^{L} S_i = \frac{S}{L} \tag{17.44}$$

which in conjunction with (17.43) gives

$$(S_i)_{ave} = \frac{1}{L}\sum_{i=1}^{L}\sum_{j=1}^{L} \mathrm{tr}\big[\boldsymbol{N}_{ij}(\boldsymbol{I}_n)\big] + \mathrm{tr}\big[\boldsymbol{W}_o\big] + \mathrm{tr}\big[\boldsymbol{K}_c\big] + \frac{L+1}{2} \tag{17.45}$$

For comparison purpose, the l_2-sensitivity measure for the SISO state-space model in (17.1) is found to be (Sections 12.2 and 12.3)

$$S_o = \mathrm{tr}\big[\boldsymbol{N}(\boldsymbol{I}_n)\big] + \mathrm{tr}\big[\boldsymbol{W}_o\big] + \mathrm{tr}\big[\boldsymbol{K}_c\big] + 1 \tag{17.46}$$

where

$$\boldsymbol{N}(\boldsymbol{I}_n) = \frac{1}{2\pi j}\oint_{|z|=1} [\boldsymbol{f}(z)\boldsymbol{g}(z)]^T \boldsymbol{f}(z^{-1})\boldsymbol{g}(z^{-1})\frac{dz}{z}$$

$$\boldsymbol{f}(z) = (z\boldsymbol{I}_n - \boldsymbol{A})^{-1}\boldsymbol{b}, \qquad \boldsymbol{g}(z) = \boldsymbol{c}(z\boldsymbol{I}_n - \boldsymbol{A})^{-1}$$

and for any $n \times n$ symmetric positive-definite matrix $\boldsymbol{P}$, the Grammian $\boldsymbol{N}(\boldsymbol{P})$ can be obtained by solving the Lyapunov equation

$$\boldsymbol{Y} = \begin{bmatrix} \boldsymbol{A} & \boldsymbol{bc} \\ 0 & \boldsymbol{A} \end{bmatrix}^T \boldsymbol{Y} \begin{bmatrix} \boldsymbol{A} & \boldsymbol{bc} \\ 0 & \boldsymbol{A} \end{bmatrix} + \begin{bmatrix} \boldsymbol{P}^{-1} & 0 \\ 0 & 0 \end{bmatrix}$$

and then taking the lower-right $n \times n$ block of $\boldsymbol{Y}$ as $\boldsymbol{N}(\boldsymbol{P})$, namely,

$$\boldsymbol{N}(\boldsymbol{P}) = \begin{bmatrix} 0 & \boldsymbol{I}_n \end{bmatrix} \boldsymbol{Y} \begin{bmatrix} 0 \\ \boldsymbol{I}_n \end{bmatrix}$$

To identify an optimal internal structure of a given IIR digital filter that achieves minimum average l_2-sensitivity, we examine a coordinate transformation defined by (17.28). When the coordinate transformation defined by (17.28) is applied to the SISO state-space model in (17.1), the new realization associated with state $\overline{\boldsymbol{x}}(k)$, denoted by $(\overline{\boldsymbol{A}}, \overline{\boldsymbol{b}}, \overline{\boldsymbol{c}}, d)_n$, is related to the original realization as

$$\begin{aligned} \overline{\boldsymbol{K}}_c &= \boldsymbol{T}^{-1}\boldsymbol{K}_c\boldsymbol{T}^{-T}, & \overline{\boldsymbol{W}}_o &= \boldsymbol{T}^T\boldsymbol{W}_o\boldsymbol{T} \\ \overline{\boldsymbol{f}}_j(z) &= \boldsymbol{T}^{-1}\boldsymbol{f}_j(z), & \overline{\boldsymbol{g}}_i(z) &= \boldsymbol{g}_i(z)\boldsymbol{T} \end{aligned} \tag{17.47}$$

and the canonical-state to block-state mapping is given by

$$(\boldsymbol{T}^{-1}\boldsymbol{A}\boldsymbol{T}, \boldsymbol{T}^{-1}\boldsymbol{b}, \boldsymbol{c}\boldsymbol{T}, d)_n \rightarrow (\boldsymbol{T}^{-1}\hat{\boldsymbol{A}}\boldsymbol{T}, \boldsymbol{T}^{-1}\hat{\boldsymbol{B}}, \hat{\boldsymbol{C}}\boldsymbol{T}, \hat{\boldsymbol{D}})_n \tag{17.48}$$

Moreover, the Grammian $\boldsymbol{N}_{ij}(\boldsymbol{I}_n)$ which is introduced in (17.39) is transformed into $\overline{\boldsymbol{N}}_{ij}(\boldsymbol{I}_n)$ as follows:

$$\begin{aligned} \overline{\boldsymbol{N}}_{ij}(\boldsymbol{I}_n) &= \frac{1}{2\pi j}\oint_{|z|=1} [\overline{\boldsymbol{f}}_j(z)\overline{\boldsymbol{g}}_i(z)]^T\,\overline{\boldsymbol{f}}_j(z^{-1})\overline{\boldsymbol{g}}_i(z^{-1})\frac{dz}{z} \\ &= \boldsymbol{T}^T\boldsymbol{N}_{ij}(\boldsymbol{P})\boldsymbol{T} \end{aligned} \tag{17.49}$$

where

$$\boldsymbol{P} = \boldsymbol{T}\boldsymbol{T}^T$$

$$\boldsymbol{N}_{ij}(\boldsymbol{P}) = \frac{1}{2\pi j}\oint_{|z|=1} [\boldsymbol{f}_j(z)\boldsymbol{g}_i(z)]^T\boldsymbol{P}^{-1}\boldsymbol{f}_j(z^{-1})\boldsymbol{g}_i(z^{-1})\frac{dz}{z}$$

It is noted that

$$\begin{aligned}\overline{\boldsymbol{f}}_j(z)\overline{\boldsymbol{g}}_i(z) &= \boldsymbol{T}^{-1}\boldsymbol{f}_j(z)\boldsymbol{g}_i(z)\,\boldsymbol{T} \\ &= \begin{bmatrix} \boldsymbol{T}^{-1} & \boldsymbol{0} \end{bmatrix} \begin{bmatrix} z\boldsymbol{I}_n - \boldsymbol{A}^L & -\boldsymbol{A}^{L-j}\boldsymbol{b}\boldsymbol{c}\boldsymbol{A}^{i-1} \\ \boldsymbol{0} & z\boldsymbol{I}_n - \boldsymbol{A}^L \end{bmatrix}^{-1} \begin{bmatrix} \boldsymbol{0} \\ \boldsymbol{T} \end{bmatrix}\end{aligned} \tag{17.50}$$

If we denote the observability Grammian of the composite system $\overline{\boldsymbol{f}}_j(z)\overline{\boldsymbol{g}}_i(z)$ in (17.50) by $\boldsymbol{Y}_{ij}$, matrix $\boldsymbol{N}_{ij}(\boldsymbol{P})$ can be obtained by solving the Lyapunov equation

$$\begin{aligned}\boldsymbol{Y}_{ij} = &\begin{bmatrix} \boldsymbol{A}^L & \boldsymbol{A}^{L-j}\boldsymbol{b}\boldsymbol{c}\boldsymbol{A}^{i-1} \\ \boldsymbol{0} & \boldsymbol{A}^L \end{bmatrix}^T \boldsymbol{Y}_{ij} \begin{bmatrix} \boldsymbol{A}^L & \boldsymbol{A}^{L-j}\boldsymbol{b}\boldsymbol{c}\boldsymbol{A}^{i-1} \\ \boldsymbol{0} & \boldsymbol{A}^L \end{bmatrix} \\ &+ \begin{bmatrix} \boldsymbol{P}^{-1} & \boldsymbol{0} \\ \boldsymbol{0} & \boldsymbol{0} \end{bmatrix}\end{aligned} \tag{17.51}$$

and then taking the lower-right $n \times n$ block of $\boldsymbol{Y}_{ij}$ as $\boldsymbol{N}_{ij}(\boldsymbol{P})$, namely,

$$\boldsymbol{N}_{ij}(\boldsymbol{P}) = \begin{bmatrix} \boldsymbol{0} & \boldsymbol{I}_n \end{bmatrix} \boldsymbol{Y}_{ij} \begin{bmatrix} \boldsymbol{0} \\ \boldsymbol{I}_n \end{bmatrix} \tag{17.52}$$

Therefore, under the coordinate transformation $\overline{x}(k) = \boldsymbol{T}^{-1}\boldsymbol{x}(k)$ defined by (17.28), the average l_2-sensitivity measure in (17.45) becomes

$$\begin{aligned}S_i(\boldsymbol{T})_{ave} = &\frac{1}{L}\sum_{i=1}^{L}\sum_{j=1}^{L} \mathrm{tr}\big[\boldsymbol{T}^T\boldsymbol{N}_{ij}(\boldsymbol{T}\boldsymbol{T}^T)\boldsymbol{T}\big] + \mathrm{tr}\big[\boldsymbol{T}^T\boldsymbol{W}_o\boldsymbol{T}\big] \\ &+ \mathrm{tr}\big[\boldsymbol{T}^{-1}\boldsymbol{K}_c\boldsymbol{T}^{-T}\big] + \frac{L+1}{2}\end{aligned} \tag{17.53}$$

For comparison purpose, it follows from (17.46) that

$$S_o(\boldsymbol{T}) = \mathrm{tr}[\boldsymbol{T}^T\boldsymbol{N}(\boldsymbol{T}\boldsymbol{T}^T)\boldsymbol{T}] + \mathrm{tr}[\boldsymbol{T}^T\boldsymbol{W}_o\boldsymbol{T}] + \mathrm{tr}[\boldsymbol{T}^{-1}\boldsymbol{K}_c\boldsymbol{T}^{-T}] + 1 \tag{17.54}$$

From (17.4b) and (17.29), it can be shown that the transfer functions $\boldsymbol{H}(z)$ in (17.34) and $H_{ij}(z)$ in (17.35) are invariant under the coordinate transformation defined by (17.28).

17.4.2 l_2-Sensitivity Minimization Subject to l_2-Scaling Constraints

17.4.2.1 Method 1: using a Lagrange function

We now consider the problem of minimizing the average l_2-sensitivity measure in (17.53) subject to l_2-scaling constraints in (17.32). Since the measure in (17.53) can be expressed in terms of matrix $\boldsymbol{P} = \boldsymbol{T}\boldsymbol{T}^T$ as

$$S_i(\boldsymbol{P})_{ave} = \frac{1}{L}\sum_{i=1}^{L}\sum_{j=1}^{L}\mathrm{tr}\big[\boldsymbol{N}_{ij}(\boldsymbol{P})\boldsymbol{P}\big] + \mathrm{tr}\big[\boldsymbol{W}_o\boldsymbol{P}\big] + \mathrm{tr}\big[\boldsymbol{K}_c\boldsymbol{P}^{-1}\big] + \frac{L+1}{2} \tag{17.55}$$

the problem we deal with is a constrained nonlinear optimization problem where the variable is matrix $\boldsymbol{P}$.

It is important that the coefficient sensitivity defined above be minimized subject to constraints so that input-to-state energy-flow is appropriately scaled which in analytic term is known as l_2-scaling. If we sum up the n l_2-scaling constraints in (17.32), then we have

$$\mathrm{tr}[\boldsymbol{T}^{-1}\boldsymbol{K}_c\boldsymbol{T}^{-T}] = \mathrm{tr}[\boldsymbol{K}_c\boldsymbol{P}^{-1}] = n \tag{17.56}$$

Consequently, the problem of minimizing (17.53) subject to the constraints in (17.32) can be *relaxed* into the following problem:

$$\begin{aligned} &\text{minimize } S_i(\boldsymbol{P})_{ave} \text{ in (17.55) with respect to } \boldsymbol{P} \\ &\text{subject to } \mathrm{tr}[\boldsymbol{K}_c\boldsymbol{P}^{-1}] = n \end{aligned} \tag{17.57}$$

We now address problem (17.57) as the first step of our solution procedure. To this end, we define the Lagrange function of the problem as

$$\begin{aligned} J(\boldsymbol{P},\lambda) = &\frac{1}{L}\sum_{i=1}^{L}\sum_{j=1}^{L}\mathrm{tr}[\boldsymbol{N}_{ij}(\boldsymbol{P})\boldsymbol{P}] + \mathrm{tr}[\boldsymbol{W}_o\boldsymbol{P}] + \mathrm{tr}[\boldsymbol{K}_c\boldsymbol{P}^{-1}] \\ &+ \frac{L+1}{2} + \lambda\left(\mathrm{tr}[\boldsymbol{K}_c\boldsymbol{P}^{-1}] - n\right) \end{aligned} \tag{17.58}$$

where λ is a Lagrange multiplier. It is well known that the solution of problem (17.57) must satisfy the Karush-Kuhn-Tucker (KKT) conditions $\partial J(\boldsymbol{P},\lambda)/\partial \boldsymbol{P} = \boldsymbol{0}$ and $\partial J(\boldsymbol{P},\lambda)/\partial\lambda = 0$ where

$$\frac{\partial J(\boldsymbol{P},\lambda)}{\partial \boldsymbol{P}} = \frac{1}{L}\sum_{i=1}^{L}\sum_{j=1}^{L}\boldsymbol{N}_{ij}(\boldsymbol{P}) + \boldsymbol{W}_o - \boldsymbol{P}^{-1}\frac{1}{L}\sum_{i=1}^{L}\sum_{j=1}^{L}\boldsymbol{M}_{ij}(\boldsymbol{P})\boldsymbol{P}^{-1}$$
$$-(\lambda+1)\boldsymbol{P}^{-1}\boldsymbol{K}_c\boldsymbol{P}^{-1} \tag{17.59}$$

$$\frac{\partial J(\boldsymbol{P},\lambda)}{\partial \lambda} = \mathrm{tr}[\boldsymbol{K}_c\boldsymbol{P}^{-1}] - n$$

The matrices $\boldsymbol{M}_{ij}(\boldsymbol{P})$ in (17.59) are obtained by solving the Lyapunov equations

$$\boldsymbol{X}_{ij} = \begin{bmatrix} \boldsymbol{A}^L & \boldsymbol{A}^{L-j}\boldsymbol{bc}\boldsymbol{A}^{i-1} \\ \boldsymbol{0} & \boldsymbol{A}^L \end{bmatrix} \boldsymbol{X}_{ij} \begin{bmatrix} \boldsymbol{A}^L & \boldsymbol{A}^{L-j}\boldsymbol{bc}\boldsymbol{A}^{i-1} \\ \boldsymbol{0} & \boldsymbol{A}^L \end{bmatrix}^T + \begin{bmatrix} \boldsymbol{0} & \boldsymbol{0} \\ \boldsymbol{0} & \boldsymbol{P} \end{bmatrix} \tag{17.60}$$

and then taking the upper-left $n \times n$ block of $\boldsymbol{X}_{ij}$ as $\boldsymbol{M}_{ij}(\boldsymbol{P})$, namely,

$$\boldsymbol{M}_{ij}(\boldsymbol{P}) = \begin{bmatrix} \boldsymbol{I}_n & \boldsymbol{0} \end{bmatrix} \boldsymbol{X}_{ij} \begin{bmatrix} \boldsymbol{I}_n \\ \boldsymbol{0} \end{bmatrix} \tag{17.61}$$

The KKT conditions in (17.59) can be expressed compactly as

$$\boldsymbol{P}\boldsymbol{F}(\boldsymbol{P})\boldsymbol{P} = \boldsymbol{G}(\boldsymbol{P},\lambda), \qquad \mathrm{tr}[\boldsymbol{K}_c\boldsymbol{P}^{-1}] = n \tag{17.62}$$

where

$$\boldsymbol{F}(\boldsymbol{P}) = \frac{1}{L}\sum_{i=1}^{L}\sum_{j=1}^{L}\boldsymbol{N}_{ij}(\boldsymbol{P}) + \boldsymbol{W}_o$$

$$\boldsymbol{G}(\boldsymbol{P},\lambda) = \frac{1}{L}\sum_{i=1}^{L}\sum_{j=1}^{L}\boldsymbol{M}_{ij}(\boldsymbol{P}) + (\lambda+1)\boldsymbol{K}_c$$

Note that the first equation in (17.62) is highly nonlinear with respect to $\boldsymbol{P}$. An effective approach to solving the first equation in (17.62) is to *relax* it into the recursive second-order matrix equation

$$\boldsymbol{P}_{k+1}\boldsymbol{F}(\boldsymbol{P}_k)\boldsymbol{P}_{k+1} = \boldsymbol{G}(\boldsymbol{P}_k,\lambda_{k+1}) \tag{17.63}$$

where $\boldsymbol{P}_k$ is assumed to be known from the previous recursion. Note that if the matrix sequence $\{\boldsymbol{P}_k\}$ converges to its limit matrix, say $\boldsymbol{P}$, then (17.63) converges to the first equation in (17.62) as k goes to infinity. The solution $\boldsymbol{P}_{k+1}$ of (17.63) is then given by

$$\boldsymbol{P}_{k+1} = \boldsymbol{F}(\boldsymbol{P}_k)^{-\frac{1}{2}} \big[\boldsymbol{F}(\boldsymbol{P}_k)^{\frac{1}{2}} \boldsymbol{G}(\boldsymbol{P}_k, \lambda_{k+1}) \boldsymbol{F}(\boldsymbol{P}_k)^{\frac{1}{2}}\big]^{\frac{1}{2}} \boldsymbol{F}(\boldsymbol{P}_k)^{-\frac{1}{2}} \quad (17.64)$$

In order to derive a recursive formula for the Lagrange multiplier λ, we use (17.62) to write

$$\mathrm{tr}\big[\boldsymbol{P}\boldsymbol{F}(\boldsymbol{P})\big] = \frac{1}{L}\sum_{i=1}^{L}\sum_{j=1}^{L} \mathrm{tr}\big[\boldsymbol{M}_{ij}(\boldsymbol{P})\boldsymbol{P}^{-1}\big] + n(\lambda + 1) \quad (17.65)$$

which naturally suggests the recursion for λ

$$\lambda_{k+1} = \frac{\mathrm{tr}\big[\boldsymbol{P}_k\boldsymbol{F}(\boldsymbol{P}_k)\big] - \dfrac{1}{L}\displaystyle\sum_{i=1}^{L}\sum_{j=1}^{L} \mathrm{tr}\big[\boldsymbol{M}_{ij}(\boldsymbol{P}_k)\boldsymbol{P}_k^{-1}\big]}{n} - 1 \quad (17.66)$$

where $\boldsymbol{P}_0$ is the initial estimate. This iteration process continues until

$$\big|S_i(\boldsymbol{P}_{k+1})_{ave} - S_i(\boldsymbol{P}_k)_{ave}\big| + \big|n - \mathrm{tr}\big[\boldsymbol{K}_c\boldsymbol{P}_{k+1}^{-1}\big]\big| < \varepsilon \quad (17.67)$$

is satisfied where $\varepsilon > 0$ is a prescribed tolerance.

If the iteration is terminated at step k, then we take $\boldsymbol{P} = \boldsymbol{P}_k$ as the solution. Since $\boldsymbol{P} = \boldsymbol{T}\boldsymbol{T}^T$, the optimal $\boldsymbol{T}$ assumes the form

$$\boldsymbol{T} = \boldsymbol{P}^{\frac{1}{2}}\boldsymbol{U} \quad (17.68)$$

where $\boldsymbol{P}^{1/2}$ is the square root of the matrix $\boldsymbol{P}$ obtained above, and $\boldsymbol{U}$ is an $n \times n$ orthogonal matrix.

As the second step of the solution procedure, we now turn our attention to the construction of the optimal coordinate transformation matrix $\boldsymbol{T}$ that solves the problem of minimizing (17.53) subject to the constraints in (17.32). To this end, we examine a procedure for determining the $n \times n$ orthogonal matrix $\boldsymbol{U}$ in (17.68) in order for the nonsingular matrix $\boldsymbol{T}$ to satisfy the l_2-scaling constraints in (17.32).

From (17.47) and (17.68), it follows that

$$\overline{\boldsymbol{K}}_c = \boldsymbol{T}^{-1}\boldsymbol{K}_c\boldsymbol{T}^{-T} = \boldsymbol{U}^T\boldsymbol{P}^{-\frac{1}{2}}\boldsymbol{K}_c\boldsymbol{P}^{-\frac{1}{2}}\boldsymbol{U} \quad (17.69)$$

In order to find an $n \times n$ orthogonal matrix $\boldsymbol{U}$ such that the matrix $\overline{\boldsymbol{K}}_c$ in (17.69) satisfies the scaling constraints in (17.32), we perform the eigenvalue-eigenvector decomposition for the symmetric positive-definite matrix $\boldsymbol{P}^{-1/2}\boldsymbol{K}_c\boldsymbol{P}^{-1/2}$ as

$$\boldsymbol{P}^{-\frac{1}{2}}\boldsymbol{K}_c\boldsymbol{P}^{-\frac{1}{2}} = \boldsymbol{R}\boldsymbol{\Theta}\boldsymbol{R}^T \tag{17.70}$$

where $\boldsymbol{\Theta} = \text{diag}\{\theta_1, \theta_2, \cdots, \theta_n\}$ with $\theta_i > 0$ and $\boldsymbol{R}$ is an orthogonal matrix. Next, an orthogonal matrix $\boldsymbol{S}$ such that

$$\boldsymbol{S}\boldsymbol{\Theta}\boldsymbol{S}^T = \begin{bmatrix} 1 & * & \cdots & * \\ * & 1 & \ddots & \vdots \\ \vdots & \ddots & \ddots & * \\ * & \cdots & * & 1 \end{bmatrix} \tag{17.71}$$

can be obtained by numerical manipulations [3, p. 278]. Using (17.69), (17.70) and (17.71), it can be readily verified that the orthogonal matrix $\boldsymbol{U} = \boldsymbol{R}\boldsymbol{S}^T$ leads to a $\overline{\boldsymbol{K}}_c$ in (17.69) whose diagonal elements are equal to unity, hence the constraints in (17.32) are now satisfied. This matrix $\boldsymbol{T}$ together with (17.68) gives the solution of the problem of minimizing (17.53) subject to the constraints in (17.32) as

$$\boldsymbol{T} = \boldsymbol{P}^{\frac{1}{2}}\boldsymbol{R}\boldsymbol{S}^T \tag{17.72}$$

17.4.2.2 Method 2: using a Quasi-Newton algorithm

Since the state-space model in (17.1) is stable and controllable, the controllability Grammian $\boldsymbol{K}_c$ is symmetric and positive-definite. Hence $\boldsymbol{K}_c^{1/2}$ satisfying $\boldsymbol{K}_c = \boldsymbol{K}_c^{1/2}\boldsymbol{K}_c^{1/2}$ is also symmetric and positive-definite.

By defining

$$\hat{\boldsymbol{T}} = \boldsymbol{T}^T\boldsymbol{K}_c^{-\frac{1}{2}} \tag{17.73}$$

the l_2-scaling constraints in (17.32) can be written as

$$(\hat{\boldsymbol{T}}^{-T}\hat{\boldsymbol{T}}^{-1})_{ii} = 1 \ \text{ for } \ i = 1, 2, \cdots, n \tag{17.74}$$

The constraints in (17.74) simply state that each column in matrix $\hat{\boldsymbol{T}}^{-1}$ must be a unit vector. Matrix $\hat{\boldsymbol{T}}^{-1}$ is assumed to have the form

$$\hat{\boldsymbol{T}}^{-1} = \left[\frac{\boldsymbol{t}_1}{||\boldsymbol{t}_1||}, \frac{\boldsymbol{t}_2}{||\boldsymbol{t}_2||}, \cdots, \frac{\boldsymbol{t}_n}{||\boldsymbol{t}_n||}\right] \tag{17.75}$$

so that (17.74) is always satisfied. Using (17.73), it follows from (17.53) that

$$S_i(\hat{\boldsymbol{T}})_{ave} = \frac{1}{L}\sum_{i=1}^{L}\sum_{j=1}^{L} \mathrm{tr}\big[\hat{\boldsymbol{T}}\hat{\boldsymbol{N}}_{ij}(\hat{\boldsymbol{T}})\hat{\boldsymbol{T}}^T\big] + \mathrm{tr}\big[\hat{\boldsymbol{T}}\hat{\boldsymbol{W}}_o\hat{\boldsymbol{T}}^T\big] + n + \frac{L+1}{2} \tag{17.76}$$

where

$$\hat{\boldsymbol{N}}_{ij}(\hat{\boldsymbol{T}}) = \boldsymbol{K}_c^{\frac{1}{2}}\boldsymbol{N}_{ij}(\boldsymbol{K}_c^{\frac{1}{2}}\hat{\boldsymbol{T}}^T\hat{\boldsymbol{T}}\boldsymbol{K}_c^{\frac{1}{2}})\boldsymbol{K}_c^{\frac{1}{2}}, \qquad \hat{\boldsymbol{W}}_o = \boldsymbol{K}_c^{\frac{1}{2}}\boldsymbol{W}_o\boldsymbol{K}_c^{\frac{1}{2}}$$

From the foregoing arguments, the problem of obtaining an $n \times n$ nonsingular matrix $\boldsymbol{T}$ which minimizes the average l_2-sensitivity $S_i(\boldsymbol{T})_{ave}$ in (17.53) subject to the l_2-scaling constraints in (17.32) can be converted into an unconstrained optimization problem of obtaining an $n \times n$ nonsingular matrix $\hat{\boldsymbol{T}}$ in (17.75) which minimizes $S_i(\hat{\boldsymbol{T}})_{ave}$ in (17.76).

We now apply a quasi-Newton algorithm [13] to minimize (17.76) with respect to matrix $\hat{\boldsymbol{T}}$ in (17.75). Let $\boldsymbol{x}$ be the column vector that collects the independent variables in matrix $\hat{\boldsymbol{T}}$, i.e.,

$$\boldsymbol{x} = (\boldsymbol{t}_1^T, \boldsymbol{t}_2^T, \cdots, \boldsymbol{t}_n^T)^T \tag{17.77}$$

Then, $S_i(\hat{\boldsymbol{T}})_{ave}$ is a function of $\boldsymbol{x}$ and is denoted by $J_o(\boldsymbol{x})$. The algorithm starts with a trivial initial point $\boldsymbol{x}_0$ obtained from an initial assignment $\hat{\boldsymbol{T}} = \boldsymbol{I}_n$. Then, in the kth iteration, a quasi-Newton algorithm updates the most recent point $\boldsymbol{x}_k$ to point $\boldsymbol{x}_{k+1}$ as

$$\boldsymbol{x}_{k+1} = \boldsymbol{x}_k + \alpha_k\boldsymbol{d}_k \tag{17.78}$$

where

$$\boldsymbol{d}_k = -\boldsymbol{S}_k\nabla J_o(\boldsymbol{x}_k), \qquad \alpha_k = arg\big[\min_{\alpha} J_o(\boldsymbol{x}_k + \alpha\boldsymbol{d}_k)\big]$$

$$\boldsymbol{S}_{k+1} = \boldsymbol{S}_k + \left(1 + \frac{\boldsymbol{\gamma}_k^T\boldsymbol{S}_k\boldsymbol{\gamma}_k}{\boldsymbol{\gamma}_k^T\boldsymbol{\delta}_k}\right)\frac{\boldsymbol{\delta}_k\boldsymbol{\delta}_k^T}{\boldsymbol{\gamma}_k^T\boldsymbol{\delta}_k} - \frac{\boldsymbol{\delta}_k\boldsymbol{\gamma}_k^T\boldsymbol{S}_k + \boldsymbol{S}_k\boldsymbol{\gamma}_k\boldsymbol{\delta}_k^T}{\boldsymbol{\gamma}_k^T\boldsymbol{\delta}_k}$$

$$\boldsymbol{S}_0 = \boldsymbol{I}, \qquad \boldsymbol{\delta}_k = \boldsymbol{x}_{k+1} - \boldsymbol{x}_k, \qquad \boldsymbol{\gamma}_k = \nabla J_o(\boldsymbol{x}_{k+1}) - \nabla J_o(\boldsymbol{x}_k)$$

In the above, $\nabla J_o(\boldsymbol{x})$ is the gradient of $J_o(\boldsymbol{x})$ with respect to $\boldsymbol{x}$, and $\boldsymbol{S}_k$ is a positive-definite approximation of the inverse Hessian matrix of $J_o(\boldsymbol{x})$.

This iteration process continues until

$$|J_o(\boldsymbol{x}_{k+1}) - J_o(\boldsymbol{x}_k)| < \varepsilon \tag{17.79}$$

is satisfied where $\varepsilon > 0$ is a prescribed tolerance. If the iteration is terminated at step k, the $\boldsymbol{x}_k$ is taken to be the solution of the minimization problem.

The implementation of (17.78) requires the computation of $\nabla J_o(\boldsymbol{x})$ in each iteration, hence the availability of a closed-form formulation to compute $\nabla J_o(\boldsymbol{x})$ will make the algorithm considerably faster relative to the evaluation of $\nabla J_o(\boldsymbol{x})$ based on numerical differentiation. To this end, note that the gradient of $J_o(\boldsymbol{x})$ with respect to t_{pq} is defined by

$$\frac{\partial J_o(\boldsymbol{x})}{\partial t_{pq}} = \lim_{\Delta \to 0} \frac{S_i(\hat{\boldsymbol{T}}_{pq})_{ave} - S_i(\hat{\boldsymbol{T}})_{ave}}{\Delta} \tag{17.80}$$

where $\hat{\boldsymbol{T}}_{pq}$ is the matrix obtained from $\hat{\boldsymbol{T}}$ with its (p, q)th component perturbed by Δ and it is follows that [14, p. 655]

$$\begin{aligned} \hat{\boldsymbol{T}}_{pq} &= \hat{\boldsymbol{T}} + \frac{\Delta \hat{\boldsymbol{T}} \boldsymbol{g}_{pq} \boldsymbol{e}_q^T \hat{\boldsymbol{T}}}{1 - \Delta \boldsymbol{e}_q^T \hat{\boldsymbol{T}} \boldsymbol{g}_{pq}} \simeq \hat{\boldsymbol{T}} + \Delta \hat{\boldsymbol{T}} \boldsymbol{g}_{pq} \boldsymbol{e}_q^T \hat{\boldsymbol{T}} \\ \boldsymbol{g}_{pq} &= -\partial \left\{ \frac{\boldsymbol{t}_q}{||\boldsymbol{t}_q||} \right\} / \partial t_{pq} = \frac{1}{||\boldsymbol{t}_q||^3} (t_{pq} \boldsymbol{t}_q - ||\boldsymbol{t}_q||^2 \boldsymbol{e}_p) \end{aligned} \tag{17.81}$$

The gradient of $J_o(\boldsymbol{x})$ with respect to t_{pq} can be evaluated using closed-form expressions as

$$\frac{\partial J_o(\boldsymbol{x})}{\partial t_{pq}} = 2(\beta_1 - \beta_2 + \beta_3) \tag{17.82}$$

where

$$\beta_1 = \frac{1}{L} \sum_{i=1}^{L} \sum_{j=1}^{L} \boldsymbol{e}_q^T \hat{\boldsymbol{T}} \hat{\boldsymbol{N}}_{ij}(\hat{\boldsymbol{T}}) \hat{\boldsymbol{T}}^T \hat{\boldsymbol{T}} \boldsymbol{g}_{pq}$$

$$\beta_2 = \frac{1}{L} \sum_{i=1}^{L} \sum_{j=1}^{L} \boldsymbol{e}_q^T \hat{\boldsymbol{T}}^{-T} \hat{\boldsymbol{M}}_{ij}(\hat{\boldsymbol{T}}) \boldsymbol{g}_{pq}, \qquad \beta_3 = \boldsymbol{e}_q^T \hat{\boldsymbol{T}} \hat{\boldsymbol{W}}_o \hat{\boldsymbol{T}}^T \hat{\boldsymbol{T}} \boldsymbol{g}_{pq}$$

$$\hat{\boldsymbol{M}}_{ij}(\hat{\boldsymbol{T}}) = \boldsymbol{K}_c^{-\frac{1}{2}} \boldsymbol{M}_{ij} (\boldsymbol{K}_c^{\frac{1}{2}} \hat{\boldsymbol{T}}^T \hat{\boldsymbol{T}} \boldsymbol{K}_c^{\frac{1}{2}}) \boldsymbol{K}_c^{-\frac{1}{2}}$$

17.4.3 l_2-Sensitivity Minimization Without Imposing l_2-Scaling Constraints

The method described above can also be utilized to minimize an average l_2-sensitivity measure in (17.53) without imposing the l_2-scaling constraints in

(17.32). This can be done by simply setting the Lagrange multiplier λ to zero in (17.58). In this case, (17.64) is changed to

$$\boldsymbol{P}_{k+1} = \boldsymbol{F}(\boldsymbol{P}_k)^{-\frac{1}{2}}\left[\boldsymbol{F}(\boldsymbol{P}_k)^{\frac{1}{2}}\boldsymbol{G}(\boldsymbol{P}_k)\boldsymbol{F}(\boldsymbol{P}_k)^{\frac{1}{2}}\right]^{\frac{1}{2}}\boldsymbol{F}(\boldsymbol{P}_k)^{-\frac{1}{2}} \tag{17.83}$$

where

$$\boldsymbol{G}(\boldsymbol{P}) = \frac{1}{L}\sum_{i=1}^{L}\sum_{j=1}^{L}\boldsymbol{M}_{ij}(\boldsymbol{P}) + \boldsymbol{K}_c$$

With an initial estimate $\boldsymbol{P}_0$, the recursive process in (17.83) continues until

$$\left|S_i(\boldsymbol{P}_{k+1})_{ave} - S_i(\boldsymbol{P}_k)_{ave}\right| < \varepsilon \tag{17.84}$$

is satisfied where $\varepsilon > 0$ is a prescribed tolerance.

If the iteration is terminated at step k, then we take $\boldsymbol{P} = \boldsymbol{P}_k$ as the solution. A coordinate transformation matrix $\boldsymbol{T}$ that minimizes the average l_2-sensitivity measure in (17.53) is then found to be

$$\boldsymbol{T} = \boldsymbol{P}^{\frac{1}{2}}\boldsymbol{U} \tag{17.85}$$

where $\boldsymbol{P}^{1/2}$ is the square root of the matrix $\boldsymbol{P}$ obtained above, and $\boldsymbol{U}$ is any $n \times n$ orthogonal matrix.

17.4.4 Numerical Experiments

A. Filter Description and Its Controllability and Observability Grammians

Consider the 4th-order Butterworth lowpass filter $(\boldsymbol{A}_o, \boldsymbol{b}_o, \boldsymbol{c}_o, d)_4$ with a narrow normalized passband of 0.05, described by

$$\boldsymbol{A}_o = \begin{bmatrix} 3.589734 & 1 & 0 & 0 \\ -4.851276 & 0 & 1 & 0 \\ 2.924053 & 0 & 0 & 1 \\ -0.663010 & 0 & 0 & 0 \end{bmatrix}, \qquad \boldsymbol{b}_o = 10^{-3}\begin{bmatrix} 0.237096 \\ 0.035885 \\ 0.216300 \\ 0.010527 \end{bmatrix}$$

$$\boldsymbol{c}_o = \begin{bmatrix} 1 & 0 & 0 & 0 \end{bmatrix}, \qquad d = 3.123898 \times 10^{-5}$$

whose numerator b and denominator a was found using MATLAB as $[b, a] =$ butter$(4, 0.05)$. Applying a coordinate transformation defined by

$$\boldsymbol{T}_o = \mathrm{diag}\{\, 0.226458,\ 0.588059,\ 0.513017,\ 0.150144 \,\}$$

to the above state-space model, we obtained an equivalent state-space model $(\boldsymbol{A}, \boldsymbol{b}, \boldsymbol{c}, d)_4$ that satisfies l_2-scaling constraints as

$$\boldsymbol{A} = \begin{bmatrix} 3.589734 & 2.596768 & 0 & 0 \\ -1.868197 & 0 & 0.872390 & 0 \\ 1.290747 & 0 & 0 & 0.292669 \\ -1.000000 & 0 & 0 & 0 \end{bmatrix}$$

$$\boldsymbol{b} = 10^{-3} \begin{bmatrix} 1.046973 & 0.061023 & 0.421624 & 0.070114 \end{bmatrix}^T$$

$$\boldsymbol{c} = \begin{bmatrix} 0.226458 & 0 & 0 & 0 \end{bmatrix}, \qquad d = 3.123898 \times 10^{-5}$$

where

$$\boldsymbol{A} = \boldsymbol{T}_o^{-1}\boldsymbol{A}_o\boldsymbol{T}_o, \qquad \boldsymbol{b} = \boldsymbol{T}_o^{-1}\boldsymbol{b}_o, \qquad \boldsymbol{c} = \boldsymbol{c}_o\boldsymbol{T}_o$$

and its controllability and observability Grammians $\boldsymbol{K}_c$ and $\boldsymbol{W}_o$ were computed by solving the Lyapunov equations

$$\boldsymbol{K}_c = \boldsymbol{A}\boldsymbol{K}_c\boldsymbol{A}^T + \boldsymbol{b}\boldsymbol{b}^T, \qquad \boldsymbol{W}_o = \boldsymbol{A}^T\boldsymbol{W}_o\boldsymbol{A} + \boldsymbol{c}^T\boldsymbol{c}$$

as

$$\boldsymbol{K}_c = \begin{bmatrix} 1.000000 & -0.999248 & 0.997433 & -0.994918 \\ -0.999248 & 1.000000 & -0.999452 & 0.998047 \\ 0.997433 & -0.999452 & 1.000000 & -0.999566 \\ -0.994918 & 0.998047 & -0.999566 & 1.000000 \end{bmatrix}$$

$$\boldsymbol{W}_o = 10^4 \begin{bmatrix} 1.063597 & 2.747677 & 2.360090 & 0.672994 \\ 2.747677 & 7.172055 & 6.224575 & 1.793652 \\ 2.360090 & 6.224575 & 5.458399 & 1.589267 \\ 0.672994 & 1.793652 & 1.589267 & 0.467539 \end{bmatrix}$$

Since $\sqrt{2}n = 5.656854$ for the filter order $n = 4$, the blocklength is set to $L = 6$ in this section.

B. Roundoff Noise

The gain of the original average roundoff noise in (17.27) was computed as

$$\frac{1}{6}\text{tr}[\boldsymbol{W}_o] + 1 = 2.360365 \times 10^4$$

which was compared with the roundoff noise gain for the SISO state-space model shown in (17.20) as

$$\text{tr}[\boldsymbol{W}_o] + 1 = 14.161690 \times 10^4$$

The optimal coordinate transformation matrix that minimizes the average roundoff noise in (17.30) was constructed using (17.33) as

$$\boldsymbol{T} = \begin{bmatrix} -0.287484 & -0.559872 & -0.450804 & -0.403517 \\ 0.257608 & 0.607739 & 0.448499 & 0.418814 \\ -0.224220 & -0.651540 & -0.442903 & -0.441924 \\ 0.190114 & 0.693107 & 0.430204 & 0.473354 \end{bmatrix}$$

and the controllability and observability Grammians were found to be

$$\begin{aligned} \overline{\boldsymbol{K}}_c &= \boldsymbol{T}^{-1}\boldsymbol{K}_c\boldsymbol{T}^{-T} \\ &= \begin{bmatrix} 1.000000 & 0.295942 & 0.341255 & 0.475526 \\ 0.295942 & 1.000000 & -0.341255 & -0.475526 \\ 0.341255 & -0.341255 & 1.000000 & 0.909588 \\ 0.475526 & -0.475526 & 0.909588 & 1.000000 \end{bmatrix} \\ \overline{\boldsymbol{W}}_o &= \boldsymbol{T}^T\boldsymbol{W}_o\boldsymbol{T} \\ &= \begin{bmatrix} 0.138885 & 0.041102 & 0.047395 & 0.066044 \\ 0.041102 & 0.138885 & -0.047395 & -0.066044 \\ 0.047395 & -0.047395 & 0.138885 & 0.126328 \\ 0.066044 & -0.066044 & 0.126328 & 0.138885 \end{bmatrix} \end{aligned}$$

The minimum gain of the average roundoff noise in (17.30) was computed as

$$\frac{1}{6}\mathrm{tr}[\boldsymbol{T}^T\boldsymbol{W}_o\boldsymbol{T}] + 1 = 1.092590$$

which was compared with the roundoff noise gain for the equivalent realization, $(\boldsymbol{T}^{-1}\boldsymbol{A}\boldsymbol{T}, \boldsymbol{T}^{-1}\boldsymbol{b}, \boldsymbol{c}\boldsymbol{T}, d)_n$, of a SISO state-space model as

$$\mathrm{tr}[\boldsymbol{T}^T\boldsymbol{W}_o\boldsymbol{T}] + 1 = 1.555541$$

C. Initial l_2-Sensitivity

The average l_2-sensitivity measure for the block-state realization described by (17.34) was computed from (17.45) as

$$(S_i)_{ave} = 40.933372 \times 10^4$$

which was compared with the l_2-sensitivity measure for the SISO state-space model shown in (17.46) as

$$S_o = 977.917589 \times 10^4$$

D. Minimization of l_2-Sensitivity Subject to l_2-Scaling Constraints

1) The Use of A Lagrange Function

The recursive matrix equation in (17.64) together with (17.66) was applied to minimize (17.58) with tolerance $\varepsilon = 10^{-8}$ in (17.67) and initial estimate $\boldsymbol{I}_4$. It took the algorithm 84 iterations to converge to the solution

$$\boldsymbol{P} = \begin{bmatrix} 0.875338 & -0.904805 & 0.932402 & -0.958480 \\ -0.904805 & 0.938156 & -0.969673 & 0.999738 \\ 0.932402 & -0.969673 & 1.005245 & -1.039528 \\ -0.958480 & 0.999738 & -1.039528 & 1.078311 \end{bmatrix}$$

and from (17.72)

$$\boldsymbol{T} = \begin{bmatrix} 0.327836 & 0.559715 & 0.663273 & 0.121038 \\ -0.313941 & -0.586064 & -0.699777 & -0.080238 \\ 0.301577 & 0.604055 & 0.740336 & 0.036271 \\ -0.294198 & -0.614717 & -0.783389 & 0.013515 \end{bmatrix}$$

was obtained. The new controllability Grammian was found to be

$$\overline{\boldsymbol{K}}_c = \begin{bmatrix} 1.000000 & 0.300486 & 0.342142 & 0.474457 \\ 0.300486 & 1.000000 & -0.342142 & -0.474457 \\ 0.342142 & -0.342142 & 1.000000 & 0.910932 \\ 0.474457 & -0.474457 & 0.910932 & 1.000000 \end{bmatrix}$$

and the average l_2-sensitivity measure for the block-state realization described by (17.34) was computed from (17.53) as

$$S_i(\boldsymbol{T})_{ave} = 8.759262$$

which was compared with the l_2-sensitivity measure for the equivalent realization, $(\overline{\boldsymbol{A}}, \overline{\boldsymbol{b}}, \overline{\boldsymbol{c}}, d)_n$, of a SISO state-space model shown in (17.54) as

$$S_o(\boldsymbol{T}) = 29.500326$$

The profiles of the average l_2-sensitivity measure $S_i(\boldsymbol{P})_{ave}$ in (17.55) and $\mathrm{tr}[\boldsymbol{K}_c\boldsymbol{P}^{-1}]$ during the first 84 iterations of the algorithm are depicted in Figures 17.4.

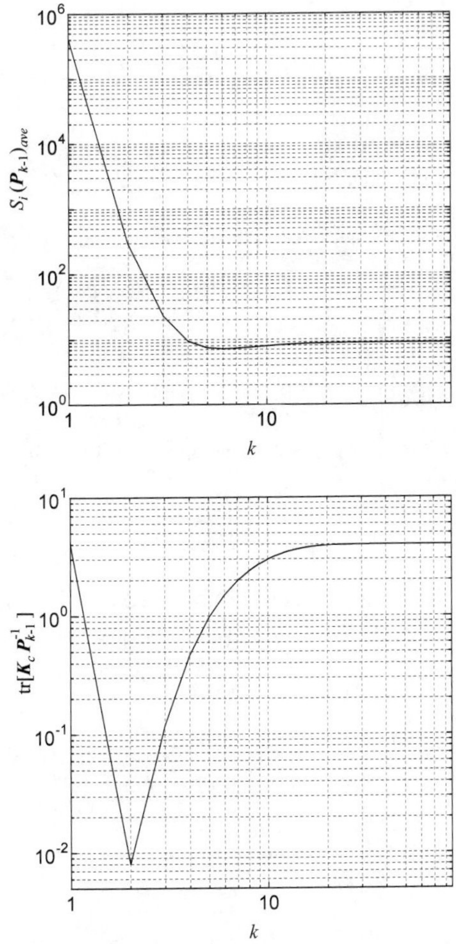

Figure 17.4 Profiles of $S_i(\boldsymbol{P})_{ave}$ and tr$[\boldsymbol{K}_c\boldsymbol{P}^{-1}]$ during the first 84 iterations.

2) The Use of A Quasi-Newton Algorithm

The quasi-Newton algorithm was applied to minimize (17.76) by choosing $\hat{\boldsymbol{T}} = \boldsymbol{I}_4$ as an initial assignment, and setting tolerance to $\varepsilon = 10^{-8}$ in (17.79). It took the algorithm 22 iterations to converge to the solution

$$\hat{\boldsymbol{T}} = \begin{bmatrix} 0.571134 & -0.415538 & -1.326551 & 0.389172 \\ 0.407792 & 0.933965 & -1.087187 & -2.120849 \\ 0.043983 & -0.117207 & 3.040058 & 2.480486 \\ -0.373909 & 0.428710 & 1.758732 & 2.561697 \end{bmatrix}$$

which is equivalent to

$$
\boldsymbol{T} = \begin{bmatrix} -0.307934 & 0.205102 & 0.403183 & -0.758913 \\ 0.350071 & -0.233366 & -0.379012 & 0.785847 \\ -0.385730 & 0.267285 & 0.350018 & -0.813980 \\ 0.414649 & -0.303751 & -0.313832 & 0.845979 \end{bmatrix}
$$

The new controllability Grammian was found to be

$$
\overline{\boldsymbol{K}}_c = \begin{bmatrix} 1.000000 & 0.609669 & 0.656806 & -0.159006 \\ 0.609669 & 1.000000 & 0.530787 & 0.197938 \\ 0.656806 & 0.530787 & 1.000000 & -0.674646 \\ -0.159006 & 0.197938 & -0.674646 & 1.000000 \end{bmatrix}
$$

and the average l_2-sensitivity measure for the block-state realization described by (17.34) was computed from (17.53) as

$$
S_i(\boldsymbol{T})_{ave} = 8.759262
$$

which was compared with the l_2-sensitivity measure for the equivalent realization, $(\overline{\boldsymbol{A}}, \overline{\boldsymbol{b}}, \overline{\boldsymbol{c}}, d)_n$, of a SISO state-space model, shown in (17.54) as

$$
S_o(\boldsymbol{T}) = 29.500303
$$

The profile of the average l_2-sensitivity measure $J_o(\boldsymbol{x})$ during the first 22 iterations of the algorithm is depicted in Figure 17.5.

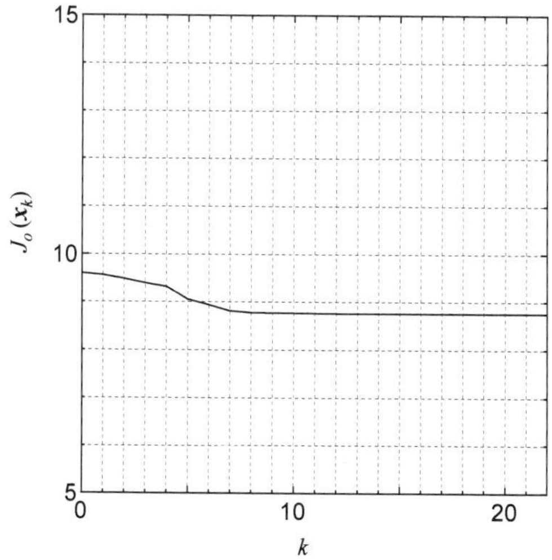

Figure 17.5 Profile of $J_o(\boldsymbol{x})$ during the first 22 iterations.

E. Minimization of l_2-Sensitivity Without Imposing l_2-Scaling Constraints

The recursive matrix equation in (17.83) was applied to minimize (17.55) with tolerance $\varepsilon = 10^{-8}$ in (17.84). Choosing $\boldsymbol{P}_0 = \boldsymbol{I}_4$ in (17.83), it took the algorithm 13 iterations to converge to the solution

$$\boldsymbol{P} = \begin{bmatrix} 2.204544 & -2.275660 & 2.342049 & -2.404601 \\ -2.275660 & 2.356583 & -2.432877 & 2.505500 \\ 2.342049 & -2.432877 & 2.519439 & -2.602761 \\ -2.404601 & 2.505500 & -2.602761 & 2.697538 \end{bmatrix}$$

which yields

$$\boldsymbol{T} = \boldsymbol{P}^{\frac{1}{2}} = \begin{bmatrix} 0.793097 & -0.756746 & 0.723513 & -0.692392 \\ -0.756746 & 0.771434 & -0.773723 & 0.768219 \\ 0.723513 & -0.773723 & 0.818655 & -0.852715 \\ -0.692392 & 0.768219 & -0.852715 & 0.949130 \end{bmatrix}$$

The new controllability Grammian was found to be

$$\overline{\boldsymbol{K}}_c = \begin{bmatrix} 0.819811 & 0.097982 & 0.106633 & -0.052902 \\ 0.097982 & 0.482067 & 0.066716 & 0.095486 \\ 0.106633 & 0.066716 & 0.146989 & -0.020279 \\ -0.052902 & 0.095486 & -0.020279 & 0.044099 \end{bmatrix}$$

and the average l_2-sensitivity measure for the block-state realization described by (17.34) was computed from (17.53) as

$$S_i(\boldsymbol{T})_{ave} = 7.190177$$

which was compared with the l_2-sensitivity measure for the equivalent realization, $(\overline{\boldsymbol{A}}, \overline{\boldsymbol{b}}, \overline{\boldsymbol{c}}, d)_n$, of a SISO state-space model shown in (17.54) as

$$S_o(\boldsymbol{T}) = 28.210150$$

The profile of the average l_2-sensitivity measure $S_i(\boldsymbol{P})_{ave}$ in (17.55) during the first 13 iterations of the algorithm is shown in Figure 17.6.

17.5 Summary

In this chapter, we have considered the block-state realization that is derived from a given SISO state-space model, and examined some of the properties of the block-state realization. Second, we have analyzed the roundoff noise in the

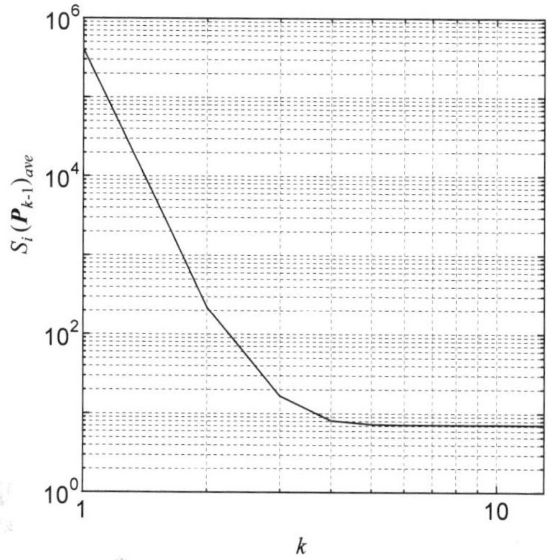

Figure 17.6 Profile of $S_i(\boldsymbol{P})_{ave}$ during the first 13 iterations.

block-state realization and minimized the average roundoff noise gain subject to l_2-scaling constraints. Third, we have analyzed l_2-sensitivity in the block-state realization and minimized the average l_2-sensitivity subject to l_2-scaling constraints where two methods have been presented. One has been based on a Lagrange function, while the other has relied on an efficient quasi-Newton algorithm. Finally, numerical experiments have been presented to demonstrate the validity and effectiveness of the techniques addressed in this chapter.

References

[1] S. Y. Hwang, "Roundoff noise in state-space digital filtering: A general analysis," *IEEE Trans. Acoust., Speech, Signal Process.*, vol. ASSP-24, no. 3, pp. 256–262, June 1976.

[2] C. T. Mullis and R. A. Roberts, "Synthesis of minimum roundoff noise fixed point digital filters," *IEEE Trans. Circuits Syst.*, vol. CAS-23, no, 9, pp. 551–562, Sept. 1976.

[3] S. Y. Hwang, "Minimum uncorrelated unit noise in state-space digital filtering," *IEEE Trans. Acoust., Speech, Signal Process.*, vol. ASSP-25, no. 4, pp. 273–281, Aug. 1977.

[4] C. W. Barnes and S. Shinnaka, "Finite word effects in block-state realization of fixed-point digital filters," *IEEE Trans. Circuits Syst.*, vol. CAS-27, no. 5, pp. 345–349, May 1980.

[5] C. W. Barnes and S. Shinnaka, "Block-shift invariance and block implementation of discrete-time filters," *IEEE Trans. Circuits Syst.*, vol. CAS-27, no. 8, pp. 667–672, Aug. 1980.

[6] J. Zeman and A. G. Lindgren, "Fast digital filters with low roundoff noise," *IEEE Trans. Circuits Syst.*, vol. CAS-28, no. 7, pp. 716–723, July 1981.

[7] W.-Y. Yan and J. B. Moore, "On L^2-sensitivity minimization of linear state-space systems," *IEEE Trans. Circuits Syst. I*, vol. 39, no. 8, pp. 641–648, Aug. 1992.

[8] T. Hinamoto, S. Yokoyama, T. Inoue, W. Zeng and W.-S. Lu, "Analysis and minimization of L_2-sensitivity for linear systems and two-dimensional state-space filters using general controllability and observability Grammians," *IEEE Trans. Circuits Syst. I*, vol. 49, no. 9, pp. 1279–1289, Sept. 2002.

[9] T. Hinamoto, H. Ohnishi and W.-S. Lu, "Minimization of L_2-sensitivity for state-space digital filters subject to L_2-dynamic-range scaling constraints," *IEEE Trans. Circuits Syst.-II*, vol. 52, no. 10, pp. 641–645, Oct. 2005.

[10] T. Hinamoto, K. Iwata and W.-S. Lu, "L_2-sensitivity Minimization of one- and two-dimensional state-space digital filters subject to L_2-scaling constraints," *IEEE Trans. Signal Processing*, vol. 54, no. 5, pp. 1804–1812, May 2006.

[11] T. Hinamoto, O. I. Omoifo and W.-S. Lu, "Realization of MIMO linear discrete-time systems with minimum l_2-sensitivity and no overflow oscillations," in *Proc. ISCAS 2006*, pp. 5215–5218.

[12] R. A. Meyer and C. S. Burrus, "A unified analysis of multirate and periodically time-varying digital filters," *IEEE Trans. Circuits Syst.*, vol. CAS-22, no. 3, pp. 162–168, Mar. 1975.

[13] R. Fletcher, *Practical Methods of Optimization*, 2nd ed., Wiley, New York, 1987.

[14] T. Kailath, *Linear System*, Englewood Cliffs, N.J.: Prentice-Hall, 1980.

Index

About the Authors

Takao Hinamoto received his B.E. degree from the Okayama University, Okayama, Japan, in 1969, and his M.E. degree from the Kobe University, Kobe, Japan, in 1971, and a Doctorate in Engineering from the Osaka University, Osaka, Japan, in 1977.

From the year 1972 to 1988, he was with the Faculty of Engineering, Kobe University and from 1979 to 1981, he was a visiting member of staff in the Department of Electrical Engineering, Queen's University, Kingston, ON, Canada. During 1988–1991, he was the Professor of Electronic Circuits in the Faculty of Engineering, Tottori University, Tottori, Japan. During 1992–2009, he was the Professor of Electronic Control in the Department of Electrical Engineering, Hiroshima University, Hiroshima, Japan. Since 2009, he has been the Professor Emeritus of Hiroshima University. His research interests include digital signal processing, system theory, and control engineering. He has published almost 450 papers in these areas. He is the coeditor and coauthor of *Two-Dimensional Signal and Image Processing* (Tokyo, Japan: SICE, 1996).

He was the Guest Editor of the special sections on *Digital Signal Processing* as well as *Adaptive Signal Processing and Its Applications* in the *IEICE Transactions on Fundamentals* in August 1998 and March 2005, respectively. He was the Co-Guest Editor of the special section on *Recent Advances in Circuits and Systems* in the July and August 2005 issues of the *IEICE Transactions on Information and Systems*. In 1997, he was the Chair of the 12th DSP Symposium held in Hiroshima, Japan, sponsored by the DSP Technical Committee of IEICE. From 1993 to 2000, he was a Senator or Member of the Board of Directors in the Society of Instrument and Control Engineers (SICE), and from 1999 to 2001 he was Chair of the Chugoku Chapter of SICE. From 2003 to 2004, he served as Chair of the DSP Technical Committee of IEICE and Chair of the Chugoku Chapter of IEICE.

From 1993 to 1995, he served as an Associate Editor of the *IEEE Transactions on Circuits and Systems II*. During 2002–2003 and 2006–2007, he served as an Associate Editor of the *IEEE Transactions on Circuits and Systems I*.

In 2004, he served as the General Chair of the 47th IEEE International Midwest Symposium on Circuits and Systems held in Hiroshima, Japan. Since 1995, he has been a Steering Committee Member of the IEEE International Midwest Symposium on Circuits and Systems. Since 1998, he has been a Digital Signal Processing Technical Committee Member in the IEEE Circuits and Systems Society. He played a leading role in establishing the Hiroshima Section of IEEE and served as Interim Chair of the Section.

He received the IEEE Third Millennium Medal in January 2000. He was elected a Fellow of the IEEE in 2001. In 2004, he was elected a Fellow of the Institute of Electronics, Information and Communication Engineers (IEICE). In 2005, he was elected a Fellow of the SICE. He became a Life Fellow of the IEEE in 2011.

Wu-Sheng Lu received his undergraduate education in mathematics from the Fudan University, Shanghai, China, during the years 1959 to 1964, and an M.S. degree in electrical engineering, and a Ph.D. in control science from the University of Minnesota, Minneapolis, USA, in 1983 and 1984, respectively.

He was a post-doctoral fellow at the University of Victoria, Victoria, B.C., Canada, in 1985, and a visiting assistant professor at the University of Minnesota from January 1986 to April 1987. He joined the Electrical and Computer Engineering Department, University of Victoria, in 1987 where he is a professor. His current research interests include analysis and design of digital filters, digital signal and image processing with a focus on sparse signal processing, and methods and applications of convex optimization. He is the co-author with A. Antoniou of *Two-Dimensional Digital Filters* (Marcel Dekker, 1992) and *Practical Optimization: Algorithms and Engineering Applications* (Springer, 2007).

He served as editor for the *Canadian Journal of Electrical and Computer Engineering* and associate editor for several journals including *IEEE Transactions on Circuits and Systems I*, *IEEE Transactions on Circuits and Systems II*, *International Journal of Multidimensional Systems and Signal Processing*, and *Journal of Circuits, Systems, and Signal Processing*.

He received several awards for his teaching at University of Victoria. He was elected a Fellow of the Engineering Institute of Canada in 1994. In 1999, he was elected a Fellow of the IEEE. He became a Life Fellow of the IEEE in 2012. He is a registered professional engineer in British Columbia, Canada.

9788793519640